Mechanism and Engineering Research on Processing, Storage and Preservation of Fresh Food

Mechanism and Engineering Research on Processing, Storage and Preservation of Fresh Food

Editors

Min Zhang
Hong-Wei Xiao

MDPI

Basel • Beijing • Wuhan • Barcelona • Belgrade • Novi Sad • Cluj • Manchester

Editors
Min Zhang
Jiangnan University
Wuxi, China

Hong-Wei Xiao
China Agricultural University
Beijing, China

Editorial Office
MDPI
St. Alban-Anlage 66
4052 Basel, Switzerland

This is a reprint of articles from the Special Issue published online in the open access journal *Foods* (ISSN 2304-8158) (available at: https://www.mdpi.com/journal/foods/special_issues/ES38423123).

For citation purposes, cite each article independently as indicated on the article page online and as indicated below:

Lastname, A.A.; Lastname, B.B. Article Title. *Journal Name* **Year**, *Volume Number*, Page Range.

ISBN 978-3-0365-9987-8 (Hbk)
ISBN 978-3-0365-9988-5 (PDF)
doi.org/10.3390/books978-3-0365-9988-5

Contents

About the Editors

Min Zhang

Prof. Min Zhang is a distinguished professor in Jiangnan University and the current director of the International Joint Laboratory (Jiangsu Province) in the field of intelligent processing and quality monitoring of fresh foods. He has been engaged in the research and teaching of fresh food processing and preservation for over 25 years and has supervised more than 200 PhD/master's degree students and post-doctoral candidates since 1997. In recent years, Prof. Zhang has paid special attention to emerging and novel food technologies including food additive manufacturing. The new technologies/products developed by his group jointly with food leading companies have been authorized by more than 230 invention patents in China and over 10 developed countries. He has authored more than 650 papers in international peer-reviewed journals as corresponding author and was featured in the list of "Chinese Highly Cited Scholars" by Elsevier (2014–now) and in the list of "Highly Cited Scientists" by Clarivate (2021–2022). His research achievements in food research have secured several important awards including the China National Scientific and Technological Progress Award in 2012, and the Arun S Mujumdar Medal in 2015.

Hong-Wei Xiao

Hong-Wei Xiao is a professor and PhD supervisor at the China Agricultural University. Prof. Xiao is one of the Associate Editors of Drying Technology, the Chief Editor of the Food and Agriproducts Processing column of IJABE, and the chairman of the 2025 Asia Pacific International Drying Conference. His research interests are innovative and emerging processing technologies for fruits, vegetables, and herbs. He has published more than 200 SCI/EI papers in prestigious journals, such as Nature, Trends in Food Science and Technology, Annual Review of Food Science and Technology, Journal of Hazardous Materials, etc. According to Google Scholar, his papers have been cited more than 9000 times, and his H-index is 52. Prof. Xiao has edited 3 professional books and has authored 12 book chapters in 10 books. He has been granted more than 40 patents. Prof. Xiao has also been featured as one of China's Highly Cited Scientists in consecutive years (as awarded by Elsevier) and as one of the World's Top 2% Scientists (awarded by Stanford University and Elsevier) since 2020.

foods

MDPI

Editorial

Mechanisms and Engineering Research on the Processing, Storage, and Preservation of Fresh Food

Min Zhang [1,*] and Hongwei Xiao [2]

[1] State Key Laboratory of Food Science and Technology, Jiangnan University, Wuxi 214122, China
[2] College of Engineering, China Agricultural University, Beijing 100083, China; xhwcaugxy@163.com
* Correspondence: min@jiangnan.edu.cn

The past few years have witnessed a significant increase in research and development activities related to advances in the processing, storage, and preservation of fresh food. This Special Issue of *Foods* aims to highlight a set of original research and review articles on the current status, challenges, and future directions of the processing, storage, and preservation of fresh food. We sincerely appreciate the active support from *Foods'* publishers and the kind invitation extended to us to serve as Guest Editors for this Special Issue.

Following an excellent response to our call for papers, we peer-reviewed a number of papers and selected 19—including 17 research articles and 2 review articles—to appear in this Special Issue. This Special Issue covers a wide range of fresh food topics as well as a variety of materials and technologies used in the field. These papers examine a variety of fresh foods ranging from fish and aquaculture (Chinese mitten crab, crucian carp, Yesso scallops), meat products (pig carcasses, chicken meat, pork mince), fruit and vegetables (fresh-cut kiwifruit, walnut, *Pleurotus eryngii*, fresh pepper fruit), processed food (pre-prepared fried rice, processed cheese) and aflatoxin in fresh food. Relatively new technologies for the preservation, processing and detection of fresh food during storage and transportation are presented in this Special Issue, including various plant essential oil components, hydrogen-rich water and slightly acidic electrolyzed water, high-quality drying, 3D and 4D functional printing, intelligent food packaging indicators for detecting freshness, and a novel detection technique for aflatoxin M1.

In the future, emerging technologies for the processing, storage, and preservation of fresh food with environmentally friendly, cost-effective, energy-saving, high-quality, intelligent-detection, and regulation characteristics will be pursued vigorously for the sustainable development of fresh food in China as well as globally. We hope that this Special Issue will further encourage research interest, engineering practice, and international communication in the field of fresh food.

We believe that all the articles published in this Special Issue will provide valuable information to scientists, engineers, researchers, and manufactures throughout the world, working in academia as well as in industry in the processing, storage, and preservation of fresh food.

Citation: Zhang, M.; Xiao, H. Mechanisms and Engineering Research on the Processing, Storage, and Preservation of Fresh Food. *Foods* **2023**, *12*, 4402. https://doi.org/10.3390/foods12244402

Received: 28 November 2023
Accepted: 29 November 2023
Published: 7 December 2023

Acknowledgments: We would like to thank all participating authors, and to thank Samuel Ariyo Okaiyeto and Li-hui Zhang for providing assistance.

Conflicts of Interest: The authors declare no conflict of interest.

List of Contributions:

1. Ye, Y.; Wang, Y.; Liu, P.; Chen, J.; Zhang, C. Uncovering the Nutritive Profiles of Adult Male Chinese Mitten Crab (E. sinensis) Harvested from the Pond and Natural Water Area of Qin Lake Based on Metabolomics. *Foods* **2023**, *12*, 2178. https://doi.org/10.3390/foods12112178.
2. Jiang, L.; Tang, J.; Huang, B.; Zhang, C.; Jiang, P.; Chen, D. Effect of Vanillin on the Anaesthesia of Crucian Carp: Effects on Physiological and Biochemical Indices, Pathology, and Volatile Aroma Components. *Foods* **2023**, *12*, 1614. https://doi.org/10.3390/foods12081614.

3. Jiang, L.; Huang, B.; Tang, J.; Jiang, P.; Chen, D.; Zhang, C. Comprehensive Analysis of Physiological, Biochemical and Flavor Characteristics Changes in Crucian Carp (Carassius auratus) under Different Concentrations of Eugenol. *Foods* **2023**, *12*, 2820. https://doi.org/10.3390/foods12152820.
4. Jiang, P.; Chen, D.; Chang, X.; Zhang, C.; Fan, X.; Qin, X. Study on the Regulation Mechanism of Quality Deterioration Due to Chilling Stress and Dry Exposure during Anhydrous Storage and Transportation of Yesso Scallop Patinopecten yessoensis. *Foods* **2023**, *12*, 2902. https://doi.org/10.3390/foods12152902.
5. Bai, H.; Zhou, G.; Liu, X. Evaluation of Temperature Uniformity in a Middle-Refrigerated Truck Loaded with Pig Carcasses. *Foods* **2023**, *12*, 1837. https://doi.org/10.3390/foods12091837.
6. Shen, Q.; Zeng, X.; Kong, L.; Sun, X.; Shi, J.; Wu, Z.; Guo, Y.; Pan, D. Research Progress of Nitrite Metabolism in Fermented Meat Products. *Foods* **2023**, *12*, 1485. https://doi.org/10.3390/foods12071485.
7. Wang, W.; Li, T.; Chen, J.; Ye, Y. Inhibition of Salmonella Enteritidis by Essential Oil Components and the Effect of Storage on the Quality of Chicken. *Foods* **2023**, *12*, 2560. https://doi.org/10.3390/foods12132560.
8. Meng, X.; Shen, Q.; Song, T.; Zhao, H.; Zhang, Y.; Ren, A.; Yang, W. Facile Fabrication of Anthocyanin-Nanocellulose Hydrogel Indicator Label for Intelligent Evaluation of Minced Pork Freshness. *Foods* **2023**, *12*, 2602. https://doi.org/10.3390/foods12132602.
9. Sun, Y.; Qiu, W.; Fang, X.; Zhao, X.; Xu, X.; Li, W. Hydrogen-Rich Water Treatment of Fresh-Cut Kiwifruit with Slightly Acidic Electrolytic Water: Influence on Antioxidant Metabolism and Cell Wall Stability. *Foods* **2023**, *12*, 426. https://doi.org/10.3390/foods12020426.
10. Tong, Q.; Zhao, W.; Guo, T.; Wang, D.; Dong, X. A Study of the Gelatin Low-Temperature Deposition Manufacturing Forming Process Based on Fluid Numerical Simulation. *Foods* **2023**, *12*, 2687. https://doi.org/10.3390/foods12142687.
11. Yang, T.; Zheng, X.; Vidyarthi, S.K.; Xiao, H.; Yao, X.; Li, Y.; Zang, Y.; Zhang, J. Artificial Neural Network Modelling and Genetic Algorithm Multiobjective Optimization of Process of Drying-Assisted Walnut Breaking. *Foods* **2023**, *12*, 1897. https://doi.org/10.3390/foods12091897.
12. Bai, J.-W.; Wang, Y.-C.; Cai, J.-R.; Zhang, L.; Dai, Y.; Tian, X.-Y.; Xiao, H.-W. Three-Dimensional Appearance and Physicochemical Properties of Pleurotus eryngii under Different Drying Methods. *Foods* **2023**, *12*, 1999. https://doi.org/10.3390/foods12101999.
13. Chen, X.; Zhang, M.; Tang, T. Microwave-Induced Rapid Shape Change of 4D Printed Vegetable-Based Food. *Foods* **2023**, *12*, 2158. https://doi.org/10.3390/foods12112158.
14. Yang, L.; Liu, X.; Lu, H.; Zhang, C.; Chen, J.; Shi, Z. Cinnamaldehyde Inhibits Postharvest Gray Mold on Pepper Fruits via Inhibiting Fungal Growth and Triggering Fruit Defense. *Foods* **2023**, *12*, 3458. https://doi.org/10.3390/foods12183458.
15. Huang, J.; Zhang, M.; Fang, Z. Perspectives on Novel Technologies of Processing and Monitoring the Safety and Quality of Prepared Food Products. *Foods* **2023**, *12*, 3052. https://doi.org/10.3390/foods12163052.
16. Deng, Y.; Zhao, G.; Cheng, K.; Shi, C.; Xiao, G. Effect of Apple Polyphenols on the Antioxidant Activity and Structure of Three-Dimensional Printed Processed Cheese. *Foods* **2023**, *12*, 1731. https://doi.org/10.3390/foods12081731.
17. Tong, Q.; Meng, Y.; Tong, Y.; Wang, D.; Dong, X. The Effect of Nozzle Temperature on the Low-Temperature Printing Performance of Low-Viscosity Food Ink. *Foods* **2023**, *12*, 2666. https://doi.org/10.3390/foods12142666.
18. Li, X.; Zhang, M.; Mo, H.; Li, H.; Xu, D.; Hu, L. The Ultrasensitive Detection of Aflatoxin M1 Using Gold Nanoparticles Modified Electrode with Fe3+ as a Probe. *Foods* **2023**, *12*, 2521. https://doi.org/10.3390/foods12132521.
19. Xu, D.; Huang, M.; Lei, J.; Song, H.; Hu, L.; Mo, H. Auricularia auricular Adsorbs Aflatoxin B1 and Ameliorates Aflatoxin B1-Induced Liver Damage in Sprague Dawley Rats. *Foods* **2023**, *12*, 2644. https://doi.org/10.3390/foods12142644.

foods

MDPI

Article

Cinnamaldehyde Inhibits Postharvest Gray Mold on Pepper Fruits via Inhibiting Fungal Growth and Triggering Fruit Defense

Lifei Yang [1,†], Xiaoli Liu [1,2,†], Haiyan Lu [2], Cunzheng Zhang [2], Jian Chen [2] and Zhiqi Shi [2,*]

1 Hexian New Countryside Development Research Institute, College of Horticulture, Nanjing Agricultural University, Nanjing 210095, China; lfy@njau.edu.cn (L.Y.); 2017104077@njau.edu.cn (X.L.)

2 Laboratory for Food Quality and Safety-State Key Laboratory Cultivation Base of Ministry of Science and Technology, Institute of Food Safety and Nutrition, Jiangsu Academy of Agricultural Sciences, Nanjing 210014, China; haiyanlu@jaas.ac.cn (H.L.); zcz@jaas.ac.cn (C.Z.); chenjian@jaas.ac.cn (J.C.)

* Correspondence: shizhiqi@jaas.ac.cn

† These authors contributed equally to this work.

Abstract: Gray mold infected with *Botrytis cinerea* frequently appears on fruits and vegetables throughout the supply chain after harvest, leading to economic losses. Biological control of postharvest disease with phytochemicals is a promising approach. CA (cinnamaldehyde) is a natural phytochemical with medicinal and antimicrobial activity. This study evaluated the effect of CA in controlling *B. cinerea* on fresh pepper fruit. CA inhibited *B. cinerea* growth in vitro significantly in a dose- (0.1–0.8 mM) and time-dependent (6–48 h) manner, with an EC_{50} (median effective concentration) of 0.5 mM. CA induced the collapse and breakdown of the mycelia. CA induced lipid peroxidation resulting from ROS (reactive oxygen species) accumulation in mycelia, further leading to cell leakage, evidenced by increased conductivity in mycelia. CA induced mycelial glycerol accumulation, resulting in osmotic stress possibly. CA inhibited sporulation and spore germination resulting from ROS accumulation and cell death observed in spores. Spraying CA at 0.5 mM induced a defense response in fresh pepper fruits, such as the accumulation of defense metabolites (flavonoid and total phenols) and an increase in the activity of defense enzymes (PAL, phenylalanine ammonia lyase; PPO, polyphenol oxidase; POD, peroxidase). As CA is a type of environmentally friendly compound, this study provides significant data on the activity of CA in the biocontrol of postharvest gray mold in peppers.

Keywords: antifungal; *Botrytis cinerea*; cinnamaldehyde; gray mold; induced defense; pepper fruit

Citation: Yang, L.; Liu, X.; Lu, H.; Zhang, C.; Chen, J.; Shi, Z. Cinnamaldehyde Inhibits Postharvest Gray Mold on Pepper Fruits via Inhibiting Fungal Growth and Triggering Fruit Defense. *Foods* **2023**, *12*, 3458. https://doi.org/10.3390/foods12183458

Academic Editor: Victor Rodov

Received: 5 August 2023
Revised: 6 September 2023
Accepted: 13 September 2023
Published: 16 September 2023

1. Introduction

Postharvest disease development is a major cause of food waste and economic losses worldwide [1]. *Botrytis cinerea* is the main phytopathogenic fungi causing postharvest gray mold on most fresh fruits and vegetables [2]. Infection with *B. cinerea* results in decay and rot on fruits, leading to a decline in quality and economic losses. *B. cinerea* can survive easily in variable environmental conditions, producing spores than can be easily dispersed to invade neighboring fruits and vegetables [3]. Pepper is a widely cultivated vegetable that is economically important for fresh consumption and processing worldwide. *B. cinerea*, causing gray mold, significantly decreases the fruit quality during the transport and storage of pepper [4].

Controlling postharvest gray mold has been considered a great challenge in avoiding the considerable postharvest losses of fruits and vegetables. Several approaches have been developed to control postharvest gray mold. The first one is a controlled atmosphere and temperature. *B. cinerea* can survive and cause disease at a wide range of temperatures (0–26 °C), but this pathogen can grow and infect rapidly under favorable environmental conditions (18–24 °C; RH > 93%) [5–7]. A low temperature and dry environment can slow down the development and spread of *B. cinerea* during the storage of fruits and vegetables. The fungi can grow rapidly in the environment with an enhanced temperature and humidity. Therefore, the temperature and humidity should be precisely controlled during cold-chain transportation, which may increase the cost of long-distance transportation [8]. The second approach is heat treatment. Rinsing bell peppers with hot water at 55 °C for 12 s can disinfect peppers and prolong storage by killing pathogens and altering the host fruit physiology. However, this approach relies on specific equipment for hot water rinsing and brushing [9]. The third approach is using physical means, such as ozone, UV irradiation, or altered atmospheric conditions (hypobaric or hyperbaric treatments). These approaches can control postharvest gray mold effectively without direct contact with the fruit, but the effect always lasts for a short time [8]. The fourth approach is using fungicides. Chemical fungicides can kill the pathogens directly, but toxic fungicide residues may pose a threat to human health. Moreover, using a single fungicide for a long time may drive the occurrence of drug resistance in *B. cinerea*. Some countries even restrict the usage of conventional synthetic fungicides in postharvest disease management [8,10]. The fifth approach is the biological control of postharvest disease, which has been drawing great attention recently [11]. Among the biological methods, the use of environmentally friendly natural compounds (e.g., essential oils, plant extracts, and microbial products) is a promising approach [4,12]. These natural compounds have fungicidal activity against *B. cinerea*, but they are safe to human health and the environment. Some natural compounds even have the ability to induce resistance in fruits and vegetables, limiting the development of postharvest disease [13].

CA (cinnamaldehyde) is a phytochemical obtained from *Cinnamomum* species [14]. CA has antimicrobial and immune-regulatory activity [15,16]. CA has been applied as a potential natural food preservative due to its antifungal and antibacterial activity [17,18]. CA (3-phenylpropenal, C_9H_8O) is a type of aldehyde with one phenyl ring (Figure S1). The fungicidal activity of CA has been associated with the conjugated double bond and the length of the CH chain outside the ring [19]. The fungicidal activity of CA against plant pathogens has been documented [20,21], and the crude extracts of cinnamon can inhibit gray mold on strawberry [22]. However, we still have little knowledge about the role of CA against gray mold caused by *B. cinerea* and its mechanisms.

In this work, we studied the effect of CA on the control of gray mold on fresh pepper. First, the antifungal activity of CA against *B. cinerea* was investigated. Then, CA-induced defensive responses in pepper fruits were detected. Finally, the possible mechanisms and their significance were discussed.

2. Materials and Methods

2.1. Strain, Media, and Treatment

The monoculture of *B. cinerea* was obtained from the College of Plant Protection in Nanjing Agricultural University, China. *B. cinerea* were placed in PDA (potato dextrose agar) medium in a petri dish at 25 °C and in PDB (potato dextrose broth) medium in a shaker (150 rpm) at 25 °C. Both PDA and PDB were obtained from Nanjing Rongshengda Instrument Co., Ltd., Nanjing, China.

CA of analytical grade (>98%) was obtained from Sinopharm Chemical Reagent Co., Ltd., Shanghai, China. CA at different concentrations (0, 0.1, 0.2, 0.4, 0.6, and 0.8 mM) was added to PDA shortly before the solidification of PDA. Then, the PDA containing CA was transferred to petri dishes for the culture of *B. cinerea* for 48 h. Then, we measured the colony diameter using a ruler crossing the center of the colony circle to evaluate the radical growth of *B. cinerea* [23].

The EC_{50} (median effective concentration) of CA against *B. cinerea* was 0.5 mM based on the calculation of the colony diameter determined as above (linear regression) [24]. The colony diameter was also determined at 6, 12, 24, 36, and 48 h, respectively, under the treatment of CA at 0.5 mM. The average growth speed of *B. cinerea* was calculated using linear regression based on the changes in colony diameter in 48 h. Then, CA at 0.2, 0.5, and

0.8 mM (representing low, median, and high CA concentrations, respectively) was added to PDB for the culture of *B. cinerea* to collect mycelia for physiological measurements.

2.2. Determination of Mycelial Biomass

B. cinerea was cultured in PDB containing CA for 48 h. Then, the mycelia were collected through filter paper. The mycelia were surface-dried with paper tissue gently. The fresh weights of mycelia was determined using a laboratory balance. Then, the fresh mycelia were dried in an oven (60 °C) for 48 h, followed by measuring the mycelial dry weight [25]. Three replicates were conducted for each treatment.

2.3. Spore Production and Germination

The spores were obtained from a PDA-cultured strain according to a previously published method with slight modification [26]. *B. cinerea* was cultured in PDA in a petri dish (2 cm in diameter) for 10 days to produce sufficient spores. Then, sterilized water (2 mL) was added to the petri dish to wash the spores out. The spores were collected through four layers of filter paper followed by centrifugation at 10,000× *g* for 5 min.

CA at different concentrations (0.2, 0.5, and 0.8 mM) was added to PDA for the culture of *B. cinerea* for 6 days to produce spores. All the spores were collected to count the number of spores produced.

To test the inhibitory effect of CA on spore germination, the spore suspension (1×10^5 spores/mL) was cultured on agar medium containing CA at 0.2, 0.5, and 0.8 mM, respectively, for 12 h. Then, the number of germinated spores was counted for the calculation of the germination rate.

2.4. SEM (Scanning Electron Microscopy)

The mycelia of *B. cinerea* were collected by filtration of the culture solution with double gauze, followed by washing with phosphate buffer (0.2 M, pH 7.0). Then, we fixed the mycelia with graded ethanol and acetone according to our previously published method [23]. Then, the mycelial samples were observed with an SEM (EVO-LS10, ZEISS, Jena, Germany). The structure of mycelia was evaluated based on morphological analysis [27].

2.5. Meaaurement of Mycelial Conductivity

The mycelia (0.5 g) of *B. cinerea* after treatment (48 h) were collected and suspended in 20 mL distilled water. Then, a conductivity meter (CON510 Eutech/Oakton, Singapore) was applied to measure the electrical conductivity at 0, 5, 10, 20, 40, 60, 80, 100, 120, 140, 160, and 180 min, respectively. The final conductivity was measured after incubating the mycelia in boiled water for 5 min. The relative conductivity was calculated as follows:

$$\text{Relative conductivity } (\%) = \frac{\text{Conductivity}}{\text{Final conductivity}} \times 100$$

2.6. Meaaurement of Mycelial TBARS (Thiobarbituric Acid Reactive Substances) Content

The mycelia after treatment were collected by filtration of the culture solution with double gauze, followed by measuring the TBARS content using a detection kit (A003; Nanjing Jiancheng Bioengineering Institute, Nanjing, China) [28].

2.7. Measurement of Mycelial Glycerol Content

Approximately 0.5 g of mycelia after treatment were harvested. The mycelia were ground and suspended with 20 mL of distilled water. The mixture was water-heated at 80 °C for 15 min, followed by centrifuging (8500 rpm, 10 min). $CuSO_4$ (0.05 g/L) was mixed with the supernatant, with shaking at 100 rpm for 12 min. This allowed the reaction between glycerol and $CuSO_4$ to produce a dark blue copper–glycerol complex with specific absorbance at 630 nm. Then, the absorbance at 630 nm was recorded for the filtered mixture.

The glycerol in mycelial samples was calculated based on a standard curve of glycerol (0–0.01 g/mL) [29].

2.8. Histochemical Analysis of ROS (Reactive Oxygen Species) and Cell Death

The mycelia after treatment with CA for 48 h were washed with distilled water three times. The mycelia were stained with 1 μM DCFH-DA (2′,7′-dichlorofluorescein diacetate) at 25 °C for 20 min. The endogenous ROS were labeled with DCF fluorescence (green) and analyzed by fluorescent microscopy (ECLIPSE Series, TE2000-S, Nikon, Tokyo, Japan) [30].

For the detection of cell death, collected mycelia were stained with 5 μM PI (propidium iodide) at 25 °C for 30 min. The dead cells were labeled with PI fluorescence (red) and photographed under fluorescent microscopy [31].

The spores of *B. cinerea* were also incubated with DCFH-DA or PI, followed by fluorescence assays using flow cytometry (Accuri C6 Plus, BD Biosciences, San Jose, CA, USA) [32].

2.9. CA-Suppressed Development of B. cinerea on Pepper Fruits

Pepper seeds (Sujiao 5) were obtained from the Institute of Vegetable Crops at Jiangsu Academy of Agricultural Sciences, Nanjing, China. The peppers were grown in a greenhouse without using any pesticides. The harvested pepper fruits were collected to test the effect of CA on the control of *B. cinerea* development. Fresh pepper fruits were surface-sterilized with 0.5% NaClO for 5 min and washed with sterilized water three times. An agar plug (5 mm in diameter) taken from PDA-cultured *B. cinerea* was inoculated on the surfaces of pepper fruits. Each fruit was inoculated with five mycelial plugs with an even distribution on the fruit surface (Supplementary Figure S2). The fruits were kept in a light chamber (photoperiod of 12 h and active radiation of 200 μmol/(m^2 s)) at 25 °C, allowing the growth and infection of *B. cinerea*. Pepper fruits were sprayed with CA at 0.5 mM for 2 h before inoculating *B. cinerea*. The peppers in the control group were treated with sterilized water before inoculation. Five fruits were inoculated for each treatment. The sizes of disease lesions were measured after 7 days. For the size of each lesion (almost a circle), the diameter was evaluated by measuring it three times from different directions crossing the center of the lesion.

2.10. Assay of Defensive Enzyme Activity of Pepper Fruits

Approximately 2 g pepper fruit samples around the inoculated sites were taken for the measurement of enzyme activity. The fruit samples were homogenized with pre-chilled PBS (phosphate buffer solution, 50 mM, pH 7.0) containing 5 mM β-mercaptoethanol and 2 mM EDTA. After centrifugation at 12,000 rpm at 4 °C for 20 min, the supernatant was collected as extracts for the determination of the enzyme activity.

The POD (peroxidase) activity was measured based on guaiacol oxidation with hydrogen peroxide [33]. The reaction mixture (2 mL) consisted of PBS, 0.5% hydrogen peroxide, 0.5% guaiacol, and 20 μL of extract. Then, we recorded the absorbance at 420 nm for the calculation of POD activity using the extinction coefficient at 26.6 /mM/cm.

The PPO (polyphenol oxidase) activity was measured based on pyrocatechol oxidation [34]. The reaction mixture consisted of 0.5 mL pyrocatechol, 2 mL PBS, and 0.5 mL extract. The absorbance at 410 nm was recorded at initial mixture and after 2 min of mixture, respectively. One unit of PPO was calculated as a change of 0.01 in absorbance per min based on the fresh weight of fruit samples.

The PAL (phenylalanine ammonia lyase) activity was measured based on the lysis of L-phenylalanine [35]. The reaction mixture consisted of 2 mL PBS, 1 mL extract, and 1 mL L-phenylalanine (0.02 M). The mixture was incubated in boiled water for 1 h. After this, 0.2 mL HCl (6 M) was added to terminate the reaction. Then, the absorbance at 290 nm was recorded before and after a water bath. One unit of PAL was calculated as a change of 0.01 in absorbance per hour based on the fresh weight of fruit samples.

2.11. Measurement of Metabolites

Approximately 2 g pepper fruit samples around inoculated sites were taken to determine the content of flavonoids and total phenols. Fruit samples were ground with cold 1% HCl (dissolved in methanol), adding distilled water to obtain a total volume of 20 mL. After the reaction at 4 °C in darkness for 20 min, the mixture was centrifuged at 12,000 rpm for 20 min. We collected the supernatant to determine the flavonoid and total phenol content using a plant total phenol test kit (A143-1-1) and plant flavonoids test kit (A142-1-1), respectively (Nanjing Jiancheng Bioengineering Institute, Nanjing, China) [36].

2.12. Data Analysis

All data were presented as the mean ± SD (standard deviation) of at least three replicates. Significant differences between two treatments ($p < 0.05$ or $p < 0.01$) were compared using ANOVA (one-way analysis of variance) with an F test. Significant differences ($p < 0.05$) among multiple treatments were compared using LSD (least significant difference test). The package "corrplot" in R was applied to perform Pearson correlation analysis [37]. TBtools was used to generate heatmaps for hierarchical cluster analysis [38].

3. Results

3.1. Mycelial Growth

B. cinerea was cultured on PDA plates containing CA for 48 h to evaluate the growth of mycelia in vitro. CA at 0.1–0.8 mM inhibited the radical growth of mycelia significantly in an dose-dependent manner (Figure 1A,B). The EC_{50} of CA was 0.5 mM based on the linear regression of the colony diameter shown in Figure 1B. The MIC (minimal inhibitory concentration) of CA against the radical growth of *B. cinerea* was 1.1 mM. A time-course experiment was performed to monitor the growth of mycelia upon CA exposure at 0.5 mM. CA (0.5 mM) began to inhibit mycelial growth significantly after 12 h. The average mycelial growth speed for the control and CA treatment was 0.108 and 0.055 cm/h, respectively (Figure 1C).

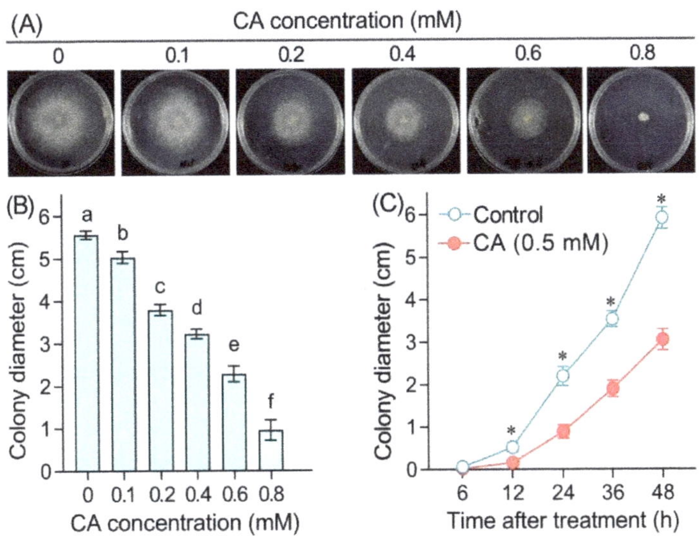

Figure 1. The antifungal effect of CA against the radical growth of *B. cinerea*. (**A**) A photograph of *B. cinerea* growing on PDA plate containing CA at different concentrations. (**B**) The colony diameter of *B. cinerea* on PDA plate. (**C**) The time-course changes in the colony diameter of *B. cinerea* under 5 mM CA treatment. Different lowercase letters in (**B**) indicate significant differences among different treatments (LSD, ANOVA, $n = 12$, $p < 0.05$). The asterisk in (**C**) indicates significant difference between control and CA treatment at each time point (ANOVA, $n = 8$, $p < 0.05$).

As CA at 0.5 mM resulted in a decrease in mycelial growth by 50%, we selected CA at 0.2, 0.5, and 0.8 mM as low, median, and high concentrations, respectively, to investigate the antifungal effect of CA against *B. cinerea* in detail. Compared to the control group, the mycelial fresh weight significantly decreased by 40.0%, 60.9%, and 83.7% upon CA addition at 0.2, 0.5, and 0.8 mM, respectively (Figure 2A). The mycelial dry weight remarkably decreased by 33.5%, 61.0%, and 92.6% upon CA addition at 0.2, 0.5, and 0.8 mM, respectively, compared to the control (Figure 2B). These results suggested that CA exposure remarkably decreased the biomass of *B. cinerea*.

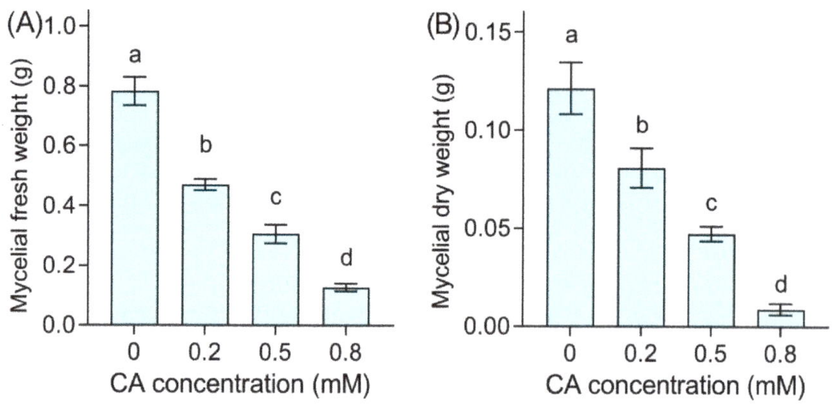

Figure 2. The effect of CA on the mycelial growth of *B. cinerea*. (**A**) Mycelial fresh weight. (**B**) Mycelial dry weight. Different lowercase letters indicate significant differences among different treatments (LSD, ANOVA, $n = 3$, $p < 0.05$).

3.2. CA Induced Mycelial Injury of *B. cinerea*

The morphology of mycelia was observed under SEM after treatment for 48 h (Figure 3). The control group without CA showed smooth and intact mycelia. CA at 0.2 mM induced distorted mycelia with minor collapse on the surface. CA at 0.5 mM resulted in much more distorted and collapsed mycelia, with fracture and shrinkage. Treatment with CA at 0.8 mM led to the almost complete breakdown of the mycelia, which shrank severely with the loss of surface integration.

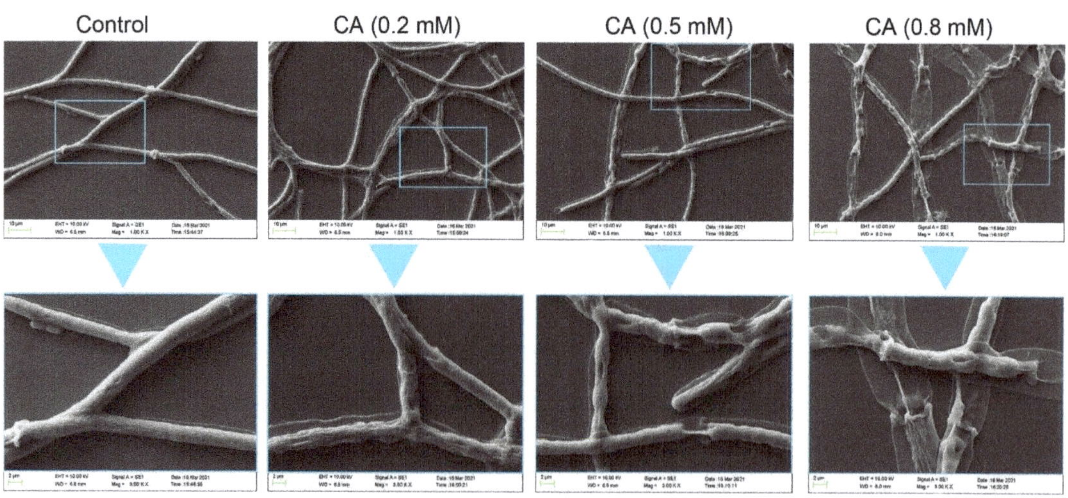

Figure 3. The effect of CA on the mycelial structure of *B. cinerea*. Upper panel and lower panel indicate the magnification of 1000× and 3000×, respectively.

To confirm the CA-induced collapse of mycelia, several physiological parameters in CA-treated mycelia were investigated. CA treatments enhanced the mycelial relative conductivity (Figure 4A), suggesting the leakage of intracellular matter. The lipid peroxidation of the plasma membrane can be indicated by the TBARS content. CA elevated the TBARS content in a dose-dependent manner (Figure 4B), indicating the occurrence of lipid peroxidation in mycelial cells upon CA exposure. CA also induced glycerol accumulation in mycelia (Figure 4C), suggesting the occurrence of an osmotic response in CA-treated mycelia.

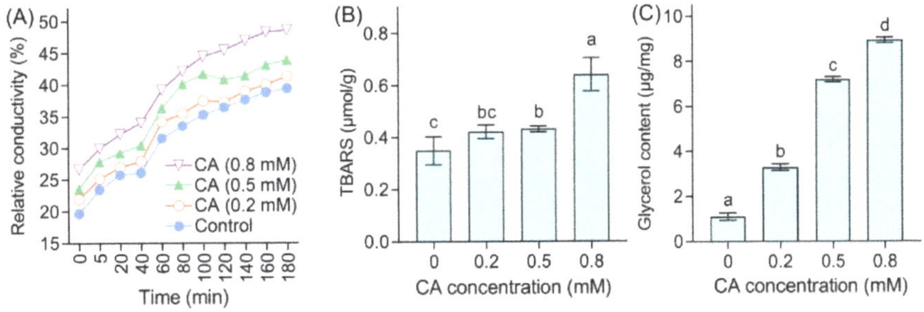

Figure 4. The effect of CA on (**A**) relative conductivity, (**B**) TBARS content, and (**C**) glycerol content in the mycelia of *B. cinerea*. Different lowercase letters in (**B**,**C**) indicate significant differences among different treatments (LSD, ANOVA, $n = 3$, $p < 0.05$).

ROS frequently attack lipids in the membrane to lead to lipid oxidation. A specific fluorescent probe, DCFH-DA, was used to detect endogenous ROS in mycelia. CA-treated mycelia showed extensive DCF fluorescence, suggesting the accumulation of ROS (Figure 5). Mycelial cell death was labeled with specific fluorescent probe PI. CA at 0.2 mM induced slight PI fluorescence in mycelia. The PI fluorescence was more extensive with the increase in CA concentration, suggesting that CA treatment led to mycelial death (Figure 5).

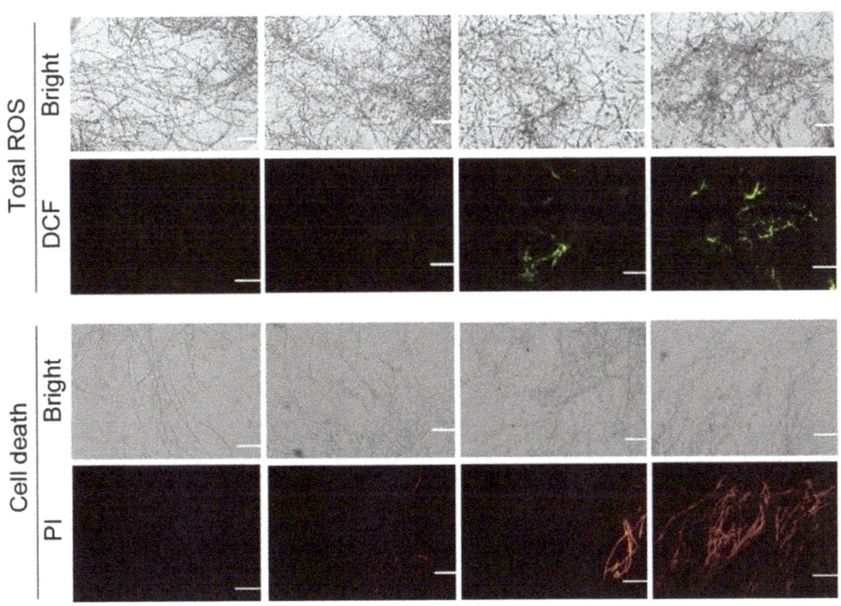

Figure 5. The effect of CA on ROS accumulation and cell death in the mycelia of *B. cinerea*.

3.3. CA Damaged the Spores of B. cinerea

CA inhibited the production of spores in a dose-dependent manner. Spore numbers decreased by 35.0%, 65.0%, and 78.3% significantly upon CA addition at 0.2, 0.5, and 0.8 mM, respectively (Figure 6A). CA also inhibited the germination of healthy spores. After culturing for 12 h, almost all of the spores germinated in the control group. However, only 54.0%, 29.7%, and 12.0% of spores germinated upon CA exposure at 0.2, 0.5, and 0.8 mM, respectively (Figure 6B).

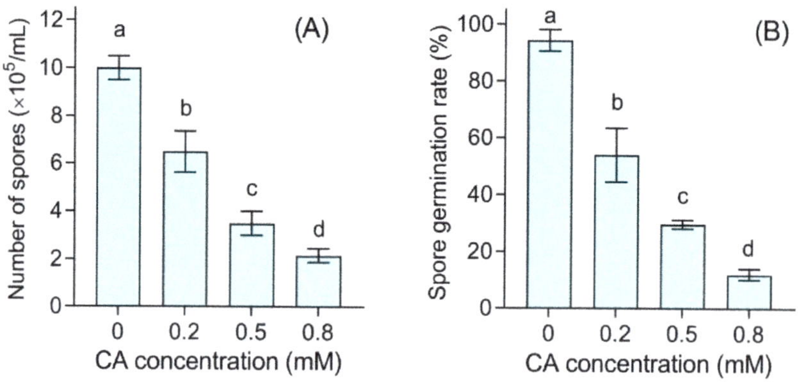

Figure 6. The effect of CA on spore generation and spore germination of *B. cinerea*. (**A**) Number of spores produced by *B. cinerea* upon CA treatment. (**B**) The germination rate of spores upon CA treatment. Different lowercase letters indicate significant differences among different treatments (LSD, ANOVA, $n = 3$, $p < 0.05$).

Flow cytometry was applied to detect DCF-labeled spores. The number of DCF-labeled spores increased under CA treatment as compared to the control (Figure 7A), suggesting that CA induced ROS accumulation in spores. The increase in PI-labeled spores also suggested that CA induced the cell death of spores (Figure 7B).

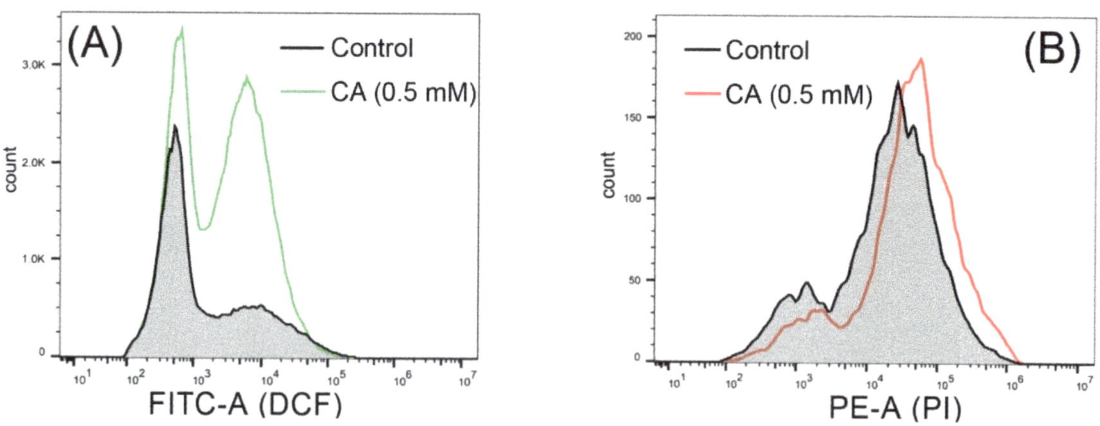

Figure 7. The effect of CA on ROS accumulation and cell death of the spores of *B. cinerea*. (**A**) DCF-labeled spores detected using flow cytometry. (**B**) PI-labeled spores detected using flow cytometry.

3.4. Hierarchical Cluster and Correlation Analysis of Physiological Parameters in CA-Treated B. cinerea

Hierarchical cluster analysis was applied to compare the changes in the parameters upon CA treatment. One cluster consisted of the parameters related to fungal growth and development, such as the mycelial diameter, mycelial fresh weight, mycelial dry weight, spore number, and spore germination. All these parameters decreased compared to the control in a dose-dependent manner (Figure 8A), suggesting the inhibitory effect of CA against *B. cinerea* growth. Glycerol content, TBARS content, and relative conductivity were clustered together. These three parameters increased with the increase in CA concentration as compared to the control (Figure 8A), suggesting that membrane injury occurred in CA-treated *B. cinerea*.

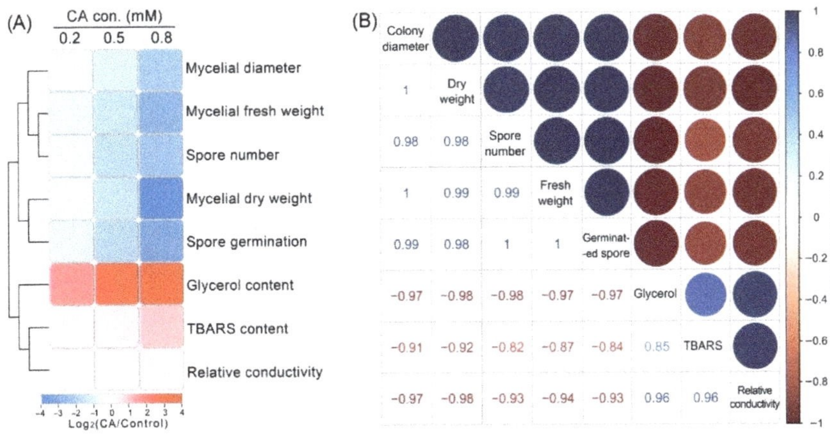

Figure 8. Cluster analysis and correlation analysis among determined physiological parameters. (**A**) Hierarchical cluster analysis of the parameters. For each parameter, the value from CA treatment is presented as log2 (fold change vs. control). (**B**) Pearson correlation analysis among different parameters.

Pearson correlation analysis was performed for all of the measured parameters. The injury indexes (glycerol content, TBARS content, and relative conductivity) were negatively correlated with the growth indexes (such as mycelial diameter, mycelial fresh weight, mycelial dry weight, spore number, and spore germination) (Figure 8B).

3.5. CA Suppresses the Infection of B. cinerea on Pepper Fruit

The above results confirmed the capability of CA in inhibiting *B. cinerea* in vitro. We further evaluated the possible role of CA in pepper fruits infected with *B. cinerea*. Inoculation with *B. cinerea* resulted in severe gray mold on pepper fruits, such as enlarged browning and decay around inoculated sites (control in Figure 9A). Pepper fruits pretreated with CA (0.5 mM) showed very slight disease symptoms (the right-hand fruit in Figure 9A). The average size of lesions was significantly decreased by 68.7% on CA-treated fruits as compared to the control (Figure 9B).

The defensive responses of pepper fruits after 12 h of *B. cinerea* inoculation were evaluated by measuring the activity of several defensive enzymes. *B. cinerea* inoculation induced a significant increase in POD activity compared to the control (neither CA nor *B. cinerea* inoculation). Pretreatment with CA induced a remarkable increase in POD activity on fruits as compared to *B. cinerea* inoculation alone (Figure 10B,C). The changes in PPO and PAL activity were similar to those of POD activity under *B. cinerea* and CA treatment.

Inoculation with *B. cinerea* significantly decreased the content of flavonoids and total phenols in pepper fruits. Pretreatment with CA effectively enhanced the content of these metabolites as compared to *B. cinerea* inoculation alone (Figure 10D,E).

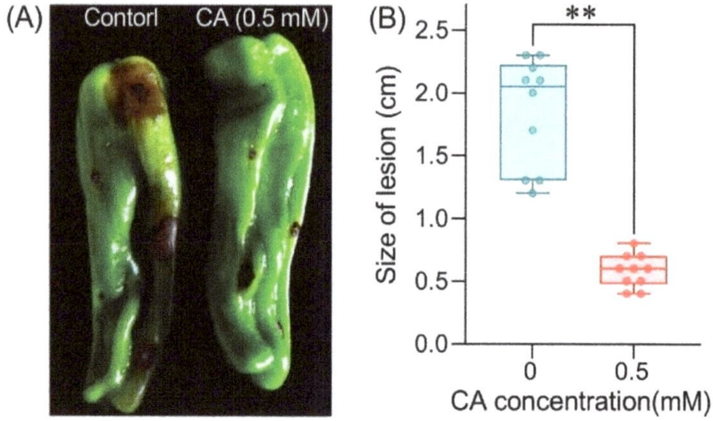

Figure 9. The effect of CA on the infection of *B. cinerea* in pepper fruits. (**A**) A photograph of the lesions on pepper fruits. (**B**) The size of lesions measured on pepper fruits. Double asterisks indicate significant differences between treatments (ANOVA, $n = 10$, $p < 0.01$).

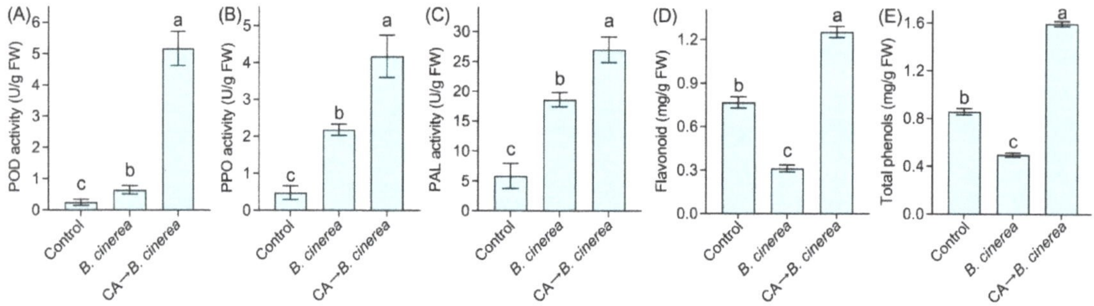

Figure 10. The effect of CA on the defensive enzyme activity and metabolites in pepper fruits infected with *B. cinerea*. (**A**) POD activity. (**B**) PPO activity. (**C**) PAL activity. (**D**) Flavonoid content. (**E**) Total phenol content. Different lowercase letters indicate significant differences among treatments (ANOVA, $n = 3$, $p < 0.05$).

4. Discussion

Applying plant extracts is an effective approach to control gray mold on vegetables and fruits [39]. CA is a small-molecule phytochemical with broad antimicrobial activity. CA can be applied as a natural food preservative [40]. In this study, we found that CA effectively controlled the gray mold on fresh pepper fruits by showing antifungal activity against *B. cinerea* and inducing defensive responses in pepper fruits, which in turn maintained pepper quality.

It has been reported that CA shows potential antifungal activity against variable phytopathogenic fungi, such as *Fusarium moniliforme*, *Verticillium fungicola*, *Sclerotinia homoeocarpa*, *Trichophyton rubrum*, and *Aspergillus fumigates* [41,42]. The antifungal activity of CA has been associated with the disruption of cell wall integrity in *Geotrichum citri-aurantii* [43]. CA damaged the mycelia of *B. cinerea* by inducing cell collapse and death. The loss of plasma membrane integrity may be one of the most important reasons, evidenced by ion leakage from mycelial cells. Plasma membrane damage is a typical mechanism behind the fungicidal activity of CA against *Alternaria alternata* [44]. The inhibition of ergosterol biosynthesis is involved in the antifungal activity of CA against *Fusarium sambucinum* [20], but we found that ROS accumulation may be another important reason. CA-induced ROS accumulation can cause lipid peroxidation, resulting in membrane damage by increasing membrane permeability in mycelial cells of *B. cinerea*. The triggering ROS accumulation is an important action of CA-induced bacterial cell death. CA-induced apoptosis-like death can be linked to the triggering of ROS accumulation in *Microcystis aeruginosa* [30]. Nox (NADPH oxidase) is an important enzyme generating ROS in eukaryotes [45]. Nox-produced ROS contributes to the cell leakage in fungi [46]. Besides damaging the cell membrane, ROS can act on mitochondrial proteins to induce fungal death [47]. CA can induce mitochondrial dysfunction in human leukemia K562 cells [48]. Therefore, further studies are needed to investigate whether and how CA triggers the Nox-ROS system to damage cell membranes and mitochondria in *B. cinerea* mycelia.

Glycerol accumulation is a typical consequence of osmotic stress in pathogenic fungi [49]. CA induced glycerol accumulation in *B. cinerea* mycelia, indicating that osmotic stress occurred. Filamentous fungi adapt to osmotic stress mainly through the HOG (high osmolarity glycerol) pathway, which consists of a set of sensors and transducers activating the biosynthesis of glycerol [50]. Fungicidal activity has been associated with osmotic stress by targeting the HOG pathway [51]. A conserved HOG pathway has been found in *B. cinerea* as well [52]. It is of interest to further understand whether the modulation of the HOG pathway contributes to the antifungal activity of CA against *B. cinerea*.

It was reported that CA-inhibited spore germination resulted in the suppression of mycelial development in *Aspergillus flavus* [40]. The antifungal activity of CA against *B. cinerea* was also related to the inhibition of sporulation and spore viability. *B. cinerea* produces large amounts of spores to finish the life cycle in order to achieve survival, dispersion, and infection among host plants [53]. Here, we found that CA significantly inhibited *B. cinerea* spore generation. This may have been due to the detrimental effect of CA on the vegetative growth of mycelia, weakening the ability of *B. cinerea* to finish its life cycle. CA also inhibited the germination of healthy spores produced by *B. cinerea*. Stress-inhibited sporulation and germination result from ROS accumulation and oxidative injury in *B. cinerea* [54]. ROS accumulation in *B. cinerea* spore may be one of the possible reasons for CA-suppressed spore viability.

Fresh pepper fruits pretreated with CA before *B. cinerea* inoculation showed decreased lesion development. This result suggested that CA may induce defense responses in pepper fruits against the invasion of *B. cinerea*. This hypothesis could be partially confirmed by the result that CA increased the activity of defensive enzymes (POD, PAL, and PPO) in pepper fruits. All of these enzymes play roles in helping host plants against pathogens [55,56]. CA induced the accumulation of flavonoids and total phenols in pepper fruits as well. These metabolites are not only important non-enzymatic components of plant defense, but also vital nutrients reflecting the quality of pepper fruits [56,57]. At present, little is known about CA-regulated plant physiology. We previously found that CA conferred cadmium tolerance by triggering endogenous Ca^{2+} in plants [58]. In addition, CA can induce defense responses in citrus fruit against the infection of *Geotrichum citri-aurantii* [21]. Moreover, CA can modulate the immune responses in fish against fungal infection [59,60]. These reports and our current results suggest an important role of CA in triggering host immunity to combat biotic or abiotic stress. However, the detailed mechanisms need further study.

Several reports suggest that CA derivatives also have effective (or even enhanced) antimicrobial activity [61–63]. In addition, antifungal agents can be encapsulated in nanoparticles in order to enhance the antifungal effect or to control the release of ingredients, which is proposed as a novel approach to control *B. cinerea* infection [64]. Encapsulation can control the release of essential oils to enhance their efficiency, which can help to decrease the dosage and possible impact of volatile flavors after treatment. Several types of encapsulated nano-CA have been developed to test the bioactivity and ingredient release [65,66]. Therefore, further efforts are needed to design and modify the structure of CA, which would help to apply CA in the biocontrol of *B. cinerea*-caused postharvest gray mold on vegetables and fruits during the supply chain in different scenarios.

5. Conclusions

In this study, we revealed that CA was capable of controlling postharvest gray mold on fresh pepper fruit by inhibiting *B. cinerea* and triggering defense responses in pepper. CA with a concentration of more than 0.2 mM effectively inhibited the growth of *B. cinerea* in vitro. The antifungal activity of CA against *B. cinerea* was closely related to ROS accumulation and oxidative injury in both mycelia and spores. Pretreatment with CA triggered defensive responses in pepper, which further prevented the infection of *B. cinerea*. The application of CA in real scenarios needs further study, but our current results extend our knowledge of the antifungal activity of CA as well as the potential application of CA in the preservation of vegetables and fruits.

Supplementary Materials: The following are available online at https://www.mdpi.com/article/10.3390/foods12183458/s1, Figure S1: The chemical structure of cinnamaldehyde, Figure S2: The schematic model for the inoculation of *B. cinerea* on the surface of pepper fruit.

Author Contributions: Conceptualization, Z.S. and J.C.; methodology, L.Y. and X.L.; investigation, L.Y., X.L., H.L. and C.Z.; writing—original draft preparation, L.Y. and X.L.; writing—review and editing, Z.S. and J.C. All authors have read and agreed to the published version of the manuscript.

Funding: This work was supported by the National Key R&D Program of China, grant number 2022YFD2100601; the Jiangsu Agricultural Science and Technology Innovation Fund, grant number CX(20)1011; and the China Agriculture Research System, grant number CARS-23-B16.

Data Availability Statement: Not applicable.

Acknowledgments: We appreciate Hao Zhou from the Central Laboratory at Jiangsu Academy of Agricultural Sciences for his kind help with the SEM analysis.

Conflicts of Interest: The authors declare no conflict of interest.

References

1. Singh, D.; Sharma, R.R. Postharvest Diseases of Fruits and Vegetables and Their Management. In *Postharvest Disinfection of Fruits and Vegetables*; Siddiqui, M.W., Ed.; Academic Press: Cambridge, MA, USA, 2018; pp. 1–52.
2. Dean, R.; van Kan, J.A.L.; Pretorius, Z.A.; Hammond-Kosack, K.E.; Di Pietro, A.; Spanu, P.D.; Rudd, J.J.; Dickman, M.; Kahmann, R.; Ellis, J.; et al. The Top 10 fungal pathogens in molecular plant pathology. *Mol. Plant Pathol.* **2012**, *13*, 414–430. [CrossRef] [PubMed]
3. Hua, L.; Yong, C.; Zhanquan, Z.; Boqiang, L.; Guozheng, Q.; Shiping, T. Pathogenic mechanisms and control strategies of *Botrytis cinerea* causing post-harvest decay in fruits and vegetables. *Food Qual. Saf.* **2018**, *2*, 111–119. [CrossRef]
4. Krasnow, C.; Ziv, C. Non-chemical approaches to control postharvest gray mold disease in bell peppers. *Agronomy* **2022**, *12*, 216. [CrossRef]
5. O'Neill, T.M.; Shtienberg, D.; Elad, Y. Effect of some host and microclimate factors on infection of tomato stems by *Botrytis cinerea*. *Plant Dis.* **1997**, *81*, 36–40. [CrossRef]
6. Tsitsigiannis, D.; Antoniou, P.; Tjamos, S.; Paplomatas, E. Major diseases of tomato, pepper and egg plant in green houses. *Eur. J. Plant Sci. Biotechnol.* **2008**, *2*, 106–124.
7. Hausbeck, M.; Moorman, G. Managing Botrytis in greenhouse-grown flower crops. *Plant Dis.* **1996**, *80*, 1212–1219. [CrossRef]
8. Romanazzi, G.; Smilanick, J.L.; Feliziani, E.; Droby, S. Integrated management of postharvest gray mold on fruit crops. *Postharvest Biol. Technol.* **2016**, *113*, 69–76. [CrossRef]
9. Fallik, E.; Grinberg, S.; Alkalai, S.; Yekutieli, O.; Wiseblum, A.; Regev, R.; Beres, H.; Bar-Lev, E. A unique rapid hot water treatment to improve storage quality of sweet pepper. *Postharvest Biol. Technol.* **1999**, *15*, 25–32. [CrossRef]
10. Schirra, M.; D'Aquino, S.; Cabras, P.; Angioni, A. Control of Postharvest Diseases of Fruit by Heat and Fungicides: Efficacy, Residue Levels, and Residue Persistence. A Review. *J. Agric. Food Chem.* **2011**, *59*, 8531–8542. [CrossRef]
11. Li, X.; Zeng, S.; Wisniewski, M.; Droby, S.; Yu, L.; An, F.; Leng, Y.; Wang, C.; Li, X.; He, M.; et al. Current and future trends in the biocontrol of postharvest diseases. *Crit. Rev. Food Sci. Nutr.* **2022**. [CrossRef]
12. Brito, C.; Hansen, H.; Espinoza, L.; Faúndez, M.; Olea, A.F.; Pino, S.; Díaz, K. Assessing the control of postharvest gray mold disease on tomato fruit using mixtures of essential oils and their respective hydrolates. *Plants* **2021**, *10*, 1719. [CrossRef]
13. Prusky, D.; Romanazzi, G. Induced resistance in fruit and vegetables: A host physiological response limiting postharvest disease development. *Annu. Rev. Phytopathol.* **2023**, *61*, 279–300. [CrossRef] [PubMed]
14. Ranasinghe, P.; Pigera, S.; Premakumara, G.S.; Galappaththy, P.; Constantine, G.R.; Katulanda, P. Medicinal properties of 'true' cinnamon (*Cinnamomum zeylanicum*): A systematic review. *BMC Complement. Altern. Med.* **2013**, *13*, 275. [CrossRef] [PubMed]

15. Ranasinghe, P.; Jayawardana, R.; Galappaththy, P.; Constantine, G.R.; de Vas Gunawardana, N.; Katulanda, P. Efficacy and safety of 'true' cinnamon (*Cinnamomum zeylanicum*) as a pharmaceutical agent in diabetes: A systematic review and meta-analysis. *Diab. Med.* **2012**, *29*, 1480–1492. [CrossRef] [PubMed]
16. Meena, V.; Sree, S.N.; Surya, P.D.V.; Sumanjali, A. A review on pharmacological activities and clinical effects of Cinnamon species. *Res. J. Pharm. Biol. Chem. Sci.* **2012**, *3*, 653–663.
17. Sun, Q.; Li, J.; Sun, Y.; Chen, Q.; Zhang, L.; Le, T. The antifungal effects of cinnamaldehyde against *Aspergillus niger* and its application in bread preservation. *Food Chem.* **2020**, *317*, 126405. [CrossRef] [PubMed]
18. Pang, D.; Huang, Z.; Li, Q.; Wang, E.; Liao, S.; Li, E.; Zou, Y.; Wang, W. Antibacterial mechanism of cinnamaldehyde: Modulation of biosynthesis of phosphatidylethanolamine and phosphatidylglycerol in *Staphylococcus aureus* and *Escherichia coli*. *J. Agric. Food Chem.* **2021**, *69*, 13628–13636. [CrossRef]
19. Xie, Y.; Huang, Q.; Wang, Z.; Cao, H.; Zhang, D. Structure-activity relationships of cinnamaldehyde and eugenol derivatives against plant pathogenic fungi. *Ind. Crops Prod.* **2017**, *97*, 388–394. [CrossRef]
20. Wei, J.; Bi, Y.; Xue, H.; Wang, Y.; Zong, Y.; Prusky, D. Antifungal activity of cinnamaldehyde against *Fusarium sambucinum* involves inhibition of ergosterol biosynthesis. *J. Appl. Microbiol.* **2020**, *129*, 256–265. [CrossRef]
21. Wu, Y.; Duan, X.; Jing, G.; Ou Yang, Q.; Tao, N. Cinnamaldehyde inhibits the mycelial growth of *Geotrichum citri-aurantii* and induces defense responses against sour rot in citrus fruit. *Postharvest Biol. Technol.* **2017**, *129*, 23–28. [CrossRef]
22. Šernaitė, L.; Rasiukevičiūtė, N.; Valiuškaitė, A. The extracts of cinnamon and clove as potential biofungicides against strawberry grey mould. *Plants* **2020**, *9*, 613. [CrossRef] [PubMed]
23. Gao, T.; Zhou, H.; Zhou, W.; Hu, L.; Chen, J.; Shi, Z. The fungicidal activity of thymol against *Fusarium graminearum* via inducing lipid peroxidation and disrupting ergosterol biosynthesis. *Molecules* **2016**, *21*, 770. [CrossRef] [PubMed]
24. Thomulka, K.W.; Abbas, C.G.; Young, D.A.; Lange, J.H. Evaluating median effective concentrations of chemicals with bioluminescent bacteria. *Bull. Environ. Contam. Toxicol.* **1996**, *56*, 446–452. [CrossRef] [PubMed]
25. Sutton, L.M.; Starzyk, M.J. Procedure and analysis of a useful method in determining mycelial dry weights from agar plates. *Appl. Microbiol.* **1972**, *24*, 1011–1012. [CrossRef] [PubMed]
26. Fernandez-San Millan, A.; Larraya, L.; Farran, I.; Ancin, M.; Veramendi, J. Successful biocontrol of major postharvest and soil-borne plant pathogenic fungi by antagonistic yeasts. *Biol. Control* **2021**, *160*, 104683. [CrossRef]
27. Zhang, J.; Hao, Y.; Lu, H.; Li, P.; Chen, J.; Shi, Z.; Xie, Y.; Mo, H.; Hu, L. Nano-thymol emulsion inhibits *Botrytis cinerea* to control postharvest gray mold on tomato fruit. *Agronomy* **2022**, *12*, 2973. [CrossRef]
28. Yang, Y.; Fan, F.; Zhuo, R.; Ma, F.; Gong, Y.; Wan, X.; Jiang, M.; Zhang, X. Expression of the laccase gene from a white rot fungus in *Pichia pastoris* can enhance the resistance of this yeast to H_2O_2-mediated oxidative stress by stimulating the glutathione-based antioxidative system. *Appl. Environ. Microbiol.* **2012**, *78*, 5845–5854. [CrossRef] [PubMed]
29. Duan, Y.; Ge, C.; Liu, S.; Chen, C.; Zhou, M. Effect of phenylpyrrole fungicide fludioxonil on morphological and physiological characteristics of *Sclerotinia sclerotiorum*. *Pestic. Biochem. Physiol.* **2013**, *106*, 61–67. [CrossRef]
30. Hu, L.; Zhou, W.; Yang, J.; Chen, J.; Yin, Y.; Shi, Z. Cinnamaldehyde Induces PCD-Like death of *Microcystis aeruginosa* via reactive oxygen species. *Water Air Soil Pollut.* **2011**, *217*, 105–113. [CrossRef]
31. Firstencel, H.; Butt, T.M.; Carruthers, R.I. A fluorescence microscopy method for determining the viability of entomophthoralean fungal spores. *J. Invertebr. Pathol.* **1990**, *55*, 258–264. [CrossRef]
32. Yao, L.; Ban, F.; Peng, S.; Xu, D.; Li, H.; Mo, H.; Hu, L.; Zhou, X. Exogenous iron induces NADPH oxidases-dependent ferroptosis in the conidia of *Aspergillus flavus*. *J. Agric. Food Chem.* **2021**, *69*, 13608–13617. [CrossRef] [PubMed]
33. Zhou, Z.S.; Huang, S.Q.; Guo, K.; Mehta, S.K.; Zhang, P.C.; Yang, Z.M. Metabolic adaptations to mercury-induced oxidative stress in roots of *Medicago sativa* L. *J. Inorg. Biochem.* **2007**, *101*, 1–9. [CrossRef] [PubMed]
34. Li, N.; Chen, F.; Cui, F.; Sun, W.; Zhang, J.; Qian, L.; Yang, Y.; Wu, D.; Dong, Y.; Jiang, J.; et al. Improved postharvest quality and respiratory activity of straw mushroom (*Volvariella volvacea*) with ultrasound treatment and controlled relative humidity. *Sci. Hortic.* **2017**, *225*, 56–64. [CrossRef]
35. Ballester, A.R.; Lafuente, M.T.; González-Candelas, L. Spatial study of antioxidant enzymes, peroxidase and phenylalanine ammonia-lyase in the citrus fruit–Penicillium digitatum interaction. *Postharvest Biol. Technol.* **2006**, *39*, 115–124. [CrossRef]
36. Ma, Q.; Hu, Y.; Dong, X.; Zhou, G.; Liu, X.; Gu, Q.; Wei, Q. Metabolic profiling and gene expression analysis unveil differences in flavonoid and lipid metabolisms between 'Huapi' kumquat (*Fortunella crassifolia* Swingle) and its wild type. *Front. Plant Sci.* **2021**, *12*, 759968. [CrossRef] [PubMed]
37. Wei, T.; Simko, V. *R Package "Corrplot": Visualization of a Correlation Matrix (Version 0.84)*; R Foundation: Vienna, Austria, 2017.
38. Chen, C.; Chen, H.; Zhang, Y.; Thomas, H.R.; Frank, M.H.; He, Y.; Xia, R. TBtools: An Integrative Toolkit Developed for Interactive Analyses of Big Biological Data. *Mol. Plant* **2020**, *13*, 1194–1202. [CrossRef] [PubMed]
39. Šernaitė, L.; Rasiukevičiūtė, N.; Valiuškaitė, A. Application of plant extracts to control postharvest gray mold and susceptibility of apple fruits to B. cinerea from different plant hosts. *Foods* **2020**, *9*, 1430. [CrossRef] [PubMed]
40. Qu, S.; Yang, K.; Chen, L.; Liu, M.; Geng, Q.; He, X.; Li, Y.; Liu, Y.; Tian, J. Cinnamaldehyde, a promising natural preservative against *Aspergillus flavus*. *Front Microbiol.* **2019**, *10*, 2895. [CrossRef]
41. Khan, M.S.; Ahmad, I. *In vitro* antifungal, anti-elastase and anti-keratinase activity of essential oils of *Cinnamomum*-, *Syzygium*- and *Cymbopogon*-species against *Aspergillus fumigatus* and *Trichophyton rubrum*. *Phytomedicine* **2011**, *19*, 48–55. [CrossRef]

42. El-Baky, N.A.; Amara, A. Recent approaches towards control of fungal diseases in plants: An updated review. *J. Fungi.* **2021**, *7*, 900. [CrossRef]
43. OuYang, Q.; Duan, X.; Li, L.; Tao, N. Cinnamaldehyde exerts its antifungal activity by disrupting the cell wall integrity of *Geotrichum citri-aurantii*. *Front Microbiol.* **2019**, *10*, 55. [CrossRef] [PubMed]
44. Xu, L.; Tao, N.; Yang, W.; Jing, G. Cinnamaldehyde damaged the cell membrane of *Alternaria alternata* and induced the degradation of mycotoxins in vivo. *Ind. Crops Prod.* **2018**, *112*, 427–433. [CrossRef]
45. Vermot, A.; Petit-Härtlein, I.; Smith, S.M.E.; Fieschi, F. NADPH Oxidases (NOX): An overview from discovery, molecular mechanisms to physiology and pathology. *Antioxidants* **2021**, *10*, 890. [CrossRef] [PubMed]
46. Wang, J.; Zhang, J.; Ma, J.; Liu, L.; Li, J.; Shen, T.; Tian, Y. The major component of cinnamon oil as a natural substitute against *Fusarium solani* on *Astragalus membranaceus*. *J. Appl. Microbiol.* **2022**, *132*, 3125–3141. [CrossRef] [PubMed]
47. Qin, G.; Liu, J.; Cao, B.; Li, B.; Tian, S. Hydrogen Peroxide Acts on Sensitive Mitochondrial Proteins to Induce Death of a Fungal Pathogen Revealed by Proteomic Analysis. *PLoS ONE* **2011**, *6*, e21945. [CrossRef] [PubMed]
48. Huang, T.-C.; Fu, H.-Y.; Ho, C.-T.; Tan, D.; Huang, Y.-T.; Pan, M.-H. Induction of apoptosis by cinnamaldehyde from indigenous cinnamon *Cinnamomum osmophloeum* Kaneh through reactive oxygen species production, glutathione depletion, and caspase activation in human leukemia K562 cells. *Food Chem.* **2007**, *103*, 434–443. [CrossRef]
49. Brown, A.J.; Budge, S.; Kaloriti, D.; Tillmann, A.; Jacobsen, M.D.; Yin, Z.; Ene, I.V.; Bohovych, I.; Sandai, D.; Kastora, S.; et al. Stress adaptation in a pathogenic fungus. *J. Exp. Biol.* **2014**, *217*, 144–155. [CrossRef]
50. Duran, R.; Cary, J.W.; Calvo, A.M. Role of the osmotic stress regulatory pathway in morphogenesis and secondary metabolism in filamentous fungi. *Toxins* **2010**, *2*, 367–381. [CrossRef]
51. Kojima, K.; Takano, Y.; Yoshimi, A.; Tanaka, C.; Kikuchi, T.; Okuno, T. Fungicide activity through activation of a fungal signalling pathway. *Mol. Microbiol.* **2004**, *53*, 1785–1796. [CrossRef]
52. Ren, W.; Liu, N.; Yang, Y.; Yang, Q.; Chen, C.; Gao, Q. The sensor proteins Bcsho1 and Bcsln1 are involved in, though not essential to, vegetative differentiation, pathogenicity and osmotic stress tolerance in *Botrytis cinerea*. *Front Microbiol.* **2019**, *10*, 328. [CrossRef]
53. Cheung, N.; Tian, L.; Liu, X.; Li, X. The destructive fungal pathogen botrytis cinerea-insights from genes studied with mutant analysis. *Pathogens* **2020**, *9*, 923. [CrossRef] [PubMed]
54. Zhao, W.; Wisniewski, M.; Wang, W.; Liu, J.; Liu, Y. Heat-induced oxidative injury contributes to inhibition of *Botrytis cinerea* spore germination and growth. *World J. Microbiol. Biotechnol.* **2014**, *30*, 951–957. [CrossRef] [PubMed]
55. Zhang, J.; Sun, X. Recent advances in polyphenol oxidase-mediated plant stress responses. *Phytochemistry* **2021**, *181*, 112588. [CrossRef] [PubMed]
56. Kaur, S.; Samota, M.K.; Choudhary, M.; Choudhary, M.; Pandey, A.K.; Sharma, A.; Thakur, J. How do plants defend themselves against pathogens-Biochemical mechanisms and genetic interventions. *Physiol. Mol. Biol. Plants* **2022**, *28*, 485–504. [CrossRef] [PubMed]
57. Batiha, G.E.; Alqahtani, A.; Ojo, O.A.; Shaheen, H.M.; Wasef, L.; Elzeiny, M.; Ismail, M.; Shalaby, M.; Murata, T.; Zaragoza-Bastida, A.; et al. Biological properties, bioactive constituents, and pharmacokinetics of some *Capsicum* spp. and Capsaicinoids. *Int. J. Mol. Sci.* **2020**, *21*, 5179. [CrossRef] [PubMed]
58. Cheng, Y.; Wang, N.; Liu, R.; Bai, H.; Tao, W.; Chen, J.; Shi, Z. Cinnamaldehyde facilitates cadmium tolerance by modulating Ca^{2+} in *Brassica rapa*. *Water Air Soil Pollut.* **2021**, *232*, 19. [CrossRef]
59. Zhou, Y.; Feng, L.; Jiang, W.-D.; Wu, P.; Liu, Y.; Jiang, J.; Kuang, S.-Y.; Tang, L.; Peng, Y.; Zhou, X.-Q. Cinnamaldehyde improved intestine immune function and alleviated inflammation associated with NF-κB pathways in grass carp (*Ctenopharyngodon idella*) after infection with *Aeromonas hydrophila*. *Aquac. Rep.* **2021**, *21*, 100837. [CrossRef]
60. Harikrishnan, R.; Devi, G.; Balasundaram, C.; Van Doan, H.; Jaturasitha, S.; Saravanan, K.; Ringø, E. Impact of cinnamaldehyde on innate immunity and immune gene expression in *Channa striatus* against *Aphanomyces invadans*. *Fish Shellfish Immunol.* **2021**, *117*, 1–16. [CrossRef]
61. Gan, Z.; Huang, J.; Chen, J.; Nisar, M.F.; Qi, W. Synthesis and antifungal activities of cinnamaldehyde derivatives against *Penicillium digitatum* causing Citrus green mold. *J. Food Qual.* **2020**, *2020*, 8898692. [CrossRef]
62. Wang, H.; Peng, Z.; Sun, H. Antifungal activities and mechanisms of *trans*-cinnamaldehyde and thymol against food-spoilage yeast *Zygosaccharomyces rouxii*. *J. Food Sci.* **2022**, *87*, 1197–1210. [CrossRef]
63. Lee, J.-E.; Jung, M.; Lee, S.-C.; Huh, M.-J.; Seo, S.-M.; Park, I.-K. Antibacterial mode of action of *trans*-cinnamaldehyde derived from cinnamon bark (*Cinnamomum verum*) essential oil against *Agrobacterium tumefaciens*. *Pestic. Biochem. Physiol.* **2020**, *165*, 104546. [CrossRef]
64. De Angelis, G.; Simonetti, G.; Chronopoulou, L.; Orekhova, A.; Badiali, C.; Petruccelli, V.; Portoghesi, F.; D'Angeli, S.; Brasili, E.; Pasqua, G.; et al. A novel approach to control *Botrytis cinerea* fungal infections: Uptake and biological activity of antifungals encapsulated in nanoparticle based vectors. *Sci. Rep.* **2022**, *12*, 7989. [CrossRef]
65. Zhang, R.; Cui, Y.; Cheng, M.; Guo, Y.; Wang, X.; Wang, J. Antifungal activity and mechanism of cinnamon essential oil loaded into mesoporous silica nanoparticles. *Ind. Crops Prod.* **2021**, *171*, 113846. [CrossRef]
66. Cionti, C.; Taroni, T.; Sabatini, V.; Meroni, D. Nanostructured oxide-based systems for the pH-triggered release of cinnamaldehyde. *Materials* **2021**, *14*, 1536. [CrossRef]

foods

MDPI

Article

Study on the Regulation Mechanism of Quality Deterioration Due to Chilling Stress and Dry Exposure during Anhydrous Storage and Transportation of Yesso Scallop *Patinopecten yessoensis*

Peihong Jiang [1,2], Dongjie Chen [2], Xiangyang Chang [1], Changfeng Zhang [2], Xiuping Fan [1] and Xiaoming Qin [1,*]

1 College of Food Science and Technology, Guangdong Ocean University, Guangdong Provincial Key Laboratory of Aquatic Products Processing and Safety, Guangdong Provincial Engineering Technology Research Center of Marine Food, Guangdong Province Engineering Laboratory for Marine Biological Products, Zhanjiang 524088, China; jiangpeihongjph@163.com (P.J.)

2 Shandong Provincial Key Laboratory of Agricultural Products Storage, Transportation and Preservation Technology, Shandong Institute of Commerce and Technology, National Engineering Research Center for Agricultural Products Logistics, Shandong Guonong Logistics Technology Co., Ltd., Jinan 250103, China

* Correspondence: xiaoming0502@21cn.com; Tel.: +86-759-2396027

Abstract: In this paper, the quality change of Yesso scallop (*Patinopecten yessoensis*) in the process of anhydrous storage and transportation after cold acclimation and induced dormancy was studied, and the regulation mechanism of quality degradation during storage and transportation in the process of gradient chilling stress and drying exposure was further explored. The results show that, when transferred from hydrous to anhydrous states, the breathing pattern of the scallops changed from aerobic to anaerobic. Their gill filaments were altered and their apparent vitality constantly declined, which was reflected by the edge shrinkage of the pallium and the direct proportions of the edge reduction rate and the stimulus response period. After being in the anhydrous state for 4 d, the AEC value dropped to 67.59%. At this time, if they were placed under hydration again, the scallops resumed a good growth state. By proteomics analysis, it was revealed that cold acclimation and dry exposure mainly led to changes in biological functions and pathways, such as mitochondrial inner membrane and ATP hydrolysis activity. In addition, it can be seen from the functional annotation and enrichment analysis of the metabolite KEGG that cold acclimation promoted the purine metabolism of scallops, while dry exposure inhibited the metabolism of saturated fatty acids. In this study, the infrared sensing mode was used for the first time, too, in order to record the heart-rate changes of the scallops during circulation, which shows that non-destructive vitality monitoring of *Lamellibranchia* is feasible.

Keywords: ecological ice temperature; chilling stress; anhydrous live preservation; regulation mechanism; *Patinopecten yessoensis*

Citation: Jiang, P.; Chen, D.; Chang, X.; Zhang, C.; Fan, X.; Qin, X. Study on the Regulation Mechanism of Quality Deterioration Due to Chilling Stress and Dry Exposure during Anhydrous Storage and Transportation of Yesso Scallop *Patinopecten yessoensis*. *Foods* **2023**, *12*, 2902. https://doi.org/10.3390/foods12152902

Academic Editor: Sidonia Martinez

Received: 25 June 2023
Revised: 22 July 2023
Accepted: 28 July 2023
Published: 30 July 2023

1. Introduction

Yesso scallop (*Patinopecten yessoensis*), as an important economic shellfish variety bred in northern China, has become the product of a leading industry that now drives the development of the local fishing economy [1,2]. Based on the dietary consumption habits of Chinese residents, vitality and freshness are the two main criteria to judge the quality of aquatic products, as well as the primary factors to consider when purchasing aquatic products. This has endowed fresh products with higher economic value due to their plump, tender, and delicious meat quality, as well as their plentiful and diversified nutrition [3]. Before going out into the market, scallops will go through four stages—that is, catching, clean-up and temporary breeding, anhydrous storage, and, finally, sale [4]. Therefore, issues such as optimizing and improving the transportation methods of aquatic products, enhancing the survival rate, reducing nutrient loss, and maintaining the original flavor quality are all urgent issues that need to be solved in industrial upgrading.

Ecological ice temperature anhydrous live preservation technology is a burgeoning method to drive aquatic animals into a dormant state at low temperatures, which can reduce their metabolic rate and thereby enable live preservation transportation to be conducted under low temperatures without the use of water [5,6]. Compared with traditional hydrous transportation, ecological ice temperature anhydrous live preservation manages to improve transportation efficiency and lower the cost of logistics, but the influence of stress factors, such as low temperature, cold acclimatization, hypoxic exposure, and desiccation, in waterless transportation on both the physiological state and the flavor quality of shellfish should be taken into consideration [7]. Many studies have shown that drastic changes in temperature can seriously affect the physiological activities, the immune function, and the disease resistance of shellfish, and can result in the death of the shellfish due to immune suppression [8]. Henzong et al. held the view that, before anhydrous live preservation, if Pacific oysters were trained in coldness at the ecological ice temperature, the effect of temperature stress on the oxidative immune system and the metabolism of energy substances could be reduced [9]. Lixin et al. simulated the vitality of scallops during circulation by analyzing the energy content, such as adenosine triphosphate, and they found that the vitality of scallops decreased for 24 h [10]. Hypoxia stress has a significant impact on the physiological metabolism and organizational structure of scallops. Indeed, Zhang et al. found that, after hypoxia treatment, scallops evinced decreased vitality, disturbed physiological and metabolic activities, and an abnormal organizational structure [11]. High temperature stress caused by increasing seawater temperature significantly affects energy metabolism, too [12]. However, there has been no systematic report on the stress regulation mechanism caused by stress, such as chilling stress and anhydrous storage. In this study, we examine the effects of environmental stress on the bodies of scallops, and we examine the effect of environmental stress in cold acclimatization and anhydrous storage and transportation in the temporary breeding stage.

2. Materials and Methods

2.1. Experimental Raw Materials

Scallops (*Patinopecten yessoensis*) cultured in net cages in Nanguang Island, Shandong, China, with a shell length of (76.30 ± 6.48) mm, a shell height of (18.97 ± 1.86) mm, and a wet weight of (59.63 ± 12.76) g, were selected as raw materials. In April 2022, immediately after being caught, they were sealed with foam boxes (ice pack cooling) and sent to Shandong Provincial Key Laboratory of Agricultural Products Storage, Transportation and Preservation Technology. Those that arrived in the laboratory in the living anhydrous hypoxia stress state were immediately placed in artificially circulated seawater for clean-up and were bred temporarily at a water temperature of (15 ± 2) °C and a salinity of 32‰. After temporary breeding, after 7 d, the scallops with strong vitality and an excellent state of health were selected to carry out the experiment.

2.2. Experimental Design and Methods

After temporary breeding in the laboratory, healthy and dynamic scallops were selected for gradient cooling and cold acclimatization to ecological ice temperature. In other words, by setting the cooling rate of the temporary breeding system to 2 °C/h, after lowering the temperature every 1 h, they were maintained at a constant temperature for 3 h. The temporary breeding temperature of 15 °C was lowered to a dormant temperature at 4 °C over 24 h and then the scallops were removed for anhydrous gas conditioning packaging. A dozen scallops were set in a group in a (335 × 220 × 175) mm foam box, with a biological ice bag placed at the bottom. Meanwhile, a temperature insulation bag was put between the ice bag and the scallops. The packed scallops were stored and transported in the low-temperature vibration chamber at 4 °C for 72 h, after which they were awakened in water in order to place the scallops after anhydrous storage and transportation into the temporary breeding system. Meanwhile, by the system program, the temperature was increased from the ecological ice temperature at a rate of 2 °C/h. After heating for 1 h, they were maintained at a constant temperature for 3 h until it returned to 15 °C, after which relevant index detection was performed.

2.3. Analysis of the Test

2.3.1. Living Health Evaluation

Edge shrinkage: The edge shrinkage rate refers to the method of Li et al. [13], where edge shrinkage rate = edge shrinkage distance/shell height × 100%.

Stimulus response time: Pointed tweezers were used to stimulate the pallium of the live scallops and the contraction response time was recorded using a stopwatch.

Heart rate: The measurement method was slightly modified according to that of Bakhmet et al. [14]. First, the scallop shell was cleaned, an infrared sensor was attached to the outside of the scallop shell near the heart, and the scallop was placed in artificial seawater at 15 °C, which would then be detected after the antennas of the pallium were fully displayed. The Power Lab instrument parameter range (2~5 V) was set with a low pass (1~10 Hz) and AC coupling. LabChart8 was utilized to record the changes in scallop heartbeat stability and to intercept the area of the 2 auricular appendages and the 1 ventricular waveform, thus obtaining a complete and smooth waveform map. The stable waveform in 10 min of detection time was then calculated and 5 scallops were checked each time, totaling 3 times in parallel.

2.3.2. H&E Staining and Histological Examination of Gill Tissues

The scallops were placed on a dissecting table to remove gill parts for H&E staining. All samples were cut into 5–6 µm thick slices. Then, H&E staining was performed for microscopic observation. Dehydration and transparency were performed in ethanol and xylene, and samples were then cover slipped. The stained sections were observed and photographed using a light microscope (Nikon, Tokyo, Japan, DS-Fi2) and spliced into a complete image using ImageJ software.

2.3.3. Nucleotides and Associated Compounds

The scallop adductor muscle was dissected, extracted, and mashed by adding 5% PCA solution in an ice bath. The pH was adjusted to 2–3.5, 4000× *g* rpm for 5 min of centrifugation in order to obtain the supernatant, which was then filtered using a 0.45 µm membrane filter and analyzed by HPLC after adding phosphate buffer. The chromatographic conditions were column symmetry C18 (4.60 mm 150 mm, 5 µm); mobile phase A: 0.05 mol/L KH_2PO_4-K_2HPO_4 (pH 6.78); mobile phase B: chromatographic methanol; detection wavelength: 259 nm; sample intake: 20 µ L; flow rate: 1 mL/min; column temperature: 40 °C; and gradient elution. According to the peak time of the standard product, the composition of the sample was qualitatively determined, and the substance content was calculated from the peak area.

The nucleotide energy charge AEC (adenylate energy charge) value is an indicator reflecting the degree of environmental stress on animals, which can reflect the freshness of scallops. AEC(%) = (2ATP + ADP)/(2(ATP + ADP + AMP)) × 100%.

2.3.4. Proteomics

The adductor muscle at different treatment stages was selected for proteomic analysis. Samples then underwent protein extraction, enzymatic peptide digestion, liquid chromatography–mass spectrometry (LC-MS) data collection, database retrieval, and bioinformatics analysis. Chromatography–mass spectrometry detection conditions were as follows—mobile phase A: formic acid aqueous solution; mobile phase B: 0.1% (*v*/*v*) formic acid solution. Peptide fragments were dissolved by liquid chromatography in mobile phase A and they were separated using an ultra-efficient liquid phase system. The liquid phase gradient setting was 0~120 min, 8~100%, and the mobile phase B flow rate was 300 nL/min.

2.3.5. Metabolomics

The adductor muscle at different treatment stages was selected for metabolomics analysis. After shock crushing and cold sonication, the samples were centrifuged at 12,000× g rmp for 10 min at 4 °C, and the supernatant was removed for machine testing. The column was a C18 column (Zorbax Eclipse C18 (1.8 µm × 2.1 mm × 100 mm), and the chromatographic separation conditions were a column temperature of 30 °C and a flow rate of 0.3 mL/min, with mobile phase compositions A (0.1% formic acid solution) and B (pure acetonitrile, gradient elution).

In positive mode, the heater temperature was 325 °C; the sheath gas flow was 45 (arbitrary units); the aux. gas flow was 15 arb; the sweep gas flow was 1 arb.; the electrospray voltage was 3.5 KV; the capillary temperature was 330 °C; and the S-Lens RF level was 55%.

In negative mode, the heater temperature was 325 °C; the sheath gas flow was 45 arb; the aux. gas flow was 15 arb; the sweep gas flow was 1 arb.; the electrospray voltage was 3.5 KV; the capillary temperature was 330 °C; and the S-Lens RF level was 55%.

For the scanning mode, full scan was m/z 100~1500 and data-dependent mass spectrometry (dd-MS2, TopN = 10), and the resolution was 120,000 (MS1) and 60,000 (MS2). The collision mode was High Energy Collision Dissociation (HCD).

2.4. Data Processing

The experimental results are expressed as the mean ± standard deviation (X ± SD), the experimental data were processed using Origin2021 and SPSS27 software and analyzed by univariate analysis, and the significance level was set as $p < 0.05$.

3. Results

3.1. The Apparent Vitality of Patinopecten yessoensis Decreased during Anhydrous Storage and Transportation

The apparent vitality of *Patinopecten yessoensis* was evaluated by determining the edge shrinkage rate and stimulus response time. During the period of purification and the temporary breeding and cold acclimatization, the pallium of the living scallops was full and had no shrinkage, and the tentacles fully extended to the shell. However, during the storage and transportation stage, as time went by, the edge shrinkage phenomenon appeared and the rate then gradually increased. By the third day of anhydrous storage and transportation, the rate reached 28.05% (Figure 1). Meanwhile, it was revealed by a Pearson correlation analysis of the pallium stimulus response time that the correlation coefficient was 0.914, showing a significant positive correlation ($p = 0.011 < 0.05$)—that is, the worse the contraction phenomenon, the worse the living state and the longer the stimulus response time (Figure 2).

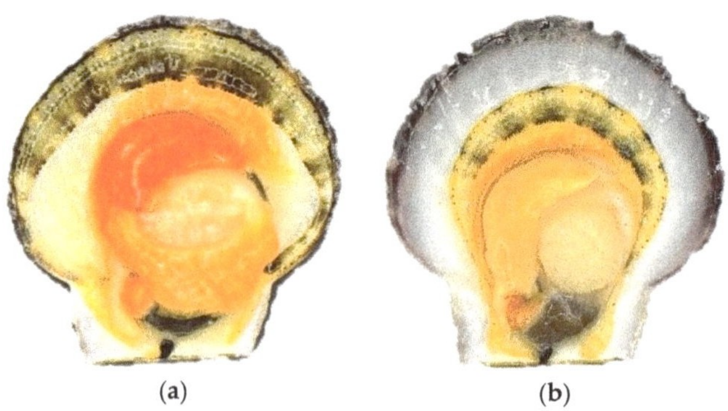

Figure 1. Living edge shrinkage of *Patinopecten yessoensis* in different periods: (**a**) temporary breeding and cold acclimatization; (**b**) anhydrous storage and transportation (3D).

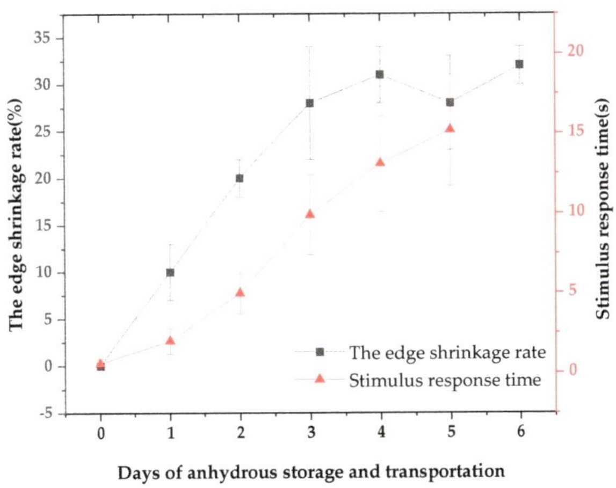

Figure 2. Changes in the apparent vitality index during anhydrous storage and transportation.

3.2. *Patinopecten yessoensis* Vitality Dropped during the Anhydrous Storage and Transportation Process

Infrared sensing was employed to monitor the heart-rate activity of scallops without doing any damage and the heart-rate cycle and intensity were recorded. The BMP value was the number of heart beats per minute. In our preliminary research, it was found that, during the 15 °C temporary breeding period, the average heart rate of scallops was 18.39 times/min. Gradient cooling was used to monitor the heart rate in real time. With the decrease in temperature, the heart rate gradually decreased and the heart rate dropped to 7.39 times/min at the cooling endpoint of 4 °C [15]. After 15 °C of rehydration, the heart rate was significantly higher than that of the control group ($p < 0.05$) (Figure 3). Among them, the heart-rate values at 24 h, 48 h, and 72 h were 23.86 ± 0.80 bpm, 28.70 ± 1.51 bpm, and 24.72 ± 0.48 bmp, respectively, which were 60.41% higher than those of the control group; the survival rate of scallops decreased after 72 h and the heart-rate measurement was thus terminated.

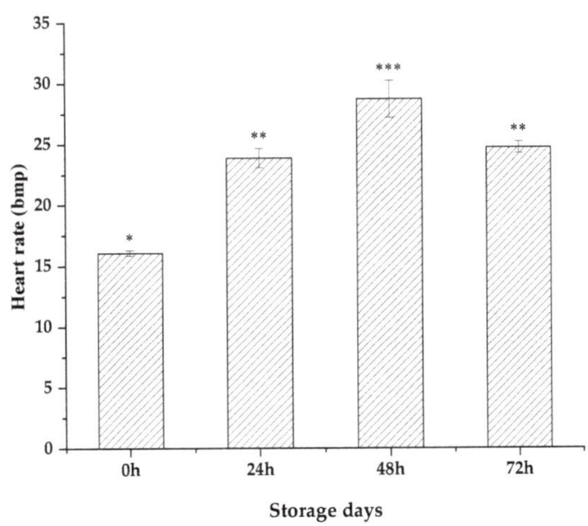

Figure 3. Heart-rate change of *Patinopecten yessoensis* after rehydration at different storage times. the significant difference was indicated by different symbos (*, ** and ***).

3.3. Changes in Microorganizational Structure of the Gills

In the comparison of the microscopic tissue sections of the gills after temporary breeding, cold acclimatization, and water storage for 3 d (see Figure 4), it can be seen that the filaments of the gills in the clean-up and temporary breeding period were uniform and neat in arrangement. After gradient cooling cold acclimatization treatment, they became shorter and thicker. After being transferred from the water environment into the anhydrous storage environment, the scallops closed their shells to retain moisture, thereby reducing their oxygen intake. As was the case from oxygen breathing to oxygen respiration, low oxygen stress led the filaments to increase in space (i.e., extend) in order to increase the contact area for air exchange. Duan et al. reported that the respiration burst of Japanese *Penaeus orientalis* in an anhydrous environment for 3 h showed that respiration weakened at 10 h and, eventually, the gill cavity movement slowed down, leading to death [16].

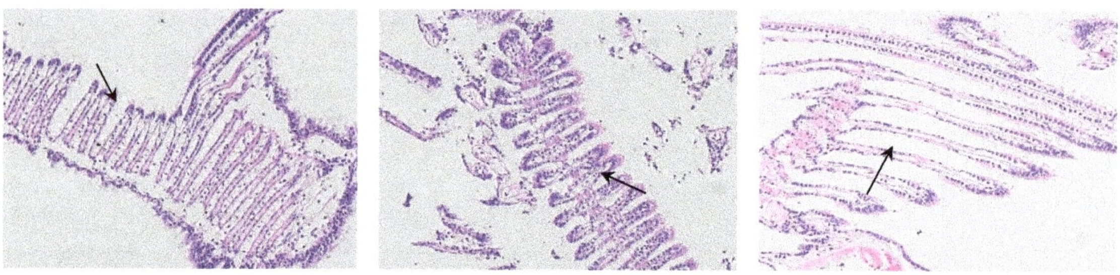

(a) (b) (c)

Figure 4. Microstructure of *Patinopecten yessoensis* gills at different stages (200), changes in gill filaments are marked in the figures: (**a**) temporary breeding stage; (**b**) cold acclimation stage; (**c**) no water storage for 3 d.

3.4. Nucleotide-Lineage Compounds

ATP provides the energy necessary to maintain normal life activities and the change in its content can better reflect the life status of shellfish [17]. In the process of anhydrous storage, the ATP content of *Patinopecten yessoensis* decreases by ladder (Figure 5). However, when the *Patinopecten yessoensis* changed from a hydrous to an anhydrous storage environment after cold acclimation, its breathing mode changed from aerobic to anaerobic and the release of energy was reduced, so the ATP showed a decreasing trend [18]. During 2–4 d of anhydrous storage and transportation, scallops were in a stress state for a long time, and they started their own stress response mechanism to mobilize energy materials and to maintain body balance, so the ATP content maintained a stable state in this stage. However, in the later period of anhydrous storage and transportation, with the gradual depletion of glycogen and other substances, the rapid ATP degradation showed a steep decline. Meanwhile, the decomposition of ATP led to an increase in ADP content. ADP is the substrate of the adenylate kinase reaction, producing one ATP and one AMP from two ADP. The changes in the ADP and AMP contents are shown in Figure 5. With the extension of anhydrous storage and transportation, the body could not maintain the relative stability of ATP content in an anoxic environment—that is to say, there was an occurrence of ADP and AMP accumulation.

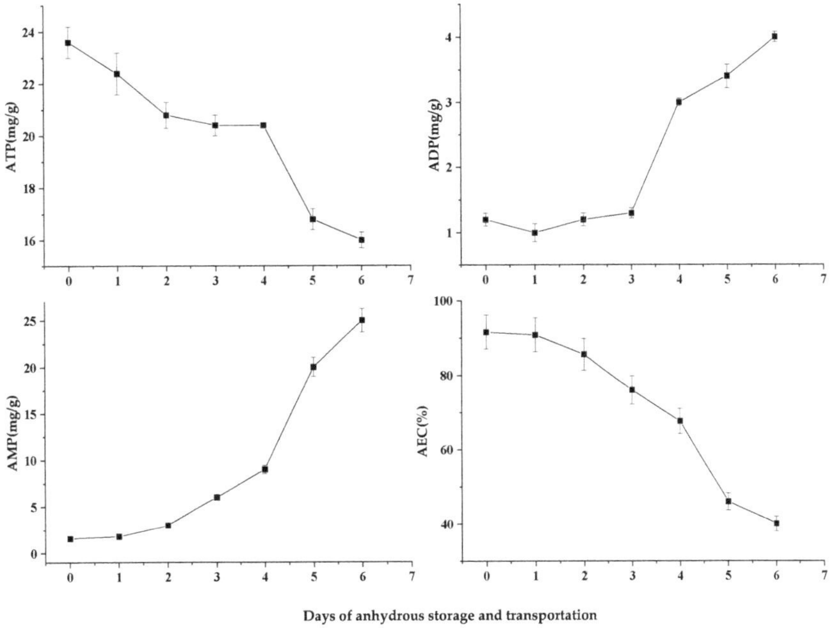

Figure 5. Changes in nucleotide series of *Patinopecten yessoensis* during anhydrous storage and transportation.

Nucleotide-energy AEC is widely used in the vitality and quality evaluation of live fish and shellfish. Maguire et al. revealed that the AEC value in scallop muscle could effectively reflect the stress intensity and state of the scallop at that time, and that the AEC value was divided into three stages to show the vitality of shellfish in different periods: 80–90% indicates a good state and reproduction being available; 50–70% represents slow growth but recoverable reproduction; and less than 50% means irreversible damage [19,20]. At 4–5 d, the AEC values were 67.59% and 46.02% (Figure 5), respectively. Therefore, to ensure a good growth state of scallops after rehydration, the storage and transportation time should be controlled within 4 d.

3.5. Environmental Stress Induced Changes in the Proteome

In order to obtain the differential protein expression changes of scallops in three different stages—fresh temporary breeding storage (XHZ), cold acclimation (LXHZ), and anhydrous live preservation storage (BHZ)—the non-labeled quantitative proteomics technique was used in this study. A search of the UniProt database identified 4548 peptides, 4134 unique peptides, and 856 proteins (Table 1). Based on the above data, a systematic bioinformatics analysis of quantitatively informative proteins was carried out, including protein annotation, significant difference analysis, annotation clustering based on significant difference, and protein interaction network analysis, in order to provide reference directions for in-depth study of the proteome.

Table 1. Protein quantitative results.

Database	Peptides	Unique Peptides	Protein Groups
Patinopecten yessoensis (Mizuhopecten yessoensis_6573)	4548	4134	856

3.5.1. GO Annotation

GO (Gene Ontology) refers to an internationally standardized classification system for gene function description, which is divided into three categories—namely, the Cellular Component, which is used to describe subcellular structure, location, and macromolecular complexes; the Molecular Function, which is used to describe the function of individual gene products; and the Biological Process, which is used to describe the biological process in which the gene-encoded products participate. According to the GO annotation results, the number of proteins corresponding to different GO entries was counted, and the annotation results of the top 10 in each major class of the GO database were drawn. As seen in Figure 6, their biological functions were mainly focused on proteolysis, translation, protein folding, phosphorylation, and other aspects, which are mainly involved in cytoplasm in terms of cellular components, membrane, ribosome, mitochondrion, and other structures. In terms of molecular functions, they were ATP binding, metal ion binding, structural constituents of ribosome, etc.

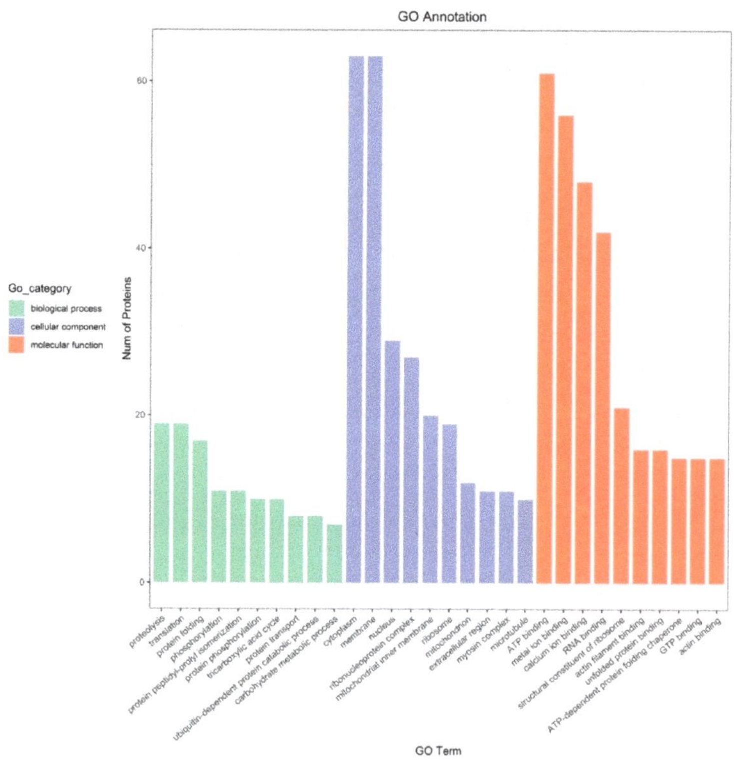

Figure 6. Bar graph of the GO annotation results.

3.5.2. COG Annotation

COG (Cluster of Orthologous Groups of proteins) refers to the protein database created and maintained by NCBI, which was constructed based on the phylogenetic relationship classification of the coding proteins of the complete genomes of bacteria, algae, and eukaryotes. Through the alignment, a certain protein sequence can be annotated to a certain COG and each cluster of COG is composed of a direct homologous sequence, so that one can speculate on the function of the sequence. The COG database is divided into 26 categories according to its functions and its annotation bar charts are drawn according to the annotation results, with the results showing that the identified proteins are mainly concentrated in post-translational modification, protein turnover, and chaperones, as well as signal transduction mechanisms, cytoskeletons, energy production and conversion, and other processes.

3.5.3. Analysis of Variance

Fresh temporary breeding (XHZ), cold acclimation (LXHZ), and anhydrous live preservation storage and transportation (BHZ) were used for between-group comparison, in which the ratio of all of the biological replicates in the comparison samples was taken as the multiple of difference (Fold Change, FC) and the protein was increased when $FC \geq 2$ ($logFC \geq 1$), while, when $FC \leq 0.5$ ($logFC \leq -1$), the protein showed downregulation of expression. Among them, the cold acclimation group upregulated 60 and 237 differentially expressed proteins compared with the transient group. It was speculated that scallops inhibited the enzyme activity by cold acclimation, so the number of downregulated expressed proteins was higher. However, the unwatered storage group raised 28 differential proteins and lowered 13 differential proteins compared with the cold acclimation group (Table 2), and Figure 7 shows volcano plots of the differential proteins.

Table 2. Differential protein quantity statistics.

Comparisons	Up	Down	All
XHZ_vs_LXHZ	60	237	297
LXHZ_vs_BHZ	28	13	41

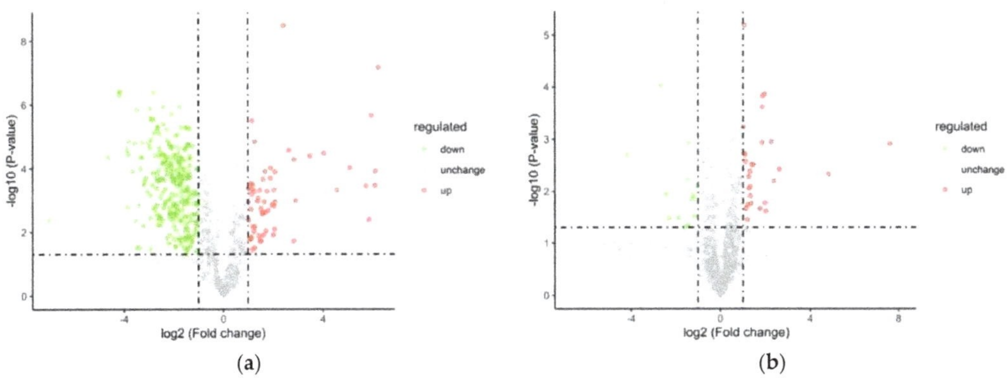

Figure 7. Volcano plots of the differential proteins: (**a**) LXHZ vs. XHZ; (**b**) BHZ vs. LXHZ.

3.5.4. GO Annotation and Enrichment Analysis of Significantly Differential Proteins

GO function significance enrichment analysis identified the GO function entries significantly enriched in the differential proteins compared to all of the identified protein backgrounds, thus providing us with the biological functions that the differential proteins are significantly associated with ($p < 0.05$). Based on the enrichment results, the bubbles of the enriched GO entries were drawn (Figure 8). The results showed that, after gradient cooling, the differential proteins were mainly enriched in mitochondrial inner membrane, ribosome, translation, actin binding, and other biological functions and pathways, while, in the transition from cold acclimation to the no water storage stage, the body mainly showed significant changes in ATP hydrolysis activity and proteolysis.

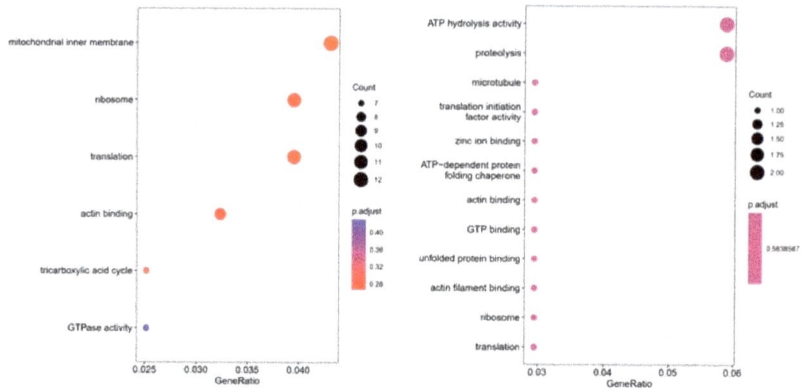

(**a**) (**b**)

Figure 8. GO enriched bubble diagram of differential proteins: (**a**) LXHZ vs. XHZ; (**b**) BHZ vs. LXHZ.

3.6. Environmental Stress Induced Changes in the Metabolome

The complex metabolic reactions and their regulation in organisms are not conducted alone, and they often form complex pathways and networks involving different genes and proteins whose mutual influence and mutual regulation eventually lead to systematic changes in the metabolome. Global metabolites in and out of the cell were analyzed qualitatively or semi-quantitatively by non-target metabolomics in order to explore the effects of cold acclimation stress and anhydrous storage on the metabolic processes of scallops. Before conducting the difference analysis, principal component analysis (PCA) of the grouped samples for the difference comparison was undertaken in order to observe the variation size between the different groups and between the samples within the group. Figure 9 shows that the three experimental samples were very significant and could be used for subsequent analysis; almost all were within the 95% confidence interval, which indicates cold acclimation (unless storage and transportation had a significant impact on scallop metabolism).

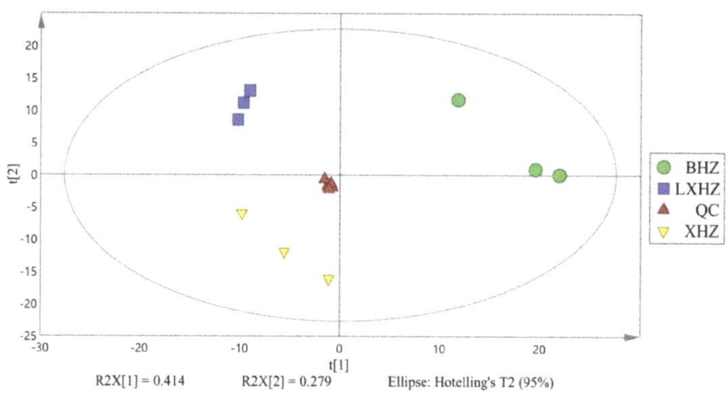

Figure 9. PCA diagram of mass spectrometry data of three scallop samples and mixed samples (XHZ—temporary breeding group, LXHZ—cold acclimation group, BHZ—anhydrous storage and transportation group, and QC—quality control mixed samples).

3.6.1. Differential Metabolite Screening

Figure 10 shows volcano plots of the differential metabolites in positive and negative ion models. Comparing the differential metabolites between the transient breeding group and the cold acclimation group, 68 differential metabolites were obtained, including adenylosuccinic acid, adenylosuccinic acid, guanosine, and dodecyl sulfate, among which 30 were upregulated products and 38 were downregulated. The cold acclimation group was stored and transported without water for 3 d, which produced 179 differential metabolites, including DL-Arginine, myristic acid, adenylosuccinic acid, Tetradecanoyl-L-Carnitine, Oleoyl-L-Carnitine, and palmitic acid, of which 85 were upregulated products and 94 were downregulated.

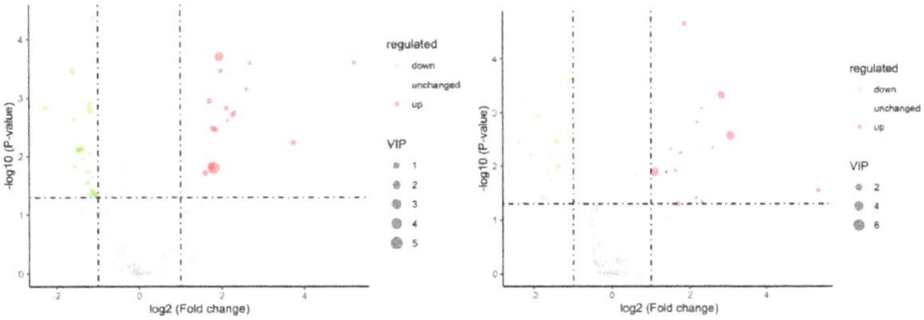

(a)

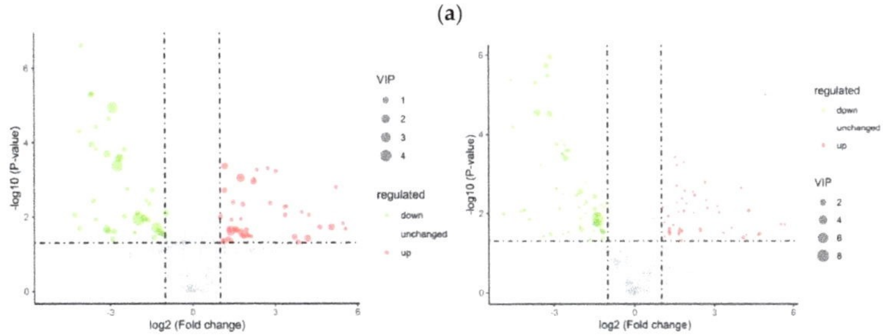

(b)

Figure 10. Volcano plots of the differential metabolite in positive and negative ion models: (**a**) LXHZ vs. XHZ; (**b**) BHZ vs. LXHZ.

3.6.2. Functional Annotation and Enrichment Analysis of the Differential Metabolite KEGG

The differential metabolites were annotated through the KEGG database and classified based on the corresponding pathway; the size of the scatter in the figure indicates the number of differential metabolites that were enriched on the pathway. As can be seen in Figures 11 and 12, the differential metabolites produced by the transient group by cold acclimation treatment mainly belonged to the purine metabolism pathway of KEGG. Cold acclimation promoted the purine metabolism of the scallops, and the differential metabolites were labeled on the pathway map showing the content of adenylosuccinate and IMP, which were 3.60 times and 3.37 times that of the transient control, respectively, while the content of guanosine decreased significantly.

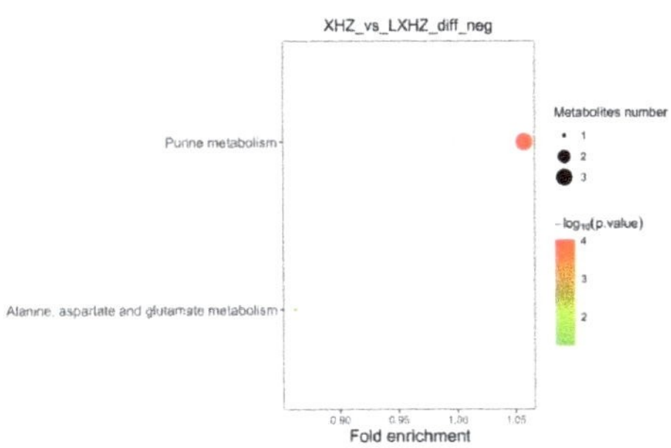

(a)

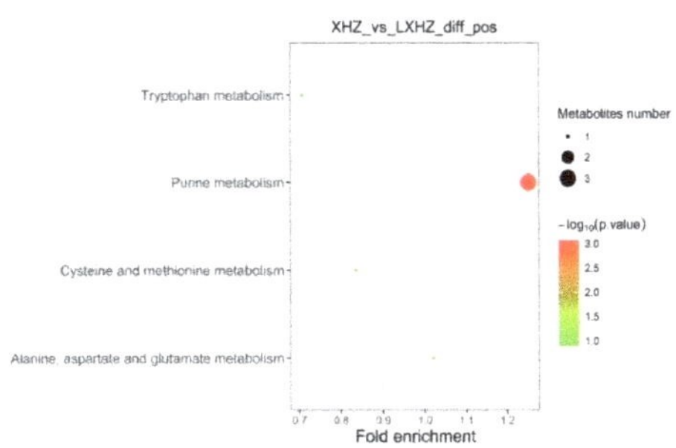

(b)

Figure 11. KEGG enrichment pathway of cold-acclimated differential metabolites: (**a**) negative ion mass spectrum; (**b**) positive ion mass spectrum.

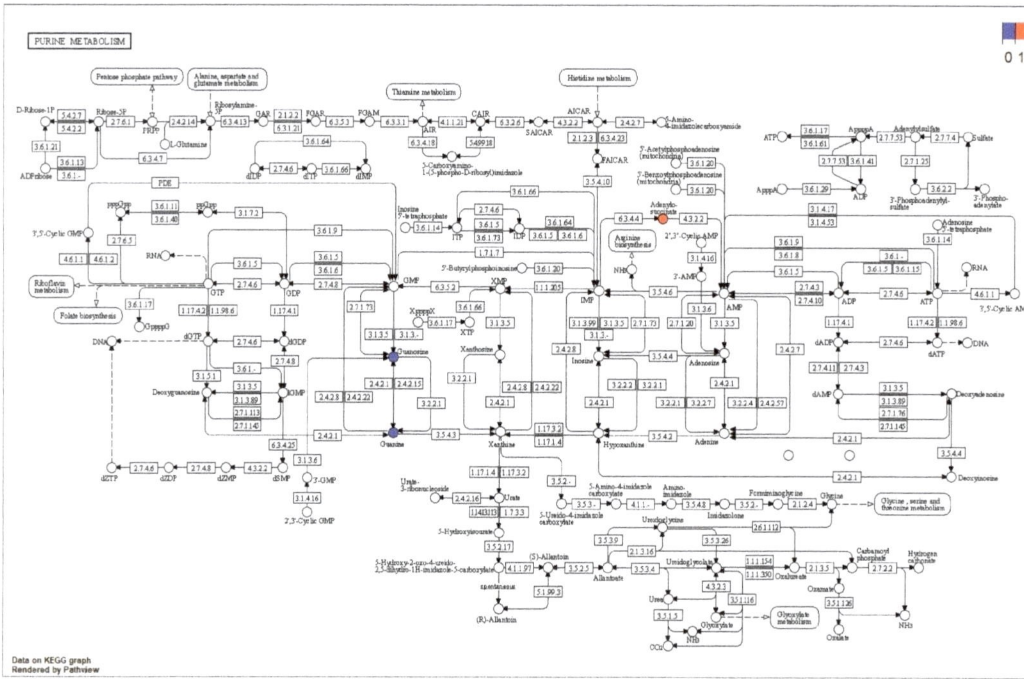

Figure 12. Metabolic pathway map of purine metabolism.

The cold acclimation group was stored and transported without water for 3 d, and the differential metabolites generated at this stage mainly belonged to the biosynthesis of unsaturated fatty acids, fatty acid biosynthesis, and purine metabolism pathways of KEGG (Figure 13). It was then speculated that low-temperature and anhydrous storage and transportation inhibited the metabolism of saturated and unsaturated fatty acids. After making the differential metabolites visible in the pathway map, tetradecanoic acid, hexadecanoic acid, icosapentaenoic acid (EPA), and palmitic acid were significantly reduced.

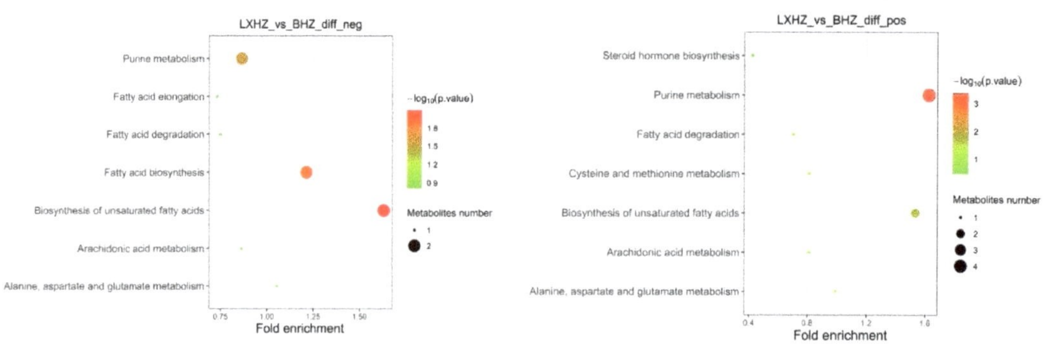

(a) (b)

Figure 13. KEGG enrichment pathway of anhydrous storage and transportation differential metabolites: (**a**) negative ion mass spectrum; (**b**) positive ion mass spectrum.

4. Discussion

The flavor quality of scallops is closely related to their physiological state. In the living market, the opening rate, closure sensitivity, and mantle shrinkage often reflect the quality of the scallop [21]. Biochemical indicators, such as pH, glycogen, ATP-related substances, and K values, are also widely used in shellfish activity evaluation methods [22,23]. The apparent vitality of *Patinopecten yessoensis* evaluated by the rate of edge shrinkage and stimulus response time was consistent with the results of Li et al. The degree of the pallium was positively correlated with the stimulus response time. The greater the degree of edge shrinkage, the longer the stimulus response time—that is, the worse the vitality state.

When stressed by an adverse external environment, living animals can produce stress and cause changes in heart rate, and non-implantable heart-rate measurement is more and more widely used in marine organisms. Mat et al. detected the rhythmic activity of Pacific oysters by means of adhesion sensing, and Dong et al. studied the temperature and the heat resistance of limpets at different temperatures through heart-rate assessment methods. Chen et al. also used the method of heart-rate assessment to evaluate the temperature and the heat resistance of abalone, which shows that the method of heart-rate assessment is becoming increasingly mature in the study of shellfish.

It is well recognized that scallops usually upregulate the expression of a large number of stress proteins in order to cope with protein damage from environmental stress and this process obviously requires the expenditure of large amounts of energy [24–29]. Energy material is often the most direct reaction to the physiological state of the scallop. ATP gradually decreased in the process of anhydrous storage and transportation of *Patinopecten yessoensis*, and this was consistent with the earlier results of Zhang et al. [30,31]. Scallops are often in a state of hypoxia during long-term transportation and severe hypoxia will have a negative impact on the general health of the scallop. The loss of ATP under hypoxia stress also leads to a decrease in blood cell adsorption capacity and phagocytosis activity [32]. In Xu's study, seven different metabolites were produced after wet storage and transportation and the pathway analysis showed that the tricarboxylic acid cycle was the most susceptible pathway [33]. Energy metabolism involves the degradation and synthesis of high-energy phosphates [34], and both GMP and IMP in this study might be related to this process. The results of proteomic GO annotation also showed that the differential proteins were mainly characterized in the tricarboxylic acid cycle, mitochondrion, and ATP, and this was consistent with the metabolome.

Author Contributions: Conceptualization, X.Q. and P.J.; methodology, X.C.; software, validation, and formal analysis, D.C.; investigation and resources, X.F.; data curation and writing—original draft preparation, P.J.; writing—review and editing, C.Z.; visualization, supervision, project administration, and funding acquisition, X.Q. All authors have read and agreed to the published version of the manuscript.

Funding: This research was funded by the National Key R&D Program of China (2021YFD2100504) and the Special Expert Program for Taishan Scholars (ts20190956).

Institutional Review Board Statement: This experiment was approved by the Animal Ethics Committee of Guangdong Ocean University (the approval code: 1122003004 and 30 November 2021).

Data Availability Statement: The data used to support the findings of this study can be made available by the corresponding author upon request.

Acknowledgments: We are grateful to Qingdao STD Standard Testing Co., Ltd. for assisting in MS analysis and bioinformatics analysis.

Conflicts of Interest: The authors declare no conflict of interest.

References

1. Yu, Z.; Li, D.; Wang, X.; Wang, Q.; Li, H.; Teng, W.; Liu, X.; Zhou, Z. Reason of massive mortality of Japanese scallop *Pationopectin yessoenisis* in raft cultivation in coastal Changhai county. *Fish. Sci.* **2019**, *38*, 420–427.
2. Wang, Q.C. Introduction of *Pationopectin yessoenisis* and the prospect of increasing culture in north China. *Fish. Sci.* **1984**, *4*, 24–27.
3. Guan, B.B.; Chen, B.; Cheng, X.H. Research on characterization and evaluation methods of aquatic products. *Fujian Anal. Test.* **2019**, *28*, 15–21. [CrossRef]
4. Pan, L.; Lin, C.; Zhang, G.; Mu, G.; Wang, Y. The influence of purification of temporary breeding and low temperature water transportation on the quality of scallop. *J. Agric. Eng.* **2017**, *33*, 301–307. [CrossRef]
5. Nie, X.; Zhang, Y.; Sun, X.; Huang, B.; Zhang, C. Key technologies for the transportation of live fish and their processes. *Fish. Mod.* **2014**, *41*, 34–39. [CrossRef]
6. Xie, J.; Wang, Q. Progress in Understanding Environmental Stress and Physiological Regulation Mechanism in Aquatic Animals during Live Transportation. *Food Sci.* **2021**, *42*, 319–325.
7. Pan, L.; Jiang, J.; Zhang, N.; Huang, D.; Gao, X.; Liu, H.; Zhang, G. Effects of different temperatures on quality of live bay scallop argopectenirradians packed by oxygenation. *Fish. Sci.* **2019**, *38*, 182–187.
8. Konstantinov, A.S.; Pushkar, V.Y.; Averyanova, O.V. Effects of fluctuations of abiotic factors on the metabolism of some hydrobionts. *Biol. Bull. Russ. Acad. Sci.* **2003**, *30*, 610–616. [CrossRef]
9. Lin, H.; Gao, J.; Liang, Z.; Fan, X.; Lin, H.; Cao, W.; Huang, Y.; Qin, X. The effect of cold stress mode on current oxidative stress and energy consumption of Pacific oysters. *J. Ocean Univ. Guangdong* **2022**, *42*, 95–103.
10. Yan, L.; Tian, Y.; Jiang, M.; Liu, R.; Xu, T. Changes of vitality and flavor characteristics of shrimp scallop in waterless transportation—Wet storage sales. *Aquat. Sci.* **2022**, *41*, 44–51. [CrossRef]
11. Zhang, W.; Lv, Z.; Zhang, Y.; Chen, J.; Li, F.; Zheng, L.; Cong, X. Influence of hypoxia stress on physiological metabolism of *Ruditapes philippinarum*. *Chin. J. Ecol.* **2014**, *33*, 2448–2453.
12. Yang, C.; Wang, X.; Zhou, K.; Jiang, D.; Shan, Y.; Wang, L.; Song, L. Effect of high temperature stress on glycogen metabolism in gills of Yesso scallop *Patinopecten yessoensis*. *Fish Shellfish Immunol.* **2023**, *138*, 108786. [CrossRef]
13. Li, Y.; Liu, J.; Zhou, J.; Zhou, Q.; Tian, Y.; Xu, T. Effects of tied-up relaying on stress-reduction and storage-stability of live dived *Patinopecten yessoensis*. *J. Fish. China* **2022**, *46*, 605–615.
14. Bakhmet, I.; Aristov, D.; Marchenko, J.; Nikolaev, K. Handling the heat: Changes in the heart rate of two congeneric blue mussel species and their hybrids in response to water temperature. *J. Sea Res.* **2022**, *185*, 102218. [CrossRef]
15. Chang, X.Y.; Jiang, P.H.; Deng, J.; Fan, X.P.; Qin, X.M. The influence of chilling stress induced dormancy on the life characteristics and nutritional quality indexes of shrimp scallop. *South China Fish. Sci.* **2023**, *19*, 129–139.
16. Duan, Y.; Zhang, J.; Dong, H.; Wang, Y.; Liu, Q.; Li, H. Effect of desiccation and resubmersion on the oxidative stress response of the kuruma shrimp *Marsupenaeus japonicus*. *Fish Shellfish Immunol.* **2016**, *49*, 91–99. [CrossRef]
17. Lee, A.-C.; Lee, K.-T. The enzyme activities of opine and lactate dehydrogenases in the gills, mantle, foot, and adductor of the hard clam Meretrix lusoria. *J. Mar. Sci. Technol.* **2011**, *19*, 4. [CrossRef]
18. Hong, H.; Regenstein, J.M.; Luo, Y. The Importance of ATP-related Compounds for the Freshness and Flavor of Post-mortem Fish and Shellfish Muscle: A Review. *CRC Crit. Rev. Food Technol.* **2015**, *57*, 1787–1798.
19. Hiltz, D.F.; Dyer, W.J. Octopine in postmortem adductor muscle of the sea scallop (*Placopecten magellanicus*). *J. Fish. Res. Board Can.* **1971**, *28*, 869–874. [CrossRef]
20. Atkinson, D.E. Energy charge of the adenylate pool as a regulatory parameter. Interaction with feedback modifiers. *Biochemistry* **1968**, *7*, 4030–4034. [CrossRef]
21. Duncan, P.F. Post-Harvest Physiology of the Scallop *Pecten maximus* (L.). Ph.D. Thesis, University of Glasgow, Glasgow, UK, 1993.
22. Vallé, M.; Malle, P.; Bouquelet, S. Evaluation of fish decomposition by liquid chromatographic assay of ATP degradation product. *J. AOAC Int.* **1998**, *81*, 571–578. [CrossRef]
23. Pacheco-Aguilar, R.; Marquez-Ríos, E.; Lugo-Sánchez, M.E.; García-Sanchez, G.; Maeda-Martínez, A.N.; Ocaño-Higuera, V.M. Postmortem changes in the adductor muscle of Pacific lions-paw scallop (*Nodipecten subnodosus*) during ice storage. *Food Chem.* **2008**, *106*, 253–259. [CrossRef]
24. Sanders, B.M.; Hope, C.; Pascoe, V.M.; Martin, L.S. Characterization of the Stress Protein in Response Two with Different Species of Collisella Limpets Tolerances Temperature. *Physiol. Zool.* **2012**, *64*, 1471–1489. [CrossRef]
25. Tomanek, L. Variation in the heat shock response and its implication for predicting the effect of global climate change on species' biogeographical distribution ranges and metabolic costs. *J. Exp. Biol.* **2010**, *213*, 971–979. [CrossRef]
26. Tomanek, L.; Somero, G.N. Interspecific- and acclimation-induced variation in levels of heat-shock proteins 70 (hsp70) and 90 (hsp90) and heat-shock transcription factor-1 (HSF1) in congeneric marine snails (genus Tegula): Implications for regulation of hsp gene expression. *J. Exp. Biol.* **2002**, *205 Pt 5*, 677–685. [CrossRef]
27. Zhang, H.; Wang, H.; Chen, H.; Wang, M.; Zhou, Z.; Qiu, L.; Wang, L.; Song, L. The transcriptional response of the Pacific oyster *Crassostrea gigas* under simultaneous bacterial and heat stresses. *Dev. Comp. Immunol. Ontog. Phylogeny Aging* **2019**, *94*, 1–10. [CrossRef]
28. Yang, C.; Gao, Q.; Liu, C.; Wang, L.; Zhou, Z.; Gong, C.; Zhang, A.; Zhang, H.; Qiu, L.; Song, L. The transcriptional response of the Pacific oyster *Crassostrea gigas* against acute heat stress. *Fish Shellfish Immunol.* **2017**, *68*, 132–143. [CrossRef]

29. Zhang, S.; Han, G.; Dong, Y. Temporal patterns of cardiac performance and genes encoding heat shock proteins and metabolic sensors of an intertidal limpet *Cellana toreuma* during sublethal heat stress. *J. Therm. Biol.* **2014**, *41*, 31–37. [CrossRef]
30. Zhang, Q.; Jiang, M.; Tian, Y.; Xu, M.; Leng, H.; Liu, J. A Relevance of Adductor Protein Solubility to ATP in Yesso Scallop *Patinopecten yessoensis*. *Fish. Sci.* **2020**, *39*, 476–482. [CrossRef]
31. Liu, J.; Liu, J.; Tian, Y.; Zhang, L.; Liu, H.; Li, D. Effects of post-harvest handling on biochemical metabolism of bottom cultured live scallop (*Patinopecten yessoensis*). *J. Fish. China* **2017**, *41*, 81–87.
32. Pampanin, D.M.; Ballarin, L.; Carotenuto, L.; Marin, M.G. Air exposure and functionality of Chamelea gallina hemocytes: Effects on haematocrit, adhesion, phagocytosis and enzyme contents. *Comp. Biochem. Phys.* **2002**, *131A*, 605–614. [CrossRef] [PubMed]
33. Xu, T.; Tian, Y.; Li, Y.; Liu, Y. UPLC-MS based metabolomic study of live scallop *Mizuhopecten yessoensis* during quality determination period. *J. Dalian Ocean Univ.* **2021**, *4*, 637–645.
34. Díaz Enrich, M.J.; Ramos Martínez, J.I.; Ibarguren Arizeta, I. Implication of guanosine 3′,5′-cyclic monophosphate, adenosine 3′,5′-cyclic monophosphate, adenosine 5′-mono-, di- and triphosphate and fructose-2,6-bisphosphate in the regulation of the glycolytic pathway in hypoxic/anoxic mussel, Mytilus galloprovincialis. *Mol. Cell. Biochem.* **2002**, *240*, 111–118. [PubMed]

foods

MDPI

Article

Comprehensive Analysis of Physiological, Biochemical and Flavor Characteristics Changes in Crucian Carp (*Carassius auratus*) under Different Concentrations of Eugenol

Lexia Jiang [1,2], Baosheng Huang [1,3,4], Jiaming Tang [1,2], Peihong Jiang [1,3], Dongjie Chen [1,3] and Changfeng Zhang [1,3,*]

1 Shandong Key Laboratory of Storage and Transportation Technology of Agricultural Products, Shandong Institute of Commerce and Technology, Jinan 250103, China; 13977695205@163.com (L.J.); 18396898921@163.com (B.H.); tang19980503@163.com (J.T.); jiangpeihongjph@163.com (P.J.); dongjie613@163.com (D.C.)
2 College of Food Science and Technology, Guangdong Ocean University, Zhanjiang 524088, China
3 National Engineering Research Center for Agricultural Products Logistics, Jinan 250103, China
4 Shandong GuoNong Logistics Technology Co., Ltd., Jinan 250103, China
* Correspondence: zcf202@163.com

Abstract: Eugenol is a widely used fishery anesthetic. This study investigated the effects of various concentrations of eugenol on blood physiological and biochemical indexes, and muscle flavor, in crucian carp (*Carassius auratus*). To determine the appropriate concentration of eugenol anesthetic for use in crucian carp transportation and production operations, we evaluated seven anesthesia groups of 20, 30, 40, 50, 60, 70, and 80 mg/L and one control group (without eugenol) to determine the effects on blood physiological and biochemical indexes, and muscle flavor. The red blood cells and platelets of crucian carp decreased significantly ($p < 0.05$) with eugenol treatment. With increasing eugenol concentration, the white blood cells and hemoglobin did not change significantly, whereas lactate dehydrogenase, alkaline phosphatase, alanine aminotransferase, and aspartate aminotransferase increased significantly ($p < 0.05$). The content of phosphorus, magnesium, and sodium increased after anesthesia, whereas the content of potassium decreased with increasing eugenol concentration. After anesthesia, the content of albumin and total protein in the serum decreased with increasing eugenol concentration ($p < 0.05$); triglyceride first increased and subsequently decreased ($p < 0.05$); blood glucose content first increased and then decreased ($p < 0.05$); and no significant difference was observed in total cholesterol content ($p > 0.05$). No significant difference was observed in muscle glycogen and liver glycogen content after eugenol anesthesia ($p > 0.05$). The eugenol-based anesthesia test did not indicate major liver histomorphological effects, but the very small number of gill sheet edema cases observed requires further study. Analysis of electronic nose data indicated that eugenol treatment affected the flavor of the fish. The anesthesia concentration of 20–80 mg/L had some effect on the physiology and biochemistry of crucian carp, thus providing a reference for the application of eugenol in crucian carp transportation and experimental research.

Keywords: eugenol; anesthesia; crucian crap; physiological and biochemical indexes; histology; flavor

Citation: Jiang, L.; Huang, B.; Tang, J.; Jiang, P.; Chen, D.; Zhang, C. Comprehensive Analysis of Physiological, Biochemical and Flavor Characteristics Changes in Crucian Carp (*Carassius auratus*) under Different Concentrations of Eugenol. *Foods* **2023**, *12*, 2820. https://doi.org/10.3390/foods12152820

Academic Editors: Min Zhang and Hong-Wei Xiao

Received: 22 June 2023
Revised: 20 July 2023
Accepted: 23 July 2023
Published: 25 July 2023

1. Introduction

Crucian carp (*Carassius auratus*) is an important freshwater aquaculture fish in China, because of its high nutritional value, fresh and tender taste, fast growth rate, and strong adaptability [1]. During transportation, crucian carp show an intense stress response. Improper operations can easily cause quality deterioration and death, thus substantially affecting breeding production and transportation. The rational use of fish anesthetics can place fish in a state of anesthesia and sedation, thereby decreasing their consumption of oxygen and energy, and metabolic rate; alleviating their stress response; and facilitating operations and transportation [2]. Simultaneously, the use of anesthetics can decrease activation of the hypothalamic-pituitary–adrenal axis [3,4], diminish the pressure and physical damage caused by operations, and help improve fish welfare.

At present, nearly 30 types of fish anesthetics are available, including ethyl m-aminobenzoate methanesulfonate (MS-222), eugenol, and carbon dioxide (CO_2) [5]. Before these drugs are used, their efficiency and safety must be assessed to ensure fish safety. MS-222 is the only anesthetic approved by the U.S. Food and Drug Administration (FDA), but its application is not safe, owing to a decrease in pH value and the formation of methanesulfonic acid [6]. The advantage of carbon dioxide over other anesthetics is that it does not require a period of withdrawal, but its anesthetic dose is difficult to control [7]. *Syzygium aromaticum* is a plant of the genus *Syzygium* in the family Myrtaceae, native to the Maluku Islands in Indonesia; its buds are generally known as cloves [8]. Cloves contain 15%–20% volatile oil, and 80% of its components are eugenol. Eugenol is a plant-derived natural compound extracted from cloves; its chemical formula is 2-methoxy-4-allylphenol [9]. Pharmacodynamic studies have indicated that eugenol has sterilization, antioxidative, anesthetic, and analgesic effects [10–12]. Local analgesia and anesthesia can be achieved by inhibiting the activity of peripheral nerves. Eugenol, as an anesthetic, inhibits the sensory center of the fish brain and elicits decreases in, or even a loss of, reflexes. Its mechanism of action is divided primarily into three stages: the tactile loss period, the excitation period, and the anesthesia period. The tactile loss period refers to the inhibitory action on the fish brain cortex. The excitation period refers to the action on the basal ganglia and cerebellum, and the anesthesia period refers to the action on the spinal cord [13,14]. The U.S. FDA lists eugenol on the Generally Recognized As Safe (GRAS) list, with an acceptable daily intake (ADI) of 2.5 mg/kg^{-1}, as recommended by the World Health Organization [15,16]. Eugenol is usually used as a dental analgesic in China. Eugenol has attracted attention in recent years because of its low cost, high efficiency, safety in live fish transportation, and good solubility in freshwater and seawater. It is widely used in fish production and in scientific experiments, such as broodstock egg collection, live fish transportation, and in vivo surgery [17]. Eugenol is advantageous because it is rapidly excreted from the blood and tissues, without inducing the body to produce mutant substances that have no effect on human health [18]. Eugenol at 40–80 mg/L has anesthetic effects on *Trachidermus fasciatus* [19], and eugenol at 20–70 mg/L has anesthetic effects on *Trachinotus ovatus* [20]. Eugenol has been reported as a fish anesthetic in *Salmon trutta* [21] and *Pterophyllum scalare* [22]. Eugenol effectively decreases stress effects in fish, and increases the transportation density and survival rate; consequently, it is frequently used in the transportation of live fish. Research on eugenol in fish has focused primarily on the anesthetic effects of anesthesia stage, time, and behavior. However, relatively few studies have assessed the specific physiological and biochemical effects in fish. Studies have observed substantial differences in the appropriate anesthetic concentrations of eugenol for different fish species.

2. Materials and Methods

2.1. Fish and Maintenance

Crucian carp were purchased from the Jinan seafood market (Jinan, Shandong, China). Healthy and active adult fish with an average body length of 16.19 ± 1.58 cm and body weight of 283.33 ± 25.16 g were selected and transported to the National Engineering Research Center for Modern Logistics of Agricultural Products. The fish were temporarily cultured at the National Agricultural Products Modern Logistics Engineering Technology Research Center with an aquatic product temperature control circulating water filtration system. The temporary culture conditions were as follows: they were transported to the Research Center for temporary breeding for 2 weeks, continuously inflated, and fed with commercial compound feed once in the morning and once in the evening every day, and treated with sewage and water exchange. During this period, fish that died or survived in poor condition were removed. Feeding was prohibited 2 days before the start of the experiment. Dissolved oxygen $\geq$ 7 mg/L, pH 7–8, and water temperature 16.30 $\pm$ 2.44 °C.

2.2. Anesthetic Agents and Preparation

Eugenol (99% purity) was obtained from Sinopharm Chemical Reagent Co., Ltd., Shanghai, China. The preparation of eugenol anesthetic followed the method of Oliveira et al. with appropriate modifications [23]. The eugenol solution was prepared half an hour before the start of the experiment. The eugenol and anhydrous ethanol (volume ratio 1:9) were mixed to prepare the stock solution. The stock solution was diluted and fully stirred to generate the required concentrations during the experiment.

2.3. Sample Collection

All animal experiments were carried out in accordance with the U.K. Animals (Scientific Procedures) Act, 1986, and the National Institutes of Health Guide for the Care and Use of Laboratory Animals (NIH Publications No. 8023, revised 1978). Anesthesia testing was performed according to the method of Favero et al. with appropriate modifications [24]. The mass concentration gradient of eugenol was set to 20, 30, 40, 50, 60, 70, and 80 mg/L, and a control without eugenol was additionally assessed. Five crucian carp were randomly placed in each tank, and no significant difference in fish size was observed among groups. Anesthesia was tested with a single-tailed experimental method; that is, each fish was used in only one experiment. In each experiment, when the fish entered the complete anesthesia period, the judgment standard of the complete anesthesia period was that the experimental fish lost the conditioned reflex ability to the external stimulation, the fish body tilted, did not struggle, the respiratory rate was decreased but regular, and the gill movement frequency was low. Blood samples were collected after crucian carp were maintained under anesthesia for 10 min. The sampling method of blood and tissue was as described by Liu et al. [25]. Five test fish were selected at each blood sampling and quickly fished out of the tank. Using a 5 mL disposable sterile syringe to draw blood from the tail vein of crucian carp, the blood volume is more than 5 mL, and each fish only draws blood once. After the blood samples were placed in a refrigerator at 4 °C for 10 min, some of them were added with EDTA anticoagulant to prevent coagulation, which was used for the determination of blood physiological indexes. The other part was centrifuged at 4 °C, 4000 r/min for 20 min, and the supernatant was stored in the refrigerator at −80 °C for the determination of serum biochemical indexes. The liver and muscle were homogenized with 9 volumes (*w/v*) of sterile saline at 4 °C and centrifuged at 6000$\times$ *g* for 20 min at 4 °C. The supernatant was collected to determine glycogen content.

2.4. Blood Biochemistry

Red blood cells (RBC), white blood cells (WBC), platelets (PLT), and hemoglobin (HGB) were measured with an XT-1800IV Sysmex se-9500 blood cell analyzer and a kit (Shenzhen Mindray Biomedical Electronics Co., Ltd., Shenzhen, China).

2.5. Serum Biochemistry

Serum lactate dehydrogenase (LDH), alkaline phosphatase (ALP), alanine aminotransferase (ALT), aspartate aminotransferase (AST), phosphorus (PO_4^{3-}), potassium (K^+), magnesium (Mg^{2+}), sodium (Na^+), glucose (GLU), albumin (ALB), total cholesterol, triglyceride (TG), and total protein (TP) were determined with BK-280 automatic biochemical analysis. The kit was purchased from Shandong Boke Biological Industry Co., Ltd., Shandong, Jinan, China.

2.6. Glycogen Content Determination

Determination of muscle glycogen content and liver mass fraction was performed with glycogen kits purchased from Nanjing Jiancheng Bioengineering Institute (Nanjing, China).

2.7. E-Nose Analysis

According to the research method of Chen et al. [26] with modifications, fish meat was ground with a tissue stirrer, and 2.0 g muscle from the control group and experimental group was weighed, then placed in 10 mL sample bottles, which were sealed. Each sample was analyzed in four replicates. The system contained 18 metal oxide sensors (Table 1). Before the experiment, the measurement parameters of the electronic nose were optimized. According to the response signal of the sensor, the measurement parameters of the electronic nose were as follows: carrier gas flow rate: 150 mL/min, headspace generation temperature: 40 °C, injection volume: 2000 μL, injection speed: 2000 μL/s, headspace generation time: 600 s, data acquisition time: 120 s, and lag time: 400 s. The samples were analyzed with a radar plot and linear discriminant analysis (LDA) to remove the individuals with large differences in the samples.

Table 1. Response characteristics of the sensors of the FOX4000 electronic nose.

Serial Number	Sensor Name	Sensor Response Characteristics
1	LY2/LG	Sensitive to gases with high oxidation capacity
2	LY2/G	Sensitive to toxic gases
3	LY2/AA	Sensitive to organic compounds
4	LY2/GH	Sensitive to toxic gases
5	LY2/gCTL	Sensitive to toxic gases
6	LY2/gCT	Sensitive to flammable gases
7	T30/1	Sensitive to organic compounds
8	P10/1	Sensitive to combustible gases
9	P10/2	Sensitive to flammable gases
10	P40/1	Sensitive to gases with high oxidation capacity
11	T70/2	Sensitive to aromatic compounds
12	PA/2	Sensitive to organic compounds, toxic gases
13	P30/1	Sensitive to combustible gases, organic compounds
14	P40/2	Sensitive to gases with high oxidation capacity
15	P30/2	Sensitive to organic compounds
16	T40/2	Sensitive to gases with high oxidation capacity
17	T40/1	Sensitive to gases with high oxidation capacity
18	TA/2	Sensitive to organic compounds

2.8. Histopathology

To further compare the effects of eugenol on the tissues of crucian carp, we examined the histological changes in the gills and liver after 10 min of eugenol anesthesia. We carefully opened the operculum and cut the gills into 5–6 μm thick slices with scissors. The gills and liver were fixed with 10% formaldehyde and washed with 4 °C PBS buffer. Subsequently, the samples were gradually dehydrated with ethanol (70–100%), rendered transparent with xylene, embedded in paraffin, and cut into 5–6 μm thick sections for H&E staining and neutral resin sealing. Optical microscopy was used for staining, observation, and imaging. Images were processed in Image Pro Plus 6.0 software.

2.9. Statistical Analysis

All data are expressed as mean ± standard deviation. Statistical analysis of the data was performed in SPSS 22.0 software (Version 22, IBM Corp., Armonk, NY, USA). Based on single-factor analysis of variance, Duncan's multiple comparison method was used for the analysis. In all cases, the minimum level of significance was set to $p < 0.05$.

3. Results

3.1. Blood and Serum Indexes in Crucian Carp

3.1.1. Blood Components

After eugenol anesthesia, the RBC concentration in each experimental group was lower than that in the control group (Figure 1A). The RBC content in the control group was $(0.25 \pm 0.01)10^{12}/L$. Significant differences were observed between the experimental groups, except the 20 mg/L group, and the control group ($p < 0.05$). The WBC content in the control group after eugenol anesthesia was $(824.15 \pm 9.03)10^9/L$ (Figure 1B). A significant difference was observed between the experimental groups, except the 20 mg/L group, and the control group ($p < 0.05$), whereas no significant difference was observed between the concentration of 30–40 mg/L ($p > 0.05$). The PLT content was lower in each experimental group ($p < 0.05$) than in the control group ($(24 \pm 1)10^9/L$; Figure 1C). A significant difference was observed between the control group and each experimental group ($p < 0.05$). The hemoglobin content of the control group was (90.3 ± 1.46) g/L. No significant difference was observed between the control group and each experimental group, except the 70 and 80 mg/L groups ($p > 0.05$; Figure 1D).

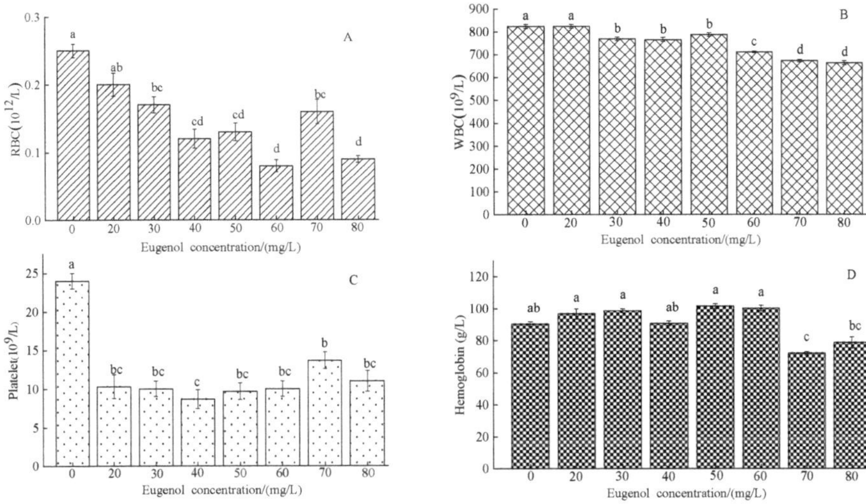

Figure 1. Changes in RBC, WBC, PLT, and HGB in crucian carp anesthetized with different concentrations of eugenol. The same lowercase letter in the graph indicates an insignificant difference ($p > 0.05$), whereas different lowercase letters indicate a significant difference ($p < 0.05$). Note: (**A**)—RBC, (**B**)—WBC, (**C**)—PLT, (**D**)—HGB.

3.1.2. The Changes in Serum Enzymes

After eugenol anesthesia, the concentration of LDH in each experimental group was higher than that in the control group (Figure 2A). The LDH content in the control group was 531.67 ± 13.67 U/L, and the highest LDH in the 80 mg/L group was 2560 ± 17 U/L. A significant difference was observed between the experimental group and the control group ($p < 0.05$). With increasing eugenol concentration, the content of ALP gradually increased (Figure 2B). The concentration of ALP in the control group was 11.32 ± 3.46 U/L. A significant difference was observed between the control group and the 60, 70, and 80 mg/L test groups ($p < 0.05$) but not between the other experimental groups ($p > 0.05$). The content of ALT in the control group was 8.7 ± 0.61 U/L (Figure 2C). The content of ALT in each experimental group was higher than that in the control group, except for that in the 20 mg/L group, which significantly differed from that in the control group ($p < 0.05$). The content of aspartate aminotransferase (AST) in the control group was 326.83 ± 12.16 U/L, whereas that in each experimental group was significantly higher ($p < 0.05$).

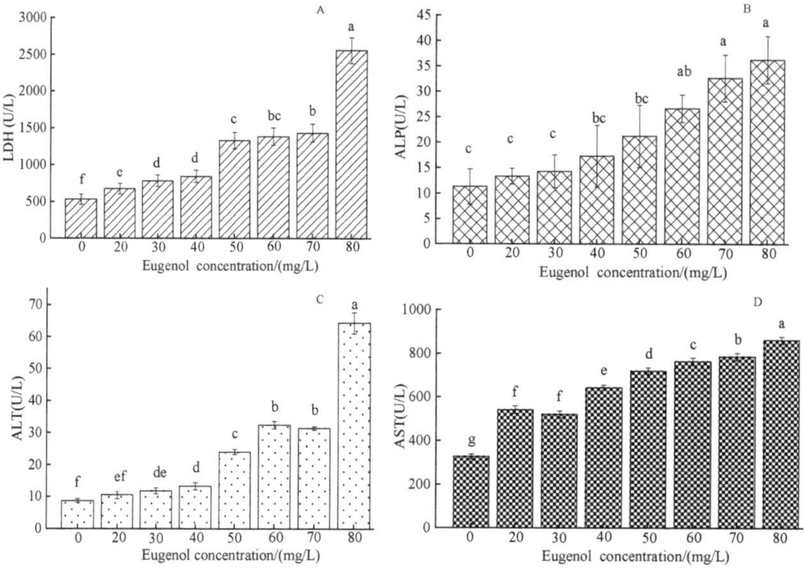

Figure 2. Changes in serum LDH, ALP, ALT, and AST content in crucian carp anesthetized with different concentrations of eugenol. Note: (**A**)—LDH, (**B**)—ALP, (**C**)—ALT, (**D**)—AST. The same lowercase letter in the graph indicates an insignificant difference ($p > 0.05$), whereas different lowercase letters indicate a significant difference ($p < 0.05$).

3.1.3. Serum Ion Content

The serum PO_4^{3-} concentration in crucian carp was 4.07 ± 0.16 mmol/L in the control group and was higher in each experimental group after anesthesia (Figure 3A), except for the 50, 60, and 70 mg/L groups, which significantly differed from the control group ($p < 0.05$). The serum K^+ concentration in crucian carp in the control group was 5.17 ± 0.17 mmol/L. The serum K^+ concentration in each experimental group after anesthesia was lower than that in the control group (Figure 3B). A significant difference in serum K^+ concentration was observed between the experimental group and the control group ($p < 0.05$). The concentration of Mg^{2+} in the serum in the control group was 1.60 ± 0.13 mmol/L, whereas that in each experimental group after anesthesia was higher (Figure 3C). A significant difference in serum Mg^{2+} concentration was observed between the experimental group and the control group ($p < 0.05$), and the concentration of Mg^{2+} in the 80 mg/L concentration group was highest, at 2.67 ± 0.08 mmol/L. The serum Na^+ concentration in the control group was 82.77 ± 4.39 mmol/L, and was higher in each experimental group after anesthesia (Figure 3D). A significant difference in serum Na^+ concentration was observed between the experimental group and the control group ($p< 0.05$), but no significant difference was observed between the 40–80 mg/L concentration groups ($p > 0.05$).

3.1.4. The Serum Concentrations of Organic Components

The concentration of ALB in the control group was 29.58 ± 3.25 g/L, whereas that in each experimental group was lower (Figure 4A). A significant difference in ALB concentration was observed between the control group and each experimental group ($p < 0.05$). The TP concentration in the control group was 52.87 ± 0.87 g/L, and a significant difference was observed between the control group and each experimental group ($p < 0.05$). The concentration of TG in the control group was 3.15 ± 0.3 mmol/L, and that in each experimental group was higher than that in each experimental group ($p < 0.05$; Figure 4B). No significant difference was observed between the concentrations of 70–80 mg/L ($p > 0.05$). The serum

GLU concentration in crucian carp in the control group was 5.9 ± 0.11 mmol/L, and was higher in each experimental group ($p < 0.05$; Figure 4D).

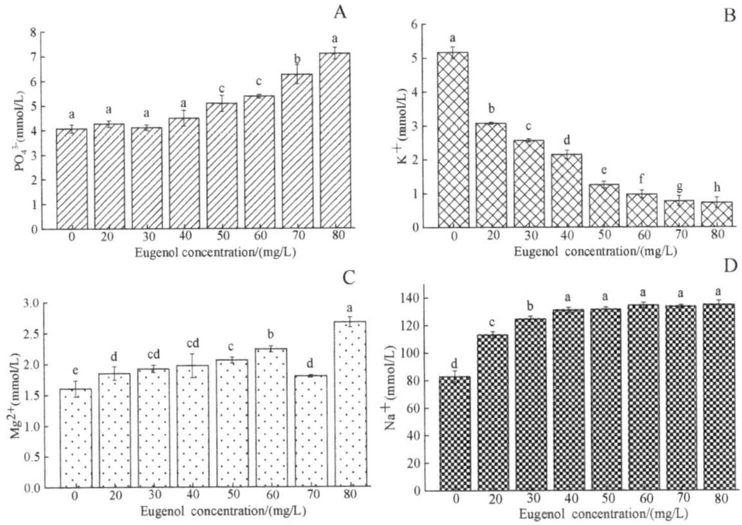

Figure 3. Changes in serum PO_4^{3-}, K^+, Mg^{2+}, and Na^+ content in crucian carp after anesthesia with different concentrations of eugenol. Notes: (**A**)—PO_4^{3-}, (**B**)—K^+, (**C**)—Mg^{2+}, (**D**)—Na^+. The same lowercase letter in the graph indicates an insignificant difference ($p > 0.05$), whereas different lowercase letters indicate a significant difference ($p < 0.05$).

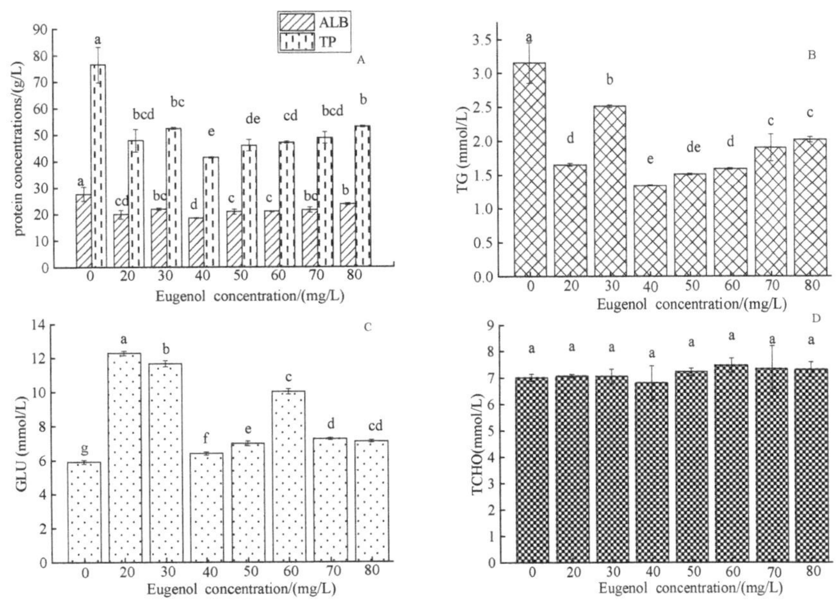

Figure 4. Changes in serum ALB, TP, TG, GLU, and TCHO in crucian carp after anesthesia with different concentrations of eugenol. Notes: (**A**)—ALB, TP, (**B**)—TG, (**C**)—GLU, (**D**)—TCHO. The same lowercase letter in the graph indicates an insignificant difference ($p > 0.05$), whereas different lowercase letters indicate a significant difference ($p < 0.05$).

3.1.5. The Glycogen Index

The content of liver glycogen in each experimental group was higher than that in the control group ($p < 0.05$). The muscle glycogen content significantly increased in crucian carp after being anesthetized with different concentrations of eugenol ($p < 0.05$; Figure 5B). The muscle glycogen content of crucian carp in the control group was 3.12 ± 0.42 mg/g, which was similar to the change trend of liver glycogen. A significant difference in muscle glycogen content was observed between the control group and each experimental group ($p < 0.05$), whereas no significant difference was observed among experimental groups except the 20 mg/L group ($p > 0.05$).

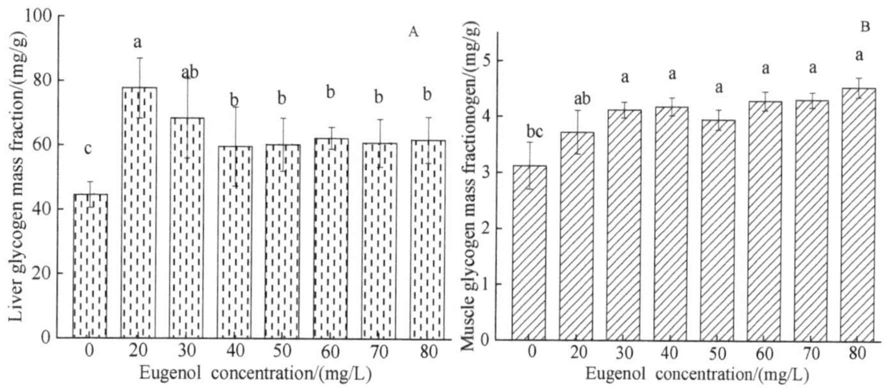

Figure 5. Changes in liver and muscle glycogen mass fraction content in crucian carp after anesthesia with different concentrations of eugenol. Notes: (**A**)—liver glycogen mass fraction, (**B**)—muscle glycogen mass fraction. The same lowercase letter in the graph indicates an insignificant difference ($p > 0.05$), whereas different lowercase letters indicate a significant difference ($p < 0.05$).

3.2. Pathology

3.2.1. Gills

The gill tissue sections of the control group and carp anesthetized with different concentrations of eugenol are shown in Figure 6. In contrast with the control group, the 20, 60, 70, and 80 mg/L eugenol-treated groups showed edema and elevation of the gill lamellae, which were not neatly arranged (neither perpendicular to the gill filaments nor parallel to one another), and the bases of the gill lamellae were elevated and not smooth. The mitochondria-rich cells were smoothly distributed at the base of the gill lamellae, with no clear elevation or detachment; the flat epithelial cells were neatly arranged on the gill lamellae; and the overall structure showed normal physiology.

3.2.2. Liver

The liver tissue sections in the control group and groups anesthetized with different concentrations of eugenol are shown in Figure 7. The liver samples of all anesthetized groups were compared with those in the control group, and no significant changes in normal liver histology were observed. Most nuclei appeared in the centers of the liver cells, and prominent heterochromatin and nucleoli were observed. In all experimental groups, no bleeding symptoms, necrotic areas, inflammation, edema, or granuloma were observed. The islets contained acinar cells and many secretory granules.

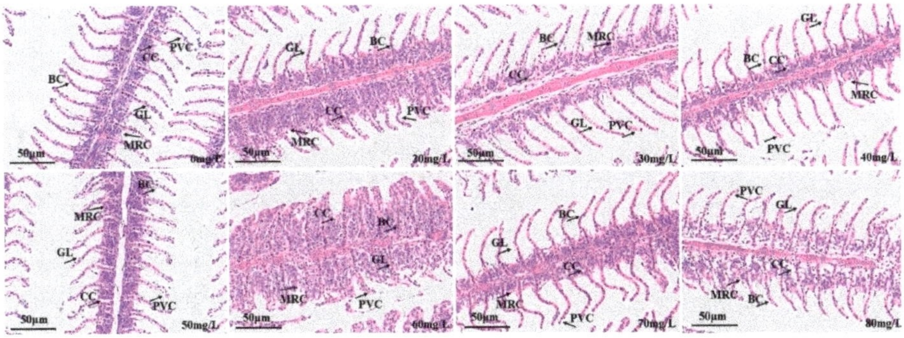

Figure 6. Effects of eugenol anesthesia on the histological structures in the liver in crucian carp. Note: BC: blood channel; GL: gill lamella; CC: chloride cell; PVC: pavement cells; MRC: mitochondrion-rich cell.

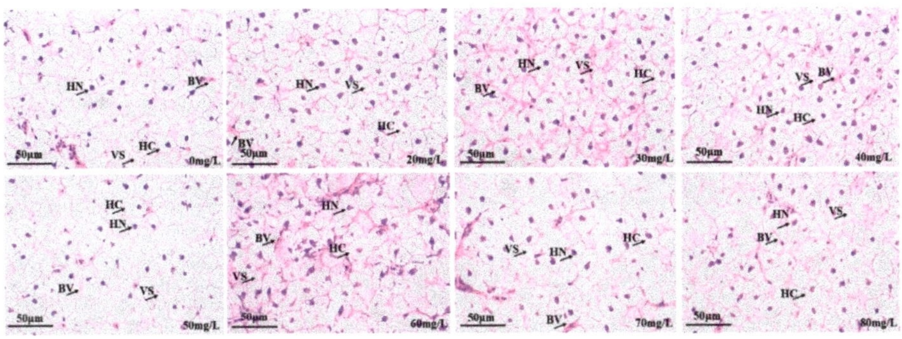

Figure 7. Effects of eugenol anesthesia on the histological structure of the liver in crucian carp. Note: VS: vacuoles; HC: hepatocytes; HN: hepatocyte nucleus; BV: blood vessels.

3.2.3. Muscle

The muscle tissue sections of the control group and the group anesthetised with different concentrations of eugenol are shown in Figure 8. Muscle is composed of many muscle fibers. In the control group, the muscle fibers were closely arranged. The muscle fiber sizes among the concentration groups differed, the muscle fiber arrangement was disordered, and the muscle fiber diameter was significantly smaller than that in the control group. In the 20–80 mg/L concentration group, the muscle fiber gap was further widened, the muscle fiber was disordered, and a clear fracture phenomenon occurred.

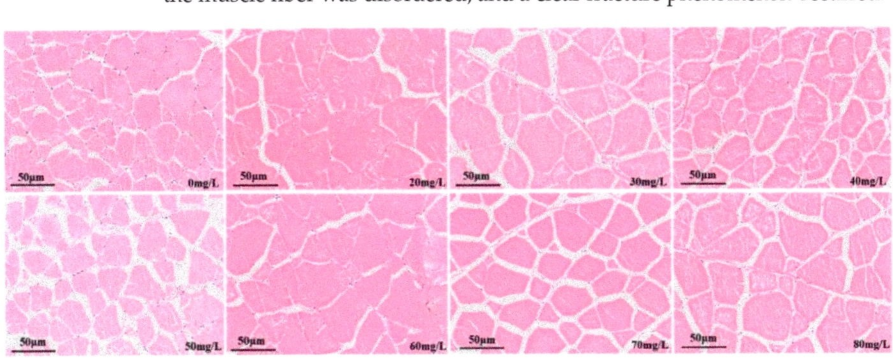

Figure 8. Effects of eugenol anesthesia on muscle structure in crucian carp.

3.3. *Response of an Electronic Nose to Volatile Flavor Compounds in Crucian Carp Anesthetized with Different Concentrations of Eugenol*

3.3.1. Radar Fingerprinting of the Effects of Different Concentrations of Eugenol on Muscle Flavor in Crucian Carp

Radar fingerprints of the effect of muscle flavour in the control group and the group anaesthetised with different concentrations of eugenol are shown in Figure 9. The aroma of muscle tissue from crucian carp anesthetized with different concentrations of eugenol was analyzed with an electronic nose. A radar map was constructed to reflect the effects of different concentrations of eugenol on volatile components in muscle. The 18 sensors of the FOX4000 electronic nose differ in their sensitivity to gas. When the sensor is in contact with gas, the ratio of relative conductivity G/G_0 is proportional to the gas concentration. When $G/G_0 > 1$, the gas concentration is high. When $G/G_0 \leq 1$, there is no response gas, or the gas concentration is below the detection limit [27]. As shown in Figure 6, the 18 sensors of the electronic nose responded to different concentrations of eugenol among the volatile flavor substances of crucian carp muscle, and the response intensity differed. The G/G_0 of T30/1, P10/1, PA/2, P30/1, and P40/2 sensors was higher than that of the other sensors. The concentration of eugenol in the control group was higher than that in all experimental groups except the 80 mg/L group. The effects of eugenol on the muscle in crucian carp in the other experimental groups were unclear, and the difference in the response signal intensity was small. The principal component analysis diagram can be visually distinguished.

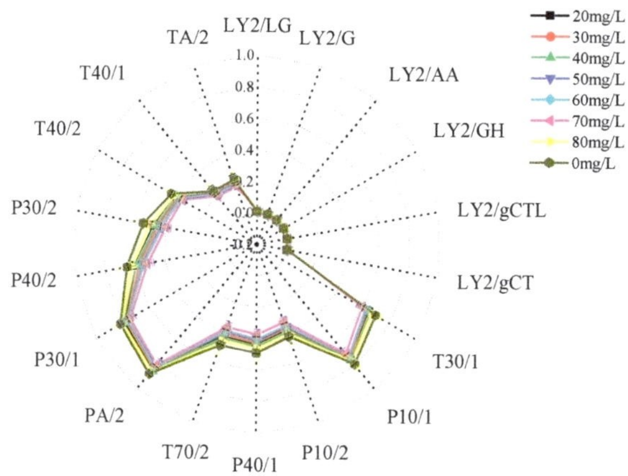

Figure 9. Radar fingerprinting of the effects of different concentrations of eugenol on muscle flavor in crucian carp.

3.3.2. LDA Analysis Based on the Electronic Nose Method

The LDA analysis of muscle based on the electronic nose method in the control group and the anesthesia group using different concentrations of eugenol in Figure 10. LDA is a statistical method that uses samples of known categories to establish a discriminant model and discriminate among unknown categories of samples [28]. The electronic nose data after LDA dimension reduction analysis are shown in Figure 7. The contribution rates of LDA1 and LDA2 were 74.2% and 14.25%, respectively, and the cumulative contribution rate was 88.45%. These two principal components essentially reflected all information characteristic of crucian carp. LDA distinguished each crucian carp muscle test group exposed to different concentrations of eugenol and the control group, and showed no overlapping area.

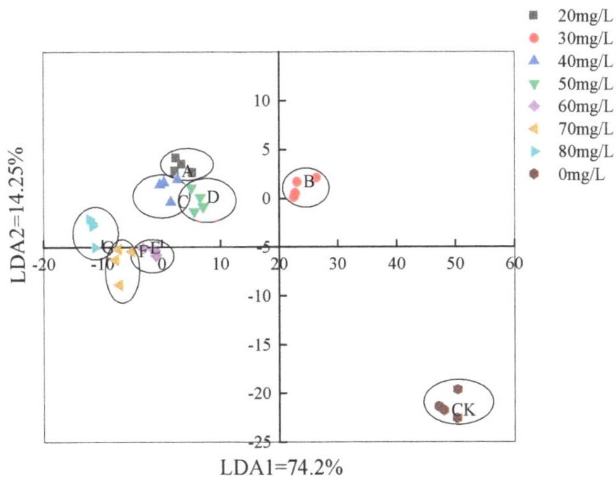

Figure 10. LDA analysis based on the electronic nose method. Note: CK: 0 mg/L, A: 20 mg/L, B: 30 mg/L, C: 40 mg/L, D: 50 mg/L, E: 60 mg/L, F: 70 mg/L, G: 80 mg/L.

4. Discussion and Conclusions

Blood physiological indexes are closely associated with metabolism, nutritional status, and diseases in fish [29]. The main role of WBC is to protect the body and resist invasion pathogens [30]. RBCs are oxygen-transporting cells in the blood, which transport fresh oxygen to the body tissue through the blood in the circulatory system in fish, and simultaneously squeeze the capillaries of the body and transport carbon dioxide from tissue to the lungs, thus allowing the fish to breathe [31]. PLT are small pieces of cytoplasm that are lysed from the cytoplasm of mature megakaryocytes in the bone marrow. These blood cells maintain hemostasis in the body and have active immunomodulatory effects. HGB is a metal protein in red blood cells, which usually transports oxygen and carbon dioxide in fish blood [32]. Liang et al. [33] analyzed the anesthetic blood indexes of clove oil on tilapia (*Oreochromis mossambicus*) and found higher content of WBC, RBC, HGB, and PLT in the blood after anesthesia than in the control group. In our study, the content of RBC, WBC, PLT, and HGB in crucian carp after anesthesia with different concentrations of eugenol was lower than that in the control group. We speculated that eugenol anesthesia might have influenced blood function in fish, in contrast to the results in tilapia research. It is speculated that the results may be caused by temperature or species differences.

Changes in serum enzymes can reflect the metabolism and material transformation of the body, and the different states of tissue structure and function, and serve as important indicators of the integrity of the cell membrane [34]. LDH is present primarily in myocardial cells, and its main function is to catalyze the oxidation of lactic acid to pyruvate in cells. An increase in LDH activity reflects the degree of damage to the kidney of the myocardial cell nucleus [35]. ALP is directly involved in the transfer and metabolism of phosphate groups in organisms, and thus plays a role in metabolism and immune protection in organisms [36]. ALT and AST are important aminotransferases widely present in animal cell mitochondria, and their content is high in hepatocytes [37]. Under normal circumstances, only small amounts of transaminases in hepatocytes are released into the blood, and consequently, the activity of transaminases in the serum is low. When the content of ALT and AST in the serum increases, hepatocytes are considered to be damaged, and ALT is an important indicator of liver damage. In general, the concentrations of LDH, ALP, ALT, and AST in the serum of crucian carp increased after eugenol anesthesia, possibly because of the synthesis and release of enzymes in the body of crucian carp in response to the stress of eugenol anesthesia, thus increasing the activity of these enzymes. This finding is consistent with research results on *Oncorhynchus mykiss* anesthetized with myrcene and eugenol [38].

Inorganic ions such as PO_4^{3-}, Mg^{2+}, K^+, and Na^+ in the serum play important roles in maintaining plasma osmotic pressure, acid–base balance, and the stability of the internal environment in fish [39]. The kidneys are associated with the metabolism of magnesium and phosphorus. When the kidneys are damaged, the level of magnesium and phosphorus metabolism in fish becomes abnormal [40]. Feng et al. [41] have found that the serum phosphorus and magnesium levels of juvenile Siberian sturgeon (*Acipenser baerii*) increase after eugenol anesthesia, similar to the results of this study. When the concentration of eugenol was 80 mg/L, the content of magnesium reached a maximum. We speculated that with increasing eugenol concentration, the damage to the fish kidneys caused by the anesthetic was aggravated, thus resulting in renal dysfunction, abnormal magnesium metabolism, and serum magnesium retention and increased level. In general, the permeability of fish gill epithelium was greatest to hydrogen ions, followed by water and Na^+. We inferred that eugenol anesthesia in crucian carp increased the permeability of the gill epithelium. Excessive Na^+ entered the blood, and its concentration significantly increased, whereas K^+ was inhibited, and its concentration significantly decreased. This finding is inconsistent with that reported by Hu et al. [42], who have found that the serum K^+ level in silver carp (*Hypophthalmichthys molitrix*) increases by four times after CO_2 anesthesia. Therefore, differences in the mechanisms of various anesthesia methods may exist among fish species, and the physiological functions of serum ions with different anesthesia methods must be further explored.

Blood GLU provides energy to support various life activities in fish and is susceptible to fluctuations in environmental factors [43]. The normal blood GLU content in fish is 2.78–12.72 mmol/L [44]. The blood GLU concentration in crucian carp in the control group was lower than that in each group after anesthesia. This finding might have been due to a decrease in blood GLU metabolism after anesthesia, thereby increasing the blood GLU concentration. The physiological functions of ALB in the blood include maintaining colloid osmotic pressure, and transporting serum calcium ions, unconjugated bilirubin, free fatty acids, and hormones [45]. After anesthesia with eugenol, the ALB value in crucian carp decreased, possibly because ALB, as a carrier of nutrients, increased consumption when dealing with the anesthesia slope, thus providing energy for the body, repairing damaged tissues, and participating in maintaining plasma colloid osmotic pressure balance. This possibility is consistent with the results of Zhu et al. [46], who observed the decreased effects of electrical anesthesia on ALB in the blood of juvenile *Coreius guichenoti*. Lipids in the blood serve as raw materials for the metabolism of liver cells, including total cholesterol, triglycerides, high-density lipoprotein, and low-density lipoprotein [47]. Total cholesterol is not only an important component of biofilm but also a prerequisite for the synthesis of bile acids, steroid hormones, vitamin D_3, and other physiologically active substances [48]. Triglyceride, total cholesterol, and total protein levels are affected by protein catabolism and hepatic glycogen decomposition. In this study, the serum triglyceride content in crucian carp in the experimental groups was lower than that in the control group, whereas the total cholesterol content in the experimental groups did not significantly differ from that in the control group, in agreement with the results of eugenol anesthesia in short-tailed bass (*Piaractus brachypomus*) [19]. The reason for the decrease may be that under anesthesia, blood lipids are involved in metabolism and are transported and decomposed. Serum GLU in fish is generally provided by liver glycogen and muscle glycogen [49]. Muscle glycogen, an important energy source for maintaining the body's metabolism, exists primarily in muscles and is an important indicator of the degree of environmental stress. Changes in muscle glycogen clearly reflect the body's adaptation to the environment [50]. The increase in muscle glycogen and liver glycogen concentration in the anesthesia group was much higher than that in the control group, possibly because of the decrease in metabolic rate and the accumulation of muscle and liver glycogen after anesthesia. No significant difference was observed in muscle glycogen and liver glycogen in each eugenol concentration group, thus indicating that different concentrations of eugenol had little effect on the muscle and liver in crucian carp.

The gills are involved in gaseous exchange, acid–base balance, ion balance, nitrogenous waste excretion, osmotic adjustment, and other multi-functional tissues [51,52]. The gills are also a biological indicator used to measure water quality [53,54] and an important means through which fish interact with the external water environment. In addition, fish gills are an important organ for the absorption of anesthetics. The gills are the main respiratory organ in fish, and are responsible for excreting metabolic wastes, such as ammonia nitrogen, and regulating osmotic pressure [55]. The fish gills are composed primarily of gill filaments and gill lamellae. The gill filaments extend vertically and have semicircular flat cystic gill lamellae, which are arranged in parallel. The main part of the gill filament contains gill cartilage, central venous sinus, and gill filament epithelial cells (composed primarily of chloride cells, mucous cells, flat cells, and undifferentiated cells). The gill lamellae are composed primarily of columnar cells, flat cells, and blood vessels (with blood cells). MS-222 has been shown to cause swelling of gill tissue [56], thereby hindering oxygen uptake by gill lamellae. Oliveira et al. [57] have found that eugenol can cause mild lamellar epithelial hyperplasia in the gill tissue in freshwater fish. Similar phenomena were also observed in this study. The liver is an important gland, and digestive and metabolic organ, in fish. It is one of the most important organs responsible for maintaining physiological functions, including bile secretion, metabolism, detoxification, and defense [58]. In this study, the different eugenol anesthesia concentrations in crucian carp did not cause liver tissue damage, in agreement with findings by Velisek et al. [59], indicating that eugenol anesthesia in carp did not cause liver or kidney tissue damage.

The electronic nose is an array based on gas sensors, which simulate the human olfactory system to capture the characteristic aroma of a sample. It has the advantages of low cost, easy operation, and high accuracy [60]. The electronic nose technology based on a bionic olfactory mechanism has developed rapidly in recent years. Li et al. [61] have quantitatively analyzed the freshness of fish meal with electronic nose technology combined with chemometrics. Zhang et al. [62] have used an electronic nose and other technologies to study the volatile flavor components and their production mechanisms in *Trachinotus blochii* fillets treated with different drying methods. Li et al. [63] used an electronic nose to detect the biochemical indexes of *Trachurus japonicus* during frozen storage, to assess its freshness. In this experiment, we observed no overlap between the control group and the groups subjected to different concentrations of eugenol anesthesia; however, overlap existed between the experimental groups, possibly because of the minor effects of eugenol on the flavor substances in crucian carp muscle. The effects of eugenol treatment on the flavor of crucian carp meat may be short-term, and the metabolism of eugenol in fish will continue to be studied in the future.

5. Conclusions

According to the changes in blood physiological and biochemical indexes in crucian carp after eugenol anesthesia, we determined that 20–80 mg/L concentrations of eugenol had relatively small effects on the physiological and biochemical indexes, but relatively greater effects on muscle, and did not result in major liver histomorphological effects. However, the edema observed in a very small number of gill lamellae requires further study to determine any potential histological effects of this anesthesia method. In fish production and scientific research, eugenol can be used reasonably in this concentration range, as appropriate. Eugenol affected the original muscle flavor of crucian carp. In the future development of fishery anesthetics, more products with less influence on flavor should be considered. This study provided a reference for the application of eugenol in crucian carp transportation and experimental research.

Author Contributions: Data Management, L.J., J.T. and D.C.; Formal Analysis, L.J.; Funding acquisition, B.H., C.Z. and P.J.; Methodology, B.H. and C.Z.; Resources, P.J. and D.C.; Writing—Original Draft, L.J.; Writing—Review and Editing, L.J. All authors have read and agreed to the published version of the manuscript.

Funding: This work was supported by the National Key R&D Program of China (2022YFD2100604, 2021YFD2100504).

Data Availability Statement: The data used to support the findings of this study can be made available by the corresponding author upon request.

Conflicts of Interest: Author Bapsheng Huang was employed by the company Shandong GuoNong Logistics Technology Co., Ltd. (Jinan, China). The remaining author declare that the research was conducted in the absence of any commercial or financial relationships that could be construed as a potential conflict of interest. The authors declare that this study received funding from Shandong GuoNong Logistics Technology Co., Ltd. The funder was not involved in the study design, collection, analysis, interpretation of data, the writing of this article of the decision to submit it for publication.

References

1. Li, L.; Li, M.; Zhu, R.; Yu, Z.; Wang, J.; Duan, J.; Wang, T.; Wu, L. Effects of β-conglycinin on growth performance, antioxidant capacity and intestinal health in juvenile golden crucian carp, *Carassius Auratus*. *Aquac. Res.* **2019**, *50*, 3231–3241. [CrossRef]
2. Palic, D.; Herolt, D.M.; Andreasen, C.B.; Menzel, B.W.; Roth, J.A. Anesthetic efficacy of tricaine methanesulfonate, metomidate and eugenol: Effects on plasma cortisol concentration and neutrophil function in fathead minnows (*Pimephales promelas* Rafinesque, 1820). *Aquaculture* **2006**, *254*, 675–685. [CrossRef]
3. Thomas, P.; Robertson, L. Plasma cortisol and glucose stress responses of red drum (*Sciaenops ocellatus*) to handling and shallow water stressors and anesthesia with MS-222, quinaldine sulfate and metomidate. *Aquaculture* **1991**, *96*, 69–86. [CrossRef]
4. Schreck, C.B.; Tort, L. The concept of stress in fish. In *Fish Physiology*; Academic Press: Cambridge, MA, USA, 2016; Volume 35, pp. 1–34.
5. Pan, H.M.; Jiang, L.X.; Zhang, C.F.; Huang, B.S. Research progress and safety evaluation of fishery anesthetics. *Preserv. Process.* **2021**, *21*, 136–143.
6. Botrel, B.M.C.; Abreu, D.C.P.; Bazana, M.J.F.; e Rosa, P.V.; Saczk, A.A. Development, optimization, and validation of the HS-SPME/GC-MS method for the residual determination of menthol in fish. *Food Anal. Methods* **2019**, *12*, 1390–1398. [CrossRef]
7. Gabriel, N.N.; Erasmus, V.N.; Namwoonde, A. Effects of different fish sizes, temperatures and concentration levels of sodium bicarbonate on anaesthesia in mozambique tilapia (*Oreochromis mossambicus*). *Aquaculture* **2020**, *529*, 735716. [CrossRef]
8. Tang, Z.L.; Zhang, J.J.; Zhou, G.Q.; Chen, S.Q.; Xu, G.C.; Xu, P.; Qiang, J.; Wang, P.P. Effects of eugenol on transport water quality and physiological indexes of blood and muscle of juvenile 'Youyu 3'. *Fish. Sci. Technol. Inf.* **2023**, *50*, 44–52.
9. Qian, W.; Sun, Z.; Wang, T.; Yang, M.; Liu, M.; Zhang, J.; Li, Y. Antimicrobial activity of eugenol against carbapenem-resistant Klebsiella pneumoniae and its effect on biofilms. *Microb. Pathog.* **2020**, *139*, 103924. [CrossRef]
10. Komala, V.V.; Ratnavathi, C.V.; Kumar, B.S.V.; Das, I.K. Inhibition of aflatoxin B_1 production by an antifungal component, eugenol in stored sorghum grains. *Food Control* **2012**, *26*, 139–146. [CrossRef]
11. He, R.; Lei, B.; Su, Y. Effectiveness of eugenol as an anesthetic for adult spotted sea bass (*Lateolabrax maculatus*). *Aquaculture* **2020**, *523*, 735180. [CrossRef]
12. Nisar, M.F.; Khadim, M.; Rafiq, M.; Chen, J.; Yang, Y.; Wan, C.C. Pharmacological properties and health benefits of eugenol: A comprehensive review. *Oxidative Med. Cell. Longev.* **2021**, *2021*, 2497354. [CrossRef]
13. Ito, M.; Murakami, K.; Yoshino, M. Antioxidant action of eugenol compounds: Role of metal ion in the inhibition of lipid peroxidation. *Food Chem. Toxicol.* **2005**, *43*, 461–466. [CrossRef]
14. Zahran, E.; Risha, E.; Rizk, A. Comparison propofol and eugenol anesthetics efficacy and effects on general health in *Nile tilapia*. *Aquaculture* **2021**, *534*, 736251. [CrossRef]
15. Anderson, W.G.; Mckinley, R.S.; Colavecchia, M. The use of clove oil as an anesthetic for rainbow trout and its effects on swimming performance. *N. Am. J. Fish. Manag.* **1997**, *17*, 301–307. [CrossRef]
16. Joint, F. *WHO Technical Report Series 934: Evaluation of Certain Food Additives*; World Health Organization: Geneva, Switzerland, 2006; pp. 49–54.
17. Wang, W.; Dong, H.; Sun, Y.; Sun, C.; Duan, Y.; Gu, Q.; Li, Y.; Xie, M.; Zhang, J. Immune and physiological responses of juvenile Chinese Sea Bass (*Lateolabrax maculatus*) to eugenol and tricaine methanesulfonate (MS-222) in gills. *Aquac. Rep.* **2020**, *18*, 100554. [CrossRef]
18. Su, M.M.; Sun, X.Q.; Yang, C.G.; Peng, X.T.; Liu, H.Y.; Cao, J.J. Research progress on application and detection methods of fishery anesthetic MS-222 and eugenol in the transportation of fresh aquatic products. *J. Food Saf. Qual. Insp.* **2015**, *6*, 25–29.
19. Hu, W.J.; Wang, C.H.; Feng, G.P.; Zhuang, P.; Zheng, Y.P.; Ji, Q. Anesthetic effects of eugenol and MS-222 on *Trachidermus fasciatus*. *Mar. Fish.* **2023**, *45*, 10.
20. Huang, X.L.; Dai, C.; Yu, W.; Yang, J.; Yang, Y.K.; Li, T.; Li, T.; Lin, H.Z.; Huang, Z.; Sun, X.Y.; et al. Anesthetic effect of eugenol on juvenile *Trachinotus ovatus*. *J. Guangdong Ocean. Univ.* **2020**, *40*, 124–131.
21. Wang, J.L.; Wang, W.L.; Zhang, B.B.; Tang, D.M. Comparison of anesthetic effects of MS-222 and eugenol on *Yadong salmon*. *J. Gansu Agric. Univ.* **2021**, *56*, 26–32+40.
22. Tarkhani, R.; Imani, A.; Jamali, H. Anaesthetic Efficacy of eugenol on various size classes of angelfish (*Pterophyllum scalare schultze*, 1823). *Aquac. Res.* **2017**, *48*, 5263–5270. [CrossRef]

23. Oliveira, C.P.B.; Lemos, C.H.P.; Vidal, L.V.O. Anaesthesia with eugenol in hybrid amazon catfish (*Pseudoplatystoma reticulatum* × *Leiarius marmoratus*) handling: Biochemical and haematological responses. *Aquaculture* **2019**, *501*, 255–259. [CrossRef]
24. Favero, G.C.; Silva, W.S.; Boaventura, T.P.; Leme, F.O.P.; Luz, R.K. Eugenol or salt to mitigate stress during the transport of juvenile *Lophiosilurus alexandri*, a neotropical carnivorous freshwater catfish. *Aquaculture* **2019**, *512*, 734321. [CrossRef]
25. Liu, J.; Zhang, P.; Wang, B.; Lu, Y.; Li, L.; Li, Y.; Liu, S. Evaluation of the effects of astragalus polysaccharides as immunostimulants on the immune response of crucian carp and against SVCV in vitro and in vivo. *Comp. Biochem. Physiol. Part C Toxicol. Pharmacol.* **2022**, *253*, 109249. [CrossRef]
26. Chen, D.J.; Zhang, M.G.; Nie, X.B.; Jiang, P.H.; Zhang, Y.H.; Zhang, C.F.; Ren, F. Detection of turbot quality treated by electrostatic field based on gas phase ion mobility spectrometry. *Food Sci.* **2019**, *40*, 313–319.
27. Jia, Z.; Chen, X.T.; Pan, N.; Cai, S.L.; Zhang, Y.; Liu, Z.Y. Changes of volatile flavor substances in *Takifugu bimaculatus* during cold storage. *Food Sci.* **2021**, *42*, 188–196.
28. Li, H.D.; Xu, Q.S.; Liang, Y.Z. LibPLS: An integrated library for partial least squares regression and linear discriminant analysis. *Chemom. Intell. Lab. Syst.* **2018**, *176*, 34–43. [CrossRef]
29. Qi, Q.; Liu, X.Y.; Liu, H.B.; Zhang, Y.; Zhang, Y.W.; Zhang, H.Y.; Peng, X.; Sun, D.J. Effects of compound Chinese herbal medicine on serum biochemical indexes of *Siberian sturgeon* broodstock. *J. Northeast. Agric. Univ.* **2012**, *43*, 134–138.
30. Safuan, S.N.M.; Tomari, M.R.M.; Zakaria, W.N.W. White blood cell (WBC) counting analysis in blood smear images using various color segmentation methods. *Measurement* **2018**, *116*, 543–555. [CrossRef]
31. Kumar, M.; Matoba, O.; Quan, X.; Rajput, S.K.; Morita, M.; Awatsuji, Y. Quantitative dynamic evolution of physiological parameters of RBC by highly stable digital holographic microscopy. *Opt. Lasers Eng.* **2022**, *151*, 106887. [CrossRef]
32. Wang, K.; Bian, X.; Zheng, M.; Liu, P.; Lin, L.; Tan, X. Rapid determination of hemoglobin concentration by a novel ensemble extreme learning machine method combined with near-infrared spectroscopy. *Spectrochim. Acta Part A Mol. Biomol. Spectrosc.* **2021**, *263*, 120138. [CrossRef]
33. Liang, Z.Y.; An, L.N.; Dong, Z.J.; Miao, L.H.; Xu, P.; Xie, Z. Anesthetic effect of clove oil on *Tilapia* and its effects on blood indexes and hormone levels. *J. Shanghai Ocean. Univ.* **2009**, *18*, 629–635.
34. Amenyogbe, E.; Yang, E.; Xie, R.; Huang, J.; Chen, G. Influences of indigenous isolates *Pantoea agglomerans* RCS_2 on growth, proximate analysis, haematological parameters, digestive enzyme activities, serum biochemical parameters, antioxidants activities, intestinal morphology, disease resistance, and molecular immune response in juvenile's cobia fish (*Rachycentron canadum*). *Aquaculture* **2022**, *551*, 737942.
35. Yong, W.; Huang, Y.; Li, X.; Wu, X. Changes of AST, ALP, CK and LDH-1 level in biological injury models caused by 18.4 mm rubber bullet. *Int. J. Lab. Med.* **2012**, *33*, 3.
36. Bharti, S.; Rasool, F. Analysis of the biochemical and histopathological impact of a mild dose of commercial malathion on *Channa punctatus* (Bloch) fish. *Toxicol. Rep.* **2021**, *8*, 443–455. [CrossRef] [PubMed]
37. Ni, J.H.; Zhu, X.J.; Ji, Y.P.; Zhou, B.; Wang, Y.J.; Xu, S.L.; Wang, D.L. Effects of stocking density on growth, metabolic enzyme activity and related gene expression of juvenile silver pomfret. *J. Trop. Oceanogr.* **2020**, *39*, 54–64.
38. Mirghaed, A.T.; Yasari, M.; Mirzargar, S.S.; Hoseini, S.M. Rainbow trout (*Oncorhynchus mykiss*) anesthesia with myrcene: Efficacy and physiological responses in comparison with eugenol. *Fish Physiol. Biochem.* **2018**, *44*, 919–926. [CrossRef] [PubMed]
39. Hughes, J.; Bardell, D. Determination of reference intervals for equine arterial blood-gas, acid-base and electrolyte analysis. *Vet. Anaesth. Analg.* **2019**, *46*, 765–771. [CrossRef]
40. Bijvelds, M.J.; Van, D.V.; Kolar, Z.I. Magnesium transport in freshwater teleosts. *J. Exp. Biol.* **1998**, *201*, 1981–1990. [CrossRef]
41. Feng, G.P.; Zhuang, P.; Zhang, L.Z.; Yan, S.W.; Qian, S. Effects of anesthetic eugenol on oxygen consumption rate and blood biochemical indexes of juvenile Siberian sturgeon. *J. Dalian Fish. Coll.* **2010**, *25*, 113–118.
42. Hu, X.; Ding, C.Y.; Ye, Q.; Wu, R.H.; Zhang, Q.; Li, H.; Li, Y. Effects of carbon dioxide anesthesia on physiological and biochemical indexes and expression of neural regulation related genes in silver carp. *Freshw. Fish.* **2020**, *50*, 9–14.
43. Sun, P.; Tang, B.J.; Jiang, Y.Z. Changes of serum biochemical indexes and liver antioxidant activity of large yellow croaker under stress state. *Mar. Fish.* **2020**, *45*, 634–640.
44. Gao, X.Q.; Wang, X.; Wang, X.Y.; Li, H.X.; Xu, L.; Huang, B.; Meng, X.S.; Zhang, T.; Chen, H.-B.; Xing, R.; et al. Effects of different feeding frequencies on the growth, plasma biochemical parameters, stress status, and gastric evacuation of juvenile tiger puffer fish (*Takifugu rubripes*). *Aquaculture* **2022**, *548*, 737718. [CrossRef]
45. Sun, Z.; Tan, X.; Wei, Z.; Liu, Q.; Mai, H.; Liu, Y.; Liu, B.; Zhuang, Y.; Zou, D.; Zhang, W.; et al. Effects of dietary dandelion extract on the growth performance, serum biochemical parameters, liver histology, and immune and apoptosis-related genes expression of hybrid grouper (*Epinephelus lanceolatus*♂ × *Epinephelus fuscoguttatus*♀) at different feeding period. *Fish Shellfish. Immunol.* **2022**, *120*, 280–286.
46. Zhu, T.B.; Li, F.; Wu, X.B.; Zhu, Y.J.; Guo, W.; Yang, D.G. Effects of electrical anesthesia on behavior and blood biochemical indexes of juvenile *Coreius guichenoti*. *Sichuan J. Zool.* **2015**, *34*, 885–888.
47. Zhu, T.T.; Li, Q.; Zhu, H.Y.; Wang, H.W.; Li, E.C.; Chen, L.Q. Effects of dietary lipid levels on growth, blood biochemical parameters and antioxidant capacity of juvenile Russian sturgeon (*Acipenser gueldenstaedti*). *Mar. Fish.* **2017**, *39*, 58–67.
48. Cheng, Z.J.; Hardy, R.W. Protein and lipid sources affect cholesterol concentrations of juvenile pacific white shrimp, *Litopenaeus vannamei* (Boone). *J. Anim. Sci.* **2004**, *82*, 1136–1145. [CrossRef] [PubMed]

49. Ji, L.Q.; Jiang, K.Y.; Han, L.J.; Yuan, K.S.; Qiu, C.W.; Wang, B.J.; Liu, M.; Wang, L.; Wen, H.S. Effects of continuous cooling on metabolic function of adult turbot. *Mar. Sci.* **2014**, *38*, 46–53.
50. Li, L.; Cao, Z.D.; Fu, S.J. Changes of lactic acid, glycogen and glucose levels in juvenile catfish after exhaustive exercise. *Acta Hydrobiol. Sin.* **2007**, *6*, 880–885.
51. Evans, D.H.; Piermarini, P.M.; Choe, K.P. The multifunctional fish gill: Dominant site of gas exchange, osmoregulation, acid-base regulation, and excretion of nitrogenous waste. *Physiol. Rev.* **2005**, *85*, 97–177. [CrossRef] [PubMed]
52. Jiao, W.; Han, Q.; Xu, Y.; Jiang, H.; Xing, H.; Teng, X. Impaired immune function and structural integrity in the gills of common carp (*Cyprinus carpio* L.) caused by chlorpyrifos exposure: Through oxidative stress and apoptosis. *Fish Shellfish Immunol.* **2019**, *86*, 239–245. [CrossRef]
53. Cappello, T.; Brandao, F.; Guilherme, S.; Santos, M.A.; Maisano, M.; Mauceri, A.; Canario, J.; Pacheco, M.; Pereira, P. Insights into the mechanisms underlying mercury-induced oxidative stress in gills of wild fish (*Liza aurata*) combining 1H NMR metabolomics and conventional biochemical assays. *Sci. Total Environ.* **2016**, *548*, 13–24. [CrossRef]
54. De Domenico, E.; Mauceri, A.; Giordano, D.; Maisano, M.; Giannetto, A.; Parrino, V.; Natalotto, A.; D'Agata, A.; Cappello, T.; Fasulo, S. Biological responses of juvenile european sea bass (*Dicentrarchus labrax*) exposed to contaminated sediments. *Ecotoxicol. Environ. Saf.* **2013**, *97*, 114–123. [CrossRef]
55. Qu, Y.J.; Liu, Q.Q.; Wen, J.F.; Li, J.E.; Li, H. Effects of acute low temperature stress on the structure of liver, muscle and gill of juvenile *Eleutheronema tetradactylum*. *Ecoscience* **2018**, *37*, 53–59.
56. Dong, H.; Wang, W.; Duan, Y.; Li, H.; Liu, Q.; Sun, Y.; Zhang, J. Transcriptomic analysis of juvenile Chinese sea bass (*Lateolabrax maculatus*) anesthetized by MS-222 (Tricaine Methanesulfonate) and eugenol. *Fish Physiol. Biochem.* **2020**, *46*, 909–920. [CrossRef] [PubMed]
57. Oliveira, C.P.B.; Lemos, C.H.P.; Felix, S.A.; de Souza, S.A.; Albinati, A.C.L.; Lima, A.O.; Copatti, C.E. Use of eugenol for the anaesthesia and transportation of freshwater angelfish (*Pterophyllum scalare*). *Aquaculture* **2019**, *513*, 734409. [CrossRef]
58. Karapanagiotidis, I.T.; Metsoviti, M.N.; Gkalogianni, E.Z.; Psofakis, P.; Asimaki, A.; Katsoulas, N.; Papapolymerou, G.; Zarkadas, I. The effects of replacing fishmeal by *Chlorella vulgaris* and fish oil by *Schizochytrium* sp. and *Microchloropsis* gaditana blend on growth performance, feed efficiency, muscle fatty acid composition and liver histology of gilthead seabream (*Sparus aurata*). *Aquaculture* **2022**, *561*, 738709. [CrossRef]
59. Velisek, J.; Svobodova, Z.; Piackova, V.; Groch, L.; Nepejchalova, L. Effects of clove oil anaesthesia on common carp (*Cyprinus carpio* L.). *Vet. Med.* **2005**, *50*, 269–275. [CrossRef]
60. Wilson, A.D. Application of electronic-nose technologies and voc-biomarkers for the noninvasive early diagnosis of gastrointestinal diseases. *Sensors* **2018**, *18*, 2613. [CrossRef]
61. Li, P.; Niu, Z.; Shao, K.; Wu, Z. Quantitative analysis of fish meal freshness using an electronic nose combined with chemometric methods. *Measurement* **2021**, *179*, 109484. [CrossRef]
62. Zhang, J.; Cao, J.; Pei, Z.; Wei, P.; Xiang, D.; Cao, X.; Shen, X.; Li, C. Volatile flavour components and the mechanisms underlying their production in golden pompano (*Trachinotus blochii*) fillets subjected to different drying methods: A comparative study using an electronic nose, an electronic tongue and SDE-GC-MS. *Food Res. Int.* **2019**, *123*, 217–225. [CrossRef]
63. Li, H.; Wang, Y.; Zhang, J.; Li, X.; Wang, J.; Yi, S.; Zhu, W.; Xu, Y.; Li, J. Prediction of the freshness of horse mackerel (*Trachurus japonicus*) using e-nose, e-tongue, and colorimeter based on biochemical indexes analyzed during frozen storage of whole fish. *Food Chem.* **2023**, *402*, 134325. [CrossRef] [PubMed]

foods

MDPI

Article

A Study of the Gelatin Low-Temperature Deposition Manufacturing Forming Process Based on Fluid Numerical Simulation

Qiang Tong [1], Wentao Zhao [1], Tairong Guo [1], Dequan Wang [1] and Xiuping Dong [2,*]

[1] College of Mechanical Engineering and Automation, Dalian Polytechnic University, Dalian 116034, China; tongqiang.work@outlook.com (Q.T.); zwt2363038548@outlook.com (W.Z.); gtr446962334@outlook.com (T.G.); work.ty1122@outlook.com (D.W.)

[2] School of Food Science and Technology, Dalian Polytechnic University, Dalian 116034, China

* Correspondence: dlgysam@outlook.com

Abstract: Low-temperature deposition manufacturing has attracted much attention as a novel printing method, bringing new opportunities and directions for the development of biological 3D printing and complex-shaped food printing. In this article, we investigated the rheological and printing properties of gelatin solution and conducted numerical simulation and experimental research on the low-temperature extrusion process of gelatin solution. The velocity, local shear rate, viscosity, and pressure distribution of the material in the extrusion process were calculated using Comsol software. The effects of the initial temperature, inlet velocity, and print head diameter of the material on the flow field distribution and printing quality were explored. The results show that: (1) the fluidity and mechanical properties of gelatin solution vary with its concentration; (2) the initial temperature of material, inlet velocity, and print head diameter all have varying degrees of influence on the distribution of the flow field; (3) the concentration change of the material mainly affects the pressure distribution in the flow channel; (4) the greater the inlet velocity, the greater the velocity and shear rate in the flow field and the higher the temperature of the material in the outlet section; and (5) the higher the initial temperature of the gel, the lower the viscosity in the flow field. This article is of great reference value for the low-temperature 3D printing of colloidal materials that are difficult to form at room temperature.

Keywords: low-temperature forming; numerical simulation; rheological properties; printing temperature

Citation: Tong, Q.; Zhao, W.; Guo, T.; Wang, D.; Dong, X. A Study of the Gelatin Low-Temperature Deposition Manufacturing Forming Process Based on Fluid Numerical Simulation. *Foods* **2023**, *12*, 2687. https://doi.org/10.3390/foods12142687

Academic Editor: Pierre Sylvain Mirade

Received: 30 May 2023
Revised: 30 June 2023
Accepted: 10 July 2023
Published: 12 July 2023

1. Introduction

3D printing technology refers to a technique employed for creating entities with special shapes/properties by stacking materials layer by layer based on digital models, and is capable of completing multiple tedious processes in traditional machining processes with a high degree of flexibility and intelligence in the absence of a mold as an alternative to traditional processing lines, thereby reducing energy consumption and production cost [1]. There are three categories of raw materials commonly used for 3D printing, namely powder, slurry, and solid. The commonly used printing methods for powder raw materials include 3D printing technology (3DP) and selective laser sintering technology (SLS), those for solid raw materials include inkjet-forming technology (LJP) and digital light technology (SLA), and those for slurry raw materials mainly include fused deposition technology (FDM), direct-write printing technology (DIW) [2], and low-temperature deposition manufacturing (LDM). LDM is available as either plunger or screw types, depending on how extrusion pressure is provided. The gel material itself has a certain viscosity. Screw printing requires direct contact between the screw and material, making it difficult to clean during actual printing [3,4]. However, plunger printing is simple in structure without requiring contact with the material, and the stepper motor is directly connected to a piston with high accuracy [5].

Gels typically consist of large amounts of water and a three-dimensional network of cross-linked polymers. Multifunctional hydrogel inks include double-network hydrogels, magnetic hydrogels, temperature-sensitive hydrogels, and biogels [6]. Gelatin (GEL) is a typical temperature-sensitive gel, which is the product of animal collagen decomposition. It is an immune reaction-free and non-toxic natural polymer material with good biocompatibility in living organisms. It has been approved by the FDA to be widely used in the food industry as well as in biomedical fields such as pharmaceuticals, drug delivery, and tissue engineering [7]. Currently, in the food field, Yang et al. [8] have achieved chicken gelatin gel printing, and Dong et al. [9] have achieved potato starch gelatin gel printing, laying a theoretical foundation for digitally accurate design and regulation of food nutrition; in the biological field, Schuurman et al. [10] mixed methacrylate-anhydride gelatin solution with cartilage cells to prepare a multilayer lattice-structured bio-scaffold by 3D printing technology. However, all the above studies were performed from the perspective of material modification. Due to the limitations of the traditional preparation process, only hydrogels with two-dimensional or simple three-dimensional structures can be prepared, and those with high-precision complex models cannot be printed. In addition, the biological activity of materials cannot be perfectly guaranteed by the vast majority of printing technologies. Cryogenic 3D printing technology, from a process perspective, controls the temperature of the printing environment, and is not only able to print some hydrogels that are difficult to form but also effectively prevents the thermal phase transition of the material in the printing process while ensuring its biological activity [11]. Therefore, low-temperature 3D printing technology is widely used for the rapid design and manufacturing of hydrogels. Tan et al. [6] conducted a preliminary biological evaluation of 3D printing materials coated with type I collagen, poly-L-lysine and gelatin, and therefore, this article investigates the gelatin printing and forming process based on low-temperature deposition 3D printing technology from a process perspective, which has important research value.

Through the research results of Gu et al., Levett et al., and Zhao [12–15], it was found that when the concentration of gelatin aqueous solution is too high, it is not suitable for cell survival and is relatively easy to form. The research has little significance. When the concentration of gelatin aqueous solution is low, it is suitable for cell survival but is not easy to print and shape. In the food industry, gelatin is a natural food additive with a small amount of addition. Therefore, we chose gelatin aqueous solutions with concentrations of 12 wt%, 9 wt%, 6 wt%, and 3 wt% within this range to study their molding process.

The solutions of non-Newtonian fluid and nonlinear flow problems based on the finite element method (FEM) have been widely used in the additive manufacturability of material flow characteristics and is of great significance to 3D printing process optimization. The location of the printing head in a closed environment makes it very difficult to observe the changes in fluid temperature and characteristics in the 3D printing process [16].

Moreover, gelatin is sensitive to temperature; its viscosity and elasticity are sensitive to temperature changes. The ink near the nozzle outlet easily solidifies [17], resulting in nozzle blockage. Therefore, in this article we numerically simulated flow field distribution (speed, shear rate, viscosity) and temperature field distribution (temperature distribution, and the change of outlet material temperature over time) in the print head during printing using Comsol software. The mathematical model between extrusion line diameter and layer thickness provided a theoretical reference for the matching of inlet velocity and printing speed, and a self-made low-temperature 3D printing device was used in the gelatin printing process experiment, which provided a theoretical basis for the 3D printing of colloidal materials that are difficult to form at room temperature.

2. Materials and Methods

2.1. Materials

12 g, 9 g, 6 g, and 3 g of gelatin (purchased from Henan Sugar Cabinet Co., Ltd, Shangqiu, China. using food grade 180 freezing force) were, respectively, weighed and mixed with 88 g, 91 g, 94 g, and 97 g of water in a beaker to prepare 12 wt%, 9 wt%, 6 wt%, and 3 wt% gelatin aqueous solution. Subsequently, the gelatin solutions were placed in a constant temperature water bath heating pot (purchased by 600 W Qun'an Experimental Instrument Co., Ltd., Huzhou, China) for melting at a constant temperature of 40 °C, and then in a negative pressure box (Puruiqi AP-01P) for 3–5 min for defoaming treatment in an environment of −2.5 MPa. Finally, they were separately stored at 35 °C, 30 °C, and 25 °C for further use.

2.2. Measurement of Rheological Properties

The rheological properties of the samples were measured using the hybrid rheometer Discovery HR-1, and the viscosity parameters of gelatin solutions at 35 °C, 30 °C, and 25 °C were measured. Three parallel experiments were conducted on each group of samples. During the measurement, gelatin solution was dropped onto a parallel plate of the rheometer with a gap of 1 mm between the two parallel plates. Before the experiment, excess material outside the plate was scraped out and allowed to stand for 2 min until it became steady. When the temperature of the material was consistent with the preset temperature, the viscosity parameters were measured at a shear rate of 0.1–100 s^{-1} [5]. In the temperature-sensitive processes, the oscillation mode was used to test the viscosity variation curve with the temperature at 0–45 °C [18].

2.3. Finite Element Simulation

2.3.1. Simulation Scheme

Material concentration, inlet velocity, and nozzle diameter are important parameters that affect flow field distribution in the flow channel. Gelatin is sensitive to temperature, so temperature distribution and changes in the printing process are also key factors affecting printing quality. In this article, we calculated the velocity, shear rate, viscosity, and pressure distribution of materials in the extrusion process by combining the coupling experiment and simulation while controlling a single variable, and then explored the effects of different process parameters (material concentration, inlet velocity, nozzle diameter, initial temperature of gelatin solution) on the flow field distribution and printing quality in the flow channel.

2.3.2. Conditions and Assumptions

1. Assuming that the material was an incompressible fluid;
2. Assuming that there was no slip between the material and the cylinder wall during extrusion;
3. The piston moved at a constant speed at the inlet. The equivalent inlet velocity boundary conditions were applied on this boundary to replace the constant movement of the piston, with wall normal velocity $v_n = 0$ and surface velocity $v_s = 0$;
4. The material was a fluid with high viscosity, whose inertial force and gravity were much smaller than the viscous force, so gravity was ignored;
5. The pressure at the nozzle outlet was 101,325 Pa, which was 1 atm;
6. The material was fully filled in the mold and stably flowed laminar in the flow channel.

2.3.3. Fluid Viscosity Model

The viscosity of the gelatin solution decreased with the increase in shear rate, characteristic of a pseudoplastic fluid, i.e., the shear stress of the fluid τ and shear rate γ have a nonlinear correlation. The selection of appropriate fluid viscosity model is very important for studying the fluidity of a pseudoplastic fluid. The constitutive equation of a pseudoplastic fluid is [19]:

$$\eta_a = \tau/\gamma = k\gamma^{n-1} \quad (1)$$

where k is the consistency coefficient and n is the rheological index.

2.3.4. Physical Model

Figure 1a shows a three-dimensional physical model of an extrusion device for a 3D printer, consisting of a material barrel, printing nozzle, fixed block, silicone heating sleeve, resistance wire, and materials. The outer diameter of the material barrel was 17 mm, the inner diameter was 14 mm, the length from inlet to outlet was 58 mm, the diameter of the print head outlet was 1 mm, the length of the fixed block was 45 mm, and the width was 30 mm, as shown in Figure 1b, which depicts the grid diagram of the extrusion device, which was automatically generated based on the physical field, with a total of 9,893,297 units.

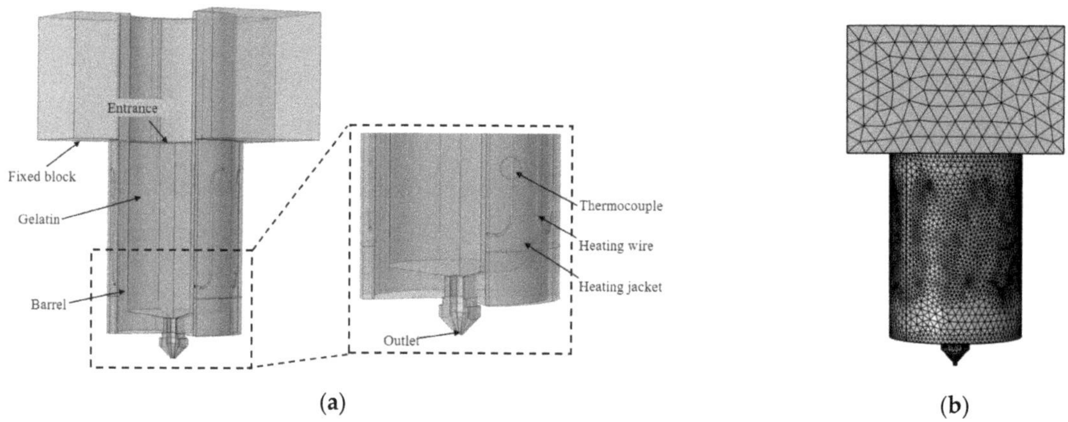

Figure 1. (**a**) 3D model of extrusion device; (**b**) Grid diagram of extrusion device.

Model Heat Transfer Analysis

The extrusion process of the gelatin solutions involved non-isothermal flow and solid heat transfer, and transient analysis of the extrusion process was conducted using Comsol software coupled with multiple physical fields. The model involved natural convection heat transfer and contact heat transfer, where natural convection heat transfer refers to the heat exchange between the air and the material in the material storage barrel, as well as that between the fixed block, the bottom of the material barrel, and the surface of the printing head and the external air, and contact heat exchange refers to the heat exchange between the material and the barrel wall, as well as that between the fixed block and the heating sleeve and the barrel wall.

The Heat Source and Its Initial Conditions

A silicone heating strip with a power of 20 W, a height of 45 mm, and a length of 106 mm was used as the heat source. A thermocouple was placed between the silicone heating strip and the outer wall of the material barrel. When the ambient temperature was 5 °C and the heating temperature of the silicone heating sleeve was 45 °C, the actual measured temperature of the outer wall of the storage barrel was 31 °C, the temperature of the inner wall of the storage barrel was 24.3 °C, and the temperature of the fixed block was 13.5 °C. Therefore, the initial temperature of the storage barrel and the printing head was set to 24.3 °C, and that of the fixed block to 13.5 °C.

Heating Condition Setting

The actual measured temperature of the outer wall of the storage barrel was 31 °C. The temperature of the thermocouple was used to replace the temperature of the outer wall of the material barrel. Using the implicit event built into Comsol, b1 = aveop1 (T) − 32 [degC] and b2 = aveop1 (T) − 30 [degC] were set. During the calculation: if b_1 > 0, stop heating; If b_2 < 0, start heating, where aveop1 (T) is the average temperature at the thermocouple, keeping the outer wall temperature of the material barrel constant at 31 °C.

2.3.5. Convective Heat Transfer Coefficient

The heat transfer in the model was mainly in the form of natural convection heat transfer, and the empirical formula for the convective heat transfer coefficient of natural convection is [20].

$$h_c = \frac{N_\mu}{L} \tag{2}$$

In formula (2), L is the characteristic length, hc is the Nusselt number, and N_μ is a dimensionless number that represents the intensity of convective heat transfer, which can be calculated by:

$$N_\mu = c(G_r P_r)^m \tag{3}$$

The material barrel was vertically placed, and the material was in the laminar flow state, so c was taken as 0.59, and m as 0.25.

$$G_r = \frac{\beta g \Delta T \rho^2 L^3}{\mu^2} \tag{4}$$

$$P_T = \frac{c_p \mu}{k} \tag{5}$$

where G_r is the Grashof number, which is dimensionless and represents the ratio of buoyancy and viscous force acting on the fluid; P_T is the Prandtl number, indicating the relationship between the temperature boundary layer and the flow boundary layer; β is the coefficient of volume change; g is the acceleration of gravity; ΔT is the temperature difference; ρ is the density of air; L is the feature length; μ is the aerodynamic viscosity; c_p is the specific heat capacity of air; and k is the thermal conductivity of the air.

2.3.6. Printing Test

A self-developed LDM (Low-Temperature Deposition Manufacturing) printer was used for process experiments, as displayed in Figure 2a. The equipment diagram is shown in Figure 2b. The piston extrusion, printing environment cooling (5–10 °C), and heating of the silicone heating sleeve (35–55 °C) were utilized, with a printing speed of 2–8 mm/s and an inlet velocity of 0.03–0.09 mm/s. The printing model is shown in Figure 3.

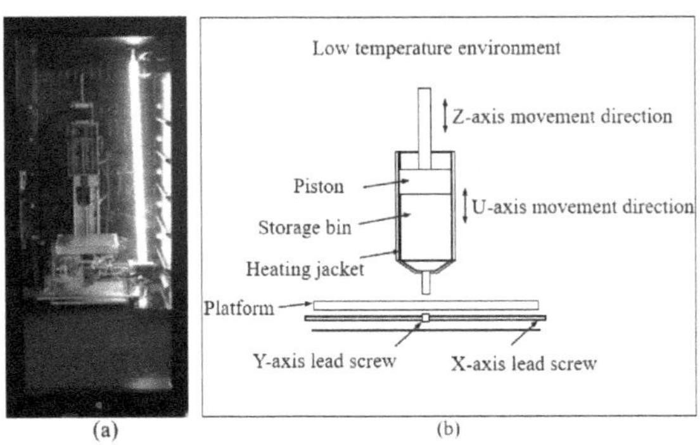

Figure 2. Schematic diagram of the structure of the cryogenic deposition 3D printer: (**a**) Equipment physical diagram; (**b**) Equipment diagram.

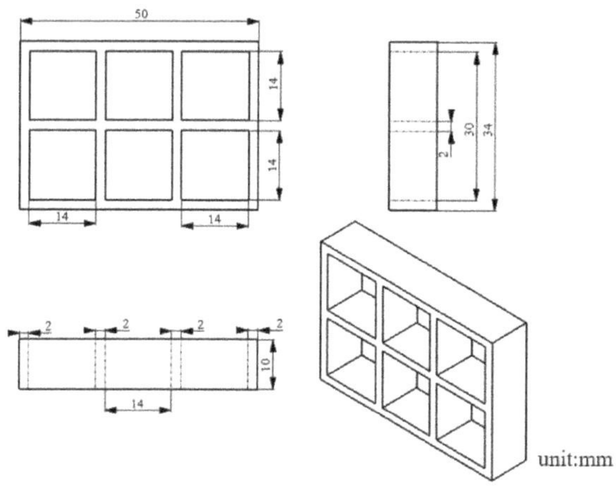

Figure 3. Schematic diagram of grid structure.

3. Results and Discussion

3.1. Rheological Properties

Rheological performance is an important indicator for evaluating the 3D printing performance of materials [21]. Materials used for extrusion 3D printing must have certain flowability or melting characteristics, as well as the ability to maintain product shape after printing is completed [22]. As shown in Figure 4a, when shear rate fluctuated from 0.1 s^{-1} to 3 s^{-1}, the apparent viscosity of the gelatin solution significantly decreased; when shear rate continued to increase from 3 s^{-1} to 100 s^{-1}, the apparent viscosity of the gelatin solution decreased slowly. The curve tended to be flat as the viscosity approached 0 Pa·s, and the gelatin solution became closer to a Newtonian fluid. Overall, the apparent viscosity decreased with the increase in shear rate, indicating that the gelatin solution exhibited shear thinning behavior and was characteristically a pseudoplastic fluid, which was conducive to the discharge of slurry through the nozzle [23]. As shown in Figure 4c,d, the viscosity of the gelatin solution followed the same trend with shear rate at different temperatures, with higher temperatures resulting in lower viscosity and shear stress at the same shear rate. As shown in Figure 4a,b, the higher the concentration of gelatin solution, the higher the viscosity and the higher the shear stress at the same shear rate.

Viscosity measurement is one of the commonly used methods to determine the gel point temperature. The viscosity mutation point was used to determine the gel point temperature. As shown in Figure 4e, the curve of 9% gelatin viscosity changes with temperature. The gelatin viscosity changes abruptly at 31 °C, so 31 °C is the gel point temperature of 9% gelatin.

The storage modulus (G′) is a measure of elastic solid materials, which reflects the mechanical strength of the specimen. Materials with high mechanical strength exhibit excellent self-supporting properties after deposition and maintain a stable shape after printing. The loss modulus (G″) is the viscous response of the stress–strain ratio under dynamic oscillation frequency [23]. Different viscoelastic behaviors were illustrated using loss tangent (Tan $\delta = G'/G''$) as a characteristic parameter. A tan δ less than 1 indicates mainly elastic properties, exhibiting solid characteristics. On the contrary, a tan δ greater than 1 indicates liquid properties [24], as shown in Figure 4f. When the temperature was greater than 31 °C, the gelatin solution tan δ was >1, and exhibited liquid properties; at a temperature less than 31 °C the gelatin solution tan δ was <1, and exhibited elastic properties.

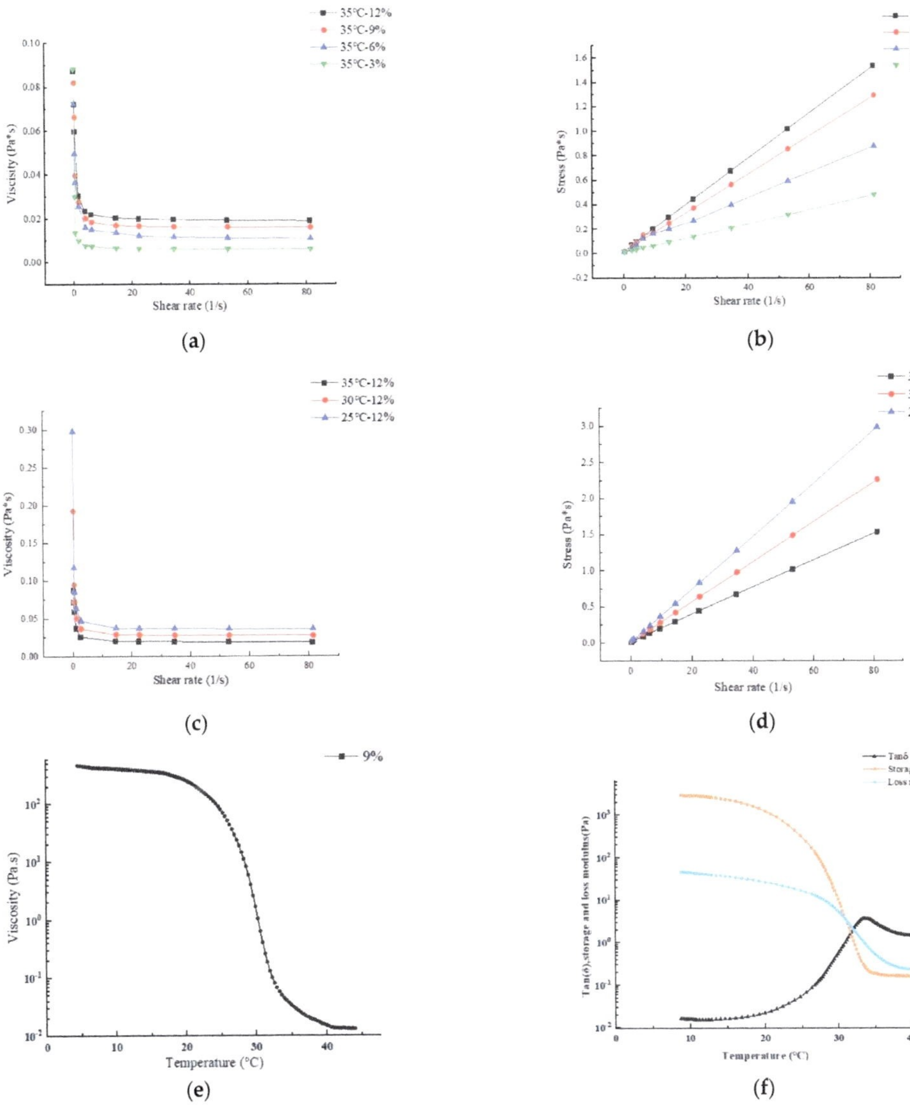

Figure 4. Material rheological properties: (**a**) Apparent viscosity versus shear rate curve of different concentrations of gelatin solution; (**b**) Stress versus shear rate curve of different concentrations of gelatin solution; (**c**) Apparent viscosity versus shear rate curve of gelatin solution at different temperatures; (**d**) Stress versus shear rate curve of gelatin solution at different temperatures; (**e**) Apparent viscosity versus temperature curve of 9% gelatin solution; (**f**) Temperature-dependent curves of storage modulus and loss modulus of 9% gelatin solution.

3.2. Analysis of the Flow Field Distribution in the Flow Channel

Taking a gelatin aqueous solution with an inlet velocity of 0.03 mm/s, an initial temperature of 30 °C, a heating jacket temperature of 45 °C, an ambient temperature of 5 °C, and a concentration of 9 wt% as an example, the flow field distribution in the flow channel was analyzed.

3.2.1. Temperature Distribution

The temperature distribution during the printing process was under the above process conditions as shown in Figure 5a. The heat exchange between the gelatin solution and the air in the barrel and the barrel wall resulted in the temperature gradient in the entire model. The temperature at the bottom of the material barrel was about 23 °C, higher than the surface temperature of the print head (about 20 °C), forming a small temperature gradient, which was conducive to the rapid cooling and solidification of materials after extrusion.

As shown in Figure 5b, the temperature at middle of the barrel was the highest, while the temperature at the fixed block was the lowest. Due to the influence of the fixed block temperature, the temperature of the gelatin solution at the inlet was relatively low.

3.2.2. Velocity Distribution

The velocity field has a direct impact on the quality of 3D printing of materials. Figures 5c and 5d, respectively, show the velocity field distribution in the flow channel and outlet under the above process conditions. It can be intuitively seen from Figure 5c that the material in the cylinder was in a low-speed state, with an average speed of 0.032 mm/s and that there was no significant change in speed. The material speed in the printing nozzle was relatively high, with an average of 1.82 mm/s. As can be seen from Figure 5d, on the same horizontal section, the fluid velocity peaked at the central axis of the nozzle and decreased from the central axis to the inner wall, with the lowest velocity near the inner wall, which was caused by the friction between the material and the solid wall [25]. These results are comparable to those found in the study by Liu et al. [26].

3.2.3. Shear Rate Distribution

Gelatin solution is a pseudoplastic fluid whose viscosity varies with shear rate. Figures 5e and 5f, respectively, show the distribution of shear rate fields in the flow channel and outlet section under the above process conditions. As shown in Figure 5e, the shear rate in the storage barrel section was lower, while that in the nozzle section was higher. As shown in Figure 5f, the shear rate in the flow channel bottomed at the central axis and increased from the central axis to the inner wall, with the shear rate higher near the inner wall, indicating that the material was subjected to significant shear force at the wall surface. These results are comparable to those of the study by Yang et al. [5].

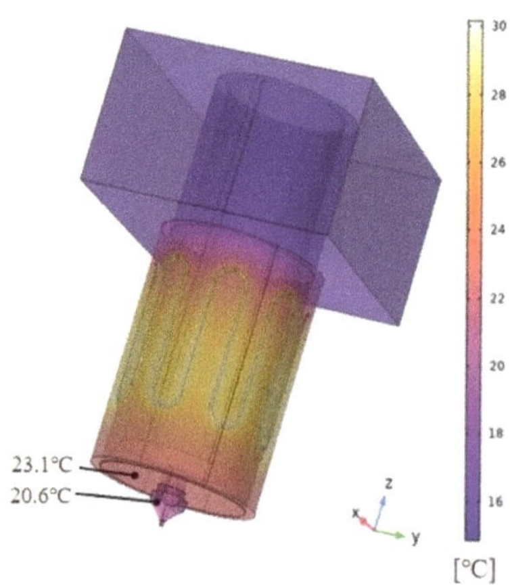

(a)

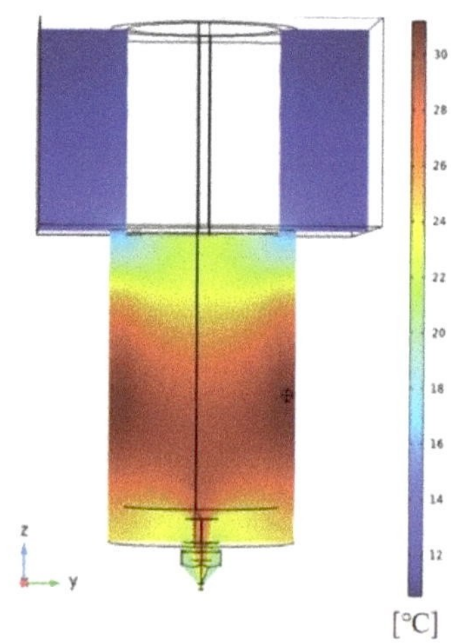

(b)

Figure 5. *Cont.*

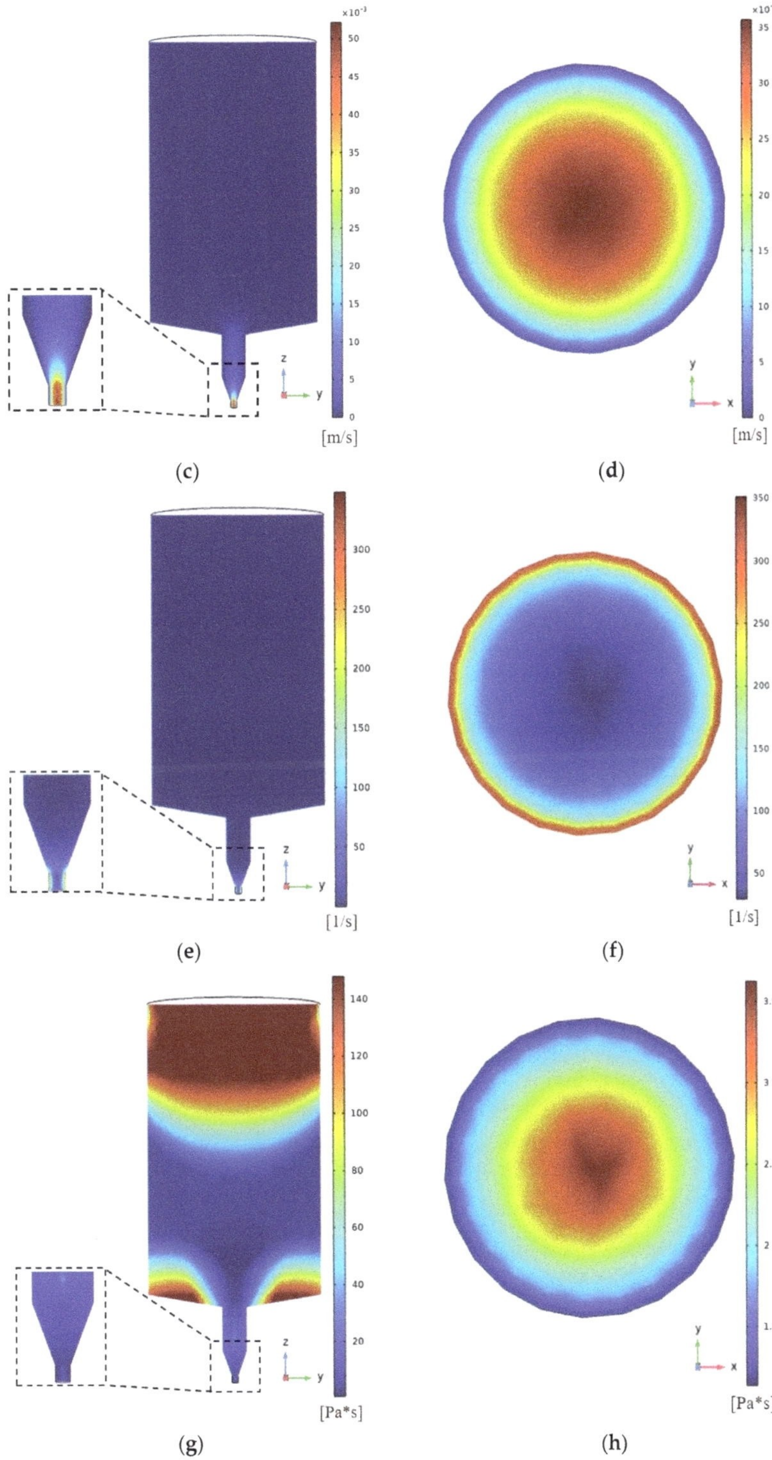

Figure 5. *Cont.*

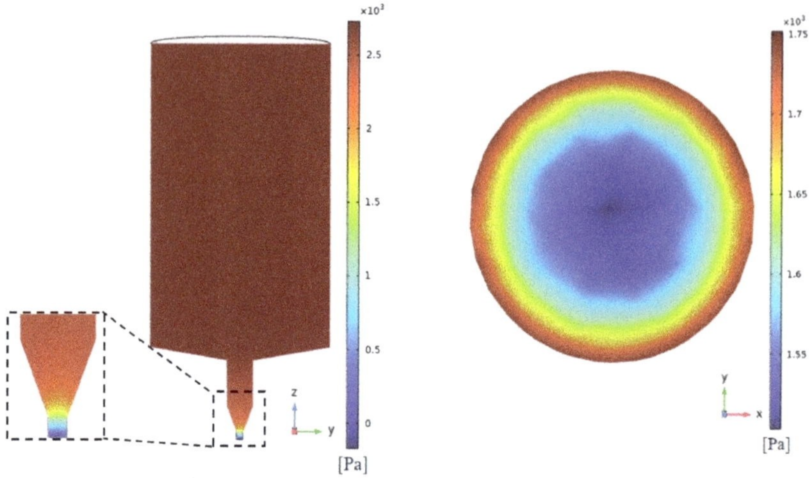

(i) (j)

Figure 5. Simulation of channel flow field distribution based on Comsol: (**a**) Three-dimensional temperature distribution; (**b**) Channel temperature distribution; (**c**) Channel velocity vertical distribution; (**d**) Outlet section velocity horizontal distribution; (**e**) Channel shear rate vertical distribution; (**f**) Outlet section shear rate horizontal distribution; (**g**) Channel viscosity vertical distribution; (**h**) Outlet section viscosity horizontal distribution; (**i**) Channel pressure vertical distribution; (**j**) Outlet section pressure horizontal distribution.

3.2.4. Viscosity Distribution

The viscosity distribution of materials is also one of the main factors affecting the forming quality of 3D printing, which is mainly reflected in the difficulty of extruding materials at the outlet. Gelatin is a thermosensitive material. When the temperature was lower than 30 °C, viscosity changed significantly with temperature. Figure 5g,h show the viscosity field distribution at the flow channel and outlet under the above process conditions, respectively. It can be intuitively seen from Figure 5g that the viscosity at the upper end of the barrel was high, and that at the center and at the bottom of material barrel was low. The viscosity at both sides was high, which was consistent with the temperature distribution at the same inlet velocity rate, indicating that temperature was one of the main factors affecting the viscosity distribution during printing.

As shown in Figure 5h, on the same horizontal section, the viscosity of the fluid decreased from the axis to the inner wall, with the lower viscosity near the wall surface. Due to the high shear rate at the nozzle outlet, the viscosity of the material dropped, indicating that the material was easier to extrude. These results are comparable to those reported by Shamsudin et al. for the effect of temperature and shear rate on viscosity distribution [27].

3.2.5. Pressure Distribution

In the continuous-extrusion 3D printing process, pressure has a significant impact on the material extrusion process [28] Figures 5i and 5j, respectively, show the pressure field distribution in the flow channel and outlet section under the above process conditions. As shown in Figure 5i, pressure in the storage barrel slightly changed, while that in the nozzle changed significantly. In addition, the pressure in the material barrel section was relatively large, while that in the nozzle section was relatively small. As shown in Figure 5j, the pressure in the flow channel increased from the axis to the inner wall, with that near the inner wall being relatively high, which was consistent with the horizontal velocity distribution inside the nozzle, both caused by the high friction between the inner wall and the material. These results are comparable to those of the study by Yang et al. [5].

3.3. Effect of Inlet Velocity on Flow Field Distribution

In the LDM process, inlet velocity was regulated by varying the velocity of the piston. The change in inlet velocity directly affected the velocity field within the flow channel, which was a key factor affecting flow channel parameters and 3D printing efficiency. Therefore, this section investigates the effect of inlet velocity on the flow field distribution within the flow channel.

As shown in Figure 6a,b, inlet velocity increased from 0.03 mm/s to 0.09 mm/s, and the average velocity at the nozzle outlet increased from 33.2 mm/s to 107.4 mm/s. The average shear rate at the outlet increased from 30 s^{-1} to 87.8 s^{-1}. The above results indicated that both velocity and shear rate increased with the increase in inlet velocity. This is comparable to the findings of Emin et al. [29].

As shown in Figure 6c, when the initial temperature was 30 °C, the temperature variation trend in the Z-axis at different inlet velocities was the same, which firstly increased and then decreased. As the inlet velocity increased, peak temperature gradually shifted from the middle part of the printing barrel to the printing nozzle.

As shown in Figure 6d, at the initial temperature of 30 °C, with the increase in time, the temperature of the gelatin solution outlet section first decreased and then shifted towards equilibrium, and the greater the inlet velocity, the slower the decrease in temperature of the gelatin solution in the outlet section, and the greater the temperature at equilibrium. The inlet velocity increased from 0.03 mm/s to 0.09 mm/s, corresponding to the increased temperature of the outlet gelatin solution from 25.1 °C to 27.1 °C, respectively.

Temperature was one of the main factors affecting the viscosity of gelatin, as shown in Figure 6e,f. The variation trend of viscosity in different inlet velocity channels was the same, with viscosity first remaining horizontal and then gradually decreasing to point A because the temperature from the inlet to point A (the highest temperature point) gradually increased due to the influence of the heating sleeve, and with the increase in temperature the viscosity of the gelatin solution decreased.

Both extrusion speed and nozzle movement speed affect 3D printing because they change the extrusion amount per unit time and length. The key problem of the continuous extrusion line in 3D printing is matching between the extrusion speed and printing speed (Figure 7). In the deposition process, the cross-section of deposited silk thread is generally taken as an elliptical flat section, which increases the contact area between the front and back deposition layers, thereby enhancing the adhesion between the two [30].

The inlet velocity of the gelatin solution is set as v_j, the cross-sectional area during extrusion as s_j, the deposition velocity of the gelatin solution as v_c, the cross-sectional area during deposition as s_c, and the volume of gelatin solution extruded within time t is Q. To avoid line breakage or accumulation, the extrusion volume of gelatin solution per unit time should be equal to the deposition volume of gelatin solution, which can be represented by Equation (6).

$$Q = v_j s_j t = vc_s c_t \tag{6}$$

Equation (7) is an expansion of Equation (6).

$$v_j \pi (d_1/2)^2 = v_c((d_2 - h)h + \pi (h/2)^2) \tag{7}$$

Equation (8) is the relationship between inlet speed and printing speed, and the K value is shown in Equation (9).

$$v_c = kv_j \tag{8}$$

$$k = \frac{\pi d_1^2}{4d_2 h - 4h^2 + \pi h^2} \tag{9}$$

$$d_2 = \frac{v_j \pi d_1^2}{4v_c h} + \frac{(4-\pi)h}{4} \tag{10}$$

where v_j is inlet velocity; v_c is deposition rate; d is the print head diameter; d_1 is the inner diameter of the storage barrel; d_2 is the width of extrusion line; and h is the thickness of deposition layer. Due to certain errors between the theoretical model and the actual situation, it was necessary to make corrections based on actual formula experiments.

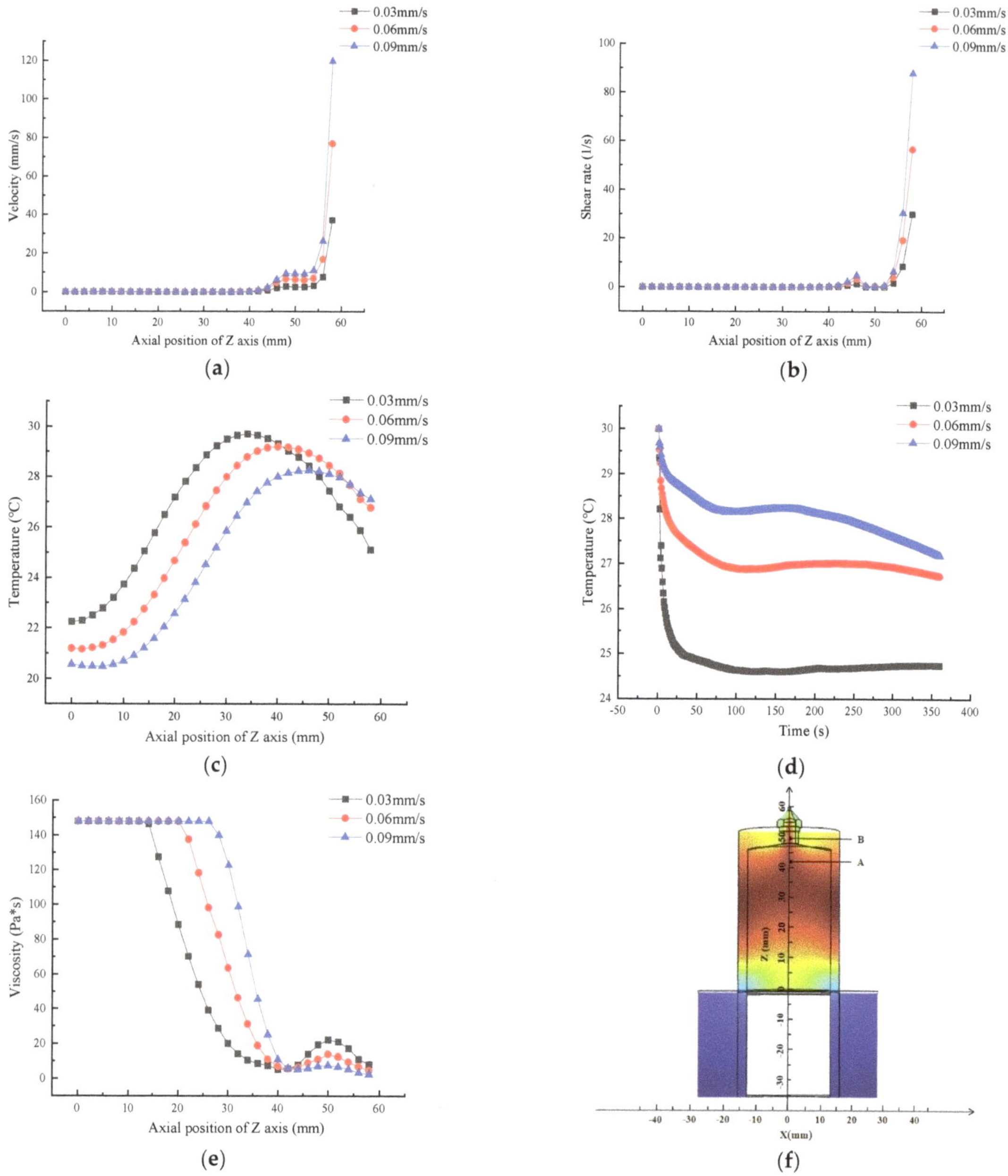

Figure 6. Effect of inlet velocity on flow field distribution based on Comsol simulation: (**a**) Velocity; (**b**) Shear rate; (**c**) Temperature; (**d**) Temperature variation curve of material in the outlet section with time; (**e**) Viscosity; (**f**) Coordinate distribution.

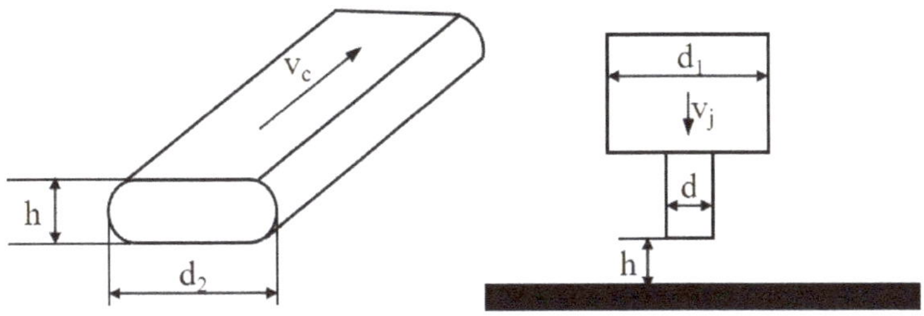

Figure 7. Line model.

Actual parameters were: h = 0.3 mm, d = 1 mm, d_1 = 14 mm, v_c = 5 mm/s. The theoretical widths of deposition lines corresponding to inlet velocities of 0.0156 mm/s, 0.0313 mm/s, 0.0469 mm/s, and 0.0625 mm/s were calculated by formula 10 to be 1.068 mm, 2.034 mm, 2.99 mm, and 3.955 mm, respectively. The actual widths of the deposition lines were measured using an electron microscope (TD-4KHT, SANQID), as shown in Figure 8, which were 1.660 mm, 2.280 mm, 2.830 mm, and 3.320 mm, respectively.

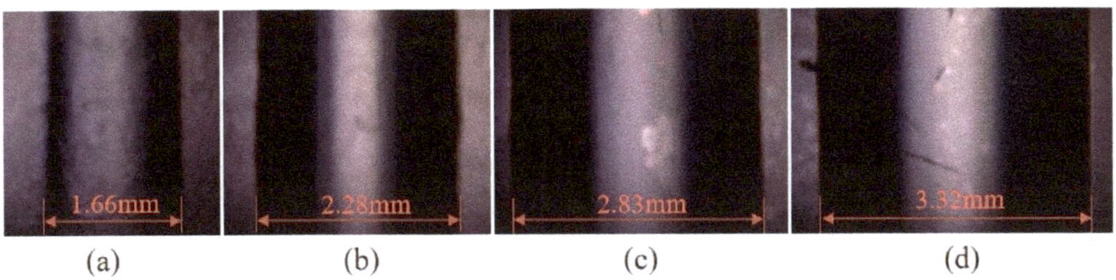

Figure 8. The effect of inlet velocity on the deposition width of wire: (**a**) 0.0156 mm/s; (**b**) 0.0313 mm/s (**c**) 0.0469 mm/s; (**d**) 0.0625 mm/s.

The relationship between the actual line width and the entrance velocity according to the linear fit of the Origin software is shown in Equation (11).

$$d_3 = 1.1415 + 35.3344v_j \tag{11}$$

The theoretical width as a function of entrance velocity is shown in Equation (12).

$$d_2 = 0.1075 + 61.529v_j \tag{12}$$

The width error between the actual width and the theoretical width is shown in Equation (13).

$$Dd_3 - d_2 = 1.0345 - 26.195v_j \tag{13}$$

The actual width minus the width error D is equal to the theoretical width, and the revised coefficient k_1 can be obtained by bringing it into Formula (10). Formula (14) is the revised velocity relationship equation.

$$v_c = k_1 v_j \tag{14}$$

$$k_1 = \frac{\pi d_1^2}{4hd_2 + (\pi - 4)h^2 - 4.14h + 104.78hv_j} \tag{15}$$

The appropriate line width was selected based on the printed model, with a wall thickness of 2 mm. Therefore, d_2 was set to 2 mm. According to Formula (14), the optimal inlet velocity should be 0.0242 mm/s when the line diameter is 2 mm and printing speed is 5 mm/s. Using this speed as a reference, the effect of inlet velocity on 3D printing was studied by controlling a single experimental variable. Grid model printing experiments were conducted with inlet velocities of 0.0156 mm/s, 0.0250 mm/s, 0.0343 mm/s, and 0.0438 mm/s at a printing speed of 5 mm/s, respectively.

As shown in Figure 9a, there was stacking in the printing model at the inlet velocity of 0.0438 mm/s, indicating that inlet velocity did not match printing speed, and that the inlet velocity was greater than its optimal value. As shown in Figure 9b, when the inlet velocity was 0.0343 mm/s there was a slight accumulation in the printing model, indicating an improvement in the mismatch degree between the inlet velocity and the printing speed, with the inlet velocity slightly higher than the optimal value. As shown in Figure 9c, when the inlet velocity was 0.0250 mm/s, the forming quality was higher and the model was smooth and flawless, indicating that the inlet velocity matched the printing speed at this time and was close to its optimal value. As shown in Figure 9d, when the inlet velocity was 0.0156 mm/s, the extrusion wire was broken, indicating that the inlet velocity did not match the printing speed and was less than its optimal value. In summary, printing quality could be guaranteed at the inlet velocity close to its optimal value at 0.0242 mm/s, indicating that the matching between inlet velocity and printing speed was an interval value. In this article, we study the relationship between extrusion speed and deposition line width by setting the printing speed. This is consistent with the research ideas of Cao et al. [31], who used a constant extrusion speed to study the relationship between printing speed and line stretching rate. The results indicate that when the printing speed remains constant within a certain range, the actual line width gradually increases with the increase of extrusion speed.

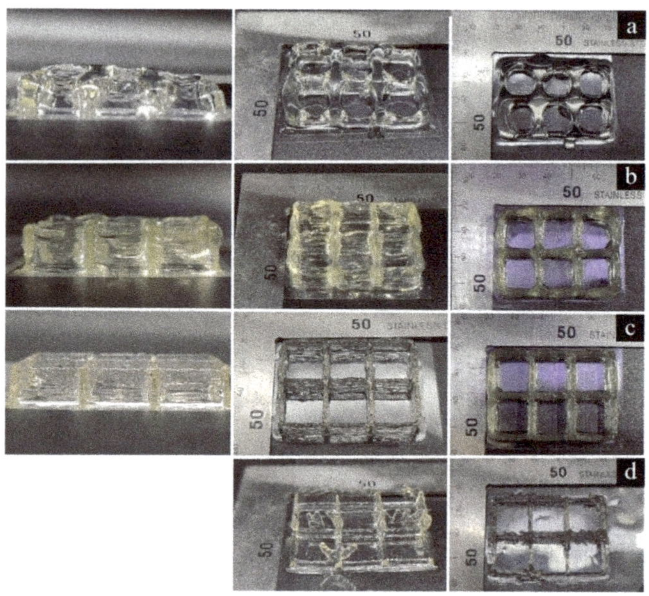

Figure 9. The effect of extrusion speed on 3D printing quality: (**a**) 0.0438 mm/s; (**b**) 0.0343 mm/s; (**c**) 0.0250 mm/s; (**d**) 0.0156 mm/s.

3.4. *Effect of Initial Gelatin Solution Temperature on Temperature Field Distribution*

During extrusion, the gelatin solution first transferred some heat to the printing head, and then exchanged heat with cold air. If the initial temperature of the gelatin solution was too high, it would take a longer time for the gelatin solution to cool down to the gel point. The gelatin solution failed to solidify rapidly after deposition, resulting in a decrease in the printing accuracy. Therefore, the initial temperature of the gelatin solution was one of the important factors affecting printing quality.

Guo et al. found that the gelatin gel process can be divided into two stages: the fast gel stage and the slow gel stage. When the temperature of 9 g/dL gelatin solution is lower than 31 °C, gelatin enters into the slow gel stage, and when the temperature is lower than 26 °C, gelatin enters into the fast gel stage [32]. Zhao et al. studied the influence of temperature on the precision of gelatin molding. The results showed that a 10 w% gelatin solution was in the form of liquid drop, pre-gelation, gelation, and super-gelation at 27, 26, 25, and 24 °C, respectively, and the lines were more uniform when it was close to gelation [15].

This means that under ideal printing conditions the gelatin solution should be at a rapid set point when extruded into the print nozzle outlet, allowing the gelatin solution to set quickly after deposition. However, in practice there is a certain difference between the temperature of the added gelatin solution and the temperature of the extruded gelatin solution, which requires consideration of the heat exchange between the gelatin solution and the barrel wall and print head during the extrusion process. So, in this section the variation of the temperature of the material in the exit section of the gelatin aqueous solution at different initial temperatures over time and the effect on print quality is explored. Figure 10 shows the change curve of gel temperature with time under the conditions of inlet velocity of 0.03 mm/s, ambient temperature of 5 °C, nozzle diameter of 1 mm and heating jacket temperature of 45 °C. As shown in Figure 10, with the initial temperature of 35 °C and 30 °C, the temperature at the exit section of the gelatin solution decreased gradually and then became stable with the increase in time. At 35 °C, the temperature of the gelatin solution decreased slowly and it took a long time to reach the equilibrium temperature. At 3 °C, the temperature of the gelatin solution quickly stabilized, getting close to the rapid setting temperature. At 25 °C, the temperature of the gelatin solution firstly dropped, then rose, and finally tended to reach the equilibrium. At 25 °C, the lowest temperature of the gelatin solution in the outlet section was 21 °C. Therefore, under the above process conditions, the optimal initial temperature of the material should be about 30 °C.

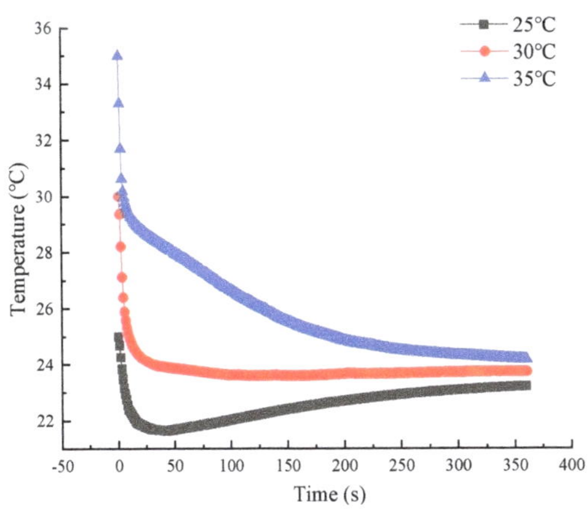

Figure 10. Temperature versus time curve of gelatin solution exit section material at different initial temperatures based on Comsol simulations.

Figure 11a shows the printing results of the gelatin solution at an initial temperature of 35 °C, which indicates that the model exhibited stacking deformation. Figure 11c shows the printing results of the gelatin solution at an initial temperature of 25 °C, accompanied by particle-like extrusion lines. Figure 11b shows the printed results of the gelatin solution at an initial temperature of 30 °C. The model has no flow or deformation phenomena, and the molding quality is high, which is consistent with the simulation results.

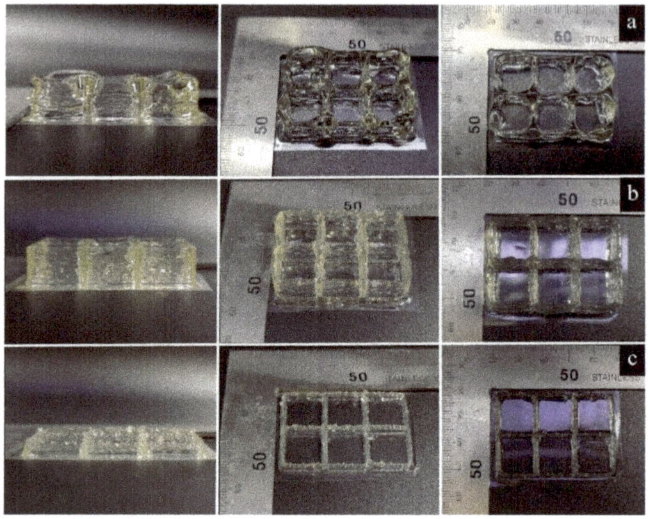

Figure 11. Gelatin printing effect at different initial temperatures: (**a**) 35 °C; (**b**) 30 °C; (**c**) 25 °C.

3.5. Effect of Material Concentration on Flow Field Distribution

At low shear rates, viscosity was related to the fluidity of the material before extrusion [8]. Therefore, it was very important to study the influence of viscosity variation with numerical values on other parameters such as shear rate and pressure field. As shown in Figure 12a,b, the distribution of velocity and shear rate fields in the flow channel did not change with material viscosity, mainly depending on inlet velocity. As shown in Figure 12c, the pressure inside the barrel remained basically constant in the Z-axis direction, and the higher the concentration, the greater the pressure inside the barrel. The pressure gradually decreased from the barrel to the outlet, and the pressure of the gelatin solution decreased more rapidly with the increase in concentration, while the pressure at the nozzle outlet basically remained unchanged. These results are comparable to the research results of Yang et al. [5].

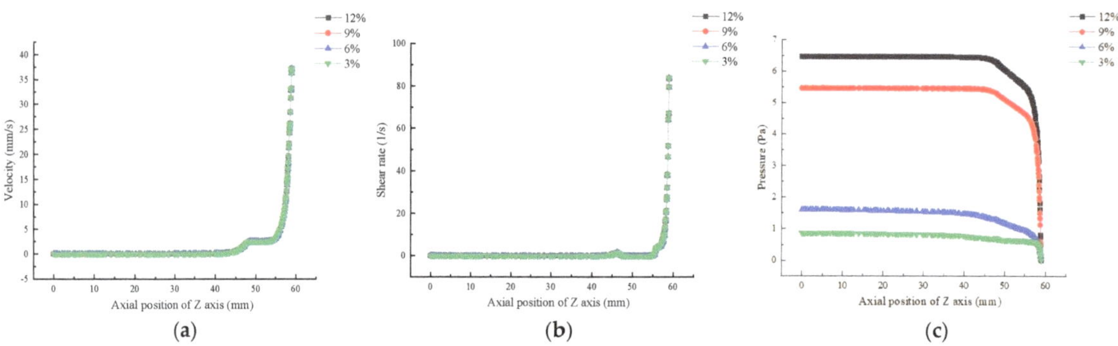

Figure 12. Effect of material concentration on flow field distribution based on Comsol simulation: (**a**) Velocity; (**b**) Shear Rate; (**c**) Pressure.

In the actual 3D printing process, the extrusion process was not instantaneously extruded to the viscosity of colloidal materials. Only when the extrusion pressure was greater than the material viscosity could it be smoothly extruded. The extrusion response time was affected by material viscosity, and the higher the viscosity, the greater the pressure required for material extrusion. The longer the extrusion response time, the greater the hysteresis, which was not conducive to printing. Therefore, pressure in the printing process could be ensured within a reasonable range by adjusting the concentration of the gelatin solution, and the material viscosity could be adjusted to an appropriate value.

The printing results of gelatin solutions with concentrations of 12%, 9%, 6%, and 3% were obtained at an ambient temperature of 5 °C, an initial temperature of 30 °C, and an inlet velocity of 0.025 mm/s, as shown in Figure 13. The storage modulus (G′) is a measure of elastic solid materials, which reflects the mechanical strength of the specimen. Materials with high mechanical strength exhibit excellent self-supporting properties after deposition and maintain a stable shape after printing.

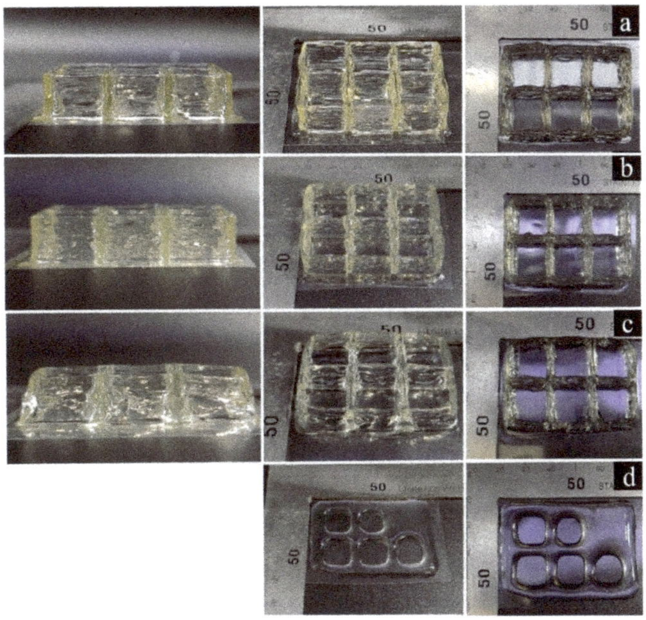

Figure 13. The effect of material concentration on 3D printing quality: (**a**) 12%; (**b**) 9%; (**c**) 6%; (**d**) 3%.

As shown in Figure 13a,b, the printing accuracy of 12% and 9% gelatin solutions is relatively high. As shown in Figure 13c, 6% gelatin solution had insufficient support performance, resulting in the serious deformation of the printing model. As shown in Figure 13d, 3% gelatin solution flowed obviously during printing, making it difficult to form. Therefore, the energy storage modulus of gelatin was proportional to the concentration of gelatin, and when the gelatin concentration was below 6% its self-supporting ability was far from being sufficient for formation. These results are comparable to the research results of He et al., who showed that in the range of 10~20% MA modified gelatin gel, the higher the concentration, the greater the Young's modulus, and the better the mechanical properties [33].

3.6. *Effect of Nozzle Diameter on Flow Field Distribution*

The nozzle diameter was a key parameter affecting printing time and accuracy. Using the nozzle diameter as a variable, we explored its impact on the parameters inside the flow channel. As shown in Figure 14a,b, there was no significant change in the velocity and shear rate in the flow channel of the printing head with different diameters in the barrel section, but they rapidly increased in the printing head section. The smaller the printing head diameter, the more rapidly the velocity and shear rate increased.

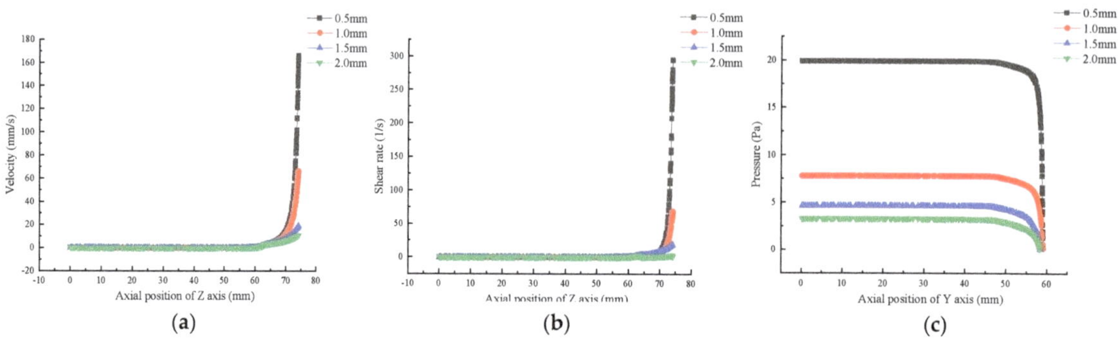

Figure 14. Effect of nozzle diameter on flow field distribution based on Comsol simulation: (**a**) Velocity; (**b**) Shear rate; (**c**) Pressure.

As shown in Figure 14c, as the nozzle diameter increased from 0.5 mm to 2.0 mm, the relative pressure inside the material barrel decreased from 19.9 Pa to 3.3 Pa, and the pressure declined significantly by about 83.5%, which indicates that the smaller the nozzle diameter, the greater the pressure required for material extrusion and the more high-pressure zones there are in the flow channel. The pressure difference between the inlet and outlet of the printing heads with diameters of 0.5 mm, 1.0 mm, 1.5 mm, and 2.0 mm was 19.9 Pa, 7.8 Pa, 4.6 Pa, and 3.3 Pa, respectively. The above results demonstrated that the smaller the nozzle diameter, the greater the pressure difference between the inlet and outlet. However, the larger the pressure difference, the more elastic potential energy is released after extrusion, which is very likely to result in expanded extrusion. Therefore, the nozzle with an appropriate diameter should be selected in the actual printing process. These results are comparable to the research results of Oyinloye et al. [34].

4. Conclusions

Simulation techniques can be used to predict the distribution of the flow field in the flow channel during the printing process, as well as the influence of process parameters on the flow field distribution. It was found that the parameter changes within the barrel during the low-temperature extrusion process were small, and the parameter changes were mainly concentrated in the nozzle section. As the inlet velocity increases, the velocity and shear rate in the flow field increase, and the material temperature in the outlet section increases. The material concentration and temperature are the main factors affecting the viscosity; as the gelatin concentration increases, its viscosity, storage modulus, and self-supporting ability increase. As the print head diameter decreases, the pressure, velocity, and shear rate in the flow channel increase.

The initial temperature of gelatin mainly affects the distribution of the temperature field in the flow channel and the temperature of the gelatin aqueous solution in the outlet section. In an ideal printing state, when the gelatin solution is extruded to the outlet of the printing nozzle, the temperature of the gelatin solution should be at the rapid condensation point temperature. When the temperature of the gelatin aqueous solution in the outlet section is lower than the rapid condensation point temperature, it can easily cause the gelatin aqueous solution to solidify in the printing head and cause extrusion blockage. When the temperature of the gelatin in the outlet section is greater than the rapid setting temperature of the gelatin, it can easily cause the print model to flow and collapse.

Through experiments and simulations, it was found that for the experimental 3D printer—under the process conditions of: environmental temperature, 5 °C; heating jacket temperature, 45 °C; nozzle diameter, 1 mm; and printing speed 5 mm/s—the molding quality was better when the inlet speed was 0.025 mm/s, the initial temperature of the gelatin solution was 30 °C, and the gelatin concentration was 9% and 12%. When the concentration of gelatin solution was below 6%, it was difficult to form due to insufficient self-supporting ability. The process parameters obtained from the experiment were within a range, which could also meet the molding process requirements when the inlet speed and the initial temperature of the gelatin solution were close to the experimental conclusion. This paper provides a certain reference value for the optimization of the low-temperature deposition 3D printing process for other food materials and the design of a low-temperature deposition 3D printer extrusion mechanism. Materials are an important factor limiting the development of 3D printing at present. In the future, the printability of more kinds of hydrogels can be explored, providing more possibilities for food and biological 3D printing.

Author Contributions: Methodology, Q.T.; software, W.Z.; verification, W.Z., T.G., D.W. and X.D.; data organization, W.Z. and T.G.; writing—original manuscript preparation, W.Z.; writing—review and editing, W.Z.; supervision, Q.T., D.W. and X.D.; funding acquisition, X.D. All authors have read and agreed to the published version of the manuscript.

Funding: This research was funded by National Key Research and Development Program of China (No. 2022YFD2100603); Basic Scientific Research Project of Institutions of Higher Learning in Liaoning Province (No. LJKMZ20220895).

Data Availability Statement: The datasets generated for this study are available on request to the corresponding author.

Conflicts of Interest: The authors declare no conflict of interest.

References

1. Sano, Y.; Matsuzaki, R.; Ueda, M.; Todoroki, A.; Hirano, Y. 3D Printing of Discontinuous and Continuous Fibre Composites Using Stereolithography. *Addit. Manuf.* **2018**, *24*, 521–527. [CrossRef]
2. Zhang, P.J. Simulation Optimization and Material Forming Research of Direct Writing 3D Device Based on Silicon Nitride Ceramics. Ph.D. Thesis, Guangdong University of Technology, Guangzhou, China, 2021.
3. Zheng, L.; Sun, W.L.; Sun, B. Influence of extrusion methods on the discharging controllability of 3D printing with viscoelastic slurry. *J. Mater. Eng.* **2020**, *48*, 134–141.
4. Liu, C.; Tong, J.; Ma, J.; Wang, D.; Xu, F.; Liu, Y.; Chen, Z.; Lao, C. Low-Temperature Deposition Manufacturing: A Versatile Material Extrusion-Based 3D Printing Technology for Fabricating Hierarchically Porous Materials. *J. Nanomater.* **2019**, *2019*, e1291067. [CrossRef]
5. Yang, F.L.; Guo, C.F.; Zhang, M. Improving 3D printing process of lemon juice gel based on fluid flow numerical simulation. *LWT* **2019**, *102*, 89–99. [CrossRef]
6. Tan, Z.; Parisi, C.; Di Silvio, L.; Dini, D.; Forte, A.E. Cryogenic 3D Printing of Super Soft Hydrogels. *Sci. Rep.* **2017**, *7*, 16293. [CrossRef]
7. Huang, H.J.; Cheng, J.S. Chemical Basis of Collagen and Gelatin Molecules and gel of Gelatin. *Sci. Technol. Gelatin* **2005**, *2*, 82–86.
8. Yang, G.H.; Han, Y. The Effect of Gelatin on the Stability of 3D Printing of Chicken Mince. *J. Food Sci.* **2021**, *43*, 51–57.
9. Liu, Z.B. *Study on Extrusion 3D Printing and Post processing Adaptability of Potato Mash and Its Starch Mixed Gel, System*; Jiangnan University: Wuxi, China, 2020.
10. Schuurman, W.; Levett, P.A.; Pot, M.W.; Van Weeren, P.R.; Dhert, W.J.A.; Hutmacher, D.W.; Melchels, F.P.W.; Klein, T.J.; Malda, J. Gelatin-Methacrylamide Hydrogels as Potential Biomaterials for Fabrication of Tissue-Engineered Cartilage Constructs: Gelatin-Methacrylamide Hydrogels as Potential Biomaterials for Fabrication. *Macromol. Biosci.* **2013**, *13*, 551–561. [CrossRef]
11. Kim, G.; Ahn, S.; Yoon, H.; Kim, Y.; Chun, W. A Cryogenic Direct-Plotting System for Fabrication of 3D Collagen Scaffolds for Tissue Engineering. *J. Mater. Chem.* **2009**, *19*, 8817–8823. [CrossRef]
12. Gu, Y.; Zhang, L.; Du, X.; Fan, Z.; Wang, L.; Sun, W.; Cheng, Y.; Zhu, Y.; Chen, C. Reversible Physical Crosslinking Strategy with Optimal Temperature for 3D Bioprinting of Human Chondrocyte-Laden Gelatin Methacryloyl Bioink. *J. Biomater. Appl.* **2018**, *33*, 609–618. [CrossRef]
13. Levett, P.A.; Melchels, F.P.W.; Schrobback, K.; Hutmacher, D.W.; Malda, J.; Klein, T.J. Chondrocyte Redifferentiation and Construct Mechanical Property Development in Single-Component Photocrosslinkable Hydrogels: Chondrocyte Redifferentiation and Construct Mechanical Property Development. *J. Biomed. Mater. Res. A* **2014**, *102*, 2544–2553. [CrossRef] [PubMed]
14. Levett, P.A.; Melchels, F.P.W.; Schrobback, K.; Hutmacher, D.W.; Malda, J.; Klein, T.J. A Biomimetic Extracellular Matrix for Cartilage Tissue Engineering Centered on Photocurable Gelatin, Hyaluronic Acid and Chondroitin Sulfate. *Acta Biomater.* **2014**, *10*, 214–223. [CrossRef] [PubMed]
15. Zhao, T.Y. *Design and Experimental Study of the Mechanical System of an Extrusion-Based Bio-3D Printer*; Harbin Institute of Technology: Harbin, China, 2019.
16. Gao, Q.; Yu, K.; Chen, F.; Lu, L.; Zhang, P. Investigation on the Temperature Distribution Uniformity of an Extrusion-Based 3D Print Head and Its Temperature Control Strategy. *Pharmaceutics* **2022**, *14*, 2108. [CrossRef]

17. Chen, X.L. Numerical Simulation of Gelatin Physical Gelatin Processes under Non-Uniform Temperature Field. Ph.D. Thesis, Shandong University, Jinan, China, 2010.
18. Kavimughil, M.; Leena, M.M.; Moses, J.A.; Anandharamakrishnan, C. Effect of Material Composition and 3D Printing Temperature on Hot-Melt Extrusion of Ethyl Cellulose Based Medium Chain Triglyceride Oleogel. *J. Food Eng.* **2022**, *329*, 111055. [CrossRef]
19. Xu, J.; Wang, N.; Severini, C. Fluent-based fluid simulation of a cigarette machine hitch gun nozzle. *Packag. J.* **2022**, *14*, 36–41. [CrossRef]
20. Gao, Q.; Lu, L.; Zhang, R.; Song, L.; Huo, D.; Wang, G. Investigation on the Thermal Behavior of an Aerostatic Spindle System Considering Multi-Physics Coupling Effect. *Int. J. Adv. Manuf. Technol.* **2019**, *102*, 3813–3823. [CrossRef]
21. Derossi, A.; Caporizzi, R.; Azzollini, D.; Severini, C. Application of 3D Printing for Customized Food. A Case on the Development of a Fruit-Based Snack for Children. *J. Food Eng.* **2018**, *220*, 65–75. [CrossRef]
22. Lille, M.; Nurmela, A.; Nordlund, E.; Metsä-Kortelainen, S.; Sozer, N. Applicability of Protein and Fiber-Rich Food Materials in Extrusion-Based 3D Printing. *J. Food Eng.* **2018**, *220*, 20–27. [CrossRef]
23. Liu, Z. Impact of Rheological Properties of Mashed Potatoes on 3D Printing. *J. Food Eng.* **2018**, *9*, 76–82. [CrossRef]
24. Eidam, D.; Kulicke, W.-M.; Kuhn, K.; Stute, R. Formation of Maize Starch Gels Selectively Regulated by the Addition of Hydrocolloids. *Starch—Stärke* **1995**, *47*, 378–384. [CrossRef]
25. Rueda, M.M.; Auscher, M.-C.; Fulchiron, R.; Périé, T.; Martin, G.; Sonntag, P.; Cassagnau, P. Rheology and Applications of Highly Filled Polymers: A Review of Current Understanding. *Prog. Polym. Sci.* **2017**, *66*, 22–53. [CrossRef]
26. Liu, Q.; Zhang, N.; Wei, W.; Hu, X.; Tan, Y.; Yu, Y.; Deng, Y.; Bi, C.; Zhang, L.; Zhang, H. Assessing the Dynamic Extrusion-Based 3D Printing Process for Power-Law Fluid Using Numerical Simulation. *J. Food Eng.* **2020**, *275*, 109861. [CrossRef]
27. Shamsudin, R.; Mohamed, I.O.; Nooi, T.S. Rheological properties of cocoa butter substitute (CBS): Effects of temperature and characteristics of fatty acids on viscocity. *J. Food Lipids* **2006**, *13*, 402–410. [CrossRef]
28. Valkenaers, H.; Jansen, D.; Voet, A.; Van Gysel, A.; Ferraris, E. Additive manufacturing for concrete a 3D printing principle. In Proceedings of the 14th Euspen International Conference, Dubrovnik, Croatia, 2–6 June 2014.
29. Emin, M.A. Analysis of the Dispersive Mixing Efficiency in a Twin-Screw Extrusion Processing of Starch Based Matrix. *J. Food Eng.* **2013**, *115*, 132–143. [CrossRef]
30. Yan, C.F. Extrusion Process of Freeze-form Extrusion Fabrication for Ceramics Material. *Bull. Am. Ceram. Soc.* **2015**, *34*, 3073–3077.
31. Cao, W.J.; Ma, Z.Y. Dimensional accuracy of wire formed by 3D low temperature deposition of sodium alginate/gelatin solution. *Mech. Eng.* **2018**, *42*, 12. [CrossRef]
32. Guo, L.; Colby, R.H.; Lusignan, C.P.; Howe, A.M. Physical Gelation of Gelatin Studied with Rheo-Optics. *Macromolecules* **2003**, *36*, 10009–10020. [CrossRef]
33. He, S.; Huang, H.J. Preliminary Study on Gelatin Hydrogels with Different Concentrations for 3D Bioprinting. *Mech. Eng.* **2021**, *4*, 148–153.
34. Oyinloye, T.M.; Yoon, W.B. Investigation of Flow Field, Die Swelling, and Residual Stress in 3D Printing of Surimi Paste Using the Finite Element Method. *Innov. Food Sci. Emerg. Technol.* **2022**, *78*, 103008. [CrossRef]

foods

MDPI

Article

The Effect of Nozzle Temperature on the Low-Temperature Printing Performance of Low-Viscosity Food Ink

Qiang Tong [1], Yuxiang Meng [1], Yao Tong [1], Dequan Wang [1] and Xiuping Dong [2,*]

[1] College of Mechanical Engineering and Automation, Dalian Polytechnic University, Dalian 116034, China; tongqiang.work@outlook.com (Q.T.); myx.work@outlook.com (Y.M.); tong0555@outlook.com (Y.T.); gtr446962334@gmail.com (D.W.)

[2] School of Food Science and Technology, Dalian Polytechnic University, Dalian 116034, China

* Correspondence: dongxp@dlpu.edu.cn

Abstract: Low-temperature food printing technology is used in many fields, such as personalized nutrition, cooking art, food design and medical nutrition. By precisely controlling the deposition temperature of the ink, a food with a finer and more controllable structure can be produced. This paper investigates the influence of nozzle temperature on printing performance via a numerical simulation and experimental research. The results indicate that the ink gradually changed from a granular state to a fLow-characteristic deposition structure when the nozzle temperature increased from 19 °C to 27 °C. When the nozzle temperature exceeded 21 °C, the ink demonstrated excellent extrusion behavior and tended to flow. The widths of the rectangular frame deposition showed no obvious changes and were 4.07 mm, 4.05 mm and 4.20 mm, respectively. The extrusion behavior of the ink showed a structural mutation in the temperature range of 19–21 °C. Its line width changed from 3.15 mm to 3.73 mm, and its deposition structure changed from a grainy shape to a normal shape. Under the influence of different environmental control capabilities, bulk structure deposition demonstrates an ideal printing performance at 21, 23 and 25 °C, and the latter temperature is more suitable in the case of large external interference. The ink flowed violently when the nozzle temperature reached 27 °C, at which point the deposit structure flowed and deformed seriously. On the other hand, evaporation losses had a strong effect on Low-viscosity ink. To reach the full potential of this promising technology, it is necessary to determine the effect of nozzle temperature on printing performance. This article provides a method for developing and applying Low-viscosity, Low-temperature food printing.

Keywords: low-temperature forming; low-viscosity ink; printing temperature; formability

Citation: Tong, Q.; Meng, Y.; Tong, Y.; Wang, D.; Dong, X. The Effect of Nozzle Temperature on the Low-Temperature Printing Performance of Low-Viscosity Food Ink. *Foods* **2023**, *12*, 2666. https://doi.org/10.3390/foods12142666

Academic Editor: Mi-Jung Choi

Received: 30 May 2023
Revised: 26 June 2023
Accepted: 7 July 2023
Published: 11 July 2023

1. Introduction

Three-dimensional food printing is a popular production technology for creating customized and personalized food, and research on food printing is expanding in both academic and commercial fields [1–3]. There are three main types of food materials currently used for printing: (1) soft materials that can maintain their support after deposition; (2) Low-melting materials that can crystallize to form layered structures; and (3) polymeric materials with shear-thinning behaviors [4–6]. The viscoelastic and mechanical properties of gel materials are determined by changes in the rheological properties of the ink [7]. In addition, the temperature range from the barrel to the nozzle outlet is closely related to the rheological behavior of the ink. A bad temperature gradient will lead to poor deposition structure and support capacity [8–10]. Therefore, based on temperature regulation, it is the focus of further research on food printing to explore the influence of the temperature distribution of gel ink at the print head on printing performance. Gel food inks usually exhibit strong colloidal characteristics at low viscosities, and are generally less affected by food materials. Taking a gelatin solution as example, this paper explores the performance characteristics of Low-viscosity food inks during printing. The gelatin solution has the typical characteristics of a pseudoplastic fluid, with high-temperature melting and Low-temperature curing points which can be used for printing. It is often used as an additive to improve ink performance, edible film preparation and bioprinting [11–13]. Compared with other food gels, it shows greater fluidity at a low viscosity. Among materials used to study the influence of nozzle domain temperature on Low-viscosity food ink, it is easier to observe the flow behavior and more representative.

Low-temperature deposition manufacturing (LDM) is a cold-cure additive manufacturing technique (AM) suitable for Low-viscosity food inks [14]. In a typical printing process, the ink needs to be melted at the right temperature to be extruded through the nozzle. The temperature is raised by a heating device on the barrel wall, and the forming environment temperature is reduced via air cooling, water cooling or a cooling environment covering the entire printer [15–17]. The cylinder has a certain diameter, and the heat transfer is carried out through the cylinder and the nozzle in contact with the cooling environment. As a result, the temperature distribution is not uniform across the center, the surface of the cylinder and the nozzle. Due to the large temperature difference between heating and cooling, the lower temperature will be concentrated in the nozzle and its upper area. This is not conducive to the extrusion of ink and changes in temperature in the nozzle. Therefore, an innovative nozzle structure is needed to regulate the temperature gradient.

Temperature changes will affect the rheology and printing properties of temperature-dependent materials [18,19]. Gelatin is a highly temperature-sensitive material that exhibits more pronounced rheological behaviors at different temperatures [20–22]. It gels at low temperatures and melts at high temperatures, with a small range of variation that is easy to observe in experiments. A large number of studies have shown that the temperature changes in the barrel and the forming environment will affect the printing performance of the ink and can regulate the state of the deposition structure [15,16,23–27]. Proper structural design can improve the temperature gradient, which further affects the mechanical properties of printing [28–32]. The ink exhibits different forming capabilities and deformation effects under different nozzle outlet temperatures and temperature gradients. Therefore, it is important to understand the underlying mechanisms of different nozzle temperatures on Low-viscosity food inks. This contributes to the development and application of 3D printing food inks.

At present, the printing process of Low-viscosity food ink, especially the influence of nozzle temperature on printing performance, has not been fully understood. We adopted an integrated research approach, combining a numerical simulation and experimental validation to obtain more comprehensive and accurate research results. The effects of nozzle temperature on the printing properties, rheological properties and macroscopic and microscopic structures of Low-viscosity food inks were systematically studied. Through the results of this study, we aim to provide a scientific basis and optimization strategy for the printing process of Low-viscosity food inks. At the same time, our research also has important engineering value for the further development and application of food printing technology. In the following chapters, we will introduce our research methods, experimental design and an analysis of our results in detail. This will further verify and explain our novel discovery and reveal the key mechanism of the effect of nozzle temperature change on printing performance.

2. Materials and Methods

2.1. Material

To determine a suitable gelatin concentration, we printed ink with gelatin contents of 4, 6, 8, 10 and 12 wt%. According to Zhao's master's thesis (2019) and our actual printing conditions, 10 wt% was used in this study [33]. The gelatin (purchased from Henan Sugar Cabinet Co., Ltd., Henan, China, using 160 bloom) and water were prepared in a mixed solution at a ratio of 1:9, placed in a constant-temperature water circulation heating pot (purchased from 600 W Qun'an Experimental Instrument Co., Ltd., Zhejiang, China) for constant-temperature melting. It was defoamed (−2.5 MPa) using an air compressor (AP-01P, Puruiqi, Beijing, China) and stored at 40 °C.

2.2. Rheological Analysis of Gelatin

The rheological properties of the gelatin were characterized using a hybrid rheometer (Discovery HR-2, TA Company, Boston, MA, USA). The liquid gelatin was placed between a 25 mm parallel plate and the platform, with a gap of 1 mm. The setup was equilibrated for 1 min to reach the desired temperature (25, 30 and 35 °C), and excess meat was scraped off of the platform. The variance in the apparent viscosity with the shear rate was recorded as the shear rate was increased from 0.1 s^{-1} to 100 s^{-1}. The elastic modulus G' and the loss modulus G'' were calculated via a temperature sweep at a frequency of 1.6 Hz, and the samples were heated from 4 °C to 40 °C within 20 min. The variance in the apparent viscosity with temperature was recorded as the temperature was increased from 4 °C to 40 °C at a shear rate of 3.12 s^{-1}. All measurements were conducted at an amplitude strain of 0.5% within the linear viscoelastic region (LVR) [9].

2.3. Configuration and Optimization of the Design of 3D Food Printing Equipment and Nozzle Structure

2.3.1. Configuration of Food 3D Printing Equipment

A 3D printer for Low-temperature food deposition was designed, as shown in Figure 1. It added ambient cooling and nozzle domain temperature control to a traditional direct writing (DW) food printer. The ambient cooling consisted of a PID controller and a cooling box that surrounded the printer as a whole, as shown in Figure 1a. Nozzle domain temperature control was achieved with a customized nozzle structure via water circulation, as shown in Figure 1b.

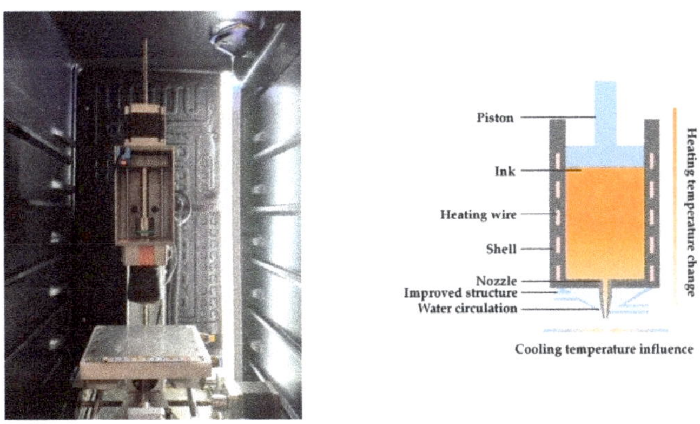

Figure 1. Configuration of 3D food printing equipment: (**a**) custom-designed Low-temperature food deposition 3D printer; (**b**) custom-made print head structure.

A temperature sensor was installed in the nozzle to accurately control its temperature. However, nozzles used for food printing generally have diameters of less than 2 mm. Considering the structural size of the temperature sensor, it was impractical to install the temperature sensor inside the nozzle. Therefore, it was installed in the gap between the self-made print head structure and the nozzle, and the temperature of the water circulation recorded by the sensor replaced the temperature at the nozzle.

2.3.2. Nozzle Structure Design

Convective heat transfer occurred between the cold air in the cooling environment and the nozzle. And the heat lost through the outer surface of the nozzle generated a thermal gradient. This phenomenon is particularly evident in Low-viscosity food printing because Low-viscosity foods are more sensitive to temperature changes. This will lead to a large deviation between the nozzle temperature and the barrel heating temperature, thus affecting the temperature control accuracy of the deposition process. Here, we designed an improved structure, as shown in Figure 1b. It was designed based on reducing heat loss and controlling temperature. The water flow temperature was heated via water circulation, and it was used to replace the thermal convection between the cold air and the nozzle. The simulation results in Section 4.1.1 show that the temperature distribution and control accuracy were improved.

The size of the optimized nozzle structure is shown in Figure 2. The inlet was above, and the outlet was below. In order to ensure a spiral flow of water around the nozzle, they were eccentrically designed. The water flowed from the inlet at a flow rate of 10 mm/s, ensuring the nozzle temperature during continuous rotation.

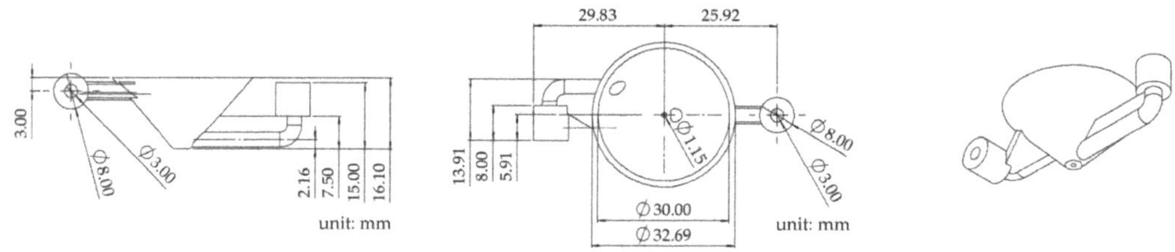

(a) (b) (c)

Figure 2. The size diagram of the optimized nozzle structure: (**a**) front view; (**b**) vertical view; (**c**) equiaxial side view.

2.4. *Finite Element Numerical Analysis*

2.4.1. Finite Element Simulation of Gelatin Fluid

Using the Heat Transfer and CFD module of COMSOL6.0 software (Comsol, Inc., Burling, MA, USA), the temperature field of the gelatin in the barrel and nozzle was analyzed via the finite element method.

2.4.2. FEM Modeling of the Print Head

The better printing conditions obtained from the early orthogonal experiments were used for analysis. In order to accurately estimate the temperature distribution of the nozzle in the printing state, a steady-state thermal simulation model of the printing state was established based on the finite element method. Furthermore, geometric features such as chamfers and threads were ignored, and resistance wire heating was regarded as an isothermal shell heat source with a temperature of 30 °C. These decisions were made to reduce the difficulty of modeling during the analysis process and improve computational efficiency. The calculation grid and thermal boundary conditions of the nozzle domain are shown in Figure 3. The outer surface of the nozzle structure was simplified as a convective heat transfer boundary.

2.4.3. Geometry and Boundary Conditions

To reduce the calculation time, the calculation range was limited to the nozzle and the barrel filled with gelatin (the part above the piston was ignored). The geometry simulation was constructed using Solidworks2018 (Solidworks, Inc., Dassault Systemes, Waltham, MA, USA). The mesh division of the simulated geometry is shown in Figure 3. Triangular mesh was used to discretize the fluid domain in the geometry. In order to ensure the calculation of turbulence, a finer physical field control grid was used. The grids of the entire model contained 3,542,847 units. The related boundary conditions are shown in Figure 3b. Flow was applied at the gelatin and water circulation inlets. The boundary conditions are listed as follows:

(1) Since it was assumed that there was no wall slip, the printing speed and water speed at the wall were zero;
(2) The pressure at the outlet of the nozzle was 101,325 Pa, which was 1 atm;
(3) The piston was considered a rigid body and did not deform. Therefore, the piston exerted a squeeze on the reservoir area at the same speed as its movement.

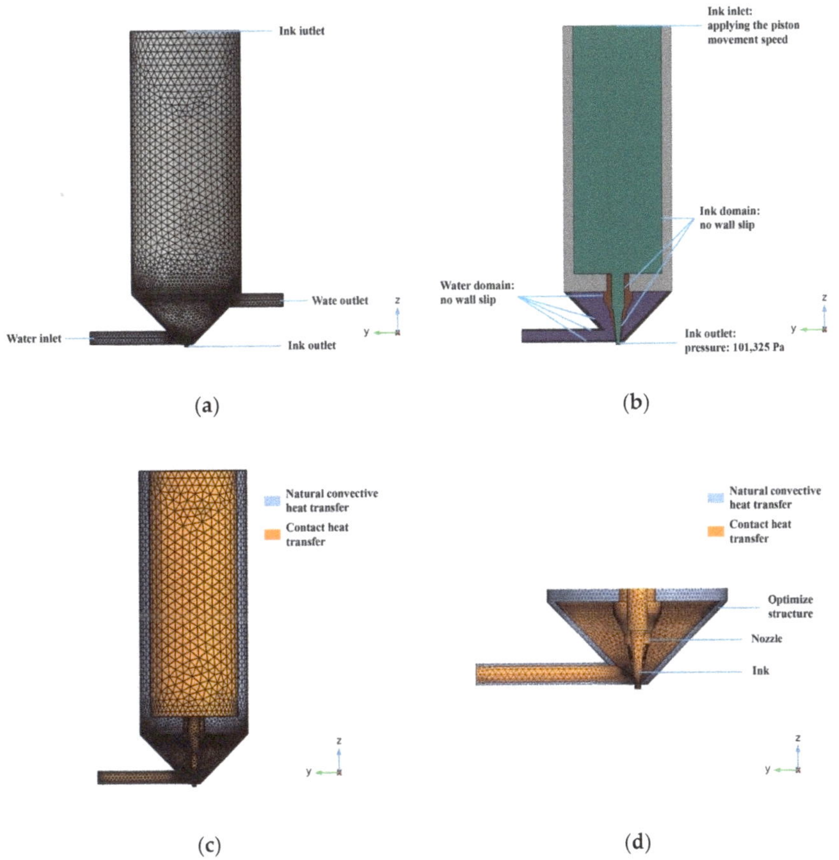

Figure 3. Schematic diagram of 3D printer nozzle and print head: (**a**) meshing of the overall model; (**b**) the essential boundary conditions; (**c**) thermal boundaries of the overall model; (**d**) meshing and heat transfer boundaries of the nozzle structure region.

2.4.4. Models and Assumptions

(1) Ink Flow Model

To analyze the flow characteristics of ink flowing through the nozzle, we evaluated the Reynolds number (Re_1) of the shear thinning flow [34]:

$$Re_1 = \frac{\rho_1 {d_1}^n {\overline{v}_1}^{2-n}}{K((3n+1)/(4n))^n 8^{n-1}} \tag{1}$$

where ρ_1 is the density, d_1 is the nozzle diameter, $\overline{v}_1$ is the average velocity of gel ink, n and K are the flow index and consistency index (Pa·s^n), respectively, derived from the rheological properties of the ink defined by the power law model [35]:

$$\eta(\dot{\gamma}) = m\dot{\gamma}^{n-1} \tag{2}$$

where η is the apparent viscosity (Pa·s) and $\dot{\gamma}$ is the local shear rate (s^{-1}). The Re can be used to identify a flow regime in a pipe, such as a laminar flow ($Re \leq 2300$), critical flow ($2000 < Re < 4000$), or turbulent flow ($Re > 4000$) [36]. The physical properties of the gel ink (90% moisture content) and the power law model coefficients are listed in Table 1.

Table 1. The geometric parameters of FEM.

Parameter	Symbol	Value	Unit
The density of the gel ink	ρ_1	1	g/cm^3
The nozzle diameter	d_1	1	mm
The average velocity of the gel ink in the nozzle	$\overline{v}_1$	0.83	mm/s
The consistency index	K	0.04	Pa·s^n
The flow index	n	0.58	
Reynolds number of the gel ink	Re_1	0.03	
The density of water	ρ_2	1	g/cm^3
The inlet diameter	d_2	3	mm
The average velocity of water	$\overline{v}_1$	10	mm/s
The dynamic viscosity of water	μ_2	10^{-3}	Pa·s
Reynolds number of water	Re_2	30	
The Prandtl number of air	P_r	0.713	

For a given shear rate, the viscosity of the fluid is only related to n, and n quantifies the response of the apparent viscosity to the shear rate. $n > 1$, $n < 1$ and $n = 1$ indicated shear thickening, shear thinning and Newtonian fluid characteristics, respectively. The printing ink studied in this paper was a non-Newtonian fluid with $n < 1$. Both exponential coefficients in the power law model were measured via the rheological tests described in Section 2.2.

Due to the low Reynolds number, the incompressible gelatin fluid was under stable, isothermal and laminar flow conditions. The mathematical model of ink flow applied in this study included the following continuity and momentum conservation equations in the vector symbol [36]:

$$\rho_1 \nabla \cdot \mathbf{u}_1 = 0 \tag{3}$$

$$\rho_1 \frac{\partial \mathbf{u}_1}{\partial t} + \rho_1 (\mathbf{u}_1 \cdot \nabla) \mathbf{u}_1 = \nabla \cdot \left[-p\mathbf{I} + \mu_1 \left(\nabla \mathbf{u}_1 + (\nabla \mathbf{u}_1)^T \right) \right] + \mathbf{F} \tag{4}$$

where $\mathbf{u}_1$ is the velocity vector, p is the hydrostatic pressure, $\mathbf{I}$ is the unit tensor, μ_1 is the dynamic viscosity of the gel ink and $\mathbf{F}$ represents other external forces. The effect of gravity was neglected.

(2) Water flow model

To analyze the flow characteristics of the water circulation part, we evaluated the Re_2 of the Newtonian fluid [37]:

$$Re_2 = \frac{\rho_2 \overline{v}_2 d_2}{\mu_2} \tag{5}$$

where ρ_2 is the density of water, $\overline{v}_2$ is the average velocity of water, d_2 is the inlet diameter and μ_2 is the dynamic viscosity of water. Table 1 shows that the Re_2 of the water cycle was small and should indicate a laminar flow. However, the water flow had an obvious vortex flow in the improved nozzle structure, which was beneficial to the heat transfer between the water body and the nozzle. The model (Low-Reynolds-number k−ε turbulence model) applied in this study of water circulation included the following continuity and momentum conservation equations in vector symbols [38]:

$$\rho_2 \frac{\partial k}{\partial t} + \rho_2 \mathbf{u}_2 \cdot \nabla k = \nabla \cdot \left(\left(\mu_2 + \frac{\mu_{2T}}{\sigma_k} \right) \nabla k \right) + P_k - \rho_2 \varepsilon \tag{6}$$

$$\rho_2\frac{\partial\varepsilon}{\partial t}+\rho_2\mathbf{u}_2\cdot\nabla\varepsilon=\nabla\cdot\left(\left(\mu_2+\frac{\mu_{2T}}{\sigma_\varepsilon}\right)\nabla\varepsilon\right)+C_{\varepsilon1}\frac{\varepsilon}{k}P_k-f_\varepsilon C_{\varepsilon2}\rho_2\frac{\varepsilon^2}{k} \quad (7)$$

$$P_k=\mu_{2T}\left(\nabla\mathbf{u}_2:\left(\nabla\mathbf{u}_2+(\nabla\mathbf{u}_2)^T\right)-\frac{2}{3}(\nabla\cdot\mathbf{u}_2)^2\right)-\frac{2}{3}\rho_2k\nabla\cdot\mathbf{u}_2 \quad (8)$$

where k is the turbulent kinetic energy, $\mathbf{u}_2$ is the velocity vector, μ_{2T} is the turbulent viscosity, σ_k is the turbulent Prandtl number for k, P_k is the production of turbulent kinetic energy, ε is the dissipation rate of turbulent kinetic energy, σ_ε is the turbulent Prandtl number for ε, f_ε, and $C_{\varepsilon1}$ and $C_{\varepsilon2}$ are constants. The effect of gravity was neglected.

(3) Assumptions

(1) Due to the slow extrusion speed and the small Reynolds number for food 3D printing, the laminar flow of the ink was adopted in the barrel;
(2) The flow in the barrel was considered to be fully developed;
(3) The influence of a small amount of gas in the barrel was ignored;
(4) The ink did not slip on the wall during printing;
(5) The friction between the piston and the storage cylinder was neglected;
(6) The ink was a homogeneous material;
(7) The ink was considered an incompressible material.

2.4.5. Heat Transfer Settings

(1) Heat Transfer Analysis

The process of extruding gelatin ink involved natural convection heat transfer and contact heat transfer. The natural convection heat transfer refers to the heat exchange between the nozzle outlet, the barrel, the optimized nozzle structure and the external air. The contact heat exchange refers to the heat exchange between the ink, barrel wall, water circulation, nozzle structure, improved structure, heating sleeve and water circulation. There was no clear thermal contact resistance between the faces of the contact part, which was considered full-contact. No thermal contact resistance was set in the model, as shown in Figure 3.

(2) Heat source for barrel heating

The heating element was installed in the heat insulation cotton as a heat source during the experiment. And the temperature of the heating element was controlled via a closed-loop temperature sensor. The temperature of the cylinder wall at a distance of 13 mm from the bottom of the cylinder remained relatively stable at different temperatures (see Section 4.1.2). Therefore, the heating element and insulation cotton were regarded as a whole as a constant temperature boundary. The temperature of the entire module was defined as 30 °C (the average value of the ink in the cartridge measured at different temperatures) during the simulation.

(3) Convective heat transfer coefficient

Using R_e and the Grashof number (G_{rL}) (two dimensionless numbers), the type of heat transfer was determined as [37]:

$$G_{rL}=\frac{g\beta(T_s-T_\infty)L^3}{\nu^2} \quad (9)$$

where g is the acceleration of gravity, β is the volumetric thermal expansion coefficient, L is the characteristic dimension and ν is the kinematic viscosity of the air. T_s and T_∞ are the temperatures of the surface air and the surrounding air. As the G_{rL}/Re^2 between the barrel, nozzle structure, heat insulation cotton and the cooling ambient air was much greater than

1, the heat transfer state between all models and the air was regarded as natural convection. The convective heat transfer coefficient (h) thus can be expressed as [29]:

$$h = \frac{0.59k}{L} G_{r_L} P_r \tag{10}$$

$$P_r = \frac{\mu_{air} c_p}{k} \tag{11}$$

where P_r is a temperature-related physical property (the Prandtl number), μ_{air} is the dynamic viscosity of air, c_p is the specific heat capacity and k is the thermal conductivity of air. P_r was defined as a 15 °C cooling environment.

3. Printing Experiment and Result Verification

A Low-temperature deposition manufacturing (LDM) food printer was used for the experiments. According to the preliminary experimental results, suitable printing conditions were selected. The diameter of the nozzle was 1.0 mm, the layer height was 0.5 mm, the ambient cooling temperature was 15 °C, the heating temperature of the barrel was 55 °C, the water circulation temperature was 19, 21, 23, 25 and 27 °C and the travel speed of the nozzle was 1.56 mm/s. The standard printing process was carried out at refrigerated temperatures to evaluate the printing performance. We experimented with the initial and optimized structures of gelatin in different nozzle temperature domains.

An electron microscope (TD-4KHT, Sanqiang Teda) was used to analyze the deposition structure changes in the gelatin photographed. We selected the same location of the deposited structure for analysis, as shown in Figure 4.

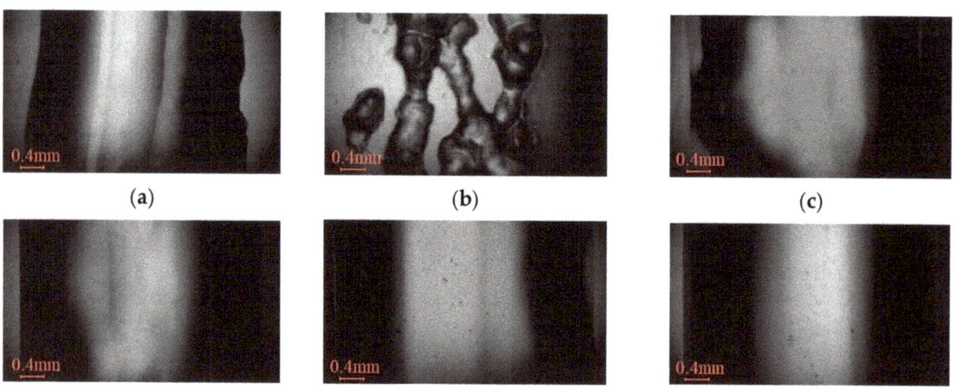

(a) (b) (c) (d) (e) (f)

Figure 4. Micrograph of sedimentary points of the unified model: (**a**) initial structure; (**b**) optimized structure at 19 °C; (**c**) 21 °C; (**d**) 23 °C; (**e**) 25 °C; (**f**) 27 °C.

4. Results and Discussion

In 3D printing Low-viscosity food, the rheological properties of the printed materials are significantly affected by temperature changes. In this section, the relationship between material printing performance and temperature is investigated and discussed by controlling the temperature of nozzle domain.

4.1. *Analysis and Experimental Verification of Simulation Results*

4.1.1. Analysis of Simulation Results

Figure 5 shows the steady-state distribution of printing temperature for the initial and optimized structures. It can be seen from the figure that under the initial conditions, the temperature peaked at 30 °C in the middle of the inner wall of the barrel and bottomed at

19 °C at the outlet of the nozzle. The temperature drop at the top of the barrel was caused by the piston being affected by the external temperature. This was the same case in the optimized structure, but it had little impact on the printing performance because the ink would be heated again during the extrusion process. Moreover, a significant temperature drop occurred at the nozzle outlet due to the convective heat transfer between the cold forming environment and the nozzle. It formed a lower and steeper temperature gradient distribution, which was not conducive to nozzle temperature control during printing.

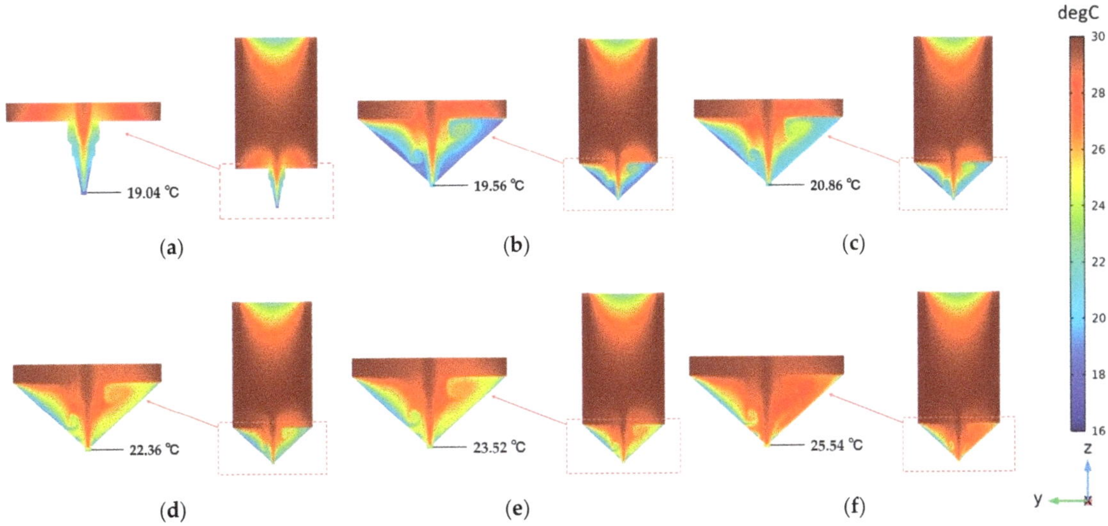

Figure 5. Steady-state temperature distribution under the initial and optimized structures: (**a**) initial structure; (**b**) optimized structure at 19 °C; (**c**) 21 °C; (**d**) 23 °C; (**e**) 25 °C; (**f**) 27 °C.

The temperature distribution of the nozzle under the optimized structure was more uniform. It had a gentler temperature drop than under the initial structure and a significant impact on the temperature of the nozzle. It can be seen that the temperature dropped only to a certain extent at the nozzle outlet. And there is a positive correlation with the water circulation temperature. This helps us indirectly characterize the nozzle outlet temperature using the water circulation temperature.

To compare the numerical difference between the simulated and actual printing processes, temperature measurements were carried out and are described in Section 4.1.2. Figure 6 shows the measured and simulated temperature curves at the nozzle outlet (Point 1) at three points under different conditions. The simulated and experimental temperatures at the nozzle outlet are also shown in the figure. And the deviations at different temperatures are 2.16, 1.66, 0.56, 0.02 and −0.26 °C, respectively. The average temperature deviation is about 0.83 °C. The mean difference from the actual value is 4.9%. Due to the simplified treatment of the model, the simulation results fit poorly at lower temperatures. Therefore, the simulation results are only used for further temperature distribution and control analysis.

4.1.2. Experimental Verification

To verify the calculation accuracy of the simulation results and the temperature improvement ability of the optimized structure, several sets of experiments were carried out to measure the actual nozzle temperature at different water circulation temperatures. As shown in Figure 7, three-point temperatures of the initial and optimized structures were measured using temperature sensors (PT100 platinum resistance temperature sensor, $\pm(0.15 + 0.002 \times |t|)$, where t is the temperature value to be measured).

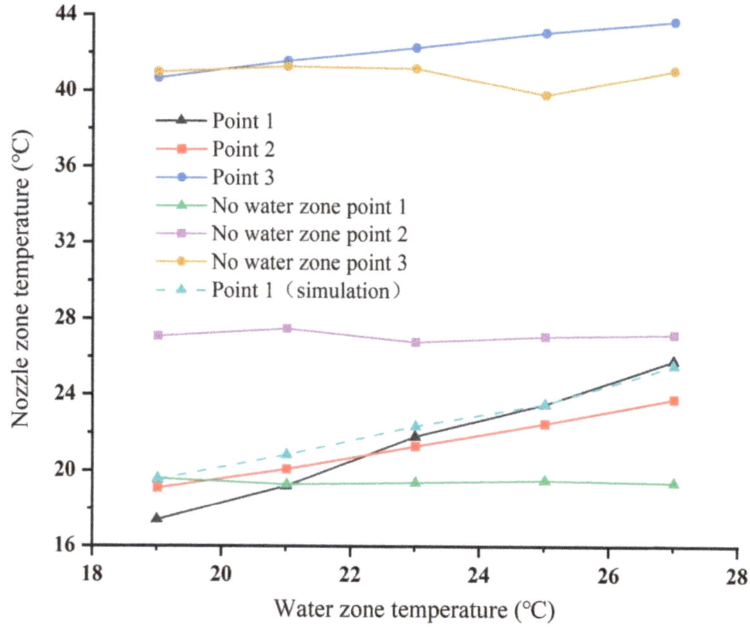

Figure 6. The temperature data measured by the sensor at three points and the simulated temperature at Point 1.

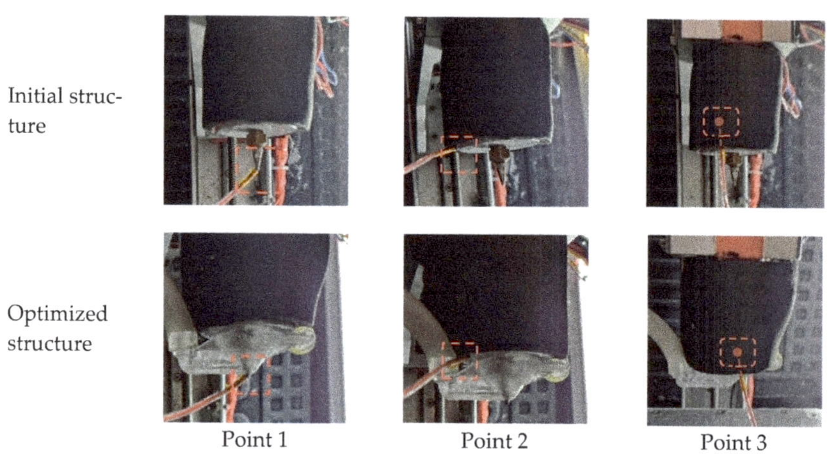

Figure 7. Temperature measurement at three points (the red box indicates the measurement location).

Figure 6 shows the temperature data measured by the sensor and simulated by Comsol software. It can be seen from the figure that with the increase in the water circulation temperature, the growth rate at Point 1 was higher than at Points 2 and 3. The temperature of Point 1 even exceeded that of Point 2 at 23 °C. This was because the influence of the temperature of the water cycle weakens with increasing height and had a weaker effect at Point 2. In addition, there was little difference between Points 1 and 3 at 19 °C whether water circulation was present or not, but there was a gap of 8 °C at Point 2. The reason for this was the difference in the dominant factors under such a condition. In the absence of water circulation, the system was dominated by the heating device, and in the presence of water circulation, it was dominated by the water circulation temperature. Figure 6 also shows the deviation range between the nozzle outlet of the optimized structure and the water circulation temperature. It was 1.6, 1.8, 1.2, 1.5 and 1.2 °C, respectively. The average deviation was 1.46 °C, indicating that the optimized nozzle structure was more effective at temperature control, and there was a strong correlation between the water body and the nozzle outlet temperature, which is consistent with the simulation results. In this paper, the temperature of the water body, which can be easily controlled, replaces the temperature of the nozzle outlet for analysis.

4.2. Rheological Properties of Gelatin

Food materials used for extrusion molding must have an appropriate viscosity, ease of extrusion and a certain degree of adhesion to avoid deformation during the deposition of structures [39,40]. The apparent viscosity of the gelatin ink with a water content of 10 wt% is shown in Figure 8a. It decreased with the increase in the shear rate, so it was a typical pseudoplastic food printing material. In addition, the viscosity also inversely proportional to the temperature and decreased sharply in the range of 25–33 °C. This meant that gelatin ink had low degree of printability at a low temperature and a high degree of fluidity at a high temperature, which was not conducive to the structure-forming process of 3D printing, as shown in Figure 8b.

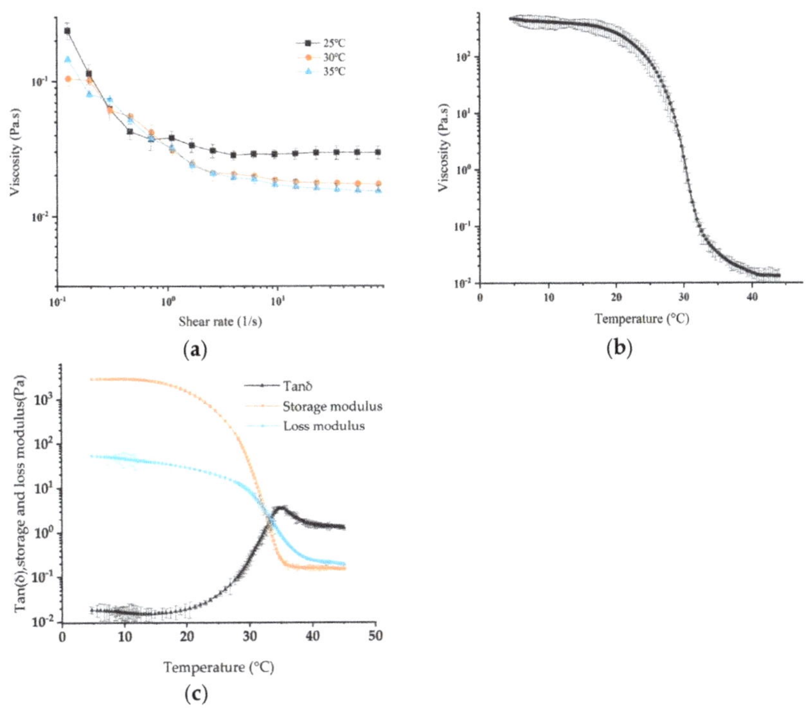

Figure 8. (**a**) The apparent viscosity of gelatin ink with a water content of 10 wt% at different temperatures as a function of shear rate; (**b**) the apparent viscosity of gelatin ink with a water content of 10 wt% as a function of temperatures; (**c**) storage modulus and loss modulus of gelatin ink with a water content of 10 wt% at different temperatures.

The storage modulus G′ and loss modulus G″ of the gelatin ink are shown in Figure 8c, which shows that they were temperature-dependent. G′ was significantly higher than G″ at low temperatures, showing a gel structure dominated by an elastic ability that made it easy to maintain the shape of the sedimentary structure [41]. With the increase in temperature, G′ and G″ gradually decreased, and the decline rate of G″ was gradually less than G′. When the gel network is depolymerized to a certain extent, the decrease rate of G' slows down until it stops, and the decrease rate of G'' gradually exceeds G'. The change in internal tension led to the transfer of intermolecular and molecular internal forces, gradually weakening the elastic part of the gel structure and enhancing its ability to flow. The loss tangent (Tanδ = G'/G'') is a characteristic parameter used to describe viscoelastic behavior. A Tanδ less than 1 shows elastic characteristics, and a Tanδ greater than 1 shows viscous characteristics [42]. At temperatures higher than 32.25 °C, G'' was greater than G', and the gelatin ink demonstrated a flow ability with liquid properties. In order to achieve successful printing, a reasonable temperature drive and precise rheological characteristics are needed.

Low-viscosity gelatin ink has an obvious fast gelling stage and a slow gelling stage. Upon cooling, gelatin undergoes physical gelation via the formation of triple helices. In the rapid gelation stage, the gelatin changes over time to show an initial rapid growth region in which a new helix is formed. In the slow gel phase, a slower growth region is present, involving spiral elongation. For gelatin gels formed in the fast-gelling regime, the rheological exponents depend on concentration but not on temperature. In addition, helical elongation occurs for a long time at each temperature and concentration for gelatin [43]. The slow gelling stage of 10 wt. % gelatin ink is large, and its slow gel time will be significantly greater than the rapid gel time. Since food printing is a rapid prototyping process, the slow gelation temperature is not suitable for printing analysis. Therefore, this paper used the rapid gelation temperature as a printing reference. It is calculated as follows [43]:

$$T_l(K) = (35 + 0.18c_0 + 273)\left(1 - \frac{0.28}{2\ln c_0 + 5.06}\right) \tag{12}$$

where c_0 is the initial concentration of gelatin (g/dL). Since the density of the ink is close to water, the initial value is 10 g/dL, and the calculated rapid gel temperature is 27.67 °C. Figure 8c shows the measured slow gelation temperature, 32.25 °C. These results are consistent with the study of Guo et al.

4.3. *Macroscopical Effect of Nozzle Temperature on Printing Performance*

4.3.1. Rectangular Frame Sedimentary Structure

The deposition results of a rectangular frame with line segments lapped in the initial and optimized structure are shown in Figure 9. Gaps and less granular deposition are present in the initial structure. The optimized structure showed a serious extrusion issue at the nozzle domain temperature of 19 °C, and the material deposited on the transparent plastic film in a granular form. This was because the material gelatinized and blocked the nozzle tip at an ambient temperature that was too low. In addition, a simple extrusion behavior was observed at the nozzle temperatures of 21, 23, 25 and 27 °C, which was caused by the strong temperature sensitivity of the Low-viscosity material. The results show that the ink underwent a strong change in its printing performance in the 2 °C range.

However, as transparent plastic film is easy to fold and bend and is not easily made completely flat, the deposited structure will show flow gaps of different sizes on the film. This is somewhat beneficial for us to judge the flow properties of materials. With the gradual increase in temperature, there were an increasing number of flow gaps in the deposition structure on the film. It was not conducive to the deposition of the base structure in the printing process.

The differences in material properties between the actual and ideal print were attributed to the shear-thinning property of the material. As a result, the material which should show elastic ability at 19–25 °C demonstrated extrusion performance. A polymer is easier to extrude than a Newtonian fluid [44]. When the ink was sprayed into the nozzle, the viscosity declined under the condition of the shear force, making it easier to extrude. As a consequence, an ink that should demonstrate good extrusion performance above 27.67 °C (a fast-gelling temperature) can achieve a good deposition structure at 21 °C.

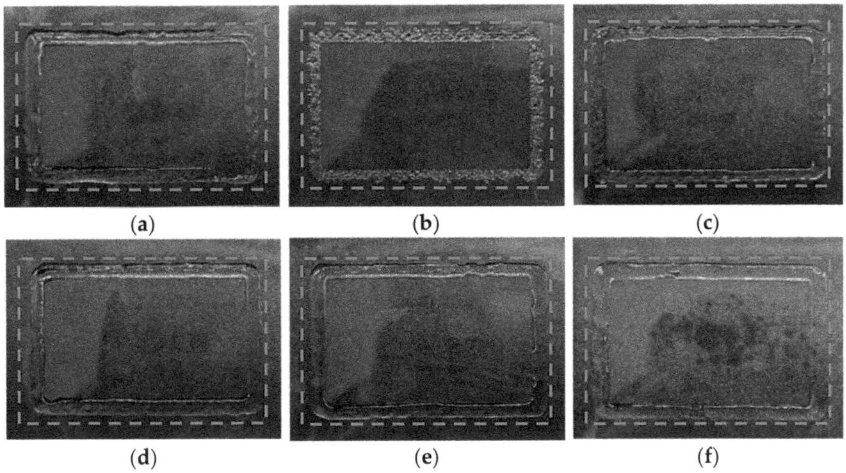

Figure 9. Initial and optimized deposition structure of rectangular frame at different nozzle domain temperatures: (a) initial structure; (b) optimized structures at 19 °C; (c) 21 °C; (d) 23 °C; (e) 25 °C; (f) 27 °C.

The flow of food materials in the nozzle was regarded as a Poiseuille flow [45]. The increase in temperature reduced viscosity but increased velocity, thus increasing the width of the deposition structure. However, the width of the deposition structure did not change significantly with the change in the nozzle temperature, as displayed in Figure 9, which was more obvious in Figure 4. This was because the viscosity of the printing material was too low and varied little with temperature at the same shear rate. This phenomenon indicated that the ink had a sudden change in response to temperature but had little effect on the width of deposition structure.

4.3.2. Cylindrical Sedimentary Structure

Figure 10 shows the cylindrical deposition structure of gelatin ink at different nozzle domain temperatures (19, 21, 23, 25 and 27 °C) and the initial conditions. Among them, the initial structure and nozzle temperature at 21 °C showed similar printing results. This was due to the fact that the nozzle temperatures in the two states ware similar (see Section 4.1), and the inks exhibited nearly identical rheological properties. The printing performance of gelatin ink varied with the change in the nozzle domain temperature. The 19 °C sedimentary structure was granular and formed a complete cylindrical sedimentary structure at the base. After printing to a certain height, there was a granular aggregation due to the uneven support layer, and the particles collapsed to the side of the cylinder after excessive aggregation, which was caused by the low fluidity of the ink at 19 °C. At 21 °C, the granular structure suddenly disappeared, which was consistent with the analysis results in the previous section. At 21–27 °C, the porous defects in the gelatin gradually disappeared until it exhibited the best printing effect at 25 °C, and the flow phenomenon occurred due to strong fluidity at 27 °C. This was different from what was described in the above section because although a better rectangular frame deposition was presented, the fluidity was poor at 21 and 23 °C. The body structure deposition was prone to influence from the external environment, leading to pore defects. The fluidity of the ink was better at 25 °C, which made up for pore defects in the printing process, and it was suitable for printing conditions that did not require precise external environment control. Therefore, it is reasonably inferred that a more accurate ink model could be obtained at 21 and 23 °C when the external environment is well controlled. If the above material conditions and printing specifications were maintained, printing performance at 21, 23 and 25 °C varied under different environmental capabilities.

Figure 10. Cylinder deposition structure of initial structure and optimized structure at different nozzle domain temperatures: (a) initial structure; (b) optimized structure at 19 °C; (c) 21 °C; (d) 23 °C; (e) 25 °C; (f) 27 °C.

4.4. Microscopic Effect of Nozzle Temperature on Deposition Structure

The micro-deposition structures of the rectangular frame under the same conditions and position for the initial and optimized structures are shown in Figure 4. The deposition width at the horizontal symmetry line of the picture was regarded as the average width of the ink at this temperature. Gelatin shows inconsistent deposition widths on different surfaces (the hydrophobic plastic material or the rough, corrugated, stainless steel surfaces, etc.). This is caused by the combined effects of uneven surface morphology, the permeability difference of gelatin, and the drying and shrinkage of gelatin. A transparent plastic film is a typical hydrophobic plastic material with good transparency and a smooth surface. Therefore, the observed width is generally more consistent than the width observed on a rough surface in the gelatin printing process. In this paper, the deposition width of the gelatin was studied using a transparent plastic film as the substrate. The deposition widths at 19, 21, 23, 25 and 27 °C were 3.15, 3.73, 4.07, 4.05 and 4.20 mm, respectively. As can be seen from the microscopic map, dark characteristics appeared on both sides and bright characteristics appeared in the middle. This is because the light source illuminated by the electron microscope deflected at the ends of the arc section, resulting in lower levels of brightness at both ends. It can be more intuitively seen from Figure 11. After evaporation, the ink arc was reduced, both ends became flat, and brightness was increased. Under the initial conditions, there were obvious shadow-like lap marks in the middle of the sedimentary structure, and it had irregular contour edges compared to the optimized structures at 21, 23, 25 and 27 °C. The optimized structure at 27 °C had a smoother surface and better shape. The reason for this situation was that the initial condition had no nozzle domain water circulation heating structure. The actual temperature at the nozzle was far lower than the charging barrel temperature but close to ambient cooling temperature. The viscosity of the gelatin material sharply increased with the decrease in temperature, which led to a decline in the passing ability of the ink at the nozzle and thus a poor deposition effect. This phenomenon was more significant at 19 °C after optimization and was accompanied by a serious granular deposition structure. In addition, lap marks gradually weakened, and edges became smoother with the increase in temperature in the range of 21–27 °C. The best effect emerged at 27 °C, but the viscosity was too low, and an obviously uneven flow appeared on the test plate, as shown in Figure 12.

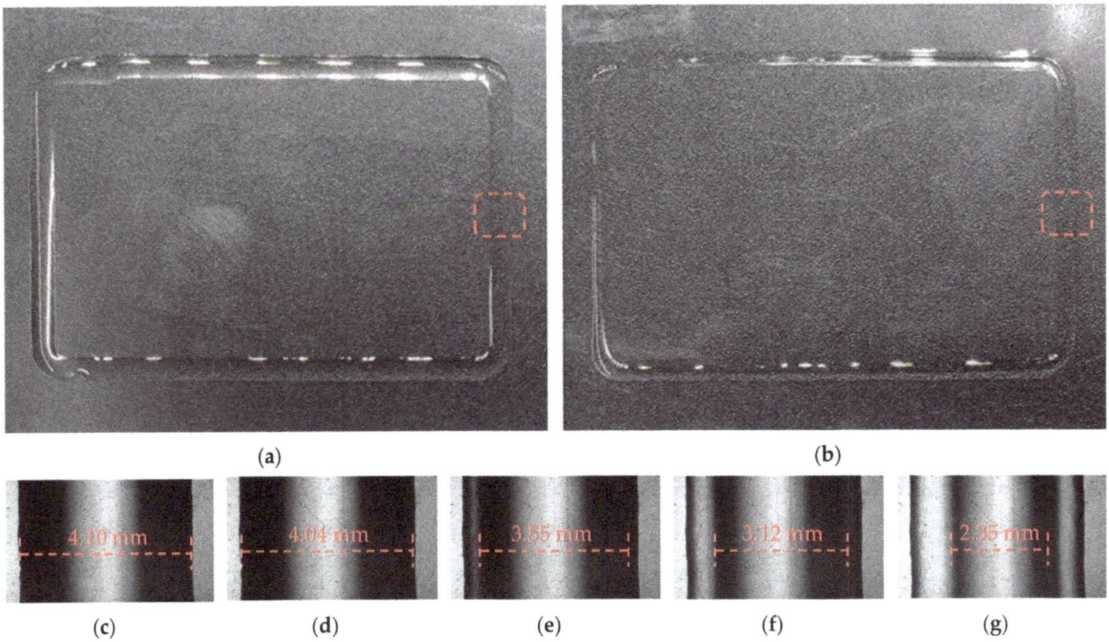

Figure 11. Deposition structures at different time under the optimal structure condition at 25 °C: (**a**) normal view at 2 min; (**b**) 10 min; (**c**) micro view at 2 min; (**d**) 4 min; (**e**) 6 min; (**f**) 8 min; (**g**) 10 min.

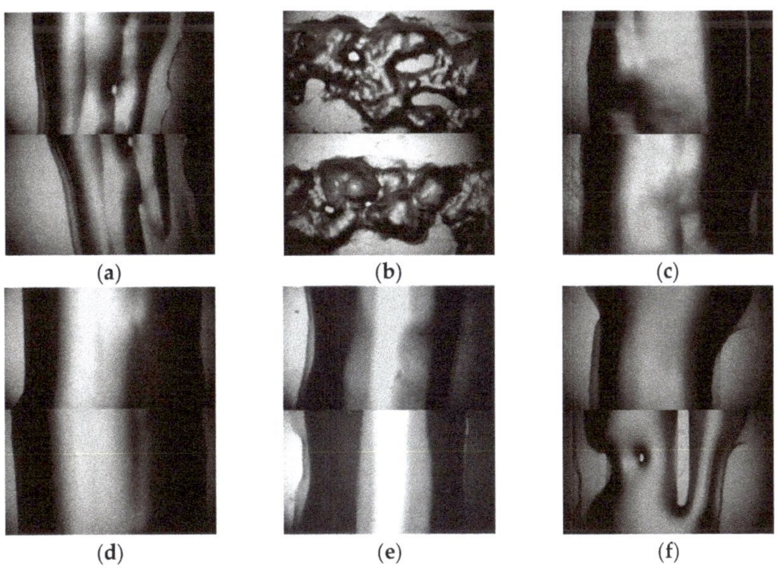

Figure 12. Typical bad spots and flows of deposition structures at different temperatures: (**a**) initial structure; (**b**) optimized structure at 19 °C; (**c**) 21 °C; (**d**) 23 °C; (**e**) 25 °C; (**f**) 27 °C.

The normal extrusion of gelatin ink was achieved under these conditions except that the deposition had obvious grainy quality at 19 °C. There was little difference in the deposition width, which was consistent with the conclusion in Section 4.3. The mutability of Low-viscosity gelatin to a temperature response was further confirmed. Considering the overall structure, lap marks and edge effects, the optimized 23 and 25 °C demonstrated better printing performance. This phenomenon illustrated the importance of nozzle domain temperature control for the overall and microscopic deposition performance in the 3D food printing process.

4.5. Effect of Evaporation on Deposition Structure of Low-Viscosity Ink

In this study, the deposition structure of Low-viscosity ink showed obvious evaporation loss and shrinkage behavior within 10 min (room temperature environment) after printing, as shown in Figure 11. This is relatively common in gelled 3D printing materials but usually occurs after a long time or in extreme environments [46]. With the increase in time, the structure width of the gelatin ink decreased after evaporation. The small-scale deformation due to the flow disappeared, showing a flatter frame structure. This phenomenon was caused by the low molecular content of the Low-viscosity ink. The reduced molecular content extended the gelation time of the gelatin ink and made it more prone to melting at room temperature. This was significantly different from conventional high-concentration food printing. Although small-scale deformation and the disappearance of lap marks were beneficial to the appearance and perception of the deposition structure, the deposition width shrank by about 55%, which was not conducive to the formation of the next layer of the structure or accurate control of the deposition structure. In the printing process, the temperature should be maintained as low as possible to inhibit the influence of evaporation loss on the printing structure. The volumetric evaporation loss of the structures that needed to be preserved at room temperature was considered in the construction of the 3D model.

5. Conclusions

The temperature in the cylinder has an important impact on the printing performance and deposition structure of extrusion 3D printing with Low-viscosity food. In order to better understand the variation mechanism and potential role of the temperature domain from the barrel to the nozzle, we developed a nozzle structure to quantify the temperature of the nozzle domain and carried out a thermal analysis simulation via the finite element method, as well as experimental verification. Through simulation and experimental comparison, a 10 wt. % gelatin ink demonstrated an ideal printing performance when exposed to varying environmental control capabilities at nozzle domain temperatures of 21 °C, 23 °C and 25 °C. In the presence of external interference, the nozzle domain temperature should be set as 25 °C to ensure excellent printing performance. The actual printable temperature is about 6 °C lower than the temperature guided by the rheological properties. The response of print performance to temperature changes was abrupt. It mutated at 19–21 °C, and the ink deposition was granular below 19 °C and normal above 21 °C. With the increase in temperature, the lap traces between the lines weakened, and the porous defects in the body structure decreased. However, it showed a strong flow capacity at 27 °C, and the flow deformation of the structure was serious. In addition, evaporation loss has a great effect on the deposition structure of the Low-viscosity ink. When stored at room temperature for 10 min, the evaporation loss occurred obviously, and the deposition width decreased by about 55%. This paper provides a certain reference for forming or printing other Low-viscosity food at low temperatures. Further research can be carried out to test the printing performances of different inks at low temperatures or on different surfaces. Future research should explore the forming performances and rheological properties of more Low-viscosity inks at low temperatures and expand the application range of Low-viscosity ink in food printing.

Author Contributions: Q.T.: writing—review and methodology; Y.M.: writing—original draft and editing and conception; Y.T.: validation and investigation; D.W.: reviewing and editing; X.D.: supervision and funding acquisition. All authors have read and agreed to the published version of the manuscript.

Funding: This research was funded by the National Key Research and Development Program of China (No. 2022YFD2100603) and Liaoning Province Ministry of Education Scientific Study Project for Institutes of Higher Learning (No. LJKMZ20220895).

Institutional Review Board Statement: Not applicable.

Informed Consent Statement: Not applicable.

Data Availability Statement: The data presented in this study are available upon request from the corresponding author.

Conflicts of Interest: The authors declare no conflict of interest.

References

1. Enfield, R.E.; Pandya, J.K.; Lu, J.; McClements, D.J.; Kinchla, A.J. The Future of 3D Food Printing: Opportunities for Space Applications. *Crit. Rev. Food Sci. Nutr.* **2022**, 1–14. [CrossRef]
2. Teng, X.; Zhang, M.; Mujumdar, A.S. 4D Printing: Recent Advances and Proposals in the Food Sector. *Trends Food Sci. Technol.* **2021**, *110*, 349–363. [CrossRef]
3. Dick, A.; Bhandari, B.; Prakash, S. 3D Printing of Meat. *Meat Sci.* **2019**, *153*, 35–44. [CrossRef]
4. Tian, H.; Wang, K.; Qiu, R.; Wang, S.; Hu, Z.; Zhao, L. Effects of Incubation Temperature on the Mechanical and Structure Performance of Beeswax-Carrageenan-Xanthan Hybrid Gelator System in 3D Printing. *Food Hydrocoll.* **2022**, *127*, 107541. [CrossRef]
5. Li, G.; Hu, L.; Liu, J.; Huang, J.; Yuan, C.; Takaki, K.; Hu, Y. A Review on 3D Printable Food Materials: Types and Development Trends. *Int. J. Food Sci. Technol.* **2022**, *57*, 164–172. [CrossRef]
6. Jiang, Q.; Zhang, M.; Mujumdar, A.S. Novel Evaluation Technology for the Demand Characteristics of 3D Food Printing Materials: A Review. *Crit. Rev. Food Sci. Nutr.* **2022**, *62*, 4669–4683. [CrossRef]
7. Nijdam, J.J.; LeCorre-Bordes, D.; Delvart, A.; Schon, B.S. A Rheological Test to Assess the Ability of Food Inks to Form Dimensionally STable 3D Food Structures. *J. Food Eng.* **2021**, *291*, 110235. [CrossRef]
8. Paolillo, M.; Derossi, A.; van Bommel, K.; Noort, M.; Severini, C. Rheological Properties, Dispensing Force and Printing Fidelity of Starchy-Gels Modulated by Concentration, Temperature and Resting Time. *Food Hydrocoll.* **2021**, *117*, 106703. [CrossRef]
9. Kavimughil, M.; Leena, M.M.; Moses, J.A.; Anandharamakrishnan, C. Effect of Material Composition and 3D Printing Temperature on Hot-Melt Extrusion of Ethyl Cellulose Based Medium Chain Triglyceride Oleogel. *J. Food Eng.* **2022**, *329*, 111055. [CrossRef]
10. Habuš, M.; Golubić, P.; Vukušić Pavičić, T.; Čukelj Mustač, N.; Voučko, B.; Herceg, Z.; Ćurić, D.; Novotni, D. Influence of Flour Type, Dough Acidity, Printing Temperature and Bran Pre-Processing on Browning and 3D Printing Performance of Snacks. *Food Bioprocess Technol.* **2021**, *14*, 2365–2379. [CrossRef]
11. Yap, K.L.; Kong, I.; Abdul Kalam Saleena, L.; Pui, L.P. 3D Printed Gelatin Film with Garcinia Atroviridis Extract. *J. Food Sci. Technol.* **2022**, *59*, 4341–4351. [CrossRef]
12. Maihemuti, A.; Zhang, H.; Lin, X.; Wang, Y.; Xu, Z.; Zhang, D.; Jiang, Q. 3D-Printed Fish Gelatin Scaffolds for Cartilage Tissue Engineering. *Bioact. Mater.* **2023**, *26*, 77–87. [CrossRef]
13. Chen, Y.; Zhang, M.; Sun, Y.; Phuhongsung, P. Improving 3D/4D Printing Characteristics of Natural Food Gels by Novel Additives: A Review. *Food Hydrocoll.* **2022**, *123*, 107160. [CrossRef]
14. Liu, C.; Tong, J.; Ma, J.; Wang, D.; Xu, F.; Liu, Y.; Chen, Z.; Lao, C. Low-Temperature Deposition Manufacturing: A Versatile Material Extrusion-Based 3D Printing Technology for Fabricating Hierarchically Porous Materials. *J. Nanomater.* **2019**, *2019*, e1291067. [CrossRef]
15. Li, Z.; Xu, M.; Wang, J.; Zhang, F. Recent Advances in Cryogenic 3D Printing Technologies. *Adv. Eng. Mater.* **2022**, *24*, 2200245. [CrossRef]
16. Warburton, L.; Lou, T.; Rubinsky, B. Development of A 3d Cryoprinter for Printing Soft Biomaterials. *Cryobiology* **2021**, *103*, 180. [CrossRef]
17. Tan, Z.; Parisi, C.; Di Silvio, L.; Dini, D.; Forte, A.E. Cryogenic 3D Printing of Super Soft Hydrogels. *Sci. Rep.* **2017**, *7*, 16293. [CrossRef]
18. Yang, N.; Feng, Y.; Su, C.; Wang, Q.; Zhang, Y.; Wei, Y.; Zhao, M.; Nishinari, K.; Fang, Y. Structure and Tribology of κ-Carrageenan Gels Filled with Natural Oil Bodies. *Food Hydrocoll.* **2020**, *107*, 105945. [CrossRef]
19. Liu, Z.; Bhandari, B.; Prakash, S.; Mantihal, S.; Zhang, M. Linking Rheology and Printability of a Multicomponent Gel System of Carrageenan-Xanthan-Starch in Extrusion Based Additive Manufacturing. *Food Hydrocoll.* **2019**, *87*, 413–424. [CrossRef]
20. Du, J.; Dai, H.; Wang, H.; Yu, Y.; Zhu, H.; Fu, Y.; Ma, L.; Peng, L.; Li, L.; Wang, Q.; et al. Preparation of High Thermal Stability Gelatin Emulsion and Its Application in 3D Printing. *Food Hydrocoll.* **2021**, *113*, 106536. [CrossRef]
21. Enrione, J.; Char, C.; Pepczynska, M.; Padilla, C.; González-Muñoz, A.; Olguín, Y.; Quinzio, C.; Iturriaga, L.; Díaz-Calderón, P. Rheological and Structural Study of Salmon Gelatin with Controlled Molecular Weight. *Polymers* **2020**, *12*, 1587. [CrossRef]
22. Poh, L.; Li, B.; Yu, W.; Narimissa, E.; Wagner, M.H. Modeling of Nonlinear Extensional and Shear Rheology of Low-Viscosity Polymer Melts. *Polym. Eng. Sci.* **2021**, *61*, 1077–1086. [CrossRef]

23. Warburton, L.; Rubinsky, B. Freezing-Modulated-Crosslinking: A Crosslinking Approach for 3D Cryoprinting. *Bioprinting* **2022**, *27*, e00225. [CrossRef]
24. Zhang, J.; Van Hooreweder, B.; Ferraris, E. T4F3: Temperature for Fused Filament Fabrication. *Prog. Addit. Manuf.* **2022**, *7*, 971–991. [CrossRef]
25. Tichý, T.; Šefl, O.; Veselý, P.; Dušek, K.; Bušek, D. Mathematical Modelling of Temperature Distribution in Selected Parts of FFF Printer during 3D Printing Process. *Polymers* **2021**, *13*, 4213. [CrossRef]
26. Solis, D.M.; Czekanski, A. The Effect of the Printing Temperature on 4D DLP Printed PNIPAM Hydrogels. *Soft Matter* **2022**, *18*, 3422–3429. [CrossRef]
27. Chen, J.; Sun, H.; Mu, T.; Blecker, C.; Richel, A.; Richard, G.; Jacquet, N.; Haubruge, E.; Goffin, D. Effect of Temperature on Rheological, Structural, and Textural Properties of Soy Protein Isolate Pastes for 3D Food Printing. *J. Food Eng.* **2022**, *323*, 110917. [CrossRef]
28. Gao, Q.; Yu, K.; Chen, F.; Lu, L.; Zhang, P. Investigation on the Temperature Distribution Uniformity of an Extrusion-Based 3D Print Head and Its Temperature Control Strategy. *Pharmaceutics* **2022**, *14*, 2108. [CrossRef]
29. Zhang, P.; Gao, Q.; Yu, K.; Yao, Y.; Lu, L. Investigation on the Temperature Control Accuracy of a Print Head for Extrusion 3D Printing and Its Improved Design. *Biomedicines* **2022**, *10*, 1233. [CrossRef]
30. El Moumen, A.; Tarfaoui, M.; Lafdi, K. Modelling of the Temperature and Residual Stress Fields during 3D Printing of Polymer Composites. *Int. J. Adv. Manuf. Technol.* **2019**, *104*, 1661–1676. [CrossRef]
31. Xu, Y.; Huang, M.; Schlarb, A.K. Print Path-Dependent Contact Temperature Dependency for 3D Printing Using Fused Filament Fabrication. *J. Appl. Polym. Sci.* **2022**, *139*, 52337. [CrossRef]
32. Martínez-Monzó, J.; Cárdenas, J.; García-Segovia, P. Effect of Temperature on 3D Printing of Commercial Potato Puree. *Food Biophys.* **2019**, *14*, 225–234. [CrossRef]
33. Zhao, T. Mechanical System Design of Extruded Biological 3D Printer and Experimental Research. Master's Thesis, Harbin Institute of Technology, Harbin, China, 2019.
34. Madlener, K.; Frey, B.; Ciezki, H.K. Generalized Reynolds Number for Non-Newtonian Fluids. In Proceedings of the Progress in Propulsion Physics; EDP Sciences: Les Ulis, France, 2009; Volume 1, pp. 237–250.
35. Albano, K.M.; Franco, C.M.L.; Telis, V.R.N. Rheological Behavior of Peruvian Carrot Starch Gels as Affected by Temperature and Concentration. *Food Hydrocoll.* **2014**, *40*, 30–43. [CrossRef]
36. Oyinloye, T.M.; Yoon, W.B. Investigation of Flow Field, Die Swelling, and Residual Stress in 3D Printing of Surimi Paste Using the Finite Element Method. *Innov. Food Sci. Emerg. Technol.* **2022**, *78*, 103008. [CrossRef]
37. Gao, Q.; Lu, L.; Zhang, R.; Song, L.; Huo, D.; Wang, G. Investigation on the Thermal Behavior of an Aerostatic Spindle System Considering Multi-Physics Coupling Effect. *Int. J. Adv. Manuf. Technol.* **2019**, *102*, 3813–3823. [CrossRef]
38. Abe, K.; Kondoh, T.; Nagano, Y. A New Turbulence Model for Predicting Fluid Flow and Heat Transfer in Separating and Reattaching Flows—I. Flow Field Calculations. *Int. J. Heat Mass Transf.* **1994**, *37*, 139–151. [CrossRef]
39. Kamlow, M.-A.; Vadodaria, S.; Gholamipour-Shirazi, A.; Spyropoulos, F.; Mills, T. 3D Printing of Edible Hydrogels Containing Thiamine and Their Comparison to Cast Gels. *Food Hydrocoll.* **2021**, *116*, 106550. [CrossRef]
40. Hussain, S.; Malakar, S.; Arora, V.K. Extrusion-Based 3D Food Printing: Technological Approaches, Material Characteristics, Printing Stability, and Post-Processing. *Food Eng. Rev.* **2022**, *14*, 100–119. [CrossRef]
41. Siacor FD, C.; Chen, Q.; Zhao, J.Y.; Han, L.; Valino, A.D.; Taboada, E.B.; Caldona, E.B.; Advincula, R.C. On the Additive Manufacturing (3D Printing) of Viscoelastic Materials and Flow Behavior: From Composites to Food Manufacturing. *Addit. Manuf.* **2021**, *45*, 102043. [CrossRef]
42. Eidam, D.; Kulicke, W.-M.; Kuhn, K.; Stute, R. Formation of Maize Starch Gels Selectively Regulated by the Addition of Hydrocolloids. *Starch-Stärke* **1995**, *47*, 378–384. [CrossRef]
43. Guo, L.; Colby, R.H.; Lusignan, C.P.; Howe, A.M. Physical Gelation of Gelatin Studied with Rheo-Optics. *Macromolecules* **2003**, *36*, 10009–10020. [CrossRef]
44. Kim, Y.H.; Shim, T.S.; Kim, J.M. Lateral Particle Migration in Shear-Thinning Colloidal Dispersion. *Korea-Aust. Rheol. J.* **2022**, *34*, 327–334. [CrossRef]
45. Yu, K.; Gao, Q.; Lu, L.; Zhang, P. A Process Parameter Design Method for Improving the Filament Diameter Accuracy of Extrusion 3D Printing. *Materials* **2022**, *15*, 2454. [CrossRef]
46. Shahmirzadi, M.R.; Gholampour, A.; Kashani, A.; Ngo, T.D. Shrinkage Behavior of Cementitious 3D Printing Materials: Effect of Temperature and Relative Humidity. *Cem. Concr. Compos.* **2021**, *124*, 104238. [CrossRef]

foods

MDPI

Article

Facile Fabrication of Anthocyanin-Nanocellulose Hydrogel Indicator Label for Intelligent Evaluation of Minced Pork Freshness

Xiangyong Meng [1], Qinqin Shen [1,2], Teng Song [1], Honglei Zhao [3], Yong Zhang [1], Aiqing Ren [4,*] and Wenbin Yang [2,*]

1 Anhui Provincial Key Laboratory of Molecular Enzymology and Mechanism of Major Diseases, Key Laboratory of Biomedicine in Gene Diseases and Health of Anhui Higher Education Institutes, College of Life Sciences, Anhui Normal University, Wuhu 241000, China
2 School of Ecology and Environment, Anhui Normal University, Wuhu 241000, China
3 Weifang Inspection and Testing Center, Weifang 261100, China
4 Institute of Food Research, Hezhou University, Hezhou 542899, China
* Correspondence: renaiqing@hzxy.edu.cn (A.R.); ywb1968@mail.ahnu.edu.cn (W.Y.)

Abstract: In order to develop a reliable and rapid method for meat freshness detection, nanocellulose (TOCNF) prepared via the TEMPO (2,2,6,6-tetramethylpiperidine oxidation) oxidation method was used as raw material to prepare hydrogels using Zn^{2+} coordination and binding. Physicochemical properties such as water absorption and porosity were analyzed. It was further used to select suitable hydrogels for the preparation of indication labels after anthocyanin adsorption, and it was applied in the freshness detection of fresh minced pork. Five percent TOCNF (*w*/*w*) aqueous solution was homogenized by high shear for 4 min, and 20% (*w*/*w*) zinc chloride solution was added to it, so that the concentration of zinc ions could reach 0.25 mol/L. After standing for 24 h, the hydrogel was obtained with good water absorption and a porous three-dimensional network structure. The activation energies of volatile base nitrogen (TVBN) and anthocyanin indicating label color changes were 59.231 kJ/mol and 69.453 kJ/mol, respectively. The difference between the two is within 25 kJ/mol, so the prepared indicator label can accurately visualize the shelf life of fresh pork.

Keywords: nanocellulose; anthocyanin; hydrogel; intelligent indicator label; freshness

Citation: Meng, X.; Shen, Q.; Song, T.; Zhao, H.; Zhang, Y.; Ren, A.; Yang, W. Facile Fabrication of Anthocyanin-Nanocellulose Hydrogel Indicator Label for Intelligent Evaluation of Minced Pork Freshness. *Foods* **2023**, *12*, 2602. https://doi.org/10.3390/foods12132602

Academic Editor: Regina Kratzer

Received: 31 May 2023
Revised: 25 June 2023
Accepted: 26 June 2023
Published: 5 July 2023

1. Introduction

With the improvement of the quality of life, people's attention to food quality and safety is increasing, and the demand for fast and effective access to food quality and freshness is also increasing [1]. To date, researchers have developed a variety of intelligent food indicator labels to monitor the status changes in packaged products, such as food freshness, spoilage and maturity, microbial counts, and product quality information such as texture and flavor [2]. The freshness indicator label uses target chemical substances such as carbon dioxide [3] or volatile nitrogen compounds [4] to dissolve in water to cause a change in pH value, thereby causing a color change on the indicator label. The color change in the label is highly sensitive and effective to humidity, but excessive humidity will cause the label to absorb water, swell, and become damaged [5]. Therefore, the hydrophilicity and stability of the material need to be considered in the preparation of the freshness indicator label. Therefore, the hydrophilicity and stability of materials are the prerequisites for the successful preparation of freshness indicator labels. Studies [6] have shown that in addition to chemical dyes such as bromothymol blue and methyl red, natural pH-sensitive dyes [2,6] such as anthocyanin, curcumin, and madder extract [7] were used to interact with highly hydrophilic chemical substances such as starch, chitosan, polyvinyl alcohol, and other substances to prepare food freshness indicator labels which have good indication effects and natural pigments with low toxicity, and the indicator labels must not pollute the food.

Hydrogel is a kind of material with a stable three-dimensional network structure made of natural or synthetic hydrophilic substances by means of thermal gelation, ionic interaction, chemical bonding or cross-linking, and it has the ability to hold a large amount of water [8]. If highly hydrophilic materials such as hydrogels are used to prepare indicator labels, the adverse effects of humidity on the sensitivity and effectiveness of color changes can be avoided, and the stability of the color response of pH-sensitive dyes to target chemicals can be improved [9]. Among them, the advantages of natural polymer-based hydrogels such as starch and cellulose compared with chemically synthesized materials in terms of biodegradability and biocompatibility have attracted more and more attention [9,10].

Oxidized nanocellulose (TOCNF), which is obtained by oxidizing the hydroxyl group at the C6 position of cellulose to a carboxyl group by using 2,2,6,6-tetramethylpiperidinium oxide (TEMPO), has a large specific surface area, strong adsorption capacity, high mechanical strength, and other characteristics [11,12]. TOCNF can not only be dispersed in the aqueous phase in the form of individual fibers to form self-assembled hydrogels, but it can also be cross-linked with counterions such as Zn^{2+} (Figure 1), Ca^{2+}, Al^{3+}, and the carboxyl groups on the surface of TOCNF to form hydrogels. Thus, the mechanical strength and stability of the hydrogel are improved [13].

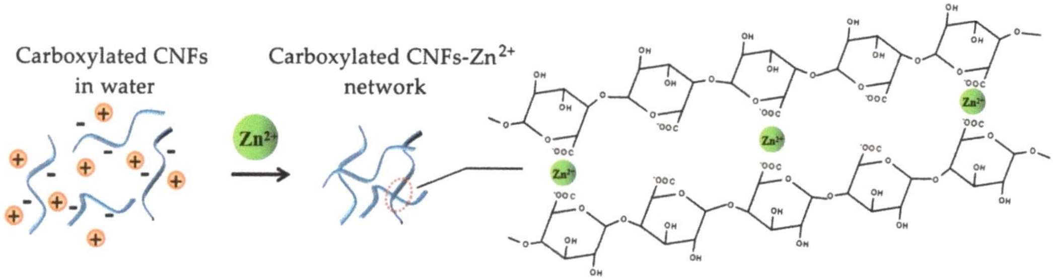

Figure 1. Schematic diagram of Zn^{2+} cross-linked with TOCNF for gelation.

Therefore, in this study, Zn^{2+} was cross-linked with TOCNF to form hydrogels, and the effects of the TOCNF addition amount, homogenization time, and Zn^{2+} addition amount on hydrogel water absorption and porosity were explored. Furthermore, a hydrogel with good water absorption was selected for the adsorption of blueberry anthocyanins, and a hydrogel-based freshness indicator label was developed and prepared. Moreover, the feasibility of using the indicator label to detect changes in the freshness of pork mince during storage was explored, and the relevant kinetic model was established by using the Arrhenius formula to evaluate the accuracy of its indicator effect, which provided a clear reference for meat products during storage. It provides a new method for intelligent monitoring of freshness in the process of transportation and in sales.

2. Materials and Methods

2.1. Materials

TEMPO oxidizes nanocellulose (TOCNF) with carboxylic acid content of 2.0 mmol/g, fiber length of 500–1000 nm and diameter of 5–10 nm was supplied by a local retailer (Tianjin Wood Spirit Biotechnology Co., Ltd., Tianjin, China). Anhydrous zinc chloride (AR, 98%, *w/w*) and aqueous ammonia (AR, 25–28%, *w/w*) were purchased from Shanghai Aladdin Reagent Company (Shanghai, China). Blueberry anthocyanins (10%, *w/w*) were purchased from Shanghai Macklin Biochemical Technology Co., Ltd. (Shanghai, China). Pork was sampled from fresh hind leg meat from RT-mart (Wuhu, China). The water used in the experiment was deionized water.

2.2. Preparation of Nanocellulose Hydrogels

The preparation of the hydrogels was performed according to the experimental method described by Lu et al. [13], as shown in Figure 2.

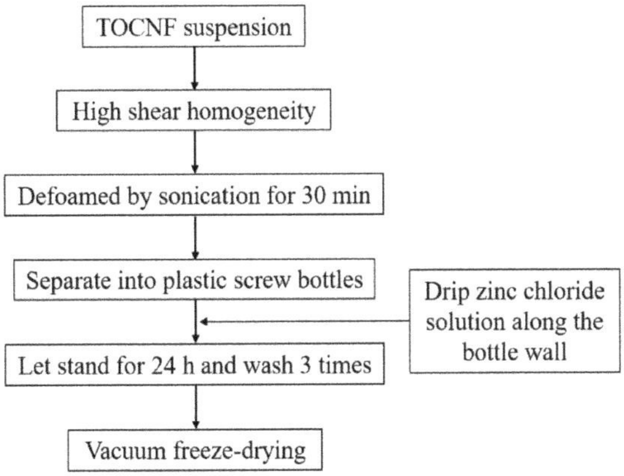

Figure 2. Preparation process of nanocellulose hydrogel.

TOCNF suspension with a certain concentration (3%, 3.5%, 4%, 4.5% and 5% (*w*/*v*)) was homogenized for a certain period of time (3, 4, 5, 6, 7 min) by a high shear homogenizer (Fluko FA25, Shanghai Fluko Technology Development Co., Ltd., Shanghai, China), followed by 30 min ultrasonic defoaming, and then transferred quantitatively to a plastic vial with an inner diameter of 25 mm. Then, a certain volume of zinc chloride solution with a concentration of 20% (*w*/*w*) and pH = 3 was added along the inner wall of the vial to maintain the content of zinc ions in the system. The total volume of zinc chloride solution and TOCNC solution is fixed at 5 mL. Zinc chloride solution, without stirring, was directly placed at 25 °C for 24 h, and then it was washed in distilled water three times to remove excess zinc ions, freeze-drying the reserve.

2.2.1. Effect of TOCNF Concentration on Water Absorption and Porosity of Hydrogels

TOCNF powder was weighed and mixed with distilled water to obtain TOCNF suspension with concentrations of 3%, 3.5%, 4%, 4.5%, and 5% (*w*/*v*), respectively. After homogenizing for 5 min (Fluko FA25, Shanghai Fluko Technology Development Co., Ltd.), the content of zinc ions in the hydrogel system was fixed at 0.25 mol/L to prepare the corresponding hydrogel, and the water absorption capacity and porosity of the hydrogel were measured.

2.2.2. Effect of Homogenization Time on Water Absorption and Porosity of Hydrogels

After the TOCNF suspension with 4% concentration was homogenized for 3 min, 4 min, 5 min, 6 min and 7 min, respectively, the content of zinc ions in the hydrogel system was fixed at 0.25 mol/L to prepare the corresponding hydrogel, and the water absorption capacity and porosity of the hydrogel were measured.

2.2.3. Effect of Zinc Ion Concentration on Water Absorption and Porosity of Hydrogels

After homogenizing the TOCNF suspension with 4% concentration for 5 min, the contents of zinc ions in the hydrogel system were 0.05 mol/L, 0.15 mol/L, 0.25 mol/L, 0.35 mol/L and 0.45 mol/L, respectively. The corresponding hydrogels were prepared, and the water absorption capacity and porosity of the hydrogels were measured.

2.3. *Determination of Water Absorption Capacity of Hydrogels*

The dried hydrogel was soaked in 10 mL distilled water for 24 h, and then the free water on the surface of the hydrogel was removed, and the hydrogel was weighed [14]. Water absorption rate is calculated according to Formula (1).

$$\text{Water absorption rate}(\%) = \frac{W_2 - W_1}{W_1} \times 100\% \quad (1)$$

where W_1 and W_2, respectively, represent the mass before and after water absorption of the dried hydrogel (g).

2.4. Determination of Hydrogel Porosity

After freeze-drying, the hydrogel was weighed, the diameter and thickness of the hydrogel were also measured, and the hydrogel was soaked in 10 mL is isopropyl alcohol solution for 24 h to fill the pores, and the mass of the hydrogel after filling the pores with isopropyl alcohol was weighed [15]. Its porosity was calculated according to Equations (2)–(4).

$$P(\%) = \frac{V_{pore}}{V_{gel} + V_{pore}} \times 100\% \quad (2)$$

$$V_{pore} = \frac{m_{wet} - m_{dry}}{\rho} \times 100\% \quad (3)$$

$$V_{gel} = \pi R^2 \times H \quad (4)$$

where V_{pore}, V_{gel}, m_{wet}, m_{dry}, R and H, respectively, represent the volume of hydrogel pore (cm^3), volume of hydrogel (cm^3), weight of hydrogel after soaking isopropyl alcohol (g), weight of freeze-dried hydrogel (g), radius of freeze-dried hydrogel (cm) and height of freeze-dried hydrogel (cm).

2.5. Morphological Characteristics of Hydrogels

2.5.1. Scanning Electron Microscope

The dried and unground hydrogel samples were sprayed with gold, and the surface morphology of the samples was observed by SEM at 10 kV accelerating voltage [16].

2.5.2. Fourier Transform Infrared Spectroscopy (FTIR)

The hydrogel samples were mixed with pure TOCNF powder and pure KBr [17]. The mixture was then ground and pressed at 100 Pa for 1 min, and the resulting transparent sheet was used for measurement. All samples were analyzed in the wavelength range of 4000~400 cm^{-1}.

2.6. Preparation and Characterization of Indicator Label

2.6.1. Response of Anthocyanins to pH

An amount of 1.0 g/L anthocyanin was obtained by dissolving 1.0 g blueberry anthocyanin in 1 L distilled water. Then, 1 mL of 1.0 g/L anthocyanin solution was added to 9 mL of solution with pH value of 1–13, stirred evenly, sealed into 25 °C incubator, and left for 30 min without light. Finally, The UV-VIS spectrum of blueberry anthocyanin solution in the pH 1–13 range was measured using a UV-VIS spectrophotometer with a scanning wavelength of 400–700 nm.

2.6.2. Response of Anthocyanins to Ammonia

Firstly, 10 mL of 1.0 g/L anthocyanin solution (pH = 5) was placed in a 20 mL glass bottle. Then, 20 mL of ammonia solution with a concentration of 0.028~0.28% (*w*/*w*) was placed in another glass bottle. The two glass bottles were placed in a 450 mL sealed jar as shown in Figure 3. After sealing, the anthocyanin solution was reacted at 25 °C for 3 h in the dark, and the UV–visible spectrum of the anthocyanin solution was measured using a UV spectrophotometer with a scanning wavelength of 400–700 nm [18].

2.6.3. Preparation of Indicator Label

Firstly, the hydrogel was prepared according to the methods described in Section 2.2. Then, the freeze-dried hydrogel was soaked in anthocyanin solution of different concentrations (0.5–2.5 g/L) for 24 h at 25 °C, the excess water was wiped off after taking it out, and it was stored in a dark place at 4 °C.

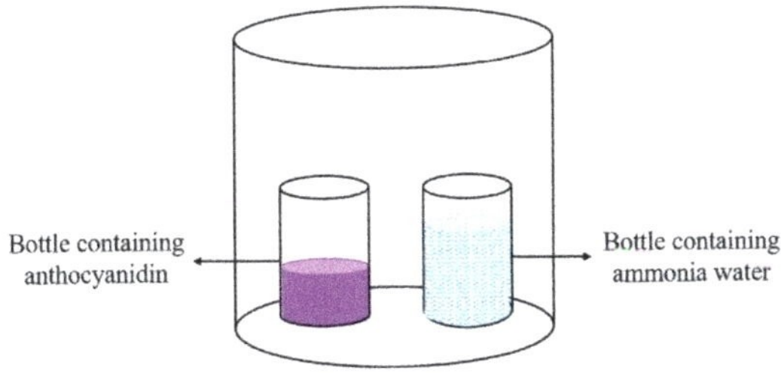

Figure 3. Schematic diagram of an experiment on the responsiveness of anthocyanins to ammonia.

2.6.4. Response of Indicator Label to Ammonia

The CIE $L^*a^*b^*$ color coordinates of the indicator label was determined with a colorimeter (HP-C220, Hanpu Photoelectric Technology Co., Ltd., Shanghai, China). In the $L^*a^*b^*$ coordinates, a^* represents the green–red color and b^* the blue–yellow color of the samples. The L^* coordinate represents lightness, where $L^* = 0$ is completely black, and $L^* = 100$ is completely white. The indicator label was placed over a 20 mL vial, with 20 mL of 0.14% ammonia added to another vial, and both bottles were placed in a 450 mL airtight jar as shown in Figure 4; after sealing and reacting in the dark at 25 °C for 3 h, the color change in the indicating label was measured. According to Formula (5), the total color difference in the indicating label before and after the reaction was calculated as follows:

$$\Delta E = \sqrt{(L - L_0)^2 + (a - a_0)^2 + (b - b_0)^2} \quad (5)$$

where L_0, a_0 and b_0 represent the initial color value of the indicating label. L, a and b represent the values after the response of the indicating label, respectively [18].

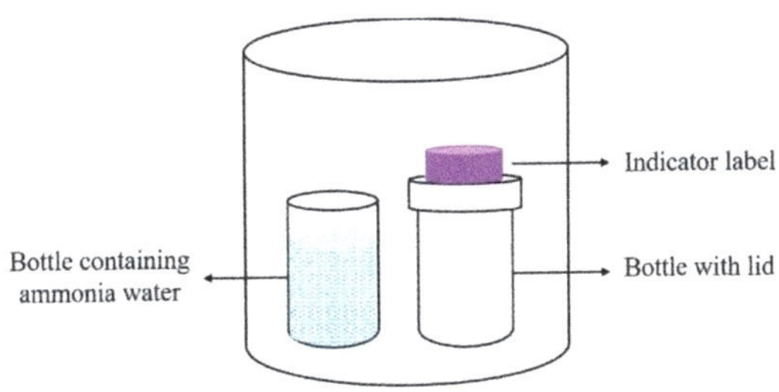

Figure 4. Schematic diagram of an experiment on the responsiveness of the indicator label to ammonia.

2.6.5. Characterization of Indicator Label

Indicator label were vacuum freeze-dried and subjected to IR and SEM analysis accorded the method as described in Section 2.5.

2.7. Application of Indicator Label in Detecting the Freshness of Minced Pork

2.7.1. Preparation of Minced Pork

Firstly, the fat and fascia on the hind legs of fresh peeled pigs were removed and minced twice using a meat grinder. Then, 50 g of pork minced meat was placed on one side of the plastic box, and the indicator label was fixed on the other side of the plastic box. After the plastic box was sealed, it was stored at 4 °C, 14 °C, 24 °C, and 34 °C, respectively. At certain intervals, the plastic box was taken out to measure the volatile base nitrogen (TVB-N) content of pork mince, as well as the L (brightness), a (redness), and b (yellowness) values of the indicator label, and analysis of the total color difference at different stages was performed according to Equation (5) (ΔE).

2.7.2. TVB-N Content Analysis

TVB-N concentration was measured according to China National Food Safety Standard methods (Determination of Total Volatile Basic Nitrogen in Food; GB/T 5009.228–2016). The results are expressed as mg of TVB-N per 100 g of pork. Finally, the freshness of minced pork was analyzed according its TVB-N content [18].

2.8. Statistical Analyses

Every experiment was performed in triplicate and average values with standard errors were reported. Using SPSS 25.0, statistical analysis was carried out according to analysis of variance (ANOVA) for which Duncan's method was adopted, and graphs were produced with origin 2019b, and different lowercase letters for the same indicator was used to indicate significant differences at 5% confidence intervals ($p < 0.05$).

3. Results and Discussion

3.1. Characteristics of Hydrogel

3.1.1. Water Absorption of Hydrogel

The anthocyanin solution needs to be fully absorbed during the preparation of the indicator labels, so the study of the water absorption ability of the hydrogel is the prerequisite for the study of the application of hydrogels [19,20]. The influence of different factors on the water absorption of hydrogel is shown in Figure 5. Figure 5A shows the influence of the TOCNF addition amount on the water absorption of hydrogel. According to the analysis, there was a very significant effect of the TOCNF addition amount on the water absorption of hydrogel ($p < 0.01$), and with the increase in the addition amount, the water absorption gradually increased, and the high shear homogenizer used in the experiment cannot effectively homogenize the solution with the addition of nanocellulose in an amount greater than 5.0%. The maximum value was 804.81 ± 112.00% at 5.0%. Figure 5B shows the effect of zinc ion addition on the water absorption of hydrogel. According to the analysis, the zinc ion addition had a very significant effect on the water absorption of hydrogel ($p < 0.01$). With the increase in zinc ion addition, the water absorption first increased and then decreased. The highest water absorption was obtained when the content of zinc ion was controlled to 0.25 mol/L in the hydrogel system. Figure 5C shows the influence of TOCNF homogenization time on the water absorption of the hydrogel. As shown in Figure 5C, there was a very significant effect of the amount of zinc ion on the water absorption of the hydrogel ($p < 0.01$). With the increase in homogenization time, the water absorption first increased and then decreased. The highest water absorption (753.68 ± 87.23%) was obtained when the TOCNF solution was homogenized for 4 min.

A large number of hydrophilic groups in the hydrogel (such as carboxyl and hydroxyl groups contained in TOCNF) could form a hydrogen bond force with water molecules to make water molecules fixed in the hydrogel's three-dimensional network structure, thus causing the hydrogel to rapidly absorb water and swell [21]. Therefore, the content of the carboxyl group increases with the increase in the amount of nanocellulose added, and the water absorption capacity of the hydrogel also increases. Zinc ion mainly increased the three-dimensional network structure inside the hydrogel and improved the stability of the structure by coordinating with the carboxyl group [22]. With the increase in the addition amount, the network structure of the hydrogel was improved, causing the improvement of the swelling performance of the hydrogel. However, the excessive zinc ion would also affect the interaction between the internal hydrophilic group carboxyl group and water and reduce the water absorption performance [22,23], making the water absorption first decline and then increase with the increase in zinc ion. During the homogenization process of TOCNF solution, cellulose was mechanically wrapped and folded to form a three-dimensional network structure with low stability. As the homogenization time increases, the effect of wrapping and folding became excessive, which also affected the subsequent addition of zinc ions that could not fully penetrate into the interior of TOCNF for coordination and binding, thereby affecting water absorption. As the homogenization time increases, the water absorption first increased and then decreased.

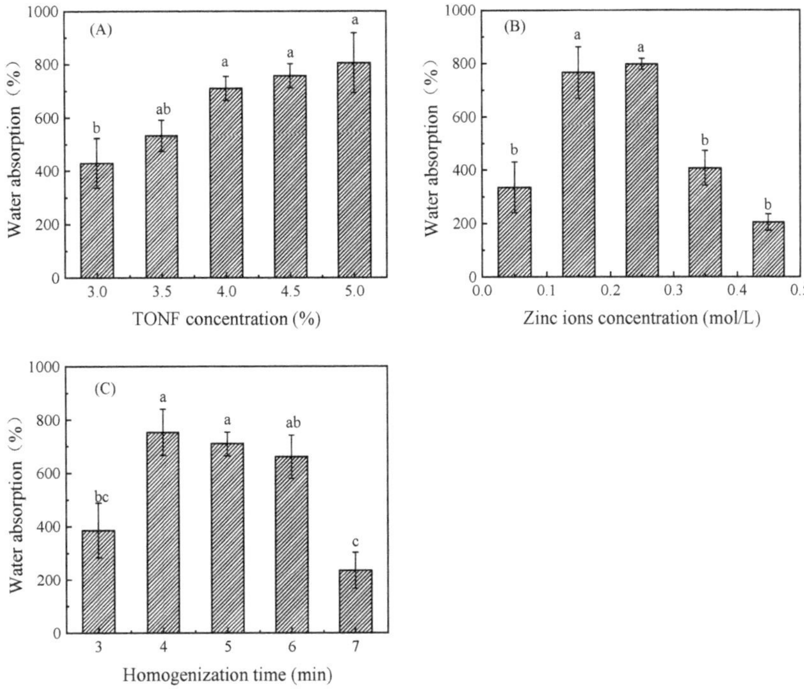

Figure 5. Water absorption properties of hydrogels induced by TONF (**A**) and zinc ions (**B**) and under different homogenization times (**C**). Each value represents the mean ± SD (n = 3). Different superscript letters (a, b, c) between columns indicate statistical differences according to Duncan's multiple comparison test ($p < 0.05$).

3.1.2. The Porosity of Hydrogel

Porosity is an important physical property of hydrogels. High porosity means that the material has a large specific surface area and high adsorption capacity, which is more conducive to the subsequent adsorption of the anthocyanin solution to prepare indicator labels [8]. The influence of different factors on the porosity of hydrogel is shown in Figure 6. As shown in the figure, the addition of TOCNF has no significant effect on the porosity of the hydrogel ($p > 0.05$). Figure 6B shows the effect of zinc ion addition on the porosity of hydrogel. As shown in Figure 6B, the amount of zinc ion added has a very significant impact on the porosity of the hydrogel ($p < 0.01$). With the increase in zinc ion addition, the porosity first increased and then decreased. The highest porosity (42.07 ± 0.59%) was achieved when the zinc ion addition was 0.25 mol/L in the hydrogel system. Figure 6C shows the effect of the TOCNF homogenization time on the porosity of hydrogel. As shown in Figure 6B, the homogenization time has a very significant effect on the porosity of the hydrogel ($p < 0.01$). As the homogenization time increases, the porosity shows a trend of first increasing and then decreasing, with the highest porosity (45.44 ± 1.05%) at 4 min of TOCNF suspension homogenization.

Under the mechanical force of the high shear homogenizer, the TOCNF solution folds and tangles between TOCNF fibers to form a three-dimensional structure. The addition of zinc ions enhanced the interaction between carboxyl groups to produce the three-dimensional structure more stable. However, excessive zinc ions induced the reduction in particle size of the hydrogel, resulting in an increase in pore density and porosity [13,22]. However, a long time of homogenization treatment leads to strong folding and entanglement effect of cellulose molecules, which leads to the inability of zinc ions to penetrate into the gel, thus reducing the stability, and the pore structure is also easily destroyed, which reduces the porosity.

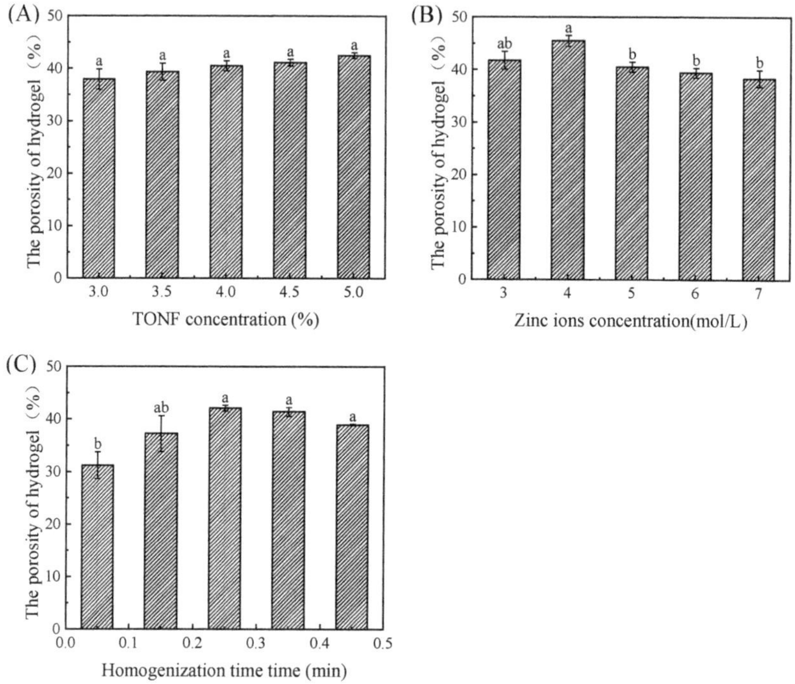

Figure 6. The porosity of hydrogels induced by TONF (**A**) and zinc ions (**B**) and under different homogenization times (**C**). Each value represents the mean ± SD ($n = 3$). Different superscript letters (a, b) between columns indicate statistical differences according to Duncan's multiple comparison test ($p < 0.05$).

3.2. Preparation of Indicator Labels

3.2.1. Response of Anthocyanins to pH

Anthocyanins exhibit different colors under different pH conditions. As shown in Figure 7A, within the pH range of 1–13, the color changes in blueberry anthocyanins are in the order of red, light pink, purple, blue purple, gray purple, and yellow brown. The color change in the solution is due to the mutual transformation of the high-molecular-weight conjugated system of anthocyanin molecules under different acid and base conditions. The four molecular structures of anthocyanins with different colors, namely, xanthate cation (red), methanol pseudobase (colorless), quinone base (blue), and chalcone (colorless), have different degrees of mutual transformation due to the increase and decrease in H^+ and OH^- ion content. Studies have demonstrated that pH value has a significant impact on the stability and color changes in anthocyanins during storage. The anthocyanin solution appeared red/pink in strong acidic conditions (pH < 2), purple/blue in a relatively neutral medium (pH 6–7), and green/yellow under alkaline conditions, which were the colors with greatest dominance (Figure 8) [24]. In addition, anthocyanin 3-glucoside achieved the highest stability at pH 8–9, while other glycosides achieved the highest stability at pH 5–7.

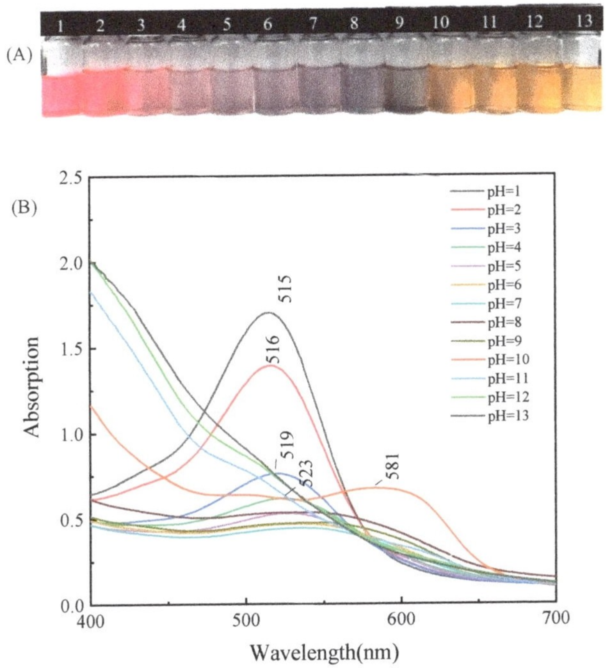

Figure 7. Color (**A**) and visible spectral absorption (**B**) characteristics of blueberry anthocyanin solution at pH 1–13.

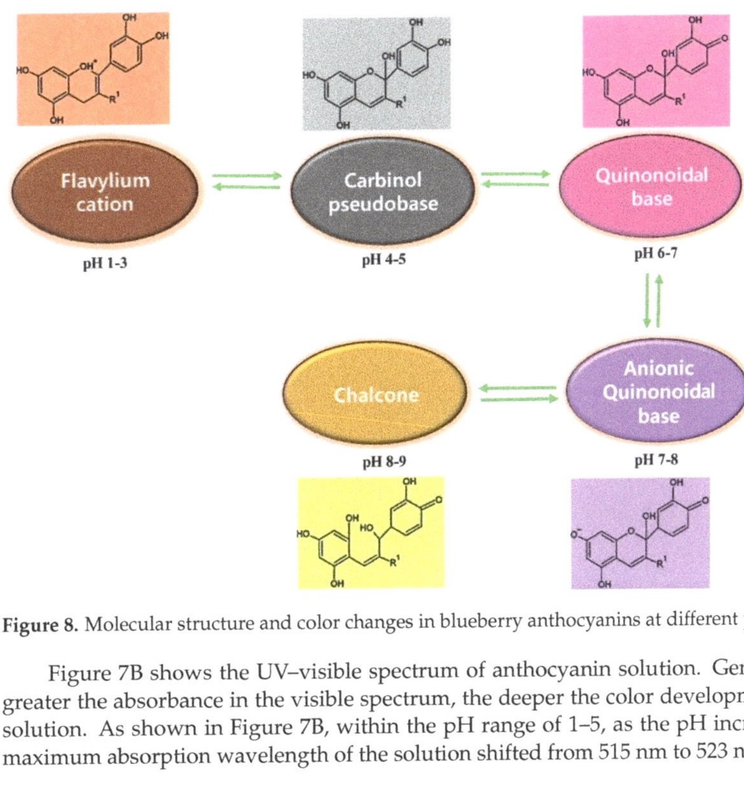

Figure 8. Molecular structure and color changes in blueberry anthocyanins at different pH.

Figure 7B shows the UV–visible spectrum of anthocyanin solution. Generally, the greater the absorbance in the visible spectrum, the deeper the color development of the solution. As shown in Figure 7B, within the pH range of 1–5, as the pH increased, the maximum absorption wavelength of the solution shifted from 515 nm to 523 nm, and the absorbance value decreased significantly. When the pH of the solution increased to 10, the maximum absorption wavelength was located near 581 nm. In the pH range of 11–13, the maximum absorption peak of the solution disappeared, and the shape of the spectrum remained unchanged. That is because, under strong alkaline conditions, the stability of anthocyanins was poor, and the central ring of the anthocyanin molecule decomposed to form chalcone. The absorbance and wavelength of blueberry anthocyanins were constantly changing due to the structural transformation of anthocyanins caused by pH changes [24].

3.2.2. Response of Anthocyanins to Ammonia

When ammonia is dissolved in water, it forms ammonium ions and hydroxide ions, leading to a change in the pH of the solution and causing a change in the color of the indicator. The color response of the anthocyanin indicator solution to different concentrations of ammonia is shown in Figure 9A. As the concentration of ammonia solution increases from 0.028% to 0.14%, the color of the solution gradually changed from bright red to purple red and then to purple. Then, as the concentration of ammonia increased to 0.28%, the color of anthocyanin remained purple. Moreover, it was observed with the naked eye that only changes in color depth were observed in anthocyanin solutions with an ammonia concentration greater than 0.14%.

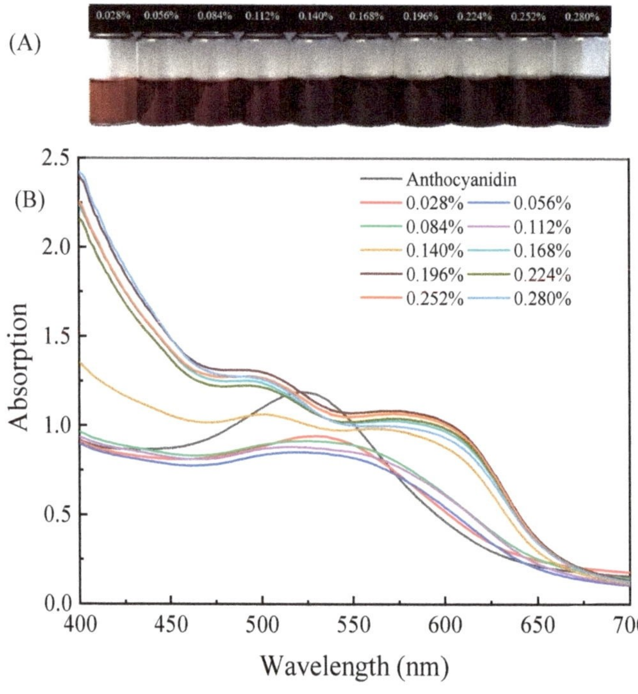

Figure 9. Color (**A**) and visible spectral absorption (**B**) characteristics of blueberry anthocyanin solution for different concentrations of ammonia.

The UV–visible absorption spectrum corresponding to the reaction between anthocyanins and ammonia is shown in Figure 9B. It was observed that an absorption peak appeared near 530 nm in a 1 g/L blueberry anthocyanin solution that was not in contact with ammonia. As the ammonia concentration increased, the absorption peak gradually shifted to the right, and the intensity of the absorption peak gradually decreased. This was also due to the solubility of ammonia in water causing an increase in pH, and the yellow salt cations gradually transformed into colorless methanol pseudobases and blue quinone-type bases. However, when the ammonia concentration increased to 0.14%, the absorption peak disappeared. It indicated that the pH of the solution was sufficient to open the central ring of anthocyanins, and the color of the anthocyanin solution also changed to deep purple [18].

As the concentration of ammonia water continued to increase, there was no significant change in the spectral pattern. Although the absorption intensity increased at 0.168% ammonia concentration, there was no significant change in absorption intensity as the ammonia concentration continued to increase. The results indicate that anthocyanin solution had a relatively high sensitivity to ammonia [18,20], and the color of anthocyanin solution varied with the change in ammonia concentration in the range of 0.028% to 0.14%. Therefore, in the subsequent preparation of indicator labels, it was considered best to use 0.14% ammonia concentration for ammonia response experiments to ensure significant color changes and reduce the waste of ammonia in the experiment.

3.2.3. The Color Difference in Color Indicator Labels

Under the action of microorganisms and enzymes, volatile nitrogen-containing compounds such as ammonia, trimethylamine, and dimethylamine are produced during the spoilage process of meat. Therefore, ammonia gas can be used to simulate the volatile nitrogen-containing compounds generated during the process of meat spoilage, in order to explore the response sensitivity of indicator labels obtained by soaking in different concentrations of anthocyanin solutions [20]. Figure 10A shows the color changes in the indicator labels on ammonia gas obtained by soaking different concentrations of anthocyanin solutions. It can be observed that the color of the indicator labels changed from purple red to blue after 3 h of exposure to ammonia gas. As the soaking concentration increases, the starting purple red color of the indicator labels becomes darker, and the blue color after the reaction ends also becomes darker. Figure 10B shows the total color difference changes in different indicator labels before and after exposure to ammonia gas. As shown in the figure, as the concentration of the anthocyanin solution used to prepare the indicator label increased, the total color difference in anthocyanins showed a tendency of first increasing and then decreasing. As the content of anthocyanins that reacted with ammonia increased, the indicator labels soaked in 1.5 g/L anthocyanin concentration has the maximum total color difference before and after contact with ammonia (19.718 $\pm$ 0.434). However, when the anthocyanin concentration continuously increased, the color change which was caused by a pH change induced by ammonia reaching saturation. Moreover, the deep red color inherent in anthocyanins also affected the color rendering effect, leading to a decreasing tendency in total color difference. Therefore, the indicator labels obtained by soaking 1.5 g/L anthocyanin solution was selected for experiments relating to the freshness detection of pork mince.

3.2.4. Characterization of Indicator Labels

Infrared Spectrum of Indicator Labels

The infrared spectra of TOCNF, anthocyanins, hydrogels, and indicator labels are shown in Figure 11. As shown in the figure, there is a strong and wide absorption peak near 3377 cm^{-1} for blueberry anthocyanins and TOCNF, mainly due to the presence of more hydroxyl groups in both anthocyanins and TOCNF. Moreover, the peak showed a significant red shift in the hydrogel prepared by TOCNF, which might be due to the increase in the number of hydrogen bonds during the preparation of the hydrogel. Compared with the single TOCNF, the infrared spectrum of the indicator labels prepared by adsorption of anthocyanins showed a blue shift. This is mainly because the addition of anthocyanins increased the hydroxyl content in the sample, which also indicated that anthocyanins were successfully adsorbed into the hydrogel [8,19]. At 1611 cm^{-1} and 1423 cm^{-1}, the characteristic peaks of TOCNF were observed, namely, the symmetric and asymmetric stretching vibration peaks of -COOH. The characteristic peaks of anthocyanins at 1627 cm^{-1} and 1515 cm^{-1} were caused by the stretching vibration of C=C on the aromatic ring skeleton of anthocyanins, and the absorption peak at 1284 cm^{-1} was caused by the stretching vibration of the pyran ring in its structure. The absorption peak at 1015cm^{-1} might be caused by the bending vibration of C-H on the aromatic ring. The blue shift in the COOH vibration peak in the hydrogel sample may be caused by the coordination bond between zinc ion and carboxyl group. Moreover, the carboxyl vibration peak in the indicator label showed a red shift, and the absorption peak at 1556 cm^{-1} indicated that anthocyanins were adsorbed on the hydrogel and affected the original coordination bond. According to FT-IR analysis, blueberry anthocyanins can attach well to hydrogels and have good compatibility with TOCNF. The performance of indicator labels is mainly determined by intermolecular forces, and the chemical composition of each substrate is not affected [9].

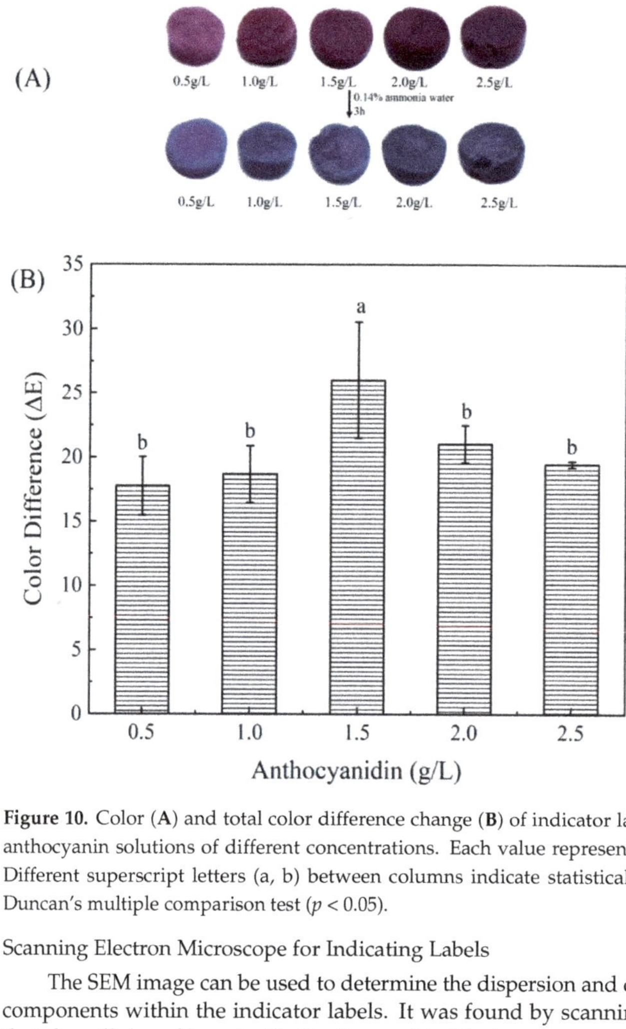

Figure 10. Color (**A**) and total color difference change (**B**) of indicator labels obtained by soaking anthocyanin solutions of different concentrations. Each value represents the mean ± SD (n = 3). Different superscript letters (a, b) between columns indicate statistical differences according to Duncan's multiple comparison test ($p < 0.05$).

Scanning Electron Microscope for Indicating Labels

The SEM image can be used to determine the dispersion and compatibility of various components within the indicator labels. It was found by scanning electron microscope that the cellulose fibers inside the freeze-dried hydrogel were coiled and wound, with an obvious pore structure, but the diameters of the holes were different, with diameters ranging from 60 to 100 μm. Therefore, the hydrogel prepared by shear homogenization and the cross-linking of zinc ion and TOCNF had good water absorption and porosity, and it has a certain application value.

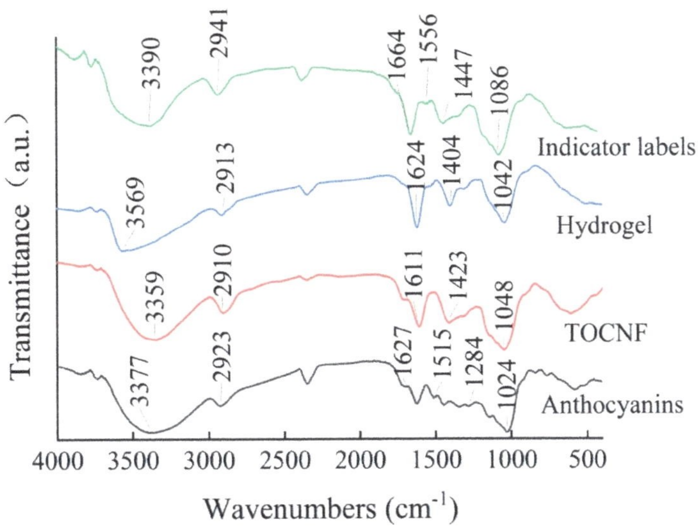

Figure 11. Infrared spectrum of hydrogel and indicator label.

Compared with the hydrogel shown in Figure 12A, the indicator labels with blueberry anthocyanin in Figure 12B showed a more compact and smooth surface, indicating that the blueberry anthocyanin and TOCNF hydrogel had good compatibility. In addition, the pore size of the hydrogel prepared with anthocyanin and TOCNF were smaller than that of the hydrogel. After the anthocyanins were adsorbed, the pore size of the hydrogel is 45 μm, and there were some small particles of anthocyanins attached to cellulose aggregates. This might be because the adsorption of anthocyanins successfully attached to the hydrogel filled the original pore structure. Moreover, the phenolic hydroxyl groups in blueberry anthocyanin molecules formed intermolecular hydrogen bonds with the hydroxyl groups in nanocellulose [8,25], reducing the phenomenon of polymer chain molecular entanglement and hydrogen bonding between cellulose molecules. Eventually, the mechanical properties of the hydrogel were affected, making the internal structure more uniform.

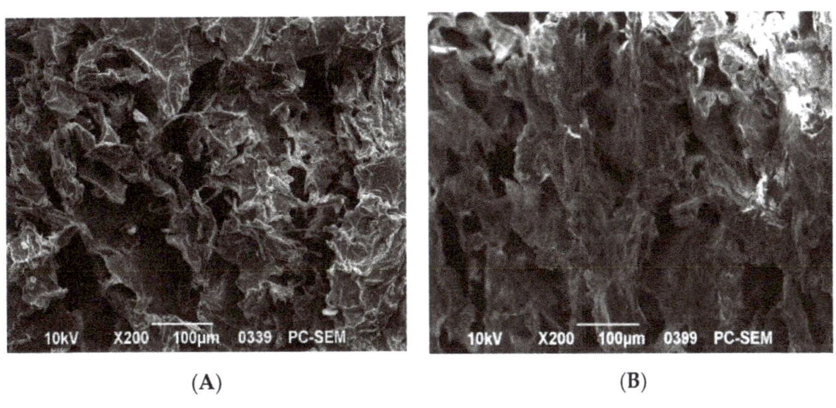

(A) (B)

Figure 12. SEM image of hydrogel (**A**) and indicator label (**B**).

3.3. Application of Indicator Labels in Detecting the Freshness of Pork Mince

3.3.1. Freshness of Pork Mince and Indicator Labels Color

The changes in the indicator label color, TVBN, and total color difference in pork mince stored at different storage temperatures (4–34 °C) are shown in Figure 13. As shown in Figure 14A–D, with the extension of storage time, the TVBN value and ΔE was constantly increasing, with an initial value of 1.90 mg/100 g for TVBN. In accordance with the GB2707-2016 'Hygienic Standard for Fresh (Frozen) Animal Meat', the TVBN value for fresh meat should be less than 15 mg/100 g, and the indicator label should be purple red at this condition. As shown in Figure 13A–D, with the extension of storage time, it was observed that the color of the indicator label changes from purple red to blue purple at different temperatures, and there were certain differences in the time and degree of color change. According to the change in total color difference between TVBN and indicator label in Figure 14A–D, it was inferred that this might be due to differences in TVBN values greater than 15 mg/100 g at different temperatures. The TVBN value at 4 °C at 192 h was 20.85 mg/100 g, which exceeded the freshness standard for pork mince. And, at this point, the minced meat ΔE was 5.88 ± 1.30, and the color was observed by the naked eye to change from initial red to light purple. Subsequently, at 240 h, the TVBN value of pork mince was 85.09 mg/100 g, ΔE was 9.09 ± 2.37, and the edge of the label was seen by the naked eye as blue purple. Similarly, the TVBN value for 72 h at a storage temperature of 14 °C was 36.46 mg/100 g, ΔE was 5.15 ± 2.13, and the edge of the indicator label could be visually observed to appear blue purple. With the extension of storage time, the blue range of the indicator label expands until it completely turns blue purple at 120 h, at which point the TVBN value is 96.72 mg/100 g. The TVBN value at 24 °C for the 36 h was 19.78 mg/100 g, and ΔE was 5.62 ± 1.89, indicating that the purple color of the label deepens, and as the storage time prolongs, the label gradually turned blue until it completely turned blue purple at 60 h. At this time, the TVBN value was 73.53 mg/100 g.

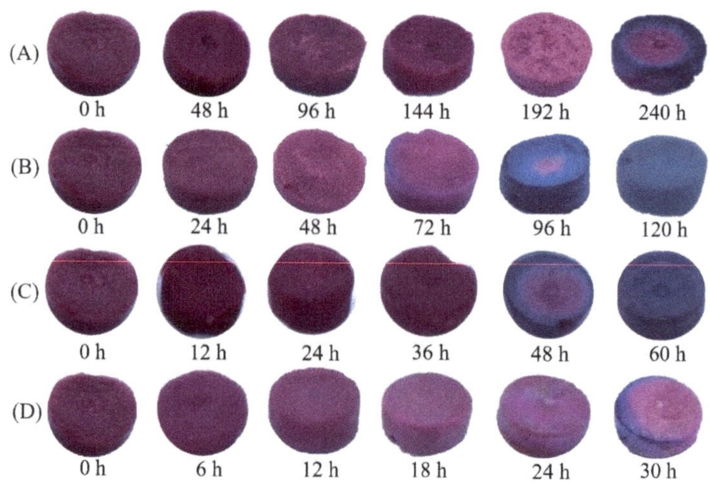

Figure 13. Color change in indicator labels (**A**–**D**) under 4, 14, 24 and 34 °C storage condition.

The TVBN value at 34 °C at 18 h was 39.68 mg/100 g, and ΔE was 15.29 ± 2.58, indicating that the surface of the label turns blue. With the extension of storage time, the blue color of the label deepened. The final TVBN value at 60 h was 133.25 mg/100 g, ΔE was 19.76 ± 1.42, and a clear blue purple color was observed on the indicator labels. There were certain differences in the types and metabolites of microorganisms at different temperatures, which resulted in differences in the range and color depth of the blue indicator labels at different temperatures. However, when the TVBN value exceeded 15 mg/100 g, the indicator labels for detecting the freshness of pork mince at different temperatures showed a change from purple red to blue purple. Some freshness indicators have been designed based on the volatile compounds released by microbial spoilage, thus resulting in the pH increase inside the package headspace. For example, an on-package indicator label was developed by Chen et al. [4] for monitoring lean pork freshness on the basis of the presence of TVB-N in the package. It was found that the lower initial pH is, the slower color change is. This colorimetric freshness indicator provides three different colors for illustrating freshness (red), medium freshness (goldenrod), and spoilage (green), thus enhancing guarantee of pork safety. In contrast, visual pH-sensing films containing curcumin and anthocyanins were developed as on-package indicator labels for monitoring fish freshness in real time [26]. This SPVA/glycerol film incorporated with curcumin and anthocyanins at a ratio of 2:8 (v/v) could provide three different colors, which were assigned to the sign of freshness, medium freshness, and spoilage for packaged fish. Furthermore, efficient prediction dynamic modeling of the freshness and freshness indicator labels of pork mince is needed to explore the correlation between the indicator labels and the quality of pork mince at different temperatures.

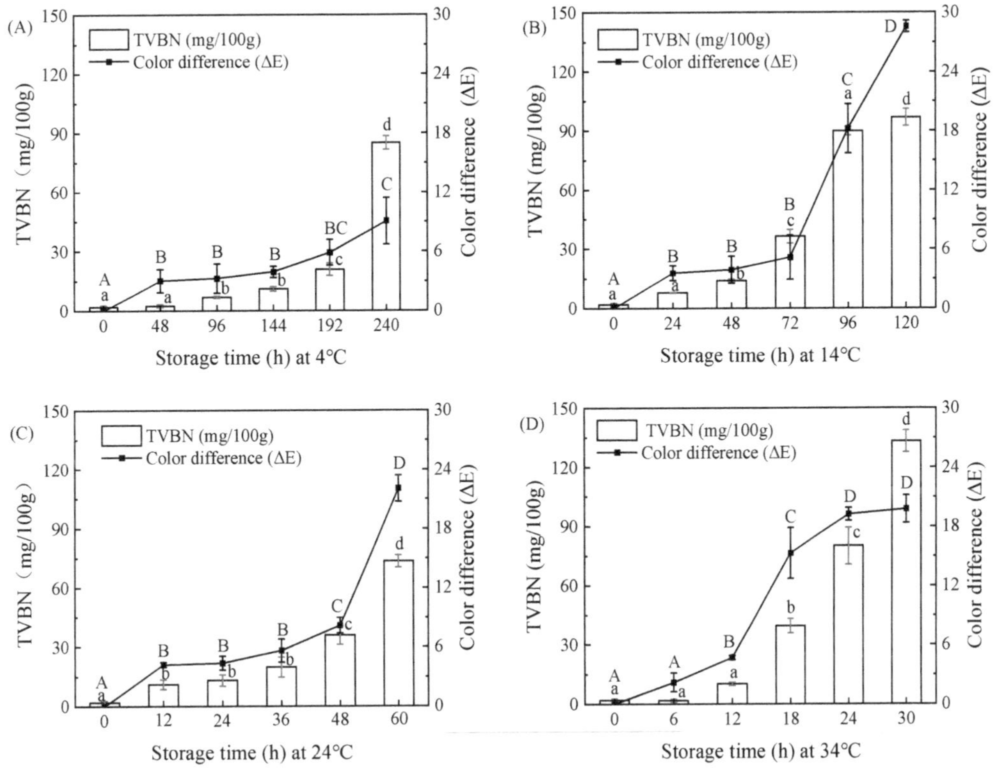

Figure 14. TVBN and total color difference in indicator labels (**A–D**) under 4, 14, 24 and 34 °C storage condition. Each value represents the mean ± SD (n = 3). Different superscript lowercase letters (a, b, c, d) between columns and capital letters (A, B, C, D) between points of curve indicate statistical differences according to Duncan's multiple comparison test ($p < 0.05$).

3.3.2. Dynamics Analysis of TVBN and Indicator Labels Color in Pork Mince

Mathematical modeling is an effective method for analyzing and predicting various biological and chemical reactions that cause changes in food quality. The Arrhenius equation is a commonly used empirical formula to represent chemical reaction rates and temperature changes, which can explore the impact of temperature on the rate of change in experimental objects. For example, the response change in the indicator labels can be modeled as a function of the response time (t) and the reaction rate constant (k) of color change. After the K value of the indicator label is determined, the activation energy (Ea) can be calculated using the Arrhenius equation. Based on the dynamic model established for pork mince and freshness indicator labels, the activation energy of pork and indicator label changes could be obtained. However, it is generally believed that the difference in activation energy between the appropriate freshness indicator label and the food should be less than 25 kJ/mol. Establishing a kinetic model can also accurately reflect the impact of temperature changes around the food on the food and the indicator label, and it can provide predictive factors for the food quality change process during the shelf life.

Analyzing the TVBN content in meat is beneficial for assessing food quality. Therefore, the changes in TVBN content of pork mince stored at 4 °C, 14 °C, 24 °C, and 34 °C at different time points were used to analyze the quality change trend of pork mince and determine its kinetic parameters. For fresh pork quality changes, the first order kinetic equation should be used to analyze the relationship between TVBN content of pork mince and time.

The trend and kinetic parameters of TVBN changes in pork mince at different temperatures are shown in Figure 15 and Tables 1 and 2. According to Figure 15, as the storage time prolongs, the color difference and TVBN of indicator labels at different storage temperatures showed a gradual upward trend, and the rate of change also varied at different temperatures. The first-order kinetic equation is suitable for analyzing the changes in food quality of fresh meat. Therefore, it is necessary to select an exponential function for linear fitting of the TVBN content changes in pork mince and the color difference in the indicator label. The response function and change rate of the TVBN and indicator color difference values of pork mince are shown in Tables 1 and 2. It was found that with the increase in temperature, the change rate constants of TVBN and total color difference were both increasing. According to Arrhenius equation, 1/RT corresponding to different temperatures was taken as the abscissa (R is the molar gas constant, also known as the ideal gas constant, with the value of 8.3144 J/mol·K; T is the thermodynamic temperature, with the unit of K), and the logarithm ln k of the change rate constant corresponding to the temperature was taken as the ordinate to draw and perform linear fitting, as shown in Figure 15. The relationship between the label change rate and temperature obtained by linear fitting is ln k (total color difference) = −68337/RT + 24.595, and the linear fitting $R^2 = 0.9587$. According to this equation, the activation energy indicating the total color difference in the label is 68.337 kJ/mol. Significantly, a means to correlate color changes to freshness of packaged foods must be established. Sun et al. [27] reported a back propagation artificial neural network model which was established using the color of the films as the input and the TVB-N content as the output. It was found that neural network model well predicted the freshness of packaged pork patties and fish balls based on color change in the films, with R^2 of 0.94–0.98.

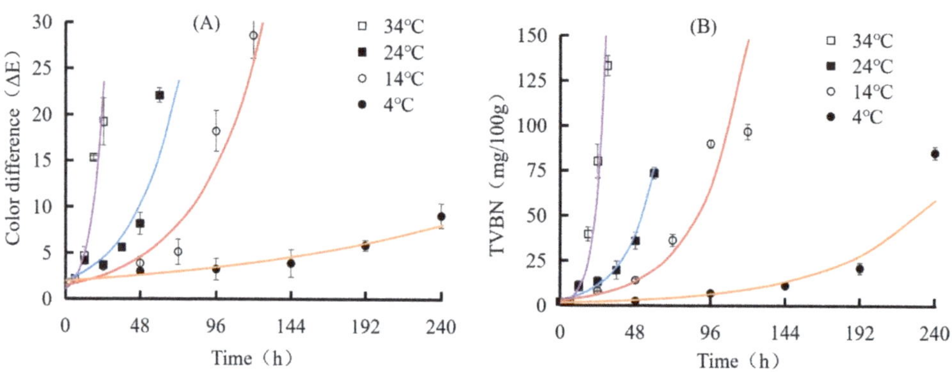

Figure 15. Changes in pork TVBN (**A**) and color of indicator label (**B**) at different temperatures.

According to the equation, the activation energies of TVBN and total color difference are 53,061 J/mol and 68,337 J/mol, respectively, with a difference of less than 25 kJ/mol. According to Taoukis' viewpoint, a good freshness indicator label that can achieve real-time monitoring of food quality should meet the activation energy of the food (i.e., TVBN content of pork mince) Ea1 and the activation energy of the indicator (total color difference before and after the indicator label) Ea2, with a difference of less than 25 kJ/mol. At four different temperatures, the color change in indicator labels was consistent with TVBN content of ground pork exceeded the spoilage threshold (greater than 15 mg/100 g). Therefore, the anthocyanin indicator labels could maintain good responsiveness at different storage temperatures. It is suitable for detecting the content of volatile nitrogen-containing odor gas produced in the process of pork ground corruption in various scenarios and for realizing real-time monitoring of meat freshness.

Table 1. Response function and change rate of pork TVBN at different temperatures.

Storage Temperature (°C)	Change Trend of TVBN	R^2	The Rate Constants	ln k
34	$y = 1.3777e^{0.162x}$	0.9471	0.162	−1.820158944
24	$y = 3.1861e^{0.0529x}$	0.9125	0.0529	−2.93935194
14	$y = 2.7466e^{0.0332x}$	0.9559	0.0332	−3.405205403
4	$y = 1.4663e^{0.0154x}$	0.9621	0.0154	−4.17338777

Table 2. Response function and change rate of color of indicator labels at different temperatures.

Storage Temperature (°C)	Change trend of ΔE	R^2	The Rate Constants (ΔE)	ln k (Color Difference)
34	$y = 1.0812e^{0.1284x}$	0.8609	0.1284	−2.0526
24	$y = 2.0219e^{0.0342x}$	0.8695	0.0342	−3.37553
14	$y = 1.4846e^{0.0238x}$	0.9607	0.0238	−3.73807
4	$y = 1.9988e^{0.0058x}$	0.9447	0.0058	−5.1499

The Arrhenius equation for TVBN and indicator color difference response factor can be obtained by fitting the equation of 1/RT and ln k, as follows (Figure 16):

$$\ln k\ (\text{TVBN}) = -53{,}061/RT + 18.792 \tag{6}$$

$$\ln k\ (\text{Color difference}) = -68{,}337/RT + 24.595 \tag{7}$$

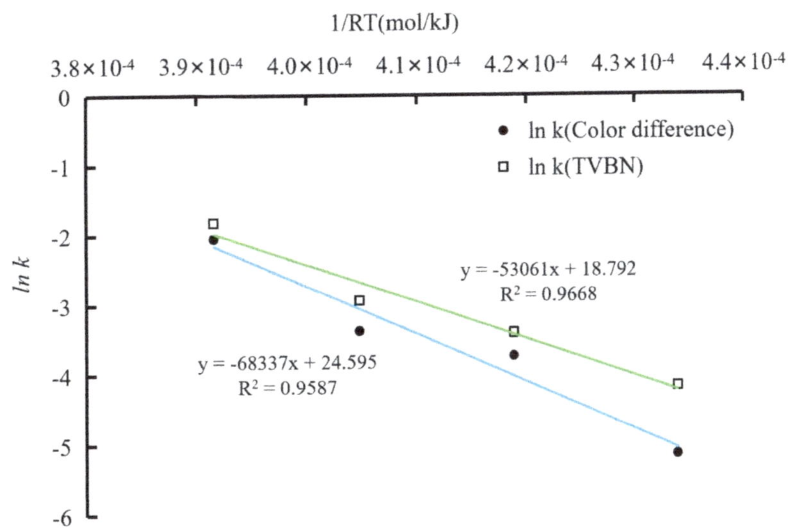

Figure 16. The ln k and 1/RT curves of pork TVBN and total color difference in indicator labels at different temperatures.

4. Conclusions

The hydrogel prepared by using zinc ion and TEMPO-oxidized nanocellulose (TOCNF) had good water absorption and a stable three-dimensional network structure. The indicator label made from the hydrogel-adsorbed blueberry anthocyanin showed obvious color change in the experiment of using ammonia gas to simulate the corruption of pork surimi, and a corresponding maximum color difference was observed before and after the hydrogel indicator labels soaked in 1.5 g/L anthocyanin solution. Applying this indicator label to the detection of pork mince freshness, it was found that when pork mince spoils, the labels changed from purple red to blue purple. Moreover, based on the Arrhenius formula, a dynamic model was established for the color difference value and TVBN change in the indicator label, and it was found that the indicator labels maintained similar color changes according to the freshness of pork mince at different temperatures. In conclusion, blueberry anthocyanin is an ideal intelligent food packaging indicator, and the hydrogel indicator label prepared by anthocyanin and nanocellulose has good application prospects in the detection of pork mince freshness.

Author Contributions: X.M.: conceptualization, writing—original draft, writing—review and editing; Q.S.: investigation, methodology, writing—original draft; T.S.: writing—review and editing; H.Z.: supervision; Y.Z.: supervision, writing—review and editing; W.Y.: conceptualization, supervision; A.R.: conceptualization, supervision, writing—review and editing. All authors have read and agreed to the published version of the manuscript.

Funding: This research was funded by the National Natural Science Foundation of China (31701663, 32160575) and Natural Science Foundation of Anhui Province (1808085MC70) And The APC was funded by Natural Science Foundation of Anhui Province (1808085MC70) and college of Life, Anhui Normal University.

Data Availability Statement: The datasets generated for this study are available on request to the corresponding author.

Acknowledgments: The authors also gratefully acknowledge the assistance from Outstanding Innovative Research Team for Molecular Enzymology and Detection in Anhui Provincial Universities (2022AH010012).

Conflicts of Interest: The authors declare no conflict of interest.

References

1. Lee, S.B.; Kim, D.H.; Jung, S.W.; Lee, S.J. Air-activation of printed time–temperature integrator: A sandwich package case study. *Food Control* **2019**, *101*, 89–96. [CrossRef]
2. Kalpana, S.; Priyadarshini, S.R.; Maria Leena, M.; Moses, J.A.; Anandharamakrishnan, C. Intelligent packaging: Trends and applications in food systems. *Trends Food Sci. Technol.* **2019**, *93*, 145–157. [CrossRef]
3. Chen, H.; Zhang, M.; Bhandari, B.; Guo, Z. Applicability of a colorimetric indicator label for monitoring freshness of fresh-cut green bell pepper. *Postharvest Biol. Technol.* **2018**, *140*, 85–92. [CrossRef]
4. Chen, H.; Zhang, M.; Bhandari, B.; Yang, C. Development of a novel colorimetric food package label for monitoring lean pork freshness. *LWT Food Sci. Technol.* **2019**, *99*, 7. [CrossRef]
5. Kim, D.; Lee, S.; Lee, K.; Baek, S.; Seo, J. Development of a pH indicator composed of high moisture-absorbing materials for real-time monitoring of chicken breast freshness. *Food Sci. Biotechnol.* **2017**, *26*, 37–42. [CrossRef]
6. Wu, C.; Sun, J.; Chen, M.; Ge, Y.; Ma, J.; Hu, Y.; Pang, J.; Yan, Z. Effect of oxidized chitin nanocrystals and curcumin into chitosan films for seafood freshness monitoring. *Food Hydrocoll.* **2019**, *95*, 308–317. [CrossRef]
7. Ezati, P.; Rhim, J. pH-responsive chitosan-based film incorporated with alizarin for intelligent packaging applications. *Food Hydrocoll.* **2020**, *102*, 105629. [CrossRef]
8. Ćorković, I.; Pichler, A.; Buljeta, I.; Šimunović, J.; Kopjar, M. Carboxymethylcellulose hydrogels: Effect of its different amount on preservation of tart cherry anthocyanins and polyphenols. *Curr. Plant Biol.* **2021**, *28*, 100222. [CrossRef]
9. Lu, P.; Yang, Y.; Liu, R.; Liu, X.; Ma, J.; Wu, M.; Wang, S. Preparation of sugarcane bagasse nanocellulose hydrogel as a colourimetric freshness indicator for intelligent food packaging. *Carbohyd. Polym.* **2020**, *249*, 116831. [CrossRef]
10. Chen, L.; Zhu, Y.; Cui, Y.; Dai, R.; Shan, Z.; Chen, H. Fabrication of starch-based high-performance adsorptive hydrogels using a novel effective pretreatment and adsorption for cationic methylene blue dye: Behavior and mechanism. *Chem. Eng. J.* **2021**, *405*, 126953. [CrossRef]

11. Jang, J.H.; Kang, H.J.; Adedeji, O.E.; Kim, G.Y.; Lee, J.K.; Kim, D.H.; Jung, Y.H. Development of a pH indicator for monitoring the freshness of minced pork using a cellulose nanofiber. *Food Chem.* **2023**, *403*, 134366. [CrossRef] [PubMed]
12. Paukkonen, H.; Kunnari, M.; Laurén, P.; Hakkarainen, T.; Uvinen, V.-V.; Oksanen, T.; Koivuniemi, R.; Yliperttula, M.; Laaksonen, T. Nanofibrillar cellulose hydrogels and reconstructed hydrogels as matrices for controlled drug release. *Int. J. Pharmaceut.* **2017**, *532*, 269–280. [CrossRef] [PubMed]
13. Lu, P.; Liu, R.; Liu, X.; Wu, M. Preparation of self-supporting bagasse cellulose nanofibrils hydrogels induced by zinc ions. *Nanomaterials* **2018**, *8*, 800. [CrossRef] [PubMed]
14. Ciuffarin, F.; Négrier, M.; Plazzotta, S.; Libralato, M.; Calligaris, S.; Budtova, T.; Manzocco, L. Interactions of cellulose cryogels and aerogels with water and oil: Structure-function relationships. *Food Hydrocoll.* **2023**, *140*, 108631. [CrossRef]
15. Guo, J.; Jin, Y.; Yang, X.; Yu, S.; Qi, J.-R. Computed microtomography and mechanical property analysis of soy protein porous hydrogel prepared by homogenizing and microbial transglutaminase cross-linking. *Food Hydrocoll.* **2013**, *31*, 220–226. [CrossRef]
16. Pilevaran, M.; Tavakolipour, H.; Naji Tabasi, S.; Elhamirad, A.H. Investigation of functional, textural, and thermal properties of soluble complex of whey protein–xanthan gum hydrogel. *J. Food Proc. Eng.* **2021**, *44*, 10. [CrossRef]
17. Ma, L.; Long, T.; Yuan, S.; Qi, P.; Han, L.; Hao, J. A pH-indicating smart tag based on porous hydrogel as food freshness sensors. *J. Colloid. Interface Sci.* **2023**, *647*, 32–42. [CrossRef]
18. Sutthasupa, S.; Padungkit, C.; Suriyong, S. Colorimetric ammonia (NH3) sensor based on an alginate-methylcellulose blend hydrogel and the potential opportunity for the development of a minced pork spoilage indicator. *Food Chem.* **2021**, *362*, 130151. [CrossRef]
19. Hanafy, N.A.N. Starch based hydrogel NPs loaded by anthocyanins might treat glycogen storage at cardiomyopathy in animal fibrotic model. *Int. J. Biol. Macromol.* **2021**, *183*, 171–181. [CrossRef]
20. Pirayesh, H.; Park, B.; Khanjanzadeh, H.; Park, H.; Cho, Y. Nanocellulose-based ammonia sensitive smart colorimetric hydrogels integrated with anthocyanins to monitor pork freshness. *Food Control* **2023**, *147*, 109595. [CrossRef]
21. Shaghaleh, H.; Hamoud, Y.A.; Xu, X.; Liu, H.; Wang, S.; Sheteiwy, M.; Dong, F.; Guo, L.; Qian, Y.; Li, P.; et al. Thermo-/pH-responsive preservative delivery based on TEMPO cellulose nanofiber/cationic copolymer hydrogel film in fruit packaging. *Int. J. Biol. Macromol.* **2021**, *183*, 1911–1924. [CrossRef] [PubMed]
22. Uyanga, K.A.; Iamphaojeen, Y.; Daoud, W.A. Effect of zinc ion concentration on crosslinking of carboxymethyl cellulose sodium-fumaric acid composite hydrogel. *Polymer* **2021**, *225*, 123788. [CrossRef]
23. Salama, H.E.; Abdel Aziz, M.S. Novel biocompatible and antimicrobial supramolecular O-carboxymethyl chitosan biguanidine/zinc physical hydrogels. *Int. J. Biol. Macromol.* **2020**, *163*, 649–656. [CrossRef] [PubMed]
24. Roy, S.; Rhim, J.W. Anthocyanin food colorant and its application in pH-responsive color change indicator films. *Crit. Rev. Food Sci.* **2021**, *61*, 2297–2325. [CrossRef] [PubMed]
25. Alizadeh-Sani, M.; Mohammadian, E.; Rhim, J.; Jafari, S.M. pH-sensitive (halochromic) smart packaging films based on natural food colorants for the monitoring of food quality and safety. *Trends Food Sci. Tech.* **2020**, *105*, 93–144. [CrossRef]
26. Chen, H.; Zhang, M.; Bhandari, B.; Yang, C. Novel pH-sensitive films containing curcumin and anthocyanins to monitor fish freshness. *Food Hydrocoll.* **2020**, *100*, 105438. [CrossRef]
27. Sun, Y.; Zhang, M.; Adhikari, B.; Devahastin, S.; Wang, H. Double-layer indicator films aided by BP-ANN-enabled freshness detection on packaged meat products. *Food Packag. Shelf* **2022**, *31*, 100808. [CrossRef]

foods

MDPI

Article

Auricularia auricular Adsorbs Aflatoxin B1 and Ameliorates Aflatoxin B1-Induced Liver Damage in Sprague Dawley Rats

Dan Xu, Minmin Huang, Jiao Lei, Hongxin Song, Liangbin Hu and Haizhen Mo *

School of Food Science and Engineering, Shaanxi University of Science and Technology, Xi'an 710021, China; xudan@sust.edu.cn (D.X.); hmm@sust.edu.cn (M.H.); leijiao1120@126.com (J.L.); songhx@sust.edu.cn (H.S.); hulb@sust.edu.cn (L.H.)
* Correspondence: mohz@sust.edu.cn

Abstract: Aflatoxin B1 (AFB1), as a class I carcinogen, poses a substantial health risk to individuals. Contamination of food sources, particularly grains and nuts, with *Aspergillus flavus* (*A. flavus*) contributes to the prevalence of AFB1. The impact of global warming has spurred research into the development of AFB1 prevention technologies. While edible fungi have shown potential in detoxifying AFB1, there is a scarcity of literature on the application of *Auricularia auricular* (*A. auricular*) in this context. This study aimed to investigate the ability and underlying mechanism of *A. auricular* mycelia to adsorb aflatoxin B1, as well as evaluate its protective effects on the AFB1-induced liver damage in SD rats. Additionally, the effects of temperature, time, pH, and reaction ratio on the adsorption rate were examined. Combining thermodynamic and kinetic data, the adsorption process was characterized as a complex mechanism primarily driven by chemical adsorption. In SD rats, the *A. auricular* mycelia exhibited alleviation of AFB1-induced liver damage. The protective effects on the liver attributed to *A. auricular* mycelia may involve a reduction in AFB1 adsorption in the intestine, mitigation of oxidative stress, and augmentation of second-phase detoxification enzyme activity. The adsorption method for AFB1 not only ensures safety and non-toxicity, but also represents a dietary regulation strategy for achieving effective defense against AFB1.

Keywords: AFB1; *Auricularia auricular*; adsorption; liver protection

Citation: Xu, D.; Huang, M.; Lei, J.; Song, H.; Hu, L.; Mo, H. *Auricularia auricular* Adsorbs Aflatoxin B1 and Ameliorates Aflatoxin B1-Induced Liver Damage in Sprague Dawley Rats. *Foods* **2023**, *12*, 2644. https://doi.org/10.3390/foods12142644

Academic Editor: Nawal Dubey

Received: 31 May 2023
Revised: 29 June 2023
Accepted: 7 July 2023
Published: 8 July 2023

1. Introduction

Aflatoxin B1 (AFB1), which is the most highly toxic mycotoxin, is predominantly produced by *Aspergillus flavus* and *Aspergillus parasiticus* [1,2], and can frequently be found in foods such as milk, rice, corn, peanuts, oils, and soybeans [3,4]. It can cause damage to our liver, lungs, heart, and kidneys, and it shows immunosuppressive properties, such as inhibiting cell-mediated immune reactions, decreasing natural killer cytolysis, and suppressing macrophage function [5–8]. Currently, it has been listed as a class I carcinogen by the World Health Organization (WHO) [9–12]. Its carcinogenic action requires exertion by metabolic activation [13]. Once adsorbed, aflatoxin undergoes metabolism by hepatic cytochrome P450 enzymes, resulting in the formation of AFB1-8,9-epoxide (AFBO) [14]. AFBO is considered pivotal in the toxicity of aflatoxins, and it undergoes an adduct reaction with DNA, causing DNA damage and genetic mutations [15].

To reduce the exposure risk of aflatoxin in the diet, numerous approaches have been investigated for the removal of AFB1. These approaches can be broadly categorized into physical, chemical, and biological control methods [16,17]. Considering safety and efficiency, biological detoxification is emerging as a favorable approach for the removal of aflatoxins [17,18]. The application of microorganisms used in the food industry to remove aflatoxins, including lactic acid bacteria, yeast, and edible fungi, has been reported in a lot of research [19–22]. It is found that AFB1 can be effectively degraded by enzymes extracted from edible fungi, such as *Armillariella tabescens*, *Pleurotus ostreatus*, *Plerotus eryngii*, and *Agaricus blazei* [23–25].

In addition, many studies have shown that AFB1 can not only be degraded by microorganisms through enzymatic degradation, but also adsorbed by the cell wall components of microorganisms. To better understand the role of fungal components in AFB1 removal activity, the ability of *Bjerkandera adusta* (*B. adusta*) to remove AFB1 was analyzed after enzymatic, physical, and chemical treatments to degrade or change the fungal cellular components. The results suggest that AFB1 was removed by its binding onto the cell wall components of *B. adusta* [26].

In the preliminary research, we studied the ability of seven edible fungi (*Lentinus edodes, Flammulina velutiper, Pleurotus ostreatus, Hericium erinaceum, Armillariella mellea, Auricularia auricular*, and *Auricularia polytricha*) to remove AFB1, and found that *A. auricular* has the strongest ability to remove AFB1, with a removal rate of 88.2%. Interestingly, the efficacy of *A. auricular* for AFB1 removal is primarily attributed to its adsorption capabilities, rather than biodegradation. Therefore, edible fungi have demonstrated potential as AFB1 adsorbents, due to their edibility and high adsorption capacity [19]. Based on the composition and structural characteristics of *A. auricular*, it contains a lot of dietary fiber that is not digested by the body and is able to adsorb heavy metals, tetracycline, and other pollutants [27–29].

Furthermore, *A. auricular* contains many pharmacologically active ingredients, which have been emphasized for use in various biological activities, such as hepatoprotective and antioxidative action [30,31]. Therefore, we speculate that *A. auricular* can effectively alleviate the liver damage caused by AFB1. On the one hand, *A. auricular* prevents AFB1 from being adsorbed into the intestine by binding the toxin; on the other hand, the active ingredients contained in *A. auricular* may activate the body's immune function and protect the liver from oxidative damage caused by AFB1.

To comprehensively assess the detoxification effect of *A. auricular* on AFB1, this research systematically investigated the impact of various parameters, such as temperature, pH, and reaction ratio, on the adsorption of AFB1 by *A. auricular in vitro*. Subsequently, the adsorption process was characterized by analyzing the kinetics and thermodynamics of adsorption. Finally, the protective effect of *A. auricular* against AFB1-induced liver injury was evaluated in an *in vivo* model.

2. Materials and Methods

2.1. Strains, Chemicals, and Reagents

A. auricular was purchased from Xixiang Edible Fungi Research Institution (Xixiang, Shaanxi Province, China) and stored at 4 °C on potato dextrose agar (PDA, BD) medium. AFB1 and dimethyl sulfoxide (DMSO) were purchased from Sigma Company (Ronkonkoma, NY, USA). All kits (Superoxide dismutase (SOD) kit, glutathione S-transferase (GST) kit, etc.) were purchased from Nanjing Jiansheng Biological Engineering Co., Ltd. (Nanjing, China). Other chemicals and reagents were purchased from Sinopharm Chemical Reagent Co., Ltd. (Shanghai, China). And 10 mg/mL AFB1 stock solution was prepared by dissolving 10 mg AFB1 powder in 1 mL of methanol solution.

2.2. Assessment of Adsorption Capacity under Different Pretreatments

A. auricular was cultured in PDA medium at 28 ± 2 °C until all over the plate. Subsequently, the mycelia were transferred to potato dextrose broth (PDB) medium, and incubated at 28 ± 2 °C under continuous shaking at 120 r/min in the absence of light, for a period of 4–7 days. The mycelia were then separated through vacuum filtration. The mycelia were divided into 4 groups, each containing 2 g, and individually combined with 2 mL of phosphate-buffered saline (PBS) buffer to form separate *A. auricular* suspensions. The four groups were subjected to different pretreatment procedures: (1) heated at 121 °C for 20 min; (2) pre-freezing at −20 °C for 12 h, followed by repeated cycles of freezing after dissolution for 12 h, repeated for 3 times; (3) freeze-drying for 12 h after pre-freezing at −20 °C; or (4) a blank control, where the *A. auricular* suspension was not subjected to any treatment. Samples with different pretreatments (the above 4 groups) were added to up to

0.4 μg/mL of AFB1, then cultured at the same conditions for 12 h. Finally, the free toxins were quantified using high-performance liquid chromatography (HPLC).

The residual AFB1 was detected according to the method by Xu [32]. AFB1 present in the 1 mL filtrate sample was extracted using 3 mL of chloroform, followed by drying using nitrogen gas. The resulting AFB1 extract was then dissolved in 100 μL of trifluoroacetic acid and 200 μL of *n*-hexane, and kept in a dark environment at 40 °C for 15 min. The derivative obtained was subsequently dried using nitrogen gas blowing, and redissolved in 1 mL of acetonitrile–water (30:70; *v*/*v*). The AFB1 derivative solution was then filtered through a 0.22 μm filter, and quantified using an Agilent HPLC 1200 liquid-chromatograph (Agilent Technologies, Palo Alto, CA, USA) equipped with a C18 column (Diamonsil®, 250 mm × 4.6 mm, 5 μm) that was conditioned at 30 °C. The mobile phase consisted of 30% acetonitrile and 70% water, with a flow rate of 1.0 mL/min. Finally, the AFB1 signal was monitored using a fluorescent detector at excitation/emission wavelengths of 360/440 nm.

2.3. Effects of Reaction Ratio (Mycelia/AFB1), pH, Temperature and Incubation Time on AFB1 Adsorption Capacity

The effects of various factors on adsorption were investigated using vacuum freeze-dried *A. auricular* powder. A suspension of *A. auricular* was prepared by mixing 0.05 g of mycelia powder with 5 mL of PBS buffer. Different initial concentrations of AFB1 (ranging from 0.0125 to 1.0 μg/mL) were added, and the mixture was cultured for 12 h. The supernatant was collected to determine the residual AFB1. In addition, experiments were conducted by adding AFB1 to a concentration of 0.4 μg/mL, and the adsorption process was studied under various pH levels (ranging from 2 to 8), temperatures (ranging from 20 to 60 °C), and co-culturing times (from 2 to 12 h).

2.4. Stability of the Adsorption

The mycelia separated from co-cultured systems were subjected to dissolution using various solvents, including sodium hydroxide (NaOH), acetonitrile (C_2H_3N), methanol-acetonitrile mixture (CH_3OH:C_2H_3N = 1:1), and 0.01 mol/L hydrochloric acid (HCl), at 30 °C for 6 h, 160 r/min. Then, the amount of AFB1 in the supernatant was defined as the desorption quantity of the binder.

The experiment was divided into two groups. In the first group, 40 μL AFB1 solution (10 mg/L) was added, then 1 mL artificial gastric juice and artificial intestinal juice were added, respectively. Then, 0.01 g of *A. auricular* mycelia powder was added, and the mixture was co-cultured 37 °C for 0.5–4 h at 200 r/min. After centrifugation at 5000 r/min for 10 min, the supernatant was collected for AFB1 content determination. In the second group, the binders obtained after 4 h of culture were mixed with 1 mL of artificial gastric juice and artificial intestinal juice, and the mixture was shaken at 37 °C for 0.5–4 h at 200 r/min. Finally, the supernatant was collected, and the quantity of AFB1 was determined using HPLC.

2.5. Adsorption Kinetics

A quantity of mycelia powder (0.05 g) was subjected to co-cultivation with AFB1 at a concentration of 0.4 μg/mL at 35 °C for 2–12 h. The adsorption of AFB1 was subsequently evaluated using the same experimental procedure. Four different kinetic models, namely the intraparticle diffusion model, pseudo-first-order kinetic model, pseudo-second-order kinetic model, and Elovich model, were chosen for fitting analysis in order to determine the most suitable kinetic model for describing the adsorption process.

$$q_t = k_p t^{0.5} \tag{1}$$

$$\ln(q_e - q_t) = \ln q_e - k_1 t \tag{2}$$

$$\frac{t}{q_t} = \frac{1}{k_2 {q_e}^2} + \frac{t}{q_e} \tag{3}$$

$$q_t = A_E + B_E \ln t \tag{4}$$

2.6. Adsorption Isotherm

A total of 0.05 g mycelia powder and PBS buffer were cultured together to form a system of 5 mL. Then, we set a range of initial concentrations of AFB1 (0.0125–0.4 μg/mL) and tested at 25–45 °C for 8 h, and separately followed the above methods to map adsorption isotherm. The quantity of AFB1 adsorbed per unit of adsorbent can be calculated using the Formula (5):

$$Q_e = \frac{(C_o - C_e) \times V}{m} \tag{5}$$

where Qe is the adsorption capacity of AFB1 (μg/Kg); Co is the concentration of AFB1 (μg/L); Ce is the equilibrium concentration of AFB1 (μg/L); V is the volume of AFB1 (mL); and M is the mass of added mycelia powder (g).

The Freundlich (6) and Langmuir empirical Formulae (7) were used to fit the adsorption results and describe the adsorption process of AFB1 by *A. auricular* mycelia:

$$Q_e = K_f \times C_e^{\frac{1}{n}} \tag{6}$$

$$Q_e = \frac{Q_m \times K_L \times C_e}{1 + K_L \times C_e} \tag{7}$$

2.7. Adsorption Thermodynamics

The relevant thermodynamic parameters are calculated by the following formulae:

$$K_{OC} = \frac{K_f}{W_{OC}} \tag{8}$$

$$\Delta G = -RT \ln K_d \tag{9}$$

$$\Delta G = \Delta H - T\Delta S \tag{10}$$

$$K_d = \frac{Q_e}{Q_c} = \frac{C_s}{C_e} \times \frac{V}{M} \tag{11}$$

$$\ln K_d = \frac{\Delta S}{R} - \frac{\Delta H}{R} \times \frac{1}{T} \tag{12}$$

2.8. The In Vivo Evaluation of the Intervention Effect of A. auricular on AFB1-Induced Liver Injury

After a period of 7 days of adaptive feeding, 5-week-old male Sprague Dawley (SD) rats were provided with a standard diet and ad libitum access to water. The rats were then randomly assigned to 6 groups based on their body weight, with each group consisting of 6 rats. The groups included the blank control group (B), which received a normal diet without AFB1 and mycelia powder; the positive control group (P), which received a 2 mg/kg AFB1 solution via oral gavage at 9 a.m. on the first day; the high dose negative control group (HC), which received a daily oral administration of 500 mg/kg BW; the high dose mycelia powder experimental group (HT), which received an AFB1-mycelia powder solution (AFB1 2 mg/kg BW, mycelia powder 500 mg/kg BW) via oral administration at 9 a.m. on the first day; the low dose negative control group (LC), which received a daily oral administration of mycelia powder solution (100 mg/kg BW); and the low dose mycelia powder experimental group (LT), which received an AFB1-mycelia powder solution (AFB1 2 mg/kg BW, mycelia powder 100 mg/kg BW) via oral administration at 9 a.m. on the first day.

Following the aforementioned protocol, the SD rats were fed a standard diet for three additional days. After an overnight fasting period, the rats were weighed the next morning, and anesthesia was induced by injecting 0.3 mL/100 g BW of 10% chloral hydrate solution. Blood samples were promptly collected after anesthesia, with 2 mL of blood stored in 5 mL anticoagulation tubes and accelerating tubes. The samples were then transported to the hospital for liver function assessment, including ALT, AST, γ-GT, T-Bil, D-Bil, I-Bil, TP, and ALB. A portion of the liver tissue was homogenized to measure enzyme activity, including antioxidant enzymes such as superoxide dismutase (SOD), glutathione peroxidase (GST-PX), and catalase (CAT). The levels of MDA were measured following the instructions provided in the MDA kit. Additionally, liver tissue was sectioned into 10 μm slices and stained with HE (hematoxylin-eosin) for further analysis. The histopathological procedure followed the methodology described in a previous study [33]. Fresh liver tissues were carefully dissected into pieces measuring 1 cm × 1 cm. These tissue samples were then immersed in neutral formalin for a minimum of 24 h, and subsequently processed for paraffin embedding. Sections with a thickness of 10 μm were obtained from the embedded tissues. To facilitate histopathological examination, the slides were stained using hematoxylin and eosin (H&E). Microscopic analyses of the liver sections from all experimental groups were performed using a Nikon E100 microscope from Japan.

2.9. Statistics Analysis

The experiments were repeated three times, and date analysis was performed using SPSS 19.0 software (International Business Machines Corporation, Armonk, NY, USA); $p \leq 0.05$ or less was considered significant.

3. Results

3.1. Characterization of AFB1 Adsorption by A. auricular Mycelia

3.1.1. Assessment of Adsorption Capacity under Different Pretreatments

Based on the previous experimental results in our laboratory regarding the detoxification of toxins by *A. auricula*, it was observed that *A. auricular* mycelia exhibited enhanced AFB1 adsorption capability. As shown in Figure 1, compared with control group, the three different pre-treatment methods of the mycelia resulted in improved AFB1 adsorption (≤20%). However, no significant differences were observed among the three treated groups. Notably, freeze-dried mycelia demonstrated the highest adsorption rate (71.2%). Freeze-drying preserves the high biological activity of the mycelia and offers the benefits of convenient storage and processing.

3.1.2. Effects of Reaction Ratio (Mycelia/AFB1), pH, Temperature and Incubation Time on AFB1 Adsorption Capacity

Furthermore, we investigated the effects of various factors on the adsorption, including the reaction mass ratio (mycelia/AFB1), pH, temperature, and incubation time (Figure 2A–D). Figure 2A showed that, as the reaction ratio increases, the adsorption rate will show an upward trend, and will reach equilibrium under a certain ratio. When the reaction ratio (mycelia/AFB1) was $1{:}2.5 \times 10^{-6}$, the adsorption quantity and adsorption rate could sustain a steady state. It showed that the adsorption has reached saturation, and the adsorption rate could still reach 81.5%.

It was observed from Figure 2B that the adsorption rate could reach the highest (75.2%) at pH 7.0, while the adsorption is less effective under acidic conditions. With the pH shifts towards alkaline, there is a slight increase in the adsorption rate, albeit within a range of approximately 20%.

The effect of temperature is shown in Figure 2C. While there were no significant variations in the adsorption rate at different temperatures, it was observed that the adsorption rate tends to be higher when the temperature approaches the human body temperature. Specifically, at 30 °C, the adsorption rate reaches 82%, surpassing the rates observed under other conditions.

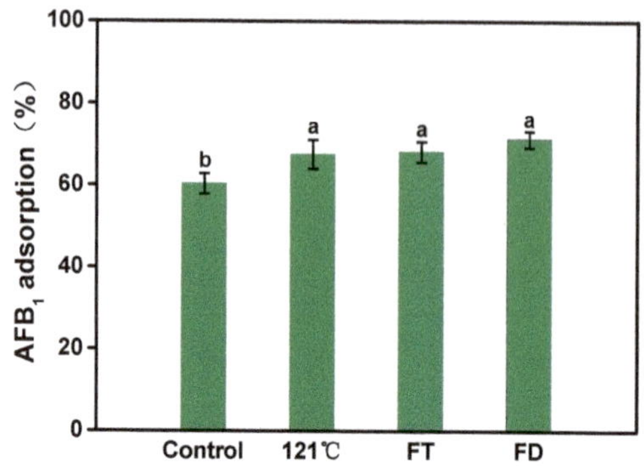

Figure 1. The effect of mycelia pretreatments on AFB1 adsorption. The abbreviations FT and FD correspond to repeated freeze–thaw and vacuum freeze-drying treatments, respectively, while the control group represents wet mycelia without any treatment. The values are presented as means ± standard deviation (SD); a and b indicate significant differences between different components or treatments ($p < 0.05$).

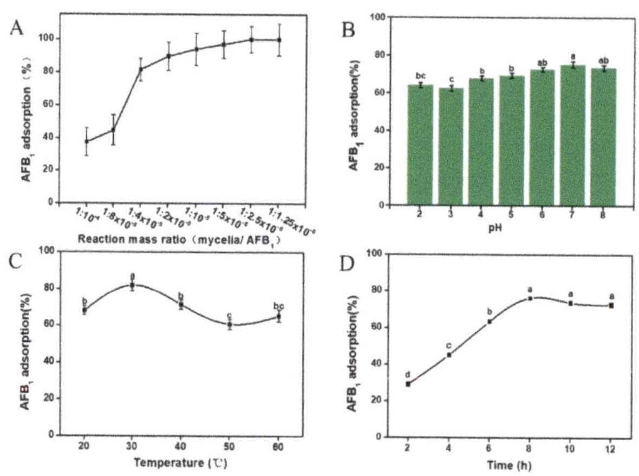

Figure 2. AFB1 adsorption capacity of *A. auricular* mycelia: effect of reaction mass ratio (mycelia/AFB1) (**A**), effect of incubation pH (**B**), effect of incubation temperature (**C**), and effect of incubation time (**D**). Values are expressed as means ± SD; a, b, c and d indicate significant differences between different treatments ($p < 0.05$).

As shown in Figure 2D, the adsorption process of AFB1 by *A. auricular* mycelia exhibited a slow progression, with approximately 30% of AFB1 being adsorbed after 2 h of mixing with the mycelia. The relationship between the adsorption rate and incubation time followed a "J"-shaped growth curve. Within 8 h, the adsorption rate exhibited rapid growth, ultimately reaching equilibrium at a rate of 78.4%.

3.1.3. Stability Evaluation of AFB1 Adsorption

The stability evaluation of the adsorption was shown in Figure 3. There was no significant difference in the desorption capacity between different eluents (Figure 3A). Following elution with seven different eluents, the residual AFB1 content in the compound was between 89.6% and 94.8%, and the elution of AFB1 to the binders were all less than 11%. The results indicated that the binders were stable.

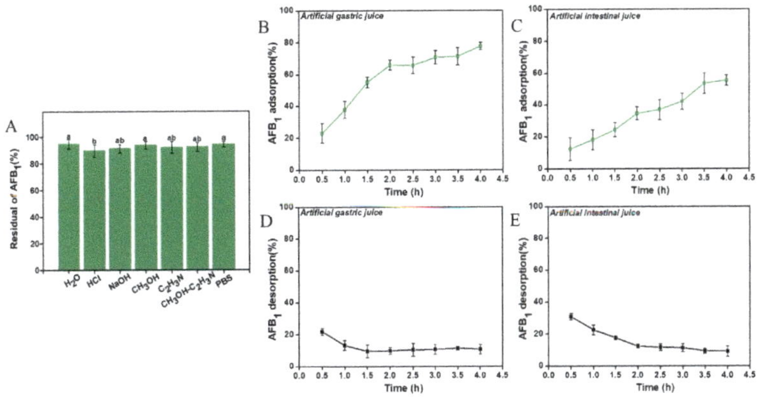

Figure 3. AFB1 adsorption stability of *A. auricular* mycelia: the residual AFB1 content after washing with different eluents (**A**), AFB1 adsorption and its desorption rate in artificial gastric juice (**B**,**D**), and AFB1 adsorption and its desorption rate in artificial intestinal juice (**C**,**E**). Values are expressed as means ± SD; a, b, and ab indicate significant differences between different treatments ($p < 0.05$).

In simulated vitro experiments, we found that with the extension of the treatment time, the AFB1 adsorption rate gradually increased, indicating that the mycelia powder has the capability to adsorb AFB1 in gastric juice and intestinal juice (Figure 3B,C). This suggests that *A. auricular* mycelia can reduce the bioavailability of AFB1 through adsorption. Our findings demonstrate the stability of the mycelia powder–AFB1 binders in the in vitro environment. From Figure 3D,E, the addition of artificial gastrointestinal juice resulted in minimal desorption of AFB1 from the binders. Hence, we believe that *A. auricular* mycelia has promising potential for application in the prevention and control of AFB1 *in vivo*.

3.2. Analysis of Adsorption Behavior

3.2.1. Adsorption Kinetics

The curve of the adsorption amount of *A. auricular* mycelia for AFB1 with adsorption time is showed in Figure 4A. The adsorption amount exhibited an increasing trend from 2 to 8 h, reaching equilibrium at 8 h (0.358 μg/kg). To analyze the adsorption process, four kinetic models, namely the intra-particle diffusion model, pseudo-first-order kinetic model, pseudo-second-order kinetic model, and Elovich model, were employed.

Figure 4B–E shows the fitting line of the kinetic models, and the four model fitting parameters are listed in Table 1. The pseudo second-order kinetic equation exhibits a significantly higher fitting value (0.9976) compared to the other models, suggesting that it provides a better description of the adsorption process of AFB1 by *A. auricular* mycelia.

Table 1. The kinetics model fitting parameters of AFB1 adsorbed by *A. auricular*.

Fitting Model	Formula	R^2
In-particle diffusion equation	$Qt = 0.06\ t^{0.5}$	0.8870
Pseudo first-order kinetic equation	$\ln(q_e - q_t) = \ln q_e - 0.16\ t$	0.9856
Pseudo second-order kinetic equation	$t/qt = 1/13.77\ qe^2 + 1/qe$	0.9976
Elovich equation	$Qe = 0.09 + 0.02\ \ln t$	0.9618

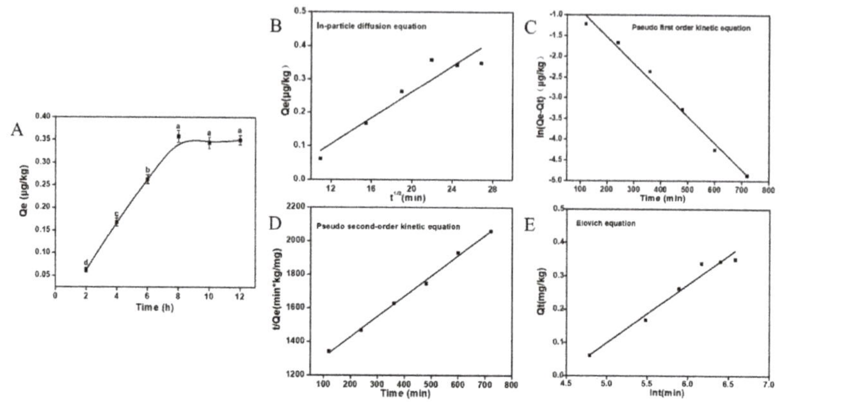

Figure 4. Kinetics of adsorption: the kinetics curve of AFB1 adsorbed by *A. auricula* mycelia (**A–E**) represent the kinetics model fitting curve of AFB1 adsorbed by *A. auricula* mycelia, using four kinetic models including the intra-particle diffusion model, pseudo-first-order kinetic model, pseudo-second-order kinetic model, and Elovich model. Values are expressed as means ± SD; a, b, c, and d indicate significant differences between different treatments ($p < 0.05$).

3.2.2. Adsorption Isotherm

The adsorption isotherm of Qe (adsorption amount of mycelia) with Ce (equilibrium concentration of AFB1) is shown in Figure 5A. The equilibrium adsorption amount of mycelia demonstrates a significant increase with the rise in the initial concentration of AFB1. At the same time, the adsorption of AFB1 by mycelia showed a decreasing trend as the temperature increased from 25 °C to 45 °C, with the highest adsorption observed at 35 °C. Furthermore, we used the Freundlich equation and the Langmuir equation to fit data of adsorption isotherm.

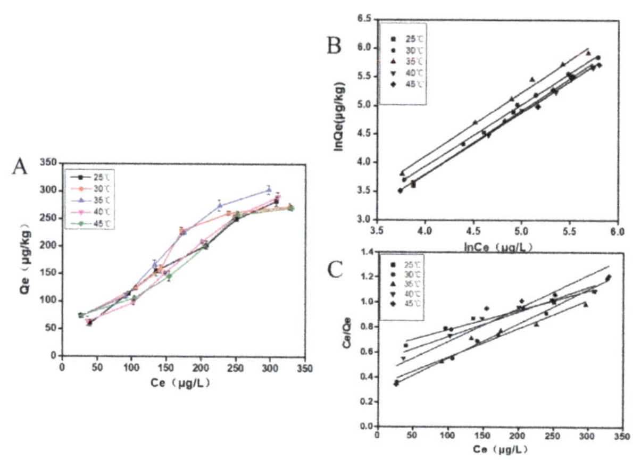

Figure 5. Adsorption behavior of *A. auricula* mycelia on AFB1: the adsorption isotherm of AFB1 (**A**), the adsorption isotherm of Freundlich equation fitting AFB1 adsorbed by *A. auricula* mycelia (**B**), and the adsorption isotherm of Langmuir equation fitting AFB1 adsorbed by *A. auricula* mycelia (**C**). Values are expressed as means ± SD.

The fitting line of the isothermal adsorption models were shown in Figure 5B,C, and the corresponding parameters are listed in Tables 2 and 3. The analysis revealed that the Freundlich model was more suitable for describing the adsorption isotherm, due to its high R^2 ($R^2 > 0.99$ at different temperatures). The results suggest that the adsorption process of AFB1 by *A. auricular* mycelia follows the heterogeneous adsorption of multi-molecular layers, as described by the Freundlich equation.

Table 2. The adsorption isotherm parameters of the Freundlich equation fitting AFB1 adsorbed by *A. auricular*.

Temperature (K)	K_f	$1/n$	Freundlich Formula	R^2
298	0.11	0.02	Qe = 0.11 × Ce0.02	0.9981
303	0.22	0.04	Qe = 0.22 × Ce0.04	0.9927
308	0.24	0.05	Qe = 0.24 × Ce0.05	0.9901
313	0.17	0.03	Qe = 0.17 × Ce0.03	0.9951
318	0.11	0.02	Qe = 0.11 × Ce0.02	0.9979

Table 3. The adsorption isotherm parameters of the Langmuir equation model.

Temperature (K)	K_L	Langmuir Formula	R^2
298	0.01	Qe = 0.278 × Ce/(1 + 0.01 Ce)	0.9943
303	0.02	Qe = 0.625 × Ce/(1 + 0.02 Ce)	0.9680
308	0.04	Qe = 0.333 × Ce/(1 + 0.04 Ce)	0.9462
313	0.09	Qe = 0.419 × Ce/(1 + 0.09 Ce)	0.8551
318	0.04	Qe = 0.527 × Ce/(1 + 0.04 Ce)	0.9464

3.2.3. Adsorption Thermodynamics

Adsorption thermodynamic parameters are listed in Table 4. It can be observed that, as the temperature increases, the value of Koc first rose and then decreased. It indicated that the adsorption of *A. auricular* mycelia on AFB1 increases with temperature, exhibiting a trend of first rising and then falling. These findings are consistent with the results obtained from the previous fitting of the Freundlich equation.

Table 4. The thermodynamics parameters of AFB1 adsorbed by *A. auricular*.

Temperature (K)	Koc	ΔG (KJ/mol)	ΔH (KJ/mol)	ΔS (KJ/mol)
298	10.59	−22.43	−6.31	0.05
303	21.59	−21.35	−6.31	0.07
308	24.38	−36.85	−6.31	0.10
313	17.27	−50.39	−6.31	0.11
318	11.13	−43.82	−6.31	0.09

According to $\Delta G < 0$, it can be judged that the adsorption process is spontaneous and is affected by temperature. The absolute value of ΔG initially increases and then decreases as the temperature rises. This indicates that the adsorption process reaches its maximum at 40 °C, and slows down beyond that temperature.

It can be judged by $\Delta H = -6.31$ that the adsorption reaction process is exothermic. It suggests that an increase in temperature is not favorable for the progress of the reaction. This is consistent with the previous single-factor result of incubation temperature, and in line with the result of $\Delta G < 0$.

The ΔS value represents the variation in adsorption entropy, reflecting the degree of disorder between the solid and liquid phases during the reaction. $\Delta S > 0$ values indicate an increase in disorder. With increasing temperature, the ΔS value initially rises and then decreases. This trend suggests that higher temperatures have a beneficial impact on the reaction progress, promoting increased disorder. However, once the maximum value is reached, the adsorption reaction proceeds unfavorably, leading to a decrease in disorder.

3.3. *Evaluation of the Intervention Effect of* A. auricular *on AFB1-Induced Liver Injury*

In this study, *A. auricular* mycelia showed an excellent ability to adsorb AFB1 *in vitro*. To investigate the effect of AFB1 adsorption by mycelia *in vivo*, SD rats were fed with mycelia, AFB1, or a mixture of both. The results showed that the addition of mycelia improved liver function and restored antioxidant levels in SD rats. Overall, *A. auricular* mycelia exhibited a protective effect against AFB1-induced liver damage.

The assessment of liver function was conducted, and the findings are presented in Figure 6A. There were significant differences between the negative control group, the positive control group, and the experimental group, as can be seen in Figure 6A. Obviously, rats in the positive control group exhibited abnormal liver function indicators due to AFB1 induction. The levels of TP and ALB significantly decreased, indicating a decline in liver reserve function. The elevated levels of ALT and AST suggested severe liver cell damage. The varying levels of T-Bil, D-Bil, and I-Bil indicated impairment in liver excretory function. The presence of high γ-GT levels in the positive control group, which is typically produced by the liver, indicated abnormal or pathological liver conditions. In summary, AFB1 infection led to abnormal liver function in the experimental rats. However, the experimental group showed mitigated liver damage compared to the negative control group. Notably, this protective effect was observed to be dose-dependent on the ingestion of mycelia powder.

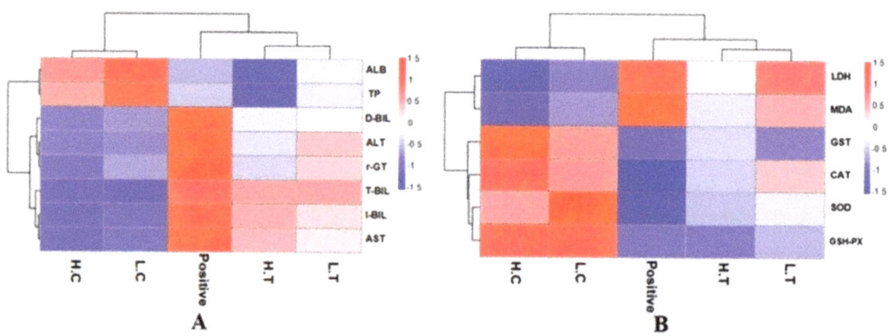

Figure 6. Heat map of liver functions, lipid peroxidation and antioxidant levels. Positive (P) represents the positive control group, rats infected with AFB1; LC and HC represent low-dose *Auricularia auricula* mycelia-fed and high-dose mycelia-fed normal SD rats, respectively; LT and HT represent the rats treated with mixed toxins and different doses of mycelia powder. (**A**) liver function related index; (**B**) lipid peroxidation and antioxidant levels.

Similarly, the protective effect of mycelia on SD rats is also evident in Figure 6B. In the positive control group, an increase in MDA and LDH levels was observed, indicating AFB1-induced lipid peroxidation in the liver of rats. The levels of GST, CAT, SOD, and GSH-PX, which are associated with *in vivo* oxidative status, significantly decreased. This reduction indicated a decrease in the antioxidant capacity of rats, and an imbalance in the regulation mechanism of endogenous enzymes or non-enzymatic factors. Such conditions are unfavorable for rats to counteract external substance interference. Conversely, in the experimental group of rats fed with a mixture of mycelia, a significant increase in the activity of endogenous antioxidant enzymes or non-enzymatic factors was observed, accompanied by a decrease in MDA and LDH levels. These findings suggest that *A. auricular mycelia* alleviated AFB1-induced lipid peroxidation in SD rats and prevented the accumulation of reactive oxygen species.

Figure 7 presents histological sections of rat livers stained with HE. The liver structure in the control group (A, B, and C, without AFB1) appeared normal, with no significant differences observed among A, B, and C. These findings indicate that the addition of mycelia powder does not induce liver damage, and is considered safe. In contrast, the positive control group exhibited evident lesions, characterized by pronounced congestion of hepatic sinusoids and a light purple appearance throughout the section. Hepatocytes showed disordered arrangement, nucleus loss, cytoplasmic outflow, and noticeable infiltration of inflammatory cells in the portal area. In liver tissue sections b and c, the overall liver structure was mostly normal, with some remaining nucleus loss. However, the pathological changes were significantly alleviated, with section c demonstrating a liver state closer to normal. In summary, *A. auricular* mycelia effectively mitigated liver damage caused by AFB1, and the protective effect improved with increasing mycelia dosage. Moreover, when fed mycelia powder alone, it did not induce liver damage and exhibited a certain level of liver protection.

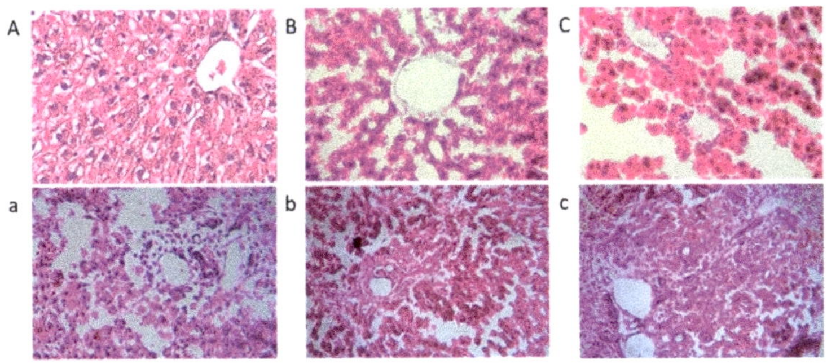

Figure 7. The liver histopathological slices observation of rats; (**a**–**c**) represent the positive group, the low dose experimental group, and the high dose experimental group, respectively, and (**A**–**C**) are their corresponding control groups.

4. Discussion

Due the high toxicity and exposure risk associated with AFB1, many methods are developed to mitigate its adverse effects. Among them, the use of microbials to remove AFB1 from materials has garnered significant attention, particularly in the feed industry [34]. A lot of microbial species, such as *Lactobacillus rhamnosus*, *Accharomyces cerevisiae*, and *Leurotus eryngii*, have been identified for their ability to remove AFB1, achieving an average clearance rate of 86% [19,35–37]. Interestingly, we discovered that the mycelia of *A. auricular*, a popular edible fungus widely cultivated in Southeast Asia and recognized for its potential pharmaceutical value, exhibited a higher efficiency in removing AFB1, with a clearance rate of 88.2% [38–40]. Under optimal conditions, the AFB1 clearance rate reached 98.3%. Therefore, *A. auricular* mycelia offer a promising, safe, and appealing approach to combat aflatoxin contamination in food [41,42]. This study was designed to investigate the adsorption effect of *A. auricular* mycelia on AFB1, establish its safety, and explore the underlying mechanism.

This study provides evidence that *A. auricular* mycelia exhibit robust adsorption capacity for AFB1. The adsorption rate remains consistently high, exceeding 60%, at both lower (20 °C) and higher (121 °C) temperatures. The influence of pH on adsorption is minimal, with a minimum adsorption rate of 60%. Alkaline conditions favor adsorption, with a notable adsorption rate of 75.2% at pH 8. Furthermore, despite a decrease in the adsorption rate in artificial intestinal juice and artificial gastric juice, it still reaches 55.5% and 77.6%, respectively, confirming the ability of *A. auricular* mycelia to adsorb AFB1 within the human gastrointestinal tract. In the case of AFB1-poisoned rats, *A. auricular* mycelia significantly alleviated liver damage and reduced oxidative stress and lipid peroxidation levels. Moreover, the adsorption process demonstrates remarkable stability, with the highest desorption rate observed at 10.4% after treatment with organic reagents. In artificial intestinal juice and gastric juice, the binder maintains high stability, exhibiting desorption rates of 10.7% and 9.1%, respectively.

Based on the findings from our experiment, we proposed the mechanism of *A. auricular* mycelia against AFB1-induced liver damage. In addition to its direct role in reducing AFB1 adsorption in the gastrointestinal tract, similar to other enterosorption therapies, *A. auricular* mycelia also exhibit the ability to attenuate AFB1-induced oxidative stress upon entering the body, and to enhance the activity of phase II detoxification enzymes.

The freeze-dried mycelia demonstrate a higher adsorption rate, which can be attributed to their elevated mycelial activity, and they are also more convenient for storage and application in food processing [43]. Owing to the high stability of this adsorption, different interpretations of adsorption behaviors emerged. According to the adsorption kinetic curve and thermodynamic research, we consider this adsorption as a form of steady chemical adsorption. However, the adsorption process aligns with empirical formulas, indicating a characteristic of heterogeneous multi-molecular layer adsorption, typically associated with physical adsorption. Considering the porosity and complexity of *A. auricular* mycelia, we propose that the adsorption process involves two phases: physical adsorption and chemical adsorption.

The hepatoprotective effects of *A. auricular* mycelia on rats have been successfully validated, confirming our initial hypothesis that *A. auricular* mycelia can alleviate AFB1-induced oxidative stress reactions and promote liver recovery. What sets it apart is that, in addition to post-digestion detoxification, *A. auricular* mycelia can directly reduce the adsorption of AFB1 in the gastrointestinal tract, thereby intercepting the initial step of toxin metabolism. Interestingly, the addition of *A. auricular* mycelia not only exhibited no toxicity in normal rats, but also provided favorable conditions for their growth, while exerting a protective effect on the liver. We speculate that this could be attributed to the nutritional and medicinal value inherent in *A. auricular* mycelia.

5. Conclusions

A. auricular mycelia exhibit the ability to adsorb AFB1 and mitigate AFB1-induced liver damage in rats. Its liver-protective effect is attributed to multiple mechanisms, including a reduction in AFB1 bioavailability through stable adsorption, attenuation of AFB1-induced oxidative stress, and enhancement of phase II detoxification enzyme activity. Furthermore, *A. auricular* mycelia contribute to the normalization of liver function indicators. The adsorption process is considered to involve both physical and chemical adsorption phenomena. In conclusion, *A. auricular* mycelia demonstrate the capability to adsorb AFB1, thereby alleviating AFB1-induced liver damage and providing liver protection. This adsorption ability remains effective even under extreme conditions, suggesting its potential for industrial applications. With its edibility and medicinal value taken into account, *A. auricular* mycelia hold promise as a potential food supplement or health-care product for the ultimate prevention of AFB1 and the promotion of liver health.

The research conducted in this study provides theoretical evidence for the edible fungi removal of AFB1. However, the adsorption of AFB1 by *A. auricula* is a complex biological reaction, and further investigation is needed to elucidate the adsorption mechanisms. Specifically, this study examined the adsorption of AFB1 by different *A. auricula* tissue structures. To further understand the adsorption mechanisms, a more in-depth investigation of the binding sites for AFB1 adsorption by *A. auricula* is warranted. Additionally, in order to clarify the impact of *A. auricula* on AFB1 adsorption and metabolism, future research should incorporate *in vivo* and *in vitro* experiments to investigate the adsorption of AFB1 in the intestine, the formation of biomarkers and parent compounds within the organism.

Author Contributions: D.X.: designed the study, formal analysis, writing—original draft, wrote the manuscript.; M.H.: performed the experiment, writing—original draft, writing—review and editing; J.L.: performed the experiment; H.S. and L.H.: formal analysis, designed the study, writing—review and editing; H.M.: supervision. All authors have read and agreed to the published version of the manuscript.

Funding: The authors disclosed receipt of the following financial support for the research and publication of this article: This work was supported by the National Key Research and Development Program of China (Grant No. 2022YFD2100601, D.X.) and Innovation Capability Support Program of Shaanxi Province (Program No. 2023-CX-TD-61).

Institutional Review Board Statement: The animal study protocol was approved by Xi'an Medical University Application for Laboratory Animal Welfare and Ethical Committee (XYLS2023005, 15 January 2023).

Data Availability Statement: The data used to support the findings of this study can be made available by the corresponding author upon request.

Conflicts of Interest: The authors declare no conflict of interest.

References

1. Rodrigues, P.; Venâncio, A.; Lima, N. Aflatoxigenic fungi and aflatoxins in portuguese almonds. *Sci. World J.* **2012**, *2012*, 471926. [CrossRef]
2. Bennett, J.W.; Klich, M. Mycotoxins. *Clin. Microbiol. Rev.* **2003**, *16*, 497–516. [CrossRef] [PubMed]
3. Kew, M.C. Aflatoxins as a cause of hepatocellular carcinoma. *J. Gastrointestin. Liver Dis.* **2013**, *22*, 305–310. [PubMed]
4. Santos, V.O.; Pelegrini, P.B.; Mulinari, F.; Lacerda, A.F.; Moura, R.S.; Cardoso, L.P.V.; Buhrer-Sekula, S.; Miller, R.N.G.; Grossi-de-Sa, M.F. Development and validation of a novel lateral flow immunoassay device for detection of aflatoxins in soy-based foods. *Anal. Methods* **2017**, *9*, 2715–2722. [CrossRef]
5. Massey, T.E.; Smith, G.B.; Tam, A.S. Mechanisms of aflatoxin B1 lung tumorigenesis. *Exp. Lung Res.* **2000**, *26*, 673–683.
6. Autrup, H.; Jorgensen, E.C.; Jensen, O. Aflatoxin B1 induced lacI mutation in liver and kidney of transgenic mice C57BL/6N: Effect of phorone. *Mutagenesis* **1996**, *11*, 69–73. [CrossRef]
7. Moon, E.Y.; Rhee, D.K.; Pyo, S. Inhibition of various functions in murine peritoneal macrophages by aflatoxin B1 exposure in vivo. *Int. J. Immunopharmacol.* **1999**, *21*, 47–58. [CrossRef]
8. Raisuddin, S.; Singh, K.; Zaidi, S.; Paul, B.; Ray, P. Immunosuppressive effects of aflatoxin in growing rats. *Mycopathologia* **1993**, *124*, 189–194. [CrossRef]
9. Hathout, A.S.; Aly, S.E. Biological detoxification of mycotoxins: A review. *Ann. Microbiol.* **2014**, *64*, 905–919. [CrossRef]
10. Do, J.H.; Choi, D.-K. Aflatoxins: Detection, toxicity, and biosynthesis. *Biotechnol. Bioprocess Eng.* **2007**, *12*, 585–593. [CrossRef]
11. Sara, D.; Watanya, J.; Flavia, G.; Consiglia, L.; Carlo, N.; Emanuela, A.; Chiara, L.; Sihem, D.; Giuseppina, A.; Achille, S.; et al. Curcumin supplementation protects broiler chickens against the renal oxidative stress induced by the dietary exposure to low levels of aflatoxin B1. *Front. Vet. Sci.* **2021**, *8*, 822227.
12. Owumi, S.E.; Kazeem, A.I.; Wu, B.C.; Ishokare, L.O.; Arunsi, U.O.; Oyelere, A.K. Apigeninidin-rich *Sorghum bicolor* (L. Moench) extracts suppress A549 cells proliferation and ameliorate toxicity of aflatoxin B1-mediated liver and kidney derangement in rats. *Sci. Rep.* **2022**, *12*, 7438. [CrossRef] [PubMed]
13. Guyonnet, D.; Belloir, C.; Suschetet, M.; Siess, M.-H.; Le Bon, A.-M. Mechanisms of protection against aflatoxin B1 genotoxicity in rats treated by organosulfur compounds from garlic. *Carcinogenesis* **2002**, *23*, 1335–1341. [CrossRef] [PubMed]
14. Eaton, D.L.; Gallagher, E.P. Mechanisms of aflatoxin carcinogenesis. *Annu. Rev. Pharmacol. Toxicol.* **1994**, *34*, 135–172. [CrossRef] [PubMed]
15. Iyer, R.S.; Coles, B.F.; Raney, K.D.; Thier, R.; Guengerich, F.P.; Harris, T.M. DNA adduction by the potent carcinogen aflatoxin B1 mechanistic studies. *J. Am. Chem. Soc.* **1994**, *116*, 1603–1609. [CrossRef]
16. Karlovsky, P.; Suman, M.; Berthiller, F.; De Meester, J.; Eisenbrand, G.; Perrin, I.; Oswald, I.P.; Speijers, G.; Chiodini, A.; Recker, T.; et al. Impact of food processing and detoxification treatments on mycotoxin contamination. *Mycotoxin Res.* **2016**, *32*, 179–205. [CrossRef]
17. Zhou, J.; Tang, L.; Wang, J.; Wang, J.S. Aflatoxin B1 disrupts gut-microbial metabolisms of short-chain fatty acids, long-chain fatty acids, and bile acids in male F344 rats. *Toxicol. Sci.* **2018**, *164*, 453–464. [CrossRef]
18. Zychowski, K.E.; Hoffmann, A.R.; Ly, H.J.; Pohlenz, C.; Buentello, A.; Romoser, A.; Gatlin, D.M.; Phillips, T.D. The effect of aflatoxin B1 on red drum (*Sciaenops ocellatus*) and assessment of dietary supplementation of NovaSil for the prevention of aflatoxicosis. *Toxins* **2013**, *5*, 1555–1573. [CrossRef]
19. Brana, M.T.; Cimmarusti, M.T.; Haidukowski, M.; Logrieco, A.F.; Altomare, C. Bioremediation of aflatoxin B1-contaminated maize by king oyster mushroom (*Pleurotus eryngii*). *PLoS ONE* **2017**, *12*, e0182574. [CrossRef]
20. El-Shiekh, H.H.; Mahdy, H.M.; El-Aaser, M.M. Bioremediation of aflatoxins by some reference fungal strains. *Pol. J. Microbiol.* **2007**, *56*, 215–223.
21. Farzaneh, M.; Shi, Z.-Q.; Ghassempour, A.; Sedaghat, N.; Ahmadzadeh, M.; Mirabolfathy, M.; Javan-Nikkhah, M. Aflatoxin B1 degradation by *Bacillus subtilis* UTBSP1 isolated from pistachio nuts of Iran. *Food Control* **2012**, *23*, 100–106. [CrossRef]
22. Gao, X.; Ma, Q.; Zhao, L.; Lei, Y.; Shan, Y.; Ji, C. Isolation of *Bacillus subtilis*: Screening for aflatoxins B-1, M-1, and G1 detoxification. *Eur. Food Res. Technol.* **2011**, *232*, 957–962. [CrossRef]
23. Schmidt, M.A.; Mao, Y.Z.; Opoku, J.; Meh, H.L. Enzymatic degradation is an effective means to reduce aflatoxin contamination in maize. *BMC Biotechnol.* **2021**, *21*, 70. [CrossRef] [PubMed]
24. Ramy, S.Y. Aflatoxin detoxification by manganese peroxidase purified from *Pleurotus ostreatus*. *Braz. J. Microbiol.* **2014**, *47*, 127–133.
25. Ali, H.S.; Badr, A.N.; Alsulami, T.; Shehata, M.G.; Youssef, M.M. Quality attributes of Sesame Butter (Tahini) fortified with lyophilized powder of edible mushroom (*Agaricus blazei*). *Foods* **2022**, *11*, 3691. [CrossRef]

26. Choo, M.J.; Hong, S.Y.; Chuang, S.H.; Om, A.S. Removal of aflatoxin B1 by edible mushroom-forming fungi and its mechanism. *Toxins* **2021**, *13*, 668. [CrossRef]
27. Igbiri, S.; Udowelle, N.A.; Ekhator, O.C.; Asomugha, R.N.; Igweze, Z.N.; Orisakwe, O.E. Edible mushrooms from niger delta, nigeria with heavy metal levels of public health concern: A human health risk assessment. *Recent Pat. Food Nutr. Agric.* **2018**, *9*, 31–41. [CrossRef]
28. Dai, Y.J.; Li, J.J.; Shan, D.X. Adsorption of tetracycline in aqueous solution by biochar derived from waste *Auricularia auricula* dregs. *Chemosphere* **2020**, *238*, 124432. [CrossRef]
29. Su, L.; Zhang, H.B.; Oh, K.; Liu, N.; Luo, Y.; Cheng, H.Y.; Zhang, G.S.; He, X.F. Activated biochar derived from spent *Auricularia auricula* substrate for the efficient adsorption of cationic azo dyes from single and binary adsorptive systems. *Water Sci. Technol.* **2021**, *84*, 101–121. [CrossRef]
30. Zhang, Y.W.; Li, X.P.; Yang, Q.H.; Zhang, C.; Song, X.L.; Wang, W.H.; Jia, L.; Zhang, J.J. Antioxidation, anti-hyperlipidaemia and hepatoprotection of polysaccharides from *Auricularia auricular* residue. *Chem. Biol. Interact.* **2021**, *335*, 109323. [CrossRef]
31. Xing, X.B.; Chitrakar, B.; Hati, S.; Xie, S.Y.; Li, H.B.; Li, C.T.; Liu, Z.B.; Mo, H.Z. Development of black fungus-based 3D printed foods as dysphagia diet: Effect of gums incorporation. *Food Hydrocoll.* **2022**, *123*, 107173. [CrossRef]
32. Xu, D.; Peng, S.R.; Guo, R.; Yao, L.S.; Mo, H.Z.; Li, H.B.; Song, H.X.; Hu, L.B. EGCG alleviates oxidative stress and inhibits aflatoxin B1 biosynthesis via MAPK signaling pathway. *Toxins* **2021**, *13*, 693. [CrossRef] [PubMed]
33. Wang, H.; Muhammad, I.; Li, W.; Sun, X.Q.; Cheng, P.; Zhang, X.Y. Sensitivity of Arbor Acres broilers and chemoprevention of aflatoxin B1-induced liver injury by curcumin, a natural potent inducer of phase-II enzymes and Nrf2. *Environ. Toxicol. Phar.* **2018**, *59*, 94–104. [CrossRef]
34. Ragoubi, C.; Quintieri, L.; Greco, D.; Mehrez, A.; Maatouk, I.; D'Ascanio, V.; Landoulsi, A.; Avantaggiato, G. Mycotoxin removal by *Lactobacillus* spp. and their application in animal liquid feed. *Toxins* **2021**, *13*, 185. [CrossRef]
35. Shetty, P.H.; Jespersen, L. *Saccharomyces cerevisiae* and lactic acid bacteria as potential mycotoxin decontaminating agents. *Trends Food Sci. Tech.* **2006**, *17*, 48–55. [CrossRef]
36. El-Nezami, H.; Kankaanpaa, P.; Salminen, S.; Ahokas, J. Ability of dairy strains of lactic acid bacteria to bind a common food carcinogen, aflatoxin B1. *Food Chem. Toxicol.* **1998**, *36*, 321–326. [CrossRef] [PubMed]
37. Rushing, B.R.; Selim, M.I. Aflatoxin B1: A review on metabolism, toxicity, occurrence in food, occupational exposure, and detoxification methods. *Food Chem. Toxicol.* **2019**, *124*, 81–100. [CrossRef]
38. Chen, H.; Wang, C.; Qi, M.; Ge, L.; Tian, Z.; Li, J.; Zhang, M.; Wang, M.; Huang, L.; Tang, X. Anti-tumor effect of *Rhaponticum uniflorum* ethyl acetate extract by regulation of peroxiredoxin1 and epithelial-to-mesenchymal transition in oral cancer. *Front. Pharmacol.* **2017**, *8*, 870. [CrossRef]
39. Wong, J.H.; Ng, T.B.; Chan, H.H.L.; Liu, Q.; Man, G.C.W.; Zhang, C.Z.; Guan, S.; Ng, C.C.W.; Fang, E.F.; Wang, H.; et al. Mushroom extracts and compounds with suppressive action on breast cancer: Evidence from studies using cultured cancer cells, tumor-bearing animals, and clinical trials. *Appl. Microbiol. Biotechnol.* **2020**, *104*, 4675–4703. [CrossRef]
40. Kang, M.A.; Jeon, Y.K.; Nam, M.J. *Auricularia auricula* increases an apoptosis in human hepatocellular carcinoma cells via a regulation of the peroxiredoxin1. *J. Food Biochem.* **2020**, *44*, e13373. [CrossRef]
41. Alberts, J.F.; Engelbrecht, Y.; Steyn, P.S.; Holzapfel, W.H.; van Zyl, W.H. Biological degradation of aflatoxin B1 by *Rhodococcus erythropolis* cultures. *Int. J. Food Microbiol.* **2006**, *109*, 121–126. [CrossRef] [PubMed]
42. Fan, Y.; Zhao, L.; Ma, Q.; Li, X.; Shi, H.; Zhou, T.; Zhang, J.; Ji, C. Effects of *Bacillus subtilis* ANSB060 on growth performance, meat quality and aflatoxin residues in broilers fed moldy peanut meal naturally contaminated with aflatoxins. *Food Chem. Toxicol.* **2013**, *59*, 748–753. [CrossRef] [PubMed]
43. Rybak, K.; Wiktor, A.; Witrowa-Rajchert, D.; Parniakov, O.; Nowacka, M. The quality of red bell pepper subjected to freeze-drying preceded by traditional and novel pretreatment. *Foods* **2021**, *10*, 226. [CrossRef] [PubMed]

foods MDPI

Article

Inhibition of *Salmonella* Enteritidis by Essential Oil Components and the Effect of Storage on the Quality of Chicken

Wu Wang *, Tingting Li, Jing Chen and Yingwang Ye

School of Food and Biological Engineering, Hefei University of Technology, Hefei 230009, China; 2021171473@mail.hfut.edu.cn (T.L.); chenjing_hfut@163.com (J.C.); yeyingwang04@126.com (Y.Y.)
* Correspondence: 1994800074@hfut.edu.cn

Abstract: This research investigates the antibacterial potential of plant essential oil components including thymol, carvacrol, citral, cinnamaldehyde, limonene, and β-pinene against *Salmonella* Enteritidis (*S.* Enteritidis). Through the determination of minimum inhibitory concentration, three kinds of natural antibacterial agents with the best inhibitory effect on *S.* Enteritidis were determined, namely thymol (128 μg/mL), carvacrol (256 μg/mL), and cinnamaldehyde (128 μg/mL). Physical, chemical, microbial, and sensory characteristics were regularly monitored on days 0, 2, 4, and 6. The findings of this study reveal that both thymol at MIC of 128 μg/mL and carvacrol at MIC of 256 μg/mL not only maintained the sensory quality of chicken, but also decreased the pH, moisture content, and TVB-N value. Additionally, thymol, carvacrol and cinnamaldehyde successfully inhibited the formation of *S.* Enteritidis biofilm, thereby minimizing the number of *S.* Enteritidis and the total aerobic plate count in chicken. Hence, thymol, carvacrol, and cinnamaldehyde have more effective inhibitory activities against *S.* Enteritidis, which can effectively prevent the spoilage of chicken and reduce the loss of its functional components.

Keywords: *Salmonella* Enteritidis; thymol; carvacrol; cinnamaldehyde

Citation: Wang, W.; Li, T.; Chen, J.; Ye, Y. Inhibition of *Salmonella* Enteritidis by Essential Oil Components and the Effect of Storage on the Quality of Chicken. *Foods* **2023**, *12*, 2560. https://doi.org/10.3390/foods12132560

Academic Editors: Min Zhang and Hong-Wei Xiao

Received: 31 May 2023
Revised: 26 June 2023
Accepted: 26 June 2023
Published: 30 June 2023

1. Introduction

In the course of the process of slaughter, circulation, sale, and processing, raw fresh meat is susceptible to contamination by *S.* Enteritidis, which can cause bacterial infections and even lead to food poisoning [1,2]. The control and elimination of *S.* Enteritidis in retail meat products poses a significant challenge. Tetracycline and sulfanilamide antibiotics are currently widely employed in poultry through drinking water and feed. However, many drug-resistant phenotypes of *Salmonella* isolates from food animals have been identified, and these are significantly higher than those found in human clinical isolates [3,4]. Consequently, the proficient surveillance of *S.* Enteritidis stands as an indispensable component in the deterrence and management of alimentary illnesses, thereby holding considerable import in the preservation of human well-being.

Plant essential oils (Eos) are considered viable alternatives to traditional antibacterial agents. The use of plant Eos as antibacterial agents not only implies greater safety for humans and a more environmentally friendly option due to their natural origin, but also represents a low risk of pathogenic microorganisms developing resistance. Thymol, carvacrol, and cinnamaldehyde have drawn considerable attention among researchers for their antibacterial potential [5,6]. Thymol, carvacrol, and cinnamaldehyde have broad-spectrum bacteriostatic effects. In the research of Qingliu Wei et al., a box-Behnken design model was selected, and 2.40 μL/mL cinnamon EO, 1.00 μL/mL cinnamon EO, and 3.50 μL/mL tea tree EO were optimized, which could effectively reduce the total aerobic plate count [7]. The antibacterial mechanism of cinnamon EO is achieved by exerting an impact on the respiratory metabolism, energy metabolism, and substance metabolism of

S. Enteritidis, thereby resulting in a decrease in bacterial growth [8]. In the exploration of the impact of EOs on the adhesion of *Salmonella* strains and HT-29 cells, Alibi noted that six types of essential oils exhibited varying degrees of inhibitory effects on biofilm formation. Cinnamon oil demonstrated a remarkable inhibition rate of 99.10%, followed by clove oil with 97.64%, thyme oil with 95.90%, rosemary oil with 79.84%, turmeric oil with 28.98%, and sage oil with 15.55% [9]. The synergistic effect of different essential oils can change the permeability of the cell membrane, the leakage of dissolved substances in cells, the denaturation of membrane proteins, and the destruction of proton kinetics, thus achieving the antibacterial effect [10–13]. In addition, the use of natural preservatives and antimicrobial agents in polymer matrices has become a new trend. Peng et al. prepared antimicrobial packaging films from agar/konjac glucomannan (KA) and carvacrol (CV) and showed that the composite film containing 2% carvacrol was the most effective in preserving freshness and could extend the shelf life of frozen chicken breasts from 5 to 9 days [14]. Lian has arrived at the conclusion that thyme EO in chitosan composite film exhibits the highest release rate in a 50% ethanol solution. As the concentration of ethanol increases, the release rate of the essential oil from the composite film slows down, while the antibacterial activity of the film improves [15]. Meanwhile, researchers have been exploring nanoencapsulation technology to address the issues posed by lipid-soluble active ingredients that are insoluble in water or that are air-unstable or that have a pungent odor [16]. Although the incorporation of thymol crystals into zeolite networks has reduced the volatility of thymol, the strong aroma of the essential oil itself may alter the flavor of food, posing potential obstacles to certain applications [17]. Sodium alginate-based cinnamaldehyde controlled-release active packaging films loaded with halloysite nanotubes have broad prospects for application in high-fat food packaging [18]. In addition, the biodegradability of composite films, preparation costs, and potential risks associated with their use in food warrant further consideration and resolution by researchers [19].

Chicken is considered to be a major source of human *Salmonella* infection. Researchers have successfully isolated bacteriophages resistant to various *Salmonella* serotypes from raw chicken skin and claws [20]. Microbial infections not only result in the loss of nutrients and functional components in chicken meat, but also produce toxins that pose a threat to human health [21]. To effectively prevent and control *S.* Enteritidis in chicken meat, the use of safe and natural antimicrobial agents such as thymol, carvacrol, and cinnamaldehyde is of utmost importance [22]. Extensive research has been conducted on the antibacterial properties and inhibitory mechanisms of using plant essential oils as a method to suppress *Salmonella* in animal feed. However, there is limited research on the application of EOs for antibacterial and preservation purposes in fresh meat products [23–25]. The ultimate purpose of this study is to investigate the inhibitory effects of thymol, carvacrol, and cinnamaldehyde on *S.* Enteritidis, as well as their potential to improve the quality of chicken meat. Therefore, this research provides sufficient theoretical and practical basis for addressing the antibacterial and preservation issues of chicken.

2. Materials and Methods

2.1. Materials

Based on our preliminary research, an *S.* Enteritidis was screened from retail meat in the Hefei area [26]. Thymol (>99.0% purity), carvacrol (99% purity), limonene (>95.0% purity), citral (97% purity), cinnamaldehyde (≥95% purity), and β-pinene (purity ≥ 95%) were purchased from Shanghai Aladdin Biochemical Technology Co., Ltd. (Shanghai, China).

Buffered peptone water (BPW), bismuth sulfite agar (BSA), brilliant green sulfa agar (BGSA), Mueller-Hinton broth (MH), plate count agar (PCA), and tryptic soy broth (TSB) were purchased from Qingdao Haibo Biotechnology Co., Ltd. (Qingdao, China).

2.2. Minimum Inhibitory Concentration(MIC) Determinations

Initially, antimicrobial agents (such as thymol, carvacrol, limonene, citral, cinnamaldehyde, and β-pinene) at different concentrations (4096, 2048, 1024, 512, 256, 128, 64, 32,

16, 8, 4, 2, and1 μg/mL, respectively) were added into sterile 96-well polystyrene plates. Antimicrobial agents were added to wells 1–13, 10 μL per well, and well 14 was utilized as a growth control without the addition of antimicrobial agents. Subsequently, bacterial suspensions with turbidity equivalent to 0.5 McFarland standard were prepared using the direct bacterial suspension method and diluted 1:1000 in Mueller-Hinton broth. Then, 100 μL suspensions of *S.* Enteritidis were added to each well, each well was sealed, and each well was incubated for 48 h in ambient air at 37 °C to determine the results. The MIC is defined as the concentration where no visible bacterial growth is observed after 48 h of incubation at 37 °C under microaerophilic conditions [27].

2.3. Sample Preparation

The chicken used in this study was obtained from the Yong-hui Supermarket in Hefei. Three essential oil components with the smallest MIC were selected, and thymol at 128 μg/mL, carvacrol at 256 μg/mL, and cinnamaldehyde at 128 μg/mL were uniformly coated on fresh chicken, while the control group was coated with anhydrous ethanol. The samples were stored at 4 °C, and samples were taken on days 0, 2, 4, and 6, respectively. Under sterile conditions, four groups of samples (25–30 g each, treated with thymol, carvacrol, and cinnamaldehyde) were immersed in *S.* Enteritidis suspension with a concentration of 10^7 CFU/g for 15 s, drained, and sampled on days 0, 2, 4, and 6.

2.4. Microbial Indicators Determinations

In the ultraclean workbench, 10 g of ground chicken was mixed with 9 mL of sterilized saline solution to create a 1:10 diluted sample. After shaking for 1 min, a series of 10-fold dilutions were performed and three appropriate dilutions were plated and incubated at 37 °C for 48 h to determine the total aerobic plate count [28].

Using the threefold dilution and most probable number (MPN) method, the amount of *S.* Enteritidis in chicken samples was determined. Specifically, 10 g of chicken meat was placed into a sterile homogenization bag containing 90 mL of BPW and homogenized for 1 min, and then 100 uL of the mixture was spread on BSA and BGSA plates, which were incubated at 37 °C for 24 h to analyze the presence of Salmonella colonies [29]. Following the procedure described in the Health Canada "Method Summary" (MFHPB-20), putative Salmonella colonies were isolated and subjected to biochemical and serological tests to confirm the identification of *S.* Enteritidis [30].

2.5. Antimicrobial Effects

A total of 20 μL of bacterial liquid (10^7 CFU/mL) of *S.* Enteritidis isolates was added into a sterile 96-well plate, and 180 μL of thymol, carvacrol, and cinnamaldehyde were added to make their final concentrations 1/4 MIC, 1/2 MIC, and MIC respectively. Six parallel wells were provided for each strain, and 200 μL of TSB broth containing no bacteria was added to the negative control well, and then it was allowed to stand at 37 °C for 12 h. Crystal violet assay was performed to evaluate biofilm formation, and the optic density at 590 nm was measured. The average value is the result of at least three tests.

The ability of bacterial strain biofilm formation is classified into four categories [31]:

$OD \leq OD_C$ is a nonbiofilm former,

$OD_C < OD \leq 2OD_C$ is a weak biofilm former,

$2OD_C < OD \leq 4OD_C$ is a moderate biofilm former,

$4OD_C < OD$ is a strong biofilm former.

Among them, the cutoff optical density value (OD_C) is defined as three standard deviations higher than the average or average OD_{AV} of the negative control.

2.6. Physical and Chemical Indexes Determinations

2.6.1. pH

By using a Five Easy Plus pH meter (Shanghai, China, Mettler Toledo Instrument Co., Ltd.), the electrode of the pH meter was directly placed in the meat tissue to determine the pH value [32]. The equipment was calibrated before measurement, and the experiment was repeated three times for each sample.

2.6.2. Moisture Content

The moisture content of chicken was determined by the AOAC (2009) standard method.

2.6.3. TVB-N

Referring to the method of Zhang et al., 10 g minced sample was added to 100 mL distilled water, homogenized for 1 min, and determined by automatic Kjeldahl nitrogen analyzer [33].

2.6.4. Color

After the sample was ground, it was spread all over the bottom of the vessel, and immediately measured by a spectrophotometer. The spectrophotometer was preheated for 15 min before use and used after calibration, and recorded as brightness (L*), redness (a*), and yellowness (b*).

2.7. Sensory Evaluation

Six people with sensory evaluation experience formed the evaluation group, including three boys and three girls. Members of the evaluation team independently evaluated the color, smell, viscosity, and overall acceptability of each sample, and could not communicate with each other. All samples were graded according to the grading criteria in Table 1.

Table 1. Standards of sensory evaluation for chicken.

Indicators	Score	Scoring Criteria
Color	8–10 points 6–7 points 6 points or fewer	Light red, glossy surface Slightly dim, slightly shiny surface Dull, surface matte
Odor	8–10 points 6–7 points 6 points or fewer	Normal smell of chicken, no peculiar smell Slightly smelly Fishy odor or ammonia odor
Stickiness	8–10 points 6–7 points 6 points or fewer	Moist, not sticky Wetter, not sticky Sticky hands
Overall acceptability	8–10 points 6–7 points 6 points or fewer	Willing to accept Acceptable Unacceptable

2.8. Statistical Analysis

The data were analyzed using one-way analysis of variance (ANOVA) and Duncan's multiple range test in SPSS 14.0 software. Statistical significance was considered when $p < 0.05$. Additionally, data visualization graphs were created using Origin 2023 software.

3. Results

3.1. MIC

The compounds, namely thymol, carvacrol, limonene, citral, cinnamaldehyde, and β-pinene, exhibited a certain degree of inhibitory effect on *S.* Enteritidis. Table 2 displays the MIC values of six essential oil constituents against *S.* Enteritidis, with thymol, carvacrol, and cinnamaldehyde exhibiting MIC values of 128, 256, and 128 μg/mL, respectively, representing the three constituents with the smallest MIC values. Generally, a smaller MIC value indicates greater inhibition of bacterial growth, particularly in relation to *S.* Enteritidis. It is worth noting that EOs mainly composed of aldehydes and phenols have higher antibacterial activity, such as cinnamaldehyde, citral, carvacrol, and thymol, followed by EOs containing terpenoid alcohols [34]. Thymol and carvacrol have been found to disrupt the bacterial biofilm, causing the breakdown of the cell membrane and leakage of intracellular contents, ultimately resulting in cell death [12]. The aldehyde groups contained in citral and cinnamaldehyde can easily penetrate the cell wall and destroy the structure of polysaccharides. In addition, the aldehyde groups can also act on protein transporters, playing a role of sterilization [35,36]. Therefore, thymol, carvacrol, and cinnamaldehyde with low MIC were chosen for the next experiment.

Table 2. Minimum inhibitory concentrations.

Essential Oil Main Ingredients	Thymol	Carvacrol	Limonene	Citral	Cinnamaldehyde	β-Pinene
MIC (μg/mL)	128	256	1024	512	128	2048

3.2. *Inhibitory Effect on S. Enteritidis Biofilm*

From Figure 1, it can be observed that thymol, carvacrol, and cinnamaldehyde all have significant inhibitory effects on the formation of *S.* Enteritidis biofilms. As shown in Figure 1A, the OD value of the thymol-treated group at MIC significantly began to decrease compared to the control group and 1/4 MIC ($p < 0.05$) after 12 h of cultivation (early stage of biofilm formation). Therefore, it can be concluded that high concentrations of thymol have an inhibitory effect on *S.* Enteritidis biofilm formation at the early stage. At 24 h (basic biofilm formation) and 48 h (mature biofilm formation), the inhibitory effect of thymol on *S.* Enteritidis biofilms became more pronounced with increasing concentrations, as the OD values decreased with increasing concentrations, indicating that the ability of thymol to inhibit *S.* Enteritidis biofilms increased with increasing concentration. As shown in Figure 1B, carvacrol at MIC had a significant inhibitory effect on *S.* Enteritidis biofilm formation throughout the entire process ($p < 0.05$) compared to the control group. It is worth noting that the OD values of 1/2 MIC and 1/4 MIC were significantly higher than those of the control group at 12 h and 24 h, indicating that low concentrations of carvacrol had a promoting effect on *S.* Enteritidis biofilm formation at the early and basic formation stages. The OD values of 1/2 MIC and MIC were significantly lower than those of the control group at 12 h and 24 h, indicating that high concentrations of carvacrol had a strong inhibitory effect on *S.* Enteritidis biofilm formation at the early and mature stages. As shown in Figure 1C, the cinnamaldehyde-treated group had a significant inhibitory effect on *S.* Enteritidis biofilms at each concentration after 24 h and 48 h ($p < 0.05$). Using lower concentrations of EO will increase the permeability of the cell membrane and eventually lead to the destruction of cell membrane structures. Nevertheless, lower concentrations of EOs do not impact the viability of the cells. Conversely, the utilization of higher concentrations of EOs leads to extensive membrane damage and complete disruption of homeostasis, which ultimately results in the death of cells. Carvacrol and thymol can prevent or interfere with biofilm formation, and they can also eradicate preformed biofilms [37,38].

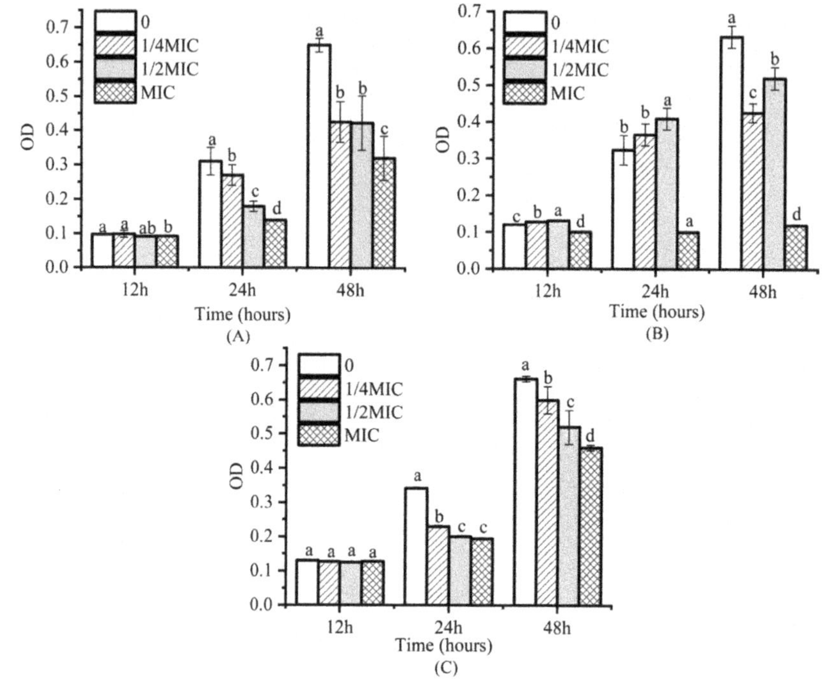

Figure 1. Inhibitory effects of thymol (**A**), carvacrol (**B**), and cinnamaldehyde (**C**) on biofilm formation in Salmonella. The different superscripts (a–d) indicate a statistically significant difference ($p < 0.05$) between different treatment groups within the same storage time.

3.3. Changes in the Total Aerobic Plate Count of Chicken during Storage

During the storage of chicken, with the increase of days, the total aerobic plate count in each groups increased significantly ($p < 0.05$), as shown in Figure 2. Among them, the total aerobic plate count in the thymol treatment group and the carvacrol treatment group was significantly lower than those in the other two groups ($p < 0.05$), which indicated that thymol and carvacrol had obvious inhibitory effects on microbial reproduction in chicken during storage. However, the antibacterial effect of the cinnamaldehyde treatment group is not ideal. There was no significant difference between the two groups on the second and fourth days of storage ($p > 0.05$).

Similar to many other antibacterial agents and food preservation techniques, the effectiveness of cinnamaldehyde in food systems has been significantly reduced. This may be attributed to the inherent properties of chicken, such as ion strength and water activity, as well as differences in composition, including protein and fat content. These factors can affect the distribution of antibacterial agents within cells, thereby further diminishing their bactericidal activity [39]. On the fourth day of storage, the total aerobic plate count in the control group reached 6.22 Lg CFU/g, which exceeded the national fresh meat standard. Thymol, carvacrol, and cinnamaldehyde were all within the standard range of fresh meat, and total aerobic plate counts were 4.98 Lg CFU/g, 4.99 Lg CFU/g, and 5.98 Lg CFU/g, respectively. Meat is generally considered to be spoiled when the total aerobic plate count in meat products exceeds 6.0 Lg CFU/g [40]. On the sixth day of storage, all the treatment groups exceeded the fresh meat standard, indicating that the chicken in the control group could be stored for about two days, while thymol, carvacrol, and cinnamaldehyde had obvious antibacterial effects, which could prolong the storage time to about four days.

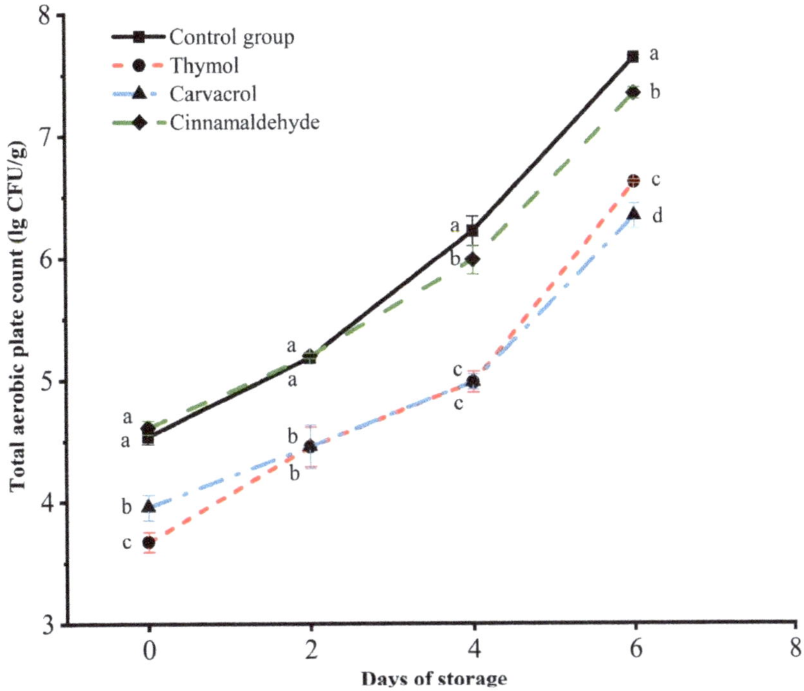

Figure 2. Effects of thymol, carvacrol, and cinnamaldehyde on total aerobic plate counts of chicken at storage. The different superscripts (a–d) indicate statistically significant differences ($p < 0.05$) between different treatment groups within the same storage time.

3.4. Changes in S. Enteritidis Count in Chicken Meat during Storage

In Figure 3, a comparison is made between the antibacterial effects of thymol, carvacrol, and cinnamaldehyde against S. Enteritidis during the storage of chicken. At day 0, the number of S. Enteritidis in each group was 4.39 Lg CFU/g, and the concentration of S. Enteritidis suspension used in the experiment was 7 Lg CFU/g. This indicates that a certain loss of *Salmonella* suspension occurred during the experimental operation, resulting in a final concentration of S. Enteritidis suspension dispersed on the surface of the chicken of 4.39 Lg CFU/g [41]. On the second day of storage, the number of S. Enteritidis in thymol and carvacrol treatment groups decreased to 4.20 Lg CFU/g and 3.96 Lg CFU/g, respectively, showing good antibacterial effect, which also proved that thymol and carvacrol had the strongest antibacterial effect in the first two days. However, the number of S. Enteritidis in the cinnamaldehyde treatment group and the control group had no significant difference in the first four days. Therefore, cinnamaldehyde's ability to inhibit S. Enteritidis in chicken was weaker than thymol and carvacrol, which was confirmed in the research of Hoffman et al. [42] Furthermore, there are researchers who have proposed an order of antibacterial effectiveness of EOs as follows: thymol > oregano > carvacrol > trans-cinnamaldehyde > eugenol. Among these, thymol has demonstrated the strongest antibacterial effect against *Salmonella*, *Escherichia coli*, and *Klebsiella pneumoniae* [43]. With the extension of storage time, the number of *Salmonella* in each group increased rapidly. On one hand, due to the volatilization of thymol and carvacrol over time, the antibacterial effects of thymol and carvacrol are weakened. On the other hand, MIC thymol and carvacrol could not completely inhibit the growth of S. Enteritidis, but increased the lag time of S. Enteritidis. Although MIC concentrations of thymol and carvacrol in culture media were effective in inhibiting the growth of S. Enteritidis, the antimicrobial effect of applying MIC concentrations of thymol and carvacrol to chicken was diminished. Hence, the use of two or more complex EOs may need to be further considered in order to achieve similar antibacterial effects in food systems as in culture-based systems.

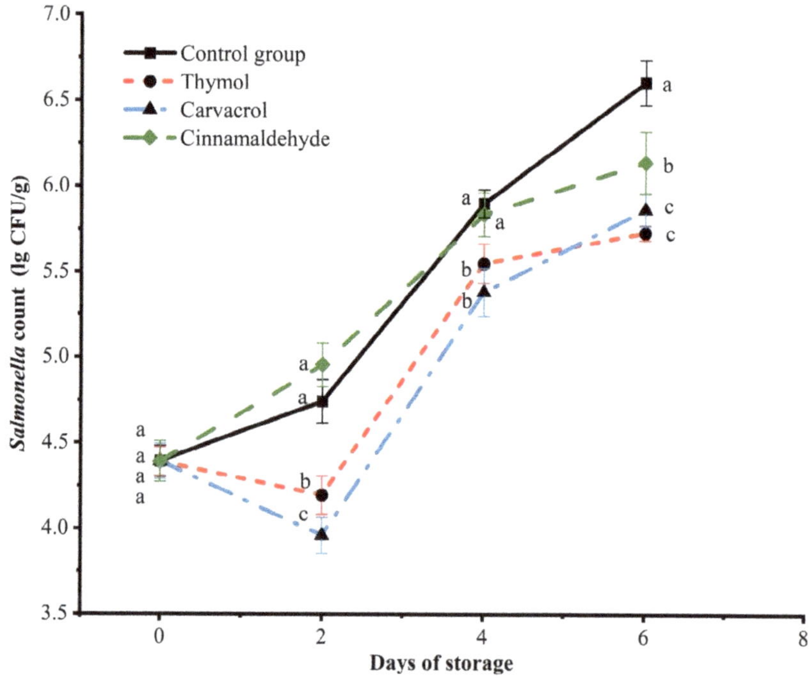

Figure 3. Effects of thymol, carvacrol, and cinnamaldehyde on *Salmonella* counts of chicken at storage. The different superscripts (a–c) indicate statistically significant differences ($p < 0.05$) between different treatment groups within the same storage time.

3.5. Changes in pH Values of Chicken during Storage

pH value is one of the most important factors affecting the quality of meat, which can affect the color, water retention capacity, flavor, tenderness, and shelf life of meat. A higher pH value is beneficial to the growth and reproduction of microorganisms, which means a short shelf life, while a lower pH value means a poor water retention ability [44]. The changes in the pH value of chicken during storage by thymol, carvacrol, and cinnamaldehyde are shown in Figure 4. The increase in microorganisms promotes the decomposition of protein in chicken meat, producing alkaline nitrogen-containing compounds such as amino acids, biogenic amines, and ammonia, which leads to aggravated meat spoilage and an increase in pH value. On the sixth day of storage, all groups experienced spoilage, indicating that chicken meat treated with thymol, carvacrol, and cinnamaldehyde had a shelf life of four days, delaying the spoilage of chicken meat. Compared with the control group, thymol, parsley phenol, and cinnamaldehyde can inhibit the growth and reproduction of microorganisms, reduce the pH value, and achieve a certain preservation effect.

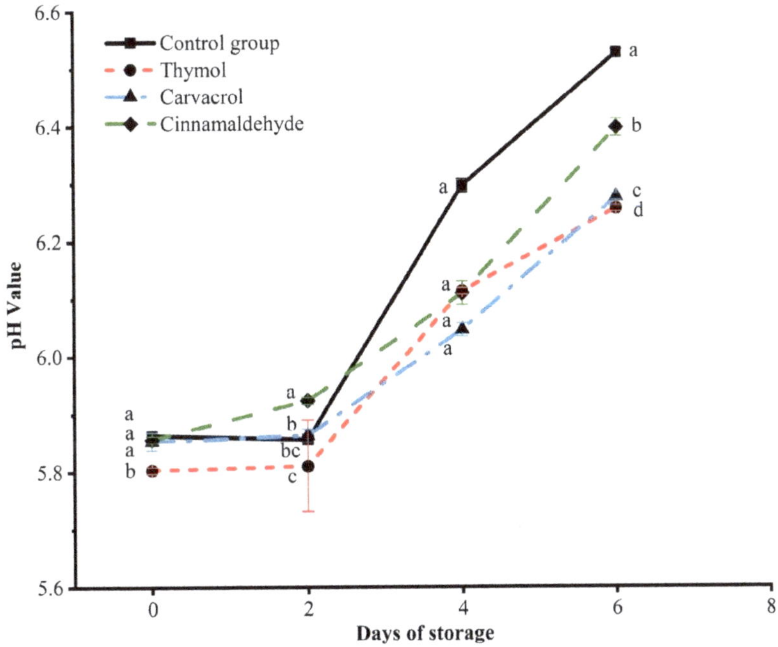

Figure 4. Effects of thymol, carvacrol, and cinnamaldehyde on the pH of chicken at storage. The different superscripts (a–d) indicate statistically significant differences ($p < 0.05$) between different treatment groups within the same storage time.

3.6. Changes of Moisture Content of Chicken during Storage

The moisture content of fresh meat is an important factor affecting sensory quality and consumer perception [45]. During storage, the water content of chicken changes smoothly, as shown in Table 3. In general, the water content of chicken in each treatment groups did not change significantly during storage ($p > 0.05$). The decrease in moisture content of the control group during the first two days of storage can be attributed to the stiff state of the chicken meat, its poor water-holding capacity, and the loss of some free water during the sample preparation process. High moisture content in meat can lead to rapid putrefaction after death, as microorganisms proliferate rapidly in meat with a water content exceeding 80% [46]. From the second day of storage, the moisture contents of the thymol, carvacrol, and cinnamaldehyde treatment groups were higher than that of the control group, indicating that these three EOs inhibited the growth of microorganisms, slowed down the decomposition of chicken protein and fat, and thereby reduced the increase in moisture content.

Table 3. Effects of thymol, carvacrol, and cinnamaldehyde on the moisture content of chicken at storage.

Parameter	Processing Groups	Days of Storage			
		0 d	2 d	4 d	6 d
Moisture content	0	0.75^{Aa}	0.74^{Bb}	0.74^{Bb}	0.74^{Bb}
	Thymol	0.75^{ABa}	0.75^{Bab}	0.76^{Aa}	0.75^{ABa}
	Carvacrol	0.75^{Aa}	0.75^{Aab}	0.76^{Aa}	0.76^{Aa}
	Cinnamaldehyde	0.74^{Ba}	0.75^{ABa}	0.75^{ABb}	0.76^{Aa}

Different superscript letters (a,b) indicate statistically significant differences ($p < 0.05$) between different treatment groups for the same storage time; different superscript letters (A,B) indicate statistically significant differences ($p < 0.05$) between different storage times for the same treatment group.

3.7. Changes in the TVB-N of Chicken during Storage

The TVB-N value is an important quality indicator to evaluate the freshness of meat. The degradation of proteins and other nitrogenous compounds leads to the accumulation of organic amines, which are often referred to as total volatile basic nitrogen (TVB-N) [47]. The changes in the TVB-N content of chicken meat during storage for thymol, carvacrol, and cinnamaldehyde are shown in Figure 5, where TVB-N values in all treatment groups showed an increasing trend with time, similar to the changes in pH and total aerobic plate count. TVB-N values were as high as 20.17 mg/100 g in the control group and 18.13 mg/100 g in the cinnamaldehyde-treated group at day four of storage, both exceeding the fresh meat standard ($\leq$15 mg/100 g), while thymol- and carvacrol-treated groups still did not exceed the standard, at 14.79 mg/100 g and 14.84 mg/100 g, respectively. The level of TVB-N in meat is related to the growth of microorganisms and the hydrolysis of protein into ammonia and amine. Chicken treated with thymol, carvacrol, and cinnamaldehyde leads to a decrease in the number of microorganisms responsible for protein degradation and the production of nonprotein nitrogen compounds including ammonia and amine. TVB-N significantly decreases compared to the control group. However, during storage, the antibacterial ability of thymol, carvacrol, and cinnamaldehyde gradually weakens, and the amplitude of TVB-N growth correspondingly increases. [48,49]. All treatment groups exceeded the standard on day six of storage, and the control group had a TVB-N value of 26.29 mg/100 g, which exceeded the standard for spoiled meat (>25 mg/100 g). The results indicate that thymol and carvacrol treatments can be effectively applied in chicken systems as natural bacterial inhibitors to extend the shelf life of chicken meat up to four days.

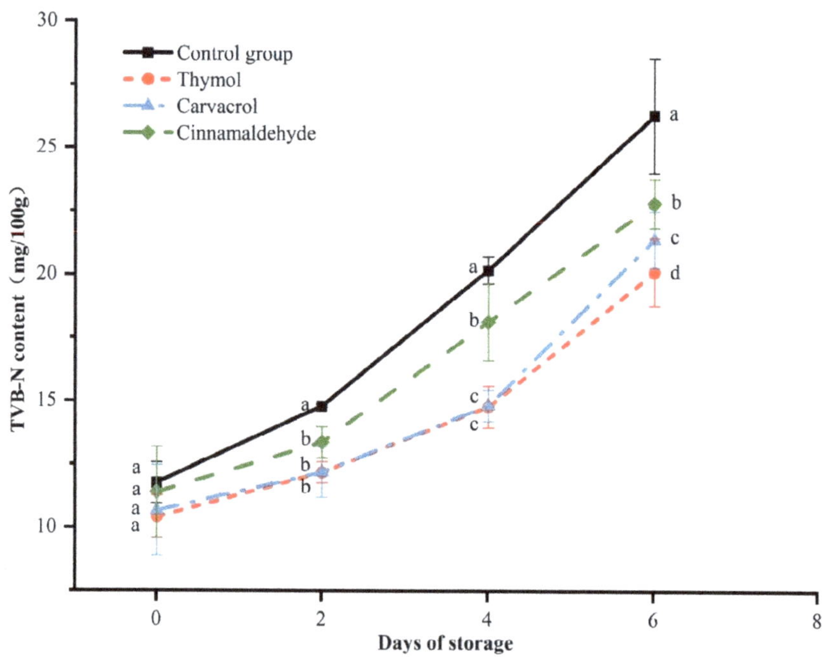

Figure 5. Effects of thymol, carvacrol, and cinnamaldehyde on the TVB-N of chicken at storage. The different superscripts (a–d) indicate statistically significant differences ($p < 0.05$) between different treatment groups within the same storage time.

3.8. Changes in Chicken Color during Storage

The color of meat is an important indicator for consumers to evaluate the freshness of meat and make a purchase decision accordingly [50]. The changes in L*, a*, and b* values on the surface of chicken during storage are shown in Table 4. The L* values of each treatment group showed a trend of increasing first and then decreasing. The increase in L* values may be due to the destruction of the protein structure caused by the enzymatic degradation of protein, which leads to more light scattering [51]. The proliferation of microorganisms accelerates the spoilage and deterioration of chicken, causing a decrease in brightness, a sticky and darkened surface, and reduced elasticity after storage for four days. In the middle storage period, the brightness of the chicken treated with thymol and carvacrol was significantly higher than that of the control group ($p < 0.05$), which may be because thymol and carvacrol inhibited the growth of microorganisms and reduced the formation of methemoglobin, indicating that thymol and carvacrol treatment can improve the brightness of chicken. With the extension of storage time, the a* value of each treatment group showed a downward trend, while the b* values showed an upward trend. From the second day of storage, the red degrees of the thymol and carvacrol treatment groups were higher than that of control group ($p < 0.05$), which slowed down the oxidation of myoglobin to methemoglobin. On the sixth day, the red value decreased rapidly, which promoted the production of methemoglobin, so the change trend of the b* value was opposite to that of the a* value.

Table 4. Effects of thymol, carvacrol, and cinnamaldehyde on color of chicken at storage.

Parameter	Processing Groups	Days of Storage			
		0 d	2 d	4 d	6 d
L* values	0	54.87 ^{Da}	56.37 ^{Cb}	57.40 ^{Bb}	55.27 ^{Ab}
	Thymol	54.61 ^{Da}	58.09 ^{Ca}	58.42 ^{Ba}	57.03 ^{Aa}
	Carvacrol	54.86 ^{Da}	57.93 ^{Ca}	58.34 ^{Ba}	56.21 ^{Aab}
	Cinnamaldehyde	54.69 ^{Da}	55.45 ^{Cc}	57.36 ^{Bb}	55.38 ^{Abc}
a* values	0	5.13 ^{Da}	4.55 ^{Cb}	4.10 ^{Bb}	3.49 ^{Ab}
	Thymol	5.12 ^{Da}	4.91 ^{Ca}	4.78 ^{Ba}	4.10 ^{Aa}
	Carvacrol	5.07 ^{Da}	4.99 ^{Ca}	4.73 ^{Ba}	3.74 ^{Ab}
	Cinnamaldehyde	5.14 ^{Da}	4.84 ^{Ca}	4.13 ^{Bb}	3.58 ^{Ab}
b* values	0	31.54 ^{Da}	32.71 ^{Ca}	34.50 ^{Ba}	35.93 ^{Aa}
	Thymol	31.14 ^{Dab}	32.12 ^{Cab}	33.11 ^{Bb}	34.60 ^{Ac}
	Carvacrol	30.92 ^{Db}	31.51 ^{Cb}	32.31 ^{Bc}	35.30 ^{Ab}
	Cinnamaldehyde	31.29 ^{Dab}	32.49 ^{Ca}	34.34 ^{Ba}	35.86 ^{Aa}

Different superscript letters (a–c) indicate statistically significant differences ($p < 0.05$) between different treatment groups for the same storage time; different superscript letters (A–D) indicate statistically significant differences ($p < 0.05$) between different storage times for the same treatment group.

3.9. Changes in the Sensory Indicators of Chicken during Storage

Table 5 illustrates the sensory changes in chicken meat during storage when treated with thymol, carvacrol, and cinnamaldehyde. Within the first two days of storage, there were no significant differences in color, odor, or overall acceptability ($p > 0.05$) among the different treatment groups. However, there was a significant difference in viscosity between the thymol group and the control group ($p < 0.05$), with only a slight stickiness observed on the surface of the chicken meat. On the fourth day of storage, there were significant differences in color, odor, viscosity, and overall acceptability ($p < 0.05$) between the thymol and carvacrol treatment groups and the control group, while only stickiness and overall acceptability ($p < 0.05$) showed significant differences between the cinnamaldehyde group and the control group, indicating a weaker preservation effect of cinnamaldehyde. The overall acceptability score of the control group was only 5.33, which was below the level that consumers would accept, with an odor score of 4, indicating a strong and unpleasant smell on the fourth day of storage, as well as a dull and sticky surface on the chicken meat. The overall acceptability score of the cinnamaldehyde group was also only 5.66, indicating a weak preservation effect, while the overall acceptability scores of the thymol and carvacrol treatment groups were both above 7, indicating that thymol and carvacrol could significantly inhibit the growth and reproduction of protein-degrading microorganisms. On the sixth day, the overall acceptability scores of all groups were around 3, indicating rapid microbial growth, which had penetrated into the chicken meat tissue and gradually decomposed to produce decay flavor substances such as dimethyl sulfide [49]. The antimicrobial properties of thymol, carvacrol, and cinnamaldehyde have been demonstrated to effectively minimize the microbial count, prolong the shelf life of products, and enhance the sensory quality of chicken.

Table 5. Effects of thymol, carvacrol, and cinnamaldehyde on the sensory parameters of chicken at storage.

Sensory Indicators	Processing Groups	0 d	2 d	4 d	6 d
Color	Control group	9.50 ± 0.84 ^{Aa}	8.67 ± 1.21 ^{Aa}	6.00 ± 0.89 ^{Bb}	4.33 ± 0.82 ^{Cb}
	Thymol	9.50 ± 0.84 ^{Aa}	9.17 ± 0.98 ^{Aa}	7.33 ± 0.82 $_{Ba}$	6.17 ± 0.75 ^{Ca}
	Carvacrol	9.33 ± 0.82 ^{Aa}	9.17 ± 0.95 ^{Aa}	7.83 ± 0.75 ^{Ba}	5.83 ± 0.75 ^{Ca}
	Cinnamaldehyde	9.33 ± 0.82 ^{Aa}	8.83 ± 0.98 ^{Aa}	6.83 ± 0.75 ^{Bab}	4.16 ± 0.75 ^{Cb}
Odor	Control group	9.67 ± 0.52 ^{Aa}	8.67 ± 1.03 ^{Aa}	4.00 ± 0.89 ^{Bb}	2.33 ± 0.82 ^{Cb}
	Thymol	9.83 ± 0.41 ^{Aa}	9.00 ± 0.89 ^{Aa}	6.50 ± 1.05 ^{Ba}	3.17 ± 0.75 ^{Cab}
	Carvacrol	9.67 ± 0.82 ^{Aa}	9.17 ± 0.75 ^{Aa}	5.67 ± 0.82 ^{Ba}	3.50 ± 0.55 ^{Ca}
	Cinnamaldehyde	9.50 ± 0.84 ^{Aa}	8.83 ± 0.75 ^{Aa}	3.83 ± 0.75 ^{Bb}	2.50 ± 0.84 ^{Cb}
Stickiness	Control group	9.83 ± 0.41 ^{Aa}	8.33 ± 0.52 ^{Bb}	6.83 ± 0.75 ^{Cb}	5.83 ± 0.75 ^{Da}
	Thymol	9.67 ± 0.52 ^{Aa}	9.50 ± 0.84 ^{Aa}	8.50 ± 1.05 ^{Ba}	6.83 ± 0.75 ^{Ca}
	Carvacrol	9.83 ± 0.41 ^{Aa}	9.33 ± 1.03 ^{ABab}	8.33 ± 1.37 ^{Ba}	6.67 ± 1.03 ^{Ca}
	Cinnamaldehyde	9.83 ± 0.41 ^{Aa}	9.17 ± 0.98 ^{ABab}	8.17 ± 0.98 ^{Ba}	6.50 ± 0.84 ^{Ca}
Overall acceptability	Control group	9.84 ± 0.41 ^{Aa}	8.33 ± 1.03 ^{Ba}	5.33 ± 0.82 ^{Cb}	2.83 ± 0.75 ^{Da}
	Thymol	9.67 ± 0.52 ^{Aa}	9.33 ± 0.82 ^{Aa}	7.17 ± 0.75 ^{Ba}	3.50 ± 1.05 ^{Ca}
	Carvacrol	9.67 ± 0.82 ^{Aa}	9.00 ± 0.89 ^{Aa}	7.00 ± 0.63 ^{Ba}	3.50 ± 0.84 ^{Ca}
	Cinnamaldehyde	9.67 ± 0.52 ^{Aa}	9.00 ± 0.89 ^{Aa}	5.67 ± 1.03 ^{Bb}	3.00 ± 1.26 ^{Ca}

Different superscript letters (a,b) indicate statistically significant differences ($p < 0.05$) between different treatment groups for the same storage time; different superscript letters (A–D) indicate statistically significant differences ($p < 0.05$) between different storage times for the same treatment group.

4. Conclusions

In this paper, the inhibitory effects of thymol, carvacrol, and cinnamaldehyde on *S.* Enteritidis during storage and on the quality characteristics of chicken meat were evaluated. The results of physicochemical analysis showed that MIC (128 μg/mL) thymol- and MIC (256 μg/mL) carvacrol-treated chicken surfaces minimized the number of *S.* Enteritidis and total aerobic plate count, reduced pH and TVB-N, and extended the shelf life from two to four days compared to MIC (128 μg/mL) cinnamaldehyde. The results of sensory analysis showed that the color, odor, viscosity, and overall acceptability of the chicken remained good in the thymol and carvacrol groups. This study confirmed that thymol and carvacrol were more effective in inhibiting *S.* Enteritidis and improving chicken quality. Based on current research findings, our future efforts will focus on the development of an antibacterial film that combines thymol, carvacrol, cinnamaldehyde, and polysaccharides, which will be applied to fresh food packaging.

Author Contributions: Conceptualization, T.L. and J.C.; methodology, T.L.; validation, T.L. and J.C.; formal analysis, J.C.; investigation, T.L.; resources, W.W.; data curation, T.L.; writing—original draft preparation, T.L. and J.C.; writing—review and editing, W.W. and Y.Y.; supervision, Y.Y.; project administration, W.W. All authors have read and agreed to the published version of the manuscript.

Funding: This work was financially supported by the National Key R&D Program of China (No. 2022YFD2100602).

Institutional Review Board Statement: Ethical review and approval were waived for this study, as no animals or humans were used within the study. All meat used in the study was purchased in commercial supermarkets.

Informed Consent Statement: Not applicable.

Data Availability Statement: The datasets generated for this study are available from the authors.

Acknowledgments: We thank Yingwang Ye from HFUT for their help during the experiment.

Conflicts of Interest: The authors declare no competing financial interest.

References

1. Jiang, Z.; Anwar, T.M.; Peng, X.; Biswas, S.; Elbediwi, M.; Li, Y.; Fang, W.; Yue, M. Prevalence and Antimicrobial Resistance of Salmonella Recovered from Pig-Borne Food Products in Henan, China. *Food Control* **2021**, *121*, 107535. [CrossRef]
2. Hofer, U. *Salmonella* Enteritidis: Chicken or Egg? *Nat. Rev. Microbiol.* **2021**, *19*, 682. [CrossRef] [PubMed]
3. Lee, K.Y.; Atwill, E.R.; Pitesky, M.; Huang, A.; Lavelle, K.; Rickard, M.; Shafii, M.; Hung-Fan, M.; Li, X. Antimicrobial Resistance Profiles of Non-Typhoidal Salmonella from Retail Meat Products in California, 2018. *Front. Microbiol.* **2022**, *13*, 835699. [CrossRef] [PubMed]
4. Food and Drug Administration [FDA]. The National Antimicrobial Resistance Monitoring System. *J. Vet. Res.* **2020**, *64*, 281–288. [CrossRef]
5. Lin, L.; Mei, C.; Shi, C.; Li, C.; Abdel-Samie, M.A.; Cui, H. Preparation and Characterization of Gelatin Active Packaging Film Loaded with Eugenol Nanoparticles and Its Application in Chicken Preservation. *Food Biosci.* **2023**, *53*, 102778. [CrossRef]
6. Hui, X.; Yan, G.; Tian, F.-L.; Li, H.; Gao, W.-Y. Antimicrobial Mechanism of the Major Active Essential Oil Compounds and Their Structure–Activity Relationship. *Med. Chem. Res.* **2017**, *26*, 442–449. [CrossRef]
7. Wei, Q.; Liu, X.; Zhao, S.; Li, S.; Zhang, J. Research Note: Preservative Effect of Compound Spices Extracts on Marinated Chicken. *Poult. Sci.* **2022**, *101*, 101778. [CrossRef]
8. Zhang, Z.; Zhao, Y.; Chen, X.; Li, W.; Wang, L.; Li, W.; Du, J.; Zhang, S. Effects of Cinnamon Essential Oil on the Physiological Metabolism of *Salmonella* Enteritidis. *Front. Microbiol.* **2022**, *13*, 1035894. [CrossRef]
9. Alibi, S.; Selma, W.B.; Mansour, H.B.; Navas, J. Activity of Essential Oils Against Multidrug-Resistant *Salmonella* Enteritidis. *Curr. Microbiol.* **2022**, *79*, 273. [CrossRef]
10. Ju, J.; Xie, Y.; Yu, H.; Guo, Y.; Cheng, Y.; Qian, H.; Yao, W. Analysis of the Synergistic Antifungal Mechanism of Eugenol and Citral. *LWT* **2020**, *123*, 109128. [CrossRef]
11. Yang, R.; Miao, J.; Shen, Y.; Cai, N.; Wan, C.; Zou, L.; Chen, C.; Chen, J. Antifungal Effect of Cinnamaldehyde, Eugenol and Carvacrol Nanoemulsion against *Penicillium digitatum* and Application in Postharvest Preservation of Citrus Fruit. *LWT* **2021**, *141*, 110924. [CrossRef]
12. Kachur, K.; Suntres, Z. The Antibacterial Properties of Phenolic Isomers, Carvacrol and Thymol. *Crit. Rev. Food Sci. Nutr.* **2020**, *60*, 3042–3053. [CrossRef] [PubMed]
13. Ma, M.; Zhao, J.; Yan, X.; Zeng, Z.; Wan, D.; Yu, P.; Xia, J.; Zhang, G.; Gong, D. Synergistic Effects of Monocaprin and Carvacrol against *Escherichia coli* O157:H7 and *Salmonella* Typhimurium in Chicken Meat Preservation. *Food Control* **2022**, *132*, 108480. [CrossRef]
14. Peng, S.; Zhang, J.; Zhang, T.; Hati, S.; Mo, H.; Xu, D.; Li, H.; Hu, L.; Liu, Z. Characterization of Carvacrol Incorporated Antimicrobial Film Based on Agar/Konjac Glucomannan and Its Application in Chicken Preservation. *J. Food Eng.* **2022**, *330*, 111091. [CrossRef]
15. Lian, H.; Shi, J.; Zhang, X.; Peng, Y. Effect of the Added Polysaccharide on the Release of Thyme Essential Oil and Structure Properties of Chitosan Based Film. *Food Packag. Shelf Life* **2020**, *23*, 100467. [CrossRef]
16. Zheng, H.; Wang, J.; You, F.; Zhou, M.; Shi, S. Fabrication, Characterization, and Antimicrobial Activity of Carvacrol-Loaded Zein Nanoparticles Using the PH-Driven Method. *IJMS* **2022**, *23*, 9227. [CrossRef]
17. Cometa, S.; Bonifacio, M.A.; Bellissimo, A.; Pinto, L.; Petrella, A.; De Vietro, N.; Iannaccone, G.; Baruzzi, F.; De Giglio, E. A Green Approach to Develop Zeolite-Thymol Antimicrobial Composites: Analytical Characterization and Antimicrobial Activity Evaluation. *Heliyon* **2022**, *8*, e09551. [CrossRef]
18. Cui, R.; Zhu, B.; Yan, J.; Qin, Y.; Yuan, M.; Cheng, G.; Yuan, M. Development of a Sodium Alginate-Based Active Package with Controlled Release of Cinnamaldehyde Loaded on Halloysite Nanotubes. *Foods* **2021**, *10*, 1150. [CrossRef] [PubMed]
19. Wang, J.; Zhao, F.; Huang, J.; Li, Q.; Yang, Q.; Ju, J. Application of Essential Oils as Slow-Release Antimicrobial Agents in Food Preservation: Preparation Strategies, Release Mechanisms and Application Cases. *Crit. Rev. Food Sci. Nutr.* **2023**, 1–26. [CrossRef]
20. Duc, H.M.; Son, H.M.; Honjoh, K.; Miyamoto, T. Isolation and Application of Bacteriophages to Reduce *Salmonella* Contamination in Raw Chicken Meat. *LWT* **2018**, *91*, 353–360. [CrossRef]
21. Zhang, Y.; Dai, J.; Ma, X.; Jia, C.; Han, J.; Song, C.; Liu, Y.; Wei, D.; Xu, H.; Qin, J.; et al. Nano-Emulsification Essential Oil of *Monarda didyma* L. to Improve Its Preservation Effect on Postharvest Blueberry. *Food Chem.* **2023**, *417*, 135880. [CrossRef] [PubMed]
22. Moon, S.H.; Waite-Cusic, J.; Huang, E. Control of *Salmonella* in Chicken Meat Using a Combination of a Commercial Bacteriophage and Plant-Based Essential Oils. *Food Control* **2020**, *110*, 106984. [CrossRef]

23. Oliveira, G.R.; Oliveira, W.K.; Andrade, C.; Melo, A.D.B.; Luciano, F.B.; Macedo, R.E.F.; Costa, L.B. Natural Antimicrobials for Control of *Salmonella* Enteritidis in Feed and in Vitro Model of the Chicken Digestive Process. *J. Anim. Physiol. Anim. Nutr.* **2019**, *103*, 756–765. [CrossRef] [PubMed]
24. Zhang, S.; Shen, Y.R.; Wu, S.; Xiao, Y.Q.; He, Q.; Shi, S.R. The Dietary Combination of Essential Oils and Organic Acids Reduces *Salmonella* Enteritidis in Challenged Chicks. *Poult. Sci.* **2019**, *98*, 6349–6355. [CrossRef]
25. Paparella, A.; Mazzarrino, G.; Chaves-López, C.; Rossi, C.; Sacchetti, G.; Guerrieri, O.; Serio, A. Chitosan Boosts the Antimicrobial Activity of *Origanum vulgare* Essential Oil in Modified Atmosphere Packaged Pork. *Food Microbiol.* **2016**, *59*, 23–31. [CrossRef]
26. Wang, W.; Chen, J.; Shao, X.; Huang, P.; Zha, J.; Ye, Y. Occurrence and Antimicrobial Resistance of *Salmonella* Isolated from Retail Meats in Anhui, China. *Food Sci. Nutr.* **2021**, *9*, 4701–4710. [CrossRef]
27. Tang, X.; Shen, Y.; Song, X.; Benghezal, M.; Marshall, B.J.; Tang, H.; Li, H. Reassessment of the Broth Microdilution Method for Susceptibility Testing of *Helicobacter pylori*. *J. Infect. Dis.* **2022**, *226*, S486–S492. [CrossRef]
28. Tu, Q.; Li, S.; Zeng, Z.; Liu, Y.; Wang, C.; Chen, S.; Hu, B.; Li, C. Cinnamon Essential Oil Liposomes Modified by Sodium Alginate-chitosan: Application in Chilled Pork Preservation. *Int. J. Food Sci. Technol.* **2023**, *58*, 939–953. [CrossRef]
29. Catford, A.; Ganz, K.; Tamber, S. Enumerative Analysis of *Salmonella* in Outbreak-Associated Breaded and Frozen Comminuted Raw Chicken Products. *J. Food Prot.* **2017**, *80*, 814–818. [CrossRef]
30. Reid, A. 2009 MFHPB-20: Isolation and Identification of Salmonella from Food and Environmental Samples. In HPB Methods for the Microbiologcal Analysis of Foods, Vol. 2. The Compendium of Analytical Methods. Health Canada, Ottawa, Ontario, Canada. *World J. Microbiol. Biotechnol.* **2020**, *36*, 24.
31. Shatila, F.; Yaşa, İ.; Yalçın, H.T. Biofilm Formation by *Salmonella* Enterica Strains. *Curr. Microbiol.* **2021**, *78*, 1150–1158. [CrossRef] [PubMed]
32. Zduńczyk, W.; Tkacz, K.; Modzelewska-Kapituła, M. The Effect of Superficial Oregano Essential Oil Application on the Quality of Modified Atmosphere-Packed Pork Loin. *Foods* **2023**, *12*, 2013. [CrossRef] [PubMed]
33. Zhang, F.; Kang, T.; Sun, J.; Wang, J.; Zhao, W.; Gao, S.; Wang, W.; Ma, Q. Improving TVB-N Prediction in Pork Using Portable Spectroscopy with Just-in-Time Learning Model Updating Method. *Meat Sci.* **2022**, *188*, 108801. [CrossRef]
34. Dhifi, W.; Bellili, S.; Jazi, S.; Bahloul, N.; Mnif, W. Essential Oils' Chemical Characterization and Investigation of Some Biological Activities: A Critical Review. *Medicines* **2016**, *3*, 25. [CrossRef]
35. Ding, J.; Dwibedi, V.; Huang, H.; Ge, Y.; Li, Y.; Li, Q.; Sun, T. Preparation and Antibacterial Mechanism of Cinnamaldehyde/Tea Polyphenol/Polylactic Acid Coaxial Nanofiber Films with Zinc Oxide Sol to *Shewanella putrefaciens*. *Int. J. Biol. Macromol.* **2023**, *237*, 123932. [CrossRef]
36. Zhang, J.; Du, C.; Li, Q.; Hu, A.; Peng, R.; Sun, F.; Zhang, W. Inhibition Mechanism and Antibacterial Activity of Natural Antibacterial Agent Citral on Bamboo Mould and Its Anti-Mildew Effect on Bamboo. *R. Soc. Open Sci.* **2021**, *8*, 202244. [CrossRef]
37. Čabarkapa, I.; Čolović, R.; Đuragić, O.; Popović, S.; Kokić, B.; Milanov, D.; Pezo, L. Anti-Biofilm Activities of Essential Oils Rich in Carvacrol and Thymol against *Salmonella* Enteritidis. *Biofouling* **2019**, *35*, 361–375. [CrossRef] [PubMed]
38. Galgano, M.; Mrenoshki, D.; Pellegrini, F.; Capozzi, L.; Cordisco, M.; Del Sambro, L.; Trotta, A.; Camero, M.; Tempesta, M.; Buonavoglia, D.; et al. Antibacterial and Biofilm Production Inhibition Activity of *Thymus vulgaris* L. Essential Oil against *Salmonella* Spp. Isolates from Reptiles. *Pathogens* **2023**, *12*, 804. [CrossRef]
39. Misiou, O.; Van Nassau, T.J.; Lenz, C.A.; Vogel, R.F. The Preservation of Listeria -Critical Foods by a Combination of Endolysin and High Hydrostatic Pressure. *Int. J. Food Microbiol.* **2018**, *266*, 355–362. [CrossRef] [PubMed]
40. Zhao, R.; Zhang, Y.; Chen, H.; Song, R.; Li, Y. Performance of Eugenol Emulsion/Chitosan Edible Coating and Application in Fresh Meat Preservation. *Food Process. Preserv.* **2022**, *46*, e16407. [CrossRef]
41. Moon, H.; Kim, N.H.; Kim, S.H.; Kim, Y.; Ryu, J.H.; Rhee, M.S. Teriyaki Sauce with Carvacrol or Thymol Effectively Controls *Escherichia coli* O157:H7, *Listeria monocytogenes*, *Salmonella typhimurium*, and Indigenous Flora in Marinated Beef and Marinade. *Meat Sci.* **2017**, *129*, 147–152. [CrossRef] [PubMed]
42. Hoffman-Pennesi, D.; Wu, C. The Effect of Thymol and Thyme Oil Feed Supplementation on Growth Performance, Serum Antioxidant Levels, and Cecal *Salmonella* Population in Broilers. *J. Appl. Poult. Res.* **2010**, *19*, 432–443. [CrossRef]
43. Liu, X.; Liu, R.; Zhao, R.; Wang, J.; Cheng, Y.; Liu, Q.; Wang, Y.; Yang, S. Synergistic Interaction Between Paired Combinations of Natural Antimicrobials Against Poultry-Borne Pathogens. *Front. Microbiol.* **2022**, *13*, 811784. [CrossRef]
44. Wen, Y.; Liu, H.; Liu, K.; Cao, H.; Mao, H.; Dong, X.; Yin, Z. Analysis of the Physical Meat Quality in Partridge (*Alectoris chukar*) and Its Relationship with Intramuscular Fat. *Poult. Sci.* **2020**, *99*, 1225–1231. [CrossRef]
45. Lin, Y.; Hu, J.; Li, S.; Hamzah, S.S.; Jiang, H.; Zhou, A.; Zeng, S.; Lin, S. Curcumin-Based Photodynamic Sterilization for Preservation of Fresh-Cut Hami Melon. *Molecules* **2019**, *24*, 2374. [CrossRef]
46. Sari, T.V.; Hasanah, U.; Harahap, R.I.H.; Zalukhu, P.; Trisna, A. Chemical and Physical Quality of Broiler Meat with Drinking Water Containing Boiled and Fermented Water of Various Cooking Spices as Phytobiotics. *IOP Conf. Ser. Earth Environ. Sci.* **2022**, *977*, 012137. [CrossRef]
47. Bekhit, A.E.-D.A.; Holman, B.W.B.; Giteru, S.G.; Hopkins, D.L. Total Volatile Basic Nitrogen (TVB-N) and Its Role in Meat Spoilage: A Review. *Trends Food Sci. Technol.* **2021**, *109*, 280–302. [CrossRef]
48. Majdinasab, M.; Niakousari, M.; Shaghaghian, S.; Dehghani, H. Antimicrobial and Antioxidant Coating Based on Basil Seed Gum Incorporated with Shirazi Thyme and Summer Savory Essential Oils Emulsions for Shelf-Life Extension of Refrigerated Chicken Fillets. *Food Hydrocoll.* **2020**, *108*, 106011. [CrossRef]

49. Hematizad, I.; Khanjari, A.; Basti, A.A.; Karabagias, I.K.; Noori, N.; Ghadami, F.; Gholami, F.; Teimourifard, R. In Vitro Antibacterial Activity of Gelatin-Nanochitosan Films Incorporated with *Zataria multiflora* Boiss Essential Oil and Its Influence on Microbial, Chemical, and Sensorial Properties of Chicken Breast Meat during Refrigerated Storage. *Food Packag. Shelf Life* **2021**, *30*, 100751. [CrossRef]
50. Tomasevic, I.; Djekic, I.; Font-i-Furnols, M.; Terjung, N.; Lorenzo, J.M. Recent Advances in Meat Color Research. *Curr. Opin. Food Sci.* **2021**, *41*, 81–87. [CrossRef]
51. Ijaz, M.; Li, X.; Zhang, D.; Hussain, Z.; Ren, C.; Bai, Y.; Zheng, X. Association between Meat Color of DFD Beef and Other Quality Attributes. *Meat Sci.* **2020**, *161*, 107954. [CrossRef] [PubMed]

foods MDPI

Article

The Ultrasensitive Detection of Aflatoxin M_1 Using Gold Nanoparticles Modified Electrode with Fe^{3+} as a Probe

Xiaobo Li [1], Miao Zhang [1], Haizhen Mo [2], Hongbo Li [2], Dan Xu [2] and Liangbin Hu [2,*]

[1] Department of Chemistry and Chemical Engineering, Henan Institute of Science and Technology, Xinxiang 453003, China; xbli@hist.edu.cn (X.L.); zmm949798@126.com (M.Z.)

[2] School of Food Science and Engineering, Shaanxi University of Science and Technology, Xi'an 710021, China; mohz@sust.edu.cn (H.M.); hongbo715@163.com (H.L.); xudan@sust.edu.cn (D.X.)

* Correspondence: hulb@sust.edu.cn; Tel.: +86-13937392557

Abstract: The increasing incidence of diseases caused by highly carcinogenic aflatoxin M_1 (AFM_1) in food demands a simple, fast, and cost-effective detection technique capable of sensitively monitoring AFM_1. Recent works predominantly focus on the electrochemical aptamer-based biosensor, which still faces challenges and high costs in experimentally identifying an efficient candidate aptamer. However, the direct electrochemical detection of AFM_1 has been scarcely reported thus far. In this study, we observed a significant influence on the electrochemical signals of ferric ions at a gold nanoparticle-modified glassy carbon electrode (AuNPs/GCE) by adding varying amounts of AFM_1. Utilizing ferricyanide as a sensitive indicator of AFM_1, we have introduced a novel approach for detecting AFM_1, achieving an unprecedentedly low detection limit of 1.6×10^{-21} g/L. Through monitoring the fluorescence quenching of AFM_1 with Fe^{3+} addition, the interaction between them has been identified at a ratio of 1:936. Transient fluorescence analysis reveals that the fluorescence quenching process is predominantly static. It is interesting that the application of iron chelator diethylenetriaminepentaacetic acid (DTPA) cannot prevent the interaction between AFM_1 and Fe^{3+}. With a particle size distribution analysis, it is suggested that a combination of AFM_1 and Fe^{3+} occurs and forms a polymer-like aggregate. Nonetheless, the mutual reaction mechanism between AFM_1 and Fe^{3+} remains unexplained and urgently necessitates unveiling. Finally, the developed sensor is successfully applied for the AFM_1 test in real samples, fully meeting the detection requirements for milk.

Keywords: aflatoxin M_1; electrochemical detection; gold nanoparticles; Fe^{3+}

Citation: Li, X.; Zhang, M.; Mo, H.; Li, H.; Xu, D.; Hu, L. The Ultrasensitive Detection of Aflatoxin M_1 Using Gold Nanoparticles Modified Electrode with Fe^{3+} as a Probe. *Foods* **2023**, *12*, 2521. https://doi.org/10.3390/foods12132521

Academic Editor: Thierry Noguer

Received: 31 May 2023
Revised: 26 June 2023
Accepted: 27 June 2023
Published: 28 June 2023

1. Introduction

Aflatoxins, produced by *Aspergillus flavus* and *Aspergillus. parasiticus*, which are commonly found in cereal grains, dairy products, beans, and nuts [1–3], are toxic compounds with a difuran ring structure. The improper storage conditions can result in contamination with aflatoxins, and the intake of aflatoxins is associated with a significant portion of hepatocellular carcinoma cases worldwide such as impaired liver function and immune response and an increase in susceptibility to infectious diseases [4–6]. These toxins can enter the bloodstream and undergo metabolization in the human body. There are four generally recognized aflatoxins designated B_1, B_2, G_1, and G_2. Aflatoxin B_1 (AFB_1) is the major mycotoxin produced by most species under culture conditions and is the most frequently studied of the four. However, the index compound of AFB1 is not carcinogenic before it is metabolically activated. AFM_1, a hydroxylated metabolite of AFB_1 in human food and animal feed, is excreted in urine and secreted in milk in mammalian species within 12 h after consumption, while its toxicity is much less known. AFM_1, in particular, is reclassified as a naturally occurring carcinogen belonging to Group 1, with the formation of DNA adducts [7]. Nonetheless, many nations have set regulatory limits for maximum allowable

AFM_1 in milk and other dairy products. In 2005, the Food and Drug Administration (FDA) set an AFM_1 action level in milk and other dairy products at 0.5 μg/L. The European Union (EU) set a much stricter standard that allows a maximum of 0.05 μg/L in milk in 2006 [7]. Turna noted that a high AFM_1 level in milk was associated with a high level of AFB_1 in animal feed, which upon consumption could harm both animal and human health [8]. Accordingly, timely monitoring of aflatoxins during biological transformation can help reduce the risk of diseases [9]. Consequently, the development of a sensitive detection method has become a prominent research focus in recent years.

Currently, the identification and quantification of aflatoxins are commonly performed using thin-layer chromatography (TLC) [10], immunoaffinity chromatography [11], high-performance liquid chromatography (HPLC) [12], and enzyme-linked immunosorbent assay (ELISA) [13]. However, these methods have drawbacks such as being time-consuming, requiring special equipment, or involving cumbersome sample pretreatment and false-positive results [14]. A reliable and promising alternative approach that offers high sensitivity, ease of operation, fast analysis, and cost-effectiveness is the electrochemical method. Recently, electrochemical aptamer-based sensors have gained significant attention for aflatoxin monitoring. However, the instability of the biological recognition piece limited the use as anticipated.

Abnous et al. [15] reported an electrochemical sensing strategy for the detection of AFB_1 based on aptamer-complementary strands of aptamer complex, forming a π-shape structure on the electrode surface, with a detection limit of 2 pg/mL. Ahmadi et al. [16] developed a pencil graphite electrode modified with reduced graphene oxide and gold nanoparticles for the detection of AFM_1, achieving a detection limit of 0.3 ng/L. Furthermore, the detection of AFM_1 has been accomplished using voltammetric biosensors, utilizing silver nanoparticles dispersed on an α-cyclodextrin-GQDs nanocomposite [17]. Aflatoxins can be electrochemically oxidized to ketone because of containing an alcohol group, which is generated by the hydrolysis of the aromatic ester group in a basic medium. This allows the direct detection of aflatoxins without the need for recognition elements or tags [18]. Gevaerd et al. [19] reported the direct determination of AFB1 in 2020 at the screen-printed electrode (SPE) modified with gold nanoparticles and graphene quantum dots (AuNPs-GQDs), which exerted an electrocatalytic effect on the oxidation of AFB_1 (shift of the oxidation peak to less positive values). The performance in ng/mL level by this approach was quite similar to those obtained with other systems where bioreceptors are used. However, the low specificity of this format of detection limits its further application for the selective determination of aflatoxins. Thus, there is still a growing demand for novel sensors that offer a simple electrode modification process, high sensitivity, low cost, and ease of use for the detection of Aflatoxins replacing commonly used biological recognition systems described above.

Aflatoxins belong to the class of coumarin compounds characterized by their difuran ring structures. Previous reports have identified coumarin compounds as fluorescent probes for various metal ions such as Mg^{2+} [20], Cu^{2+} [21,22], Zn^{2+} [23], and others. In recent years, numerous small molecule fluorescent probes for Fe^{3+} have been developed [24,25] based on a selective binding approach for example complexation or chelation. Wang et al. [26] reported a highly selective coumarin-based chemosensor for the detection of Fe^{3+} where coumarin FB displayed a high affinity to Fe^{3+} resulting in forming an FB-Fe^{3+} complex. Zhao et al. [27] specifically designed and synthesized Schiff base probes using phenanthro [9,10-d] imidazole-coumarin derivatives, demonstrating the formation of a 1:1 complex between these probes and Fe^{3+}.

Considering this, the electrochemical signals of Fe^{3+} could potentially reflect the concentration of aflatoxins. There appear to be interactions between Fe^{3+} and aflatoxins, which warrant further comprehensive investigation, particularly in understanding the nature of the interaction between Fe^{3+} and aflatoxins.

As is known, electrode modifiers having good conductivity and catalytic activity play an important role in influencing the sensitivity and capability of modified electrodes. Commonly, conducting polymers, molecularly imprinted materials and some metals such as gold, iron, silver, and palladium can be used as electrode modifier materials for enhancing the peak currents, which is necessary for determining the trace amount of analytes in real samples. In this work, we have compared the electrochemical activity of polythionine, molecularly imprinted L-cysteine, and electrodeposited gold nanoparticles (AuNPs). Among them, the uniform deposition of AuNPs onto the glassy carbon electrode (GCE) surface is well known for its ability to increase the effective area and confer the direct electron transfer between the analyte and the electrode base [28,29], making it an excellent sensing platform and giving a much better electrochemical signal towards Fe^{3+} as shown in Supplementary Figure S1. In this study, we have successfully developed an electrochemical sensor by modifying a GCE with AuNPs (AuNPs/GCE) for the detection of AFM_1. While the electrochemical synthesis of AuNPs without the requirement of an external linker or functionalizing ligand is well established, aiming for the best electrochemical performance; herein, we have particularly optimized the electrochemical sweeping methods and parameters, deposition times, and minimal solution preparation because the in situ tailoring of nanoparticle surface chemistry resulted in improved catalytic activity and selectivity. The sensor utilizes ferricyanide as a mediator, where the presence of AuNPs greatly facilitates the electrode reaction and enhances the catalytic activity towards ferricyanide. Consequently, this sensor exhibits an unprecedented lowest detectable concentration of AFM_1 over the widest linear range reported thus far.

To demonstrate the specificity of Fe^{3+} in AFM_1 detection, an immunoassay column was utilized. Remarkably, an impressive reaction ratio of AFM_1 to Fe^{3+} was obtained. The performance of the developed sensor was evaluated by measuring AFM_1 in spiked milk samples, yielding satisfactory analytical results since it is easy, quick, and does not involve developing the biological material methodology.

2. Materials and Methods

2.1. Chemicals and Apparatus

AFM_1 was obtained from Toronto Research Chemicals, while $HAuCl_4 \cdot 3H_2O$ (≥99%) was acquired from Sigma Aldrich. KCl, $K_3[Fe(CN)_6]$, $K_4[Fe(CN)_6]$, NaH_2PO_4, and Na_2HPO_4 were purchased from Aladdin Reagents. Unless otherwise specified, all reagents were used as received. Phosphate buffer solutions (PBS) were prepared by diluting 0.1 M NaH_2PO_4 and 0.1 M Na_2HPO_4 stock solutions. All solutions were prepared using double-distilled water with a resistivity of 18 MΩ·cm.

The electrochemical experiments were conducted at room temperature using a CHI900D workstation (Shanghai CH Instrument Ltd., Shanghai, China) equipped with a conventional three-electrode system. The system consisted of a glassy carbon electrode (GCE, 3.0 mm in diameter) as the working electrode, a platinum wire as the counter electrode, and an Ag/AgCl (saturated KCl) electrode as the reference electrode. Electrochemical impedance spectroscopy (EIS) measurements were performed in PBS containing 5 mM $K_3[Fe(CN)_6]/K_4[Fe(CN)_6]$ (mole ratio of 1:1) and 0.1 M KCl at room temperature. An AUTOLAB PGSTAT302N (Metrohm Auto lab B.V., Herisau, Switzerland) was used for EIS measurements, employing a formal potential of 0.2 V, 5 mV amplitude, and a frequency range from 0.1 Hz to 100 kHz. Nyquist plots were generated from the impedance data and fitted using AUTOLAB Nova 1.8.

Scanning electron microscopy (SEM) experiments were conducted using an XL30 ESEM-FEG (FEI Company, Hillsboro, OR, USA) with an acceleration voltage of 20.0 kV. Molecular fluorescence spectra were measured using an F-180 fluorescence spectroscope (Tianjin Gangdong Co., Ltd., Tianjin, China). The steady-state and transient-state fluorescence spectra were obtained using an FLS 1000 spectrometer (Edinburgh Instruments, West Lothian, UK). Particle size distribution analysis was obtained using a nanoparticle size analyzer Winner 802 (Jinan Weina Particle Instruments, Jinan, China). The enzyme-linked immunosorbent assay (ELISA) was performed using a Varioskan™ LUX (Thermo Fisher Scientific, Waltham, MA, USA).

2.2. Preparation of AuNPs/GCE

Figure 1A depicts the schematic diagram illustrating the modification of GCE with AuNPs. To achieve this, the GCE was initially polished to a mirror finish using 0.05 μm alumina slurry on a microcloth. Subsequently, it was ultrasonicated with distilled water for 1 min to ensure cleanliness. The polished and cleaned GCE was then immersed in a solution containing 5 mM HAuCl4 and 0.1 M KCl. AuNPs were formed using cyclic voltammetry (CV) in a potential range from −0.4 V to 1.2 V, with a scan rate of 10 mV/s for 20 cycles. Figure 1B displays the cyclic voltammograms, which exhibit a monotonically increasing trend of the redox waves, confirming the continuous growth of the AuNPs layer.

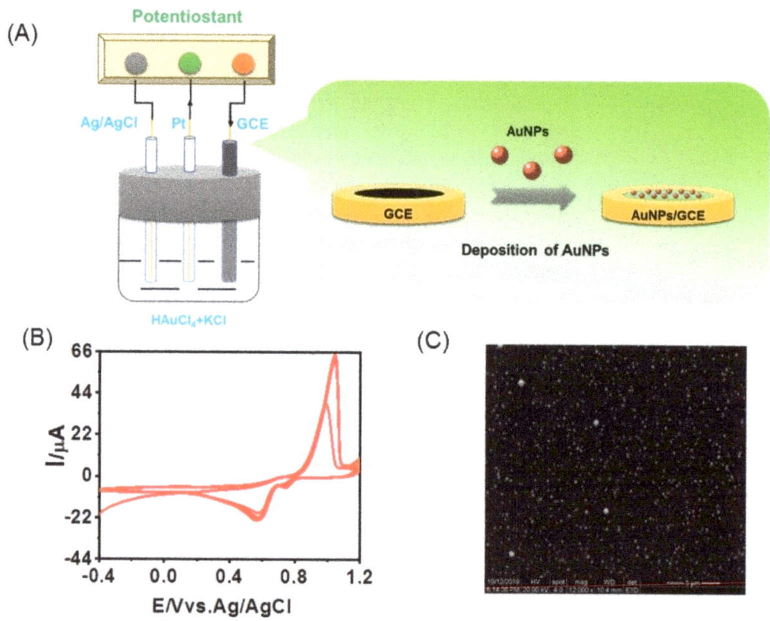

Figure 1. The preparation of AuNPs−modified glassy carbon electrode. (**A**) Schematic diagram of AuNPs modification onto glassy carbon electrode; (**B**) CVs of Au electrodeposition onto GCE in 0.1 M pH 7.0 PBS; (**C**) SEM image of AuNPs/GCE.

Following the AuNPs formation, the AuNPs/GCE was thoroughly rinsed with deionized water and transferred to a 0.1 M PBS solution at pH 7.0. The electrode was scanned until a stable voltammogram was obtained. Figure 1C shows the corresponding SEM image of the AuNPs/GCE, revealing a compact layer with uniformly distributed and smaller AuNPs, providing full coverage.

2.3. Sample Preparation

AFM_1 stock solution was prepared by dissolving an appropriate mass of AFM_1 in pH 7.0 PBS, and working solutions with different concentrations of AFM_1 were prepared by diluting the stock solution with buffer. The commercial milk was purchased from a local store. In the experimental procedure, 30 mL of the milk sample was transferred to a centrifuge tube. To remove proteins, 20% TCA (trichloroacetic acid) was added, and the mixture was centrifuged for 5 min. Subsequently, centrifugation was performed at a speed of 6000 rpm for 10 min. The resulting supernatant was then filtered through a 0.22 μm filter and passed through an immunoaffinity column. Finally, the filtered sample was subjected to testing. The possibility and reliability of the method being applied in practice were established in regard to evaluating the recovery rate in actual samples.

3. Results and Discussion

3.1. Electrochemical Characterization of the AuNPs/GCE

The electrochemical behaviors of AuNPs/GCE were investigated by performing CV (cyclic voltammetry) and EIS (electrochemical impedance spectroscopy) measurements in a 0.1 M PBS solution (pH 7.0) containing 5 mM $Fe(CN)6^{3-/4-}$. Figure 2A clearly shows that AuNPs/GCE exhibits remarkable activity and reversibility, as evidenced by the distinct peak-to-peak separation (ΔEp = ca. 78 mV at 100 $mV \cdot s^{-1}$) and enhanced peak current observed in the CVs, in comparison to the bare GCE. Furthermore, the CVs of AuNPs/GCE remained nearly constant despite variations in the number of ultrasonic cleaning cycles, indicating the high stability of the modifier layer.

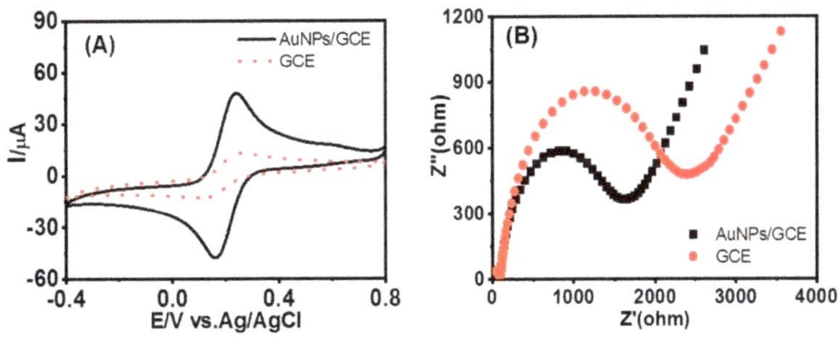

Figure 2. The improved performance of AuNPs−modified glassy carbon electrode. (**A**) CVs of ferricyanide at GCE and AuNPs/GCE in the presence of AFM_1; (**B**) Nyquist Plot (Z' vs. $-Z''$) of GCE and AuNPs/GCE.

Additionally, the Nyquist plots presented in Figure 2B, obtained from the EIS measurements, further support the superior conductivity of AuNPs/GCE when compared to the bare GCE. This enhanced conductivity contributes to the exceptional electrochemical catalytic performance exhibited by AuNPs/GCE without the use of any additional biomolecule as an electrode modifier.

3.2. Effect of pH and Scan Rates on Fe^{3+} Signals at AuNPs/GCE in the Presence of AFM_1

The impact of pH variation on the electrochemical response of Fe^{3+} at AuNPs/GCE in the presence of AFM_1 was further investigated using different buffer solutions prepared and adjusted to a pH range of 4.0 to 8.0. Protons always exert a significant impact on the reaction speed when being involved in the electrochemical reactions of organic compounds. Figure 3A illustrates the gradual increase in peak currents in differential pulse voltammetry (DPV) with an increase in pH within the range of 4–9. The peak currents reach a maximum at pH 7.0 and then decline, leading to the selection of pH 7 for subsequent experiments. This phenomenon is also aligned with the fact that AFM_1 is generally more stable in neutral pH. Notably, the absence of proton involvement in the reaction is evident as the peak potential does not exhibit a linear shift with pH.

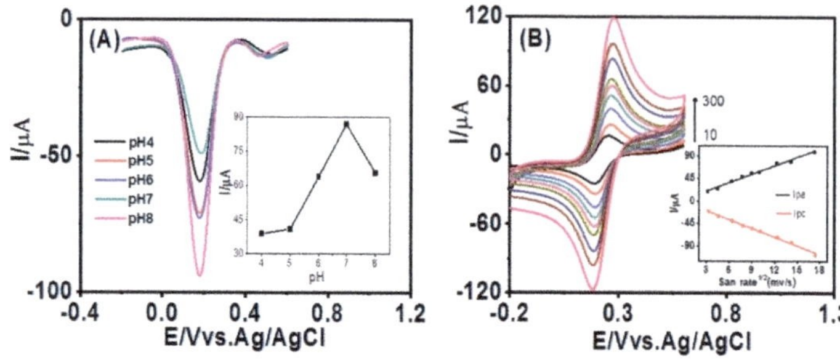

Figure 3. The effects of different pH on the sensitivity of AuNPs/GCE to Fe^{3+} and AFM_1. (**A**) DPVs of ferricyanide at AuNPs/GCE in 0.1 M PBS in the presence of AFM1 with pH ranging from 4.0 to 8.0. (**B**) CVs of ferricyanide GCE in 0.1 M pH 7.0 PBS at AuNPs/GCE in the presence of AFM_1 at various scan rates (10, 20, 40, 60, 80, 100, 150, 200, and 300 $mV{\cdot}s^{-1}$, respectively). Inset shows plots of *Ipa* and *Ipc* (μA) versus the square root of the scan rate.

Figure 3B presents the cyclic voltammograms (CVs) obtained at AuNPs/GCE with different scan rates. It is observed that both the cathodic peak current (Ipa) and anodic peak current (Ipc) are directly proportional to the square root of the scan rates within the range of 10–300 $mV{\cdot}s^{-1}$. The correlated linear equations can be expressed as Ipa (μA) = 5.63 $v^{1/2}$ ($mV{\cdot}s^{-1}$) + 2.72, and Ipc (μA) = −5.84 $v^{1/2}$ ($mV{\cdot}s^{-1}$) − 2.23, respectively, with the consistent regression coefficient (r^2) of 0.99 (inset of Figure 3B), suggesting a diffusion-controlled redox behavior of Fe^{3+} at AuNPs/GCE according to Randles–Sevcik equation instead of a surface reaction-controlled process. Moreover, the fact that peak potential is nearly independent on the scan rate suggests that the redox reaction is electrochemically reversible.

3.3. Electrochemical Response of Fe^{3+} at AuNPs/GCE in the Presence of AFM_1

Differential pulse voltammetry (DPV) is an effective and rapid electroanalytical technique with lower concentration detection limits. The electrochemical responses of Fe^{3+} were further investigated using the DPV technique, with varying concentrations of AFM_1 added to the electrolyte. Figure 4 illustrates the findings, where it can be observed that the peak currents of Fe^{3+} decrease as the concentrations of AFM_1 increase within the range of 1.6×10^{-21} to 2.5×10^{-4} g/L. A linear regression equation for AFM_1 of I (μA) = −2.34 lg [AFM_1] (g/L) + 54.261 with a correlation coefficient of 0.99879 was derived from the data. Each current response was measured three times, yielding a relative standard deviation (RSD) of 4.3%. These results clearly demonstrate the successful application of the developed sensor for ultrasensitive detection of AFM_1, with the lowest observed concentration of 1.6×10^{-21} g/L, surpassing previous reports based on the electrochemical method as listed in Table 1. When the electrode was stored in the refrigerator at 4 °C, the current response remained almost unchanged for about 2 weeks by taking advantage of a highly reliable electrode-preparing process. This raises a crucial question regarding the nature of the reaction occurring between Fe^{3+} and AFM_1, leading to significant suppression of the electrochemical signals of Fe^{3+} in the presence of AFM_1. We also confirmed the capability of the present method for the monitoring of AFB_1-NAC and AFB_1-lysine, which are another two metabolites from AFB1 as shown in Supplementary Figure S2.

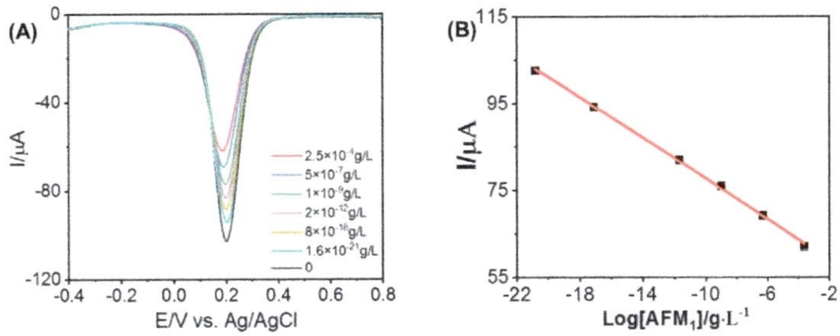

Figure 4. (**A**) DPVs profile of 5 mM ferricyanide at AuNPs/AGCE in the presence of AFM_1 at different concentrations (2.5×10^{-4}, 5×10^{-7}, 1×10^{-9}, 2×10^{-12}, 8×10^{-18}, 1.6×10^{-21} g/L, and 0, respectively). (**B**) Inset shows the plot of *Ipa* as a Logarithmic function of the concentration of AFM_1.

Table 1. Comparison of analytical performance for the electrochemical detection of AFM_1.

System	Detection Limit	Linear Range	Ref.
A-CD-GQDs-AgNPs/GCE	2 µm	0.015–25 µM	[17]
Apt-CS-AuNPs/SPGE	0.9 ng/L	2–600 ng/L	[29]
anti-AFM1/SPGE	2.5×10^{-8} g/kg	3×10^{-8}–1.6×10^{-7} g/kg	[30]
Fe_3O_4-PANi/IDE	1.98 ng/mL	6–60 ng/mL	[31]
NR/P [5]A-COOH/GCE	0.5 ng/L	5–120 ng/L	[32]
AuNPs/SPE	37 pg/mL	-	[33]
ss-HSDNA-AuNPs/GE	0.36 ng/mL	1–14 ng/mL	[34]
Fe^{3+}-AuNPs/GCE	1.6×10^{-21} g/L	1.6×10^{-21}–2.5×10^{-4} g/L	This work

Furthermore, the potential application of this method for detecting other toxins such as zearalenone (ZEA), ochratoxins (OTA), AFB_2, and AFG_1 was investigated. The DPVs obtained for these toxins are shown in Figure 5. Similar phenomena were observed for AFB_2 and AFG_1, confirming the method's suitability for selective detection of aflatoxins.

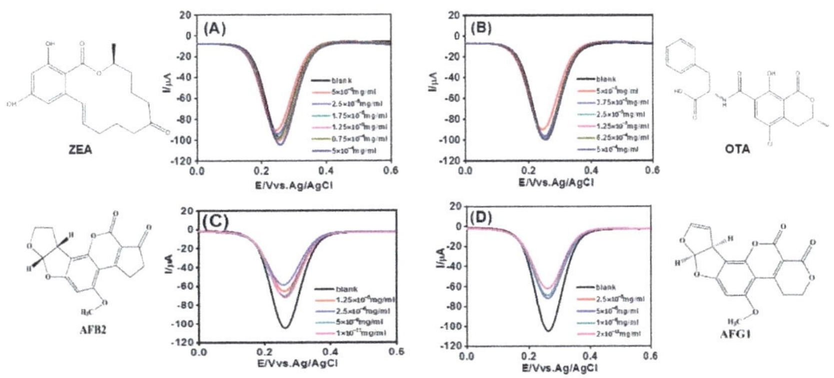

Figure 5. DPVs profile of ferricyanide at AuNPs/GCE in 0.1M PBS in the presence of ZEA (**A**), OTA (**B**), AFB_2 (**C**), and AFG_1 (**D**) at different concentrations.

3.4. Chronoamperometric Studies

The diffusion coefficient and catalytic rate constant of Fe^{3+} in the presence of AFM_1 were calculated from chronoamperometry. From the time-current curve, as shown in Figure 6A, it has been deduced that inversely linear dependency exists between the current and the square root of time as shown in Figure 6B. The slope of the linear equation could be obtained by using the Cottrell Equation:

$$I = nFAD^{1/2}C\pi^{1/2}t^{1/2}$$

where n is the number of transferred electrons, F is the Faraday constant, A is the proportion of the electrode, D is the diffusion coefficient of active substance, C is the initial molar concentration, and t is the running time. From the resulting slope, the D value was obtained to be 6.463×10^{-8} $cm^2 \cdot s^{-1}$. Chronoamperometry was also used to measure the catalytic rate constants from the following Equation:

$$Icat/Id = c^{1/2}\pi^{1/2} = \pi^{1/2}(kCt)$$

where *Icat* and *Id* were the currents of Fe^{3+} at AuNPs/GCE in the presence and absence of AFM_1 and Fe^{3+}, respectively, γ = kCt is the error function, k is the catalytic rate constant, C is the concentration of AFM_1 and Fe^{3+}, and t is the running time (s). From the slope of the *Icat/Id* vs. $t^{1/2}$ plot, as shown in Figure 6C, the k value was obtained to be 4.8×10^{-2} $cm^3 \cdot mol^{-1} \cdot s^{-1}$.

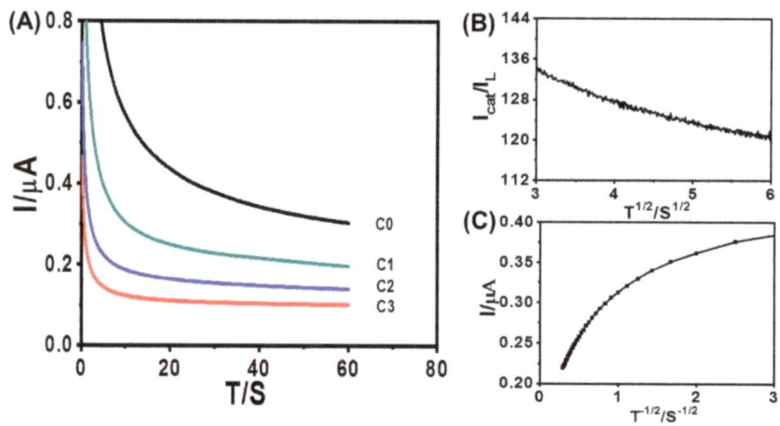

Figure 6. Amperomertric i-t Curves of AFM1 with different concentrations (C0: 0, C1: 6.25×10^{-10} g/L, C2:2.5×10^{-15} g/L, and C3:1×10^{-20} g/L) at AuNPs/GCE in the presence of 5 mM $[Fe(CN)6]^{4-}$. (**A**) Potential is 400 mV. (**B**) Dependence of Icatal/IL on $t^{1/2}$. (**C**) Dependency of transient current on $t^{-1/2}$.

3.5. Effect of Fe^{3+} Concentration on AFM_1 Fluorescence Intensity

The fluorescence intensity of AFM_1 in the presence of Fe^{3+} was observed to decrease as the solution pH values increased, as shown in Supplementary Figure S3. Although, generally, the fluorescence intensity at neutral pH is higher than that in acidic or base environments, there is competition occurring between Fe^{3+} and H^+ assumed by Patel-Sorrentino et al., for the explanation of pH effect [35]. Additionally, in order to further explore the response properties of AFM_1 to Fe^{3+}, the fluorescence titration experiment in tris buffer solution was performed with the gradual addition of Fe^{3+} to AFM_1. The concentration of AFM_1 was maintained at 1×10^{-5} M, while the concentration of Fe^{3+} was over the range from 0 to 5.12×10^{-6} M. Figure 7A demonstrates that the fluorescence intensity gradually decreases with increasing Fe^{3+} concentration; to put it another way, the addition of Fe^{3+} leads to a remarkable fluorescence quenching of AFM_1. Once the concentration reaches 8 mM, the fluorescence intensity remains constant at nearly zero and does not change because of quenching saturation. It can be assumed that they may tend to form polymer-like nano-aggregates [36]. The fitting curve is illustrated in Figure 7B, revealing a concentration ratio of AFM_1 to Fe^{3+} of approximately 1:936. Transient fluorescence analysis

(Figure 7C) confirms that the fluorescence quenching process is predominantly static, as AFM_1, Fe^{3+}, and AFM_1+Fe^{3+} exhibit high coinciding properties. The second-order fitting curve of transient fluorescence spectra by the ordinary least square method is displayed in Supplementary Figure S4. The non-linear Equation is expressed as follows:

$$Y = A1 \times \exp(-x/t1) + A2 \times \exp(-x/t2) + y0\ (r^2 = 0.99)$$

where A1 = 3155.94 ± 17.67, t1 = 1250.31 ± 14.00, A2 = 156.62 ± 11.16, t2 = 13,712.75 ± 1427.08, y0 = −3.42 ± 2.37, respectively. A calculated fluorescence lifetime of 12.402 μs by formula $t = (A1 \times t1^2) + (A2 \times t2^2)/(A1 \times t1 + A2 \times t2)$ proves that the interaction between AFM_1 and Fe^{3+} is ultrafast.

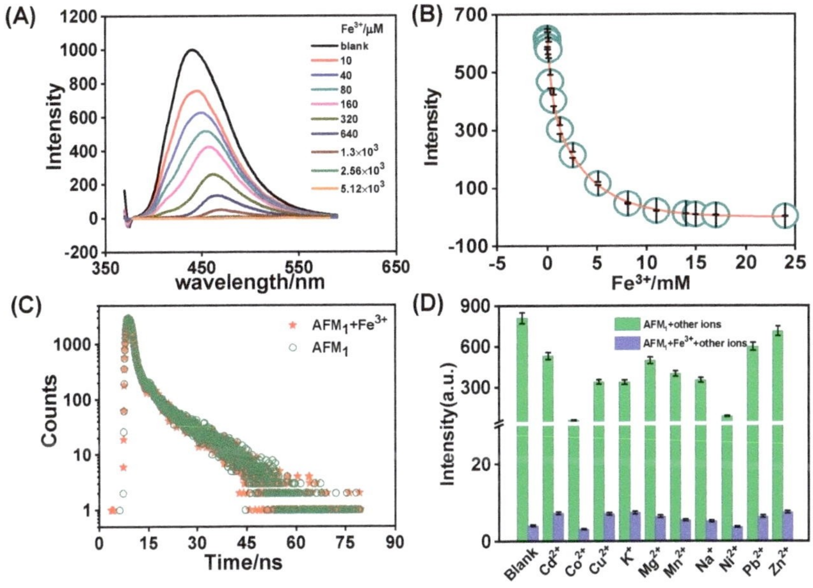

Figure 7. The specific interaction between Fe^{3+} and AFM_1. (**A**) Effects of Fe^{3+} addition on fluorescence emission of AFM_1; (**B**) Fluorescence intensity as a function of Fe^{3+} concentration; (**C**) Transient fluorescence lifetime of AFM_1, AFM_1 + Fe^{3+}; (**D**) Effects of other metal ions on the fluorescence emission of AFM_1.

To determine the specific effect of Fe^{3+}, various ions including Na^+, K^+, Zn^{2+}, Cu^{2+}, Ni^{2+}, Pb^{2+}, Cd^{2+}, Mn^{2+}, Mg^{2+}, and Co^{2+} were investigated for their interference on the fluorescence intensity of AFM_1 under the same conditions. As demonstrated in Figure 7D, none of these ions caused any significant interference even at higher concentrations reflected by negligible responses of AFM_1. The fluorescence intensity of AFM_1 was also examined in the presence of vitamin B_{12}, a water-soluble vitamin known for its metal ion content, and heme iron. Similar to the effect of Fe^{3+}, the fluorescence intensity gradually decreased with increasing concentration, as depicted in Supplementary Figure S5. These findings indicate a certain interaction between AFM_1 and Fe^{3+}.

To evaluate the intensity of this interaction, diethylenetriaminepentaacetic acid (DTPA) was employed as a competitor against AFM_1. As it was, DTPA may form a complexation with Fe^{3+} to release AFM_1 so as to observe an increase in fluorescence intensity. Despite the strong Fe^{3+}-binding ability of DTPA, the addition of DTPA (5 mM) to a mixture of AFM_1 (4 μg/mL) and Fe^{3+} (3.2 mM) unexpectedly led to a further decrease in fluorescence intensity, as depicted in Figure 8A. This suggests that DTPA can surprisingly enhance the interaction between AFM1 and Fe^{3+}. The particle size distribution analysis of AFM_1+Fe^{3+}

in the 420–600 nm range (Figure 8B) compared with Fe^{3+} and AFM_1 in the range of 500–800 nm and 400–1000 nm, respectively, further confirms their combination.

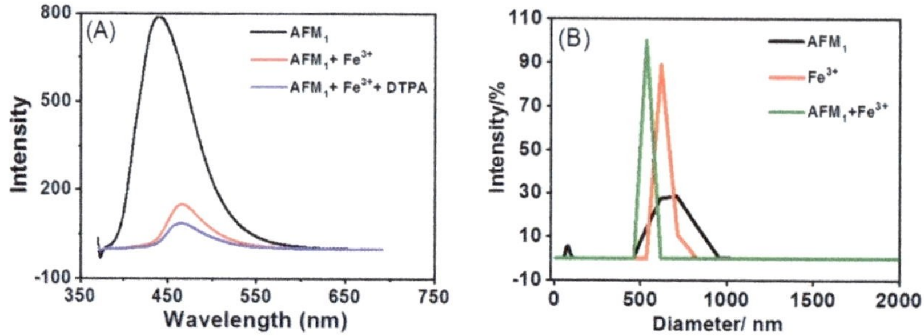

Figure 8. Evaluation on the intensity of interaction between AFM_1 and Fe^{3+}. (**A**) The Fluorescence intensity of AFM_1, AFM_1+Fe^{3+}, and AFM_1+Fe^{3+}+DTPA. (**B**) The Particle size distribution of AFM_1, Fe^{3+}, and AFM_1+Fe^{3+}.

3.6. Determination of AFM_1 in Milk

To assess the effectiveness and feasibility of the proposed method, AFM_1 levels in milk were measured. Spike and recovery experiments were conducted by measuring DPV responses in real milk samples with known concentrations of AFM_1 added. The AFM_1 concentrations in the milk samples were determined through calibration and are presented in Table 2. In all cases, good recoveries were obtained for AFM_1 varying from 92.0% to 93.9% considering the level of concentration being analyzed, which is comparatively better than those obtained from spectrofluorimetry and ELISA. These findings strongly demonstrate the practical applicability and reliability of the proposed method.

Table 2. The determination of AFM_1 in milk with three methods.

Method	Add/μM	Detected/μM	Recovery%
This work	2.96	2.78	93.9
	8.91	8.24	92.4
	23.78	22.17	93.2
Spectrofluorimetry	2.2	1.9	86
	3.16	2.5	78
	12.65	10.08	84
ELISA	2.6	2.39	92.1
	8.41	7.82	93
	20.25	18.71	92.4

4. Conclusions

In this study, we have introduced a novel approach for detecting AFM_1, utilizing the electrochemical signals of ferricyanide as a sensitive indicator of AFM_1 concentrations. Notably, we achieved an unprecedentedly low detection limit of 1.6×10^{-21} g/L, surpassing previous reports. Furthermore, the recovery rates of 92.4–93.9% obtained from real sample testing underscore the potential of this method as a reliable screening technique for AFM_1 detection in food. This approach combines the advantages of nanotechnology, supramolecular recognition techniques, and signal amplification, providing a versatile tool for monitoring aflatoxins. Ongoing investigations aim to elucidate the underlying interaction mechanism between AFM_1 and Fe^{3+}. Since most research on the electrochemical detection of aflatoxins is focused on aptamer immunosensors, this work may open new opportunities for Fe^{3+} as a probe for reversely monitoring coumarin-based small molecules. Meanwhile, this work will provide a beneficial reference for sensing of other toxins in food or pharmaceutical assays.

Supplementary Materials: The following supporting information can be downloaded at: https://www.mdpi.com/article/10.3390/foods12132521/s1, Figure S1: CV response of different modified electrodes in Fe^{3+} and AFM1; Figure S2: Effect of pH on fluorescence spectrum intensity; Figure S3: The DPV response of AFB1, AFB1-NAC and AFB1-lysine respectively at AuNPs/GCE; Figure S4: The second-order fitting curve of transient fluorescence spectra; Figure S5 Fluorescence intensity at different concentrations of heme iron and VB_{12}.

Author Contributions: Conceptualization, X.L. and M.Z.; methodology, X.L.; software, H.M.; formal analysis, X.L. and H.L.; investigation, M.Z.; resources, D.X. and L.H.; data curation, X.L.; writing—original draft preparation, X.L.; writing—review and editing, D.X. and L.H. All authors have read and agreed to the published version of the manuscript.

Funding: This work was supported by National Key R&D Program of China(2022YFD2100601), National Natural Science Foundation of China (32272278) and Innovation Capability Support Program of Shaanxi Province (Program No. 2023-CX-TD-61).

Data Availability Statement: The data used to support the findings of this study can be made available by the corresponding author upon request.

Conflicts of Interest: The authors declare no conflict interest.

References

1. Ma, H.H.; Sun, J.Z.; Zhang, Y.; Bian, C.; Xia, S.H.; Zhen, T. Label-free immunosensor based on one-step electrodeposition of chitosan-gold nanoparticles biocompatible film on Au microelectrode for determination of aflatoxin B_1 in maize. *Biosens. Bioelectron.* **2016**, *80*, 222–229. [CrossRef]
2. Malhotra, B.D.; Srivastava, S.; Ali, M.A.; Singh, C. Nanomaterial-based biosensors for food toxin detection. *Appl. Biochem. Biotechnol.* **2014**, *174*, 880–896. [CrossRef]
3. Machida, M.; Gomi, K. *Aspergillus: Molecular Biology and Genomics*; Horizon Scientific Press: Tsukuba, Japan, 2010.
4. Liu, Y.; Wu, F. Global burden of aflatoxin-induced hepatocellular carcinoma: A risk assessment. *Environ. Health Perspect.* **2010**, *118*, 818–824. [CrossRef]
5. Liu, Y.; Chang, C.C.; Marsh, G.M.; Wu, F. Population attributable risk of aflatoxin-related liver cancer: Systematic review and meta-analysis. *Eur. J. Cancer* **2012**, *48*, 2125–2136. [CrossRef] [PubMed]
6. Wu, F. Perspective: Time to face the fungal threat. *Nature* **2014**, *516*, S7. [CrossRef] [PubMed]
7. Lee, J.; Her, J.Y.; Lee, K.G. Reduction of aflatoxins (B_1, B_2, G_1, and G_2) in soybean-based model systems. *Food Chem.* **2015**, *189*, 45–51. [CrossRef]
8. Turna, N.S.; Wu, F. Aflatoxin M_1 in milk: A global occurrence, intake, & exposure assessment. *Trends Food Sci. Tech.* **2021**, *110*, 183–192.
9. Turner, N.W.; Subrahmanyam, S.; Piletsky, S.A. Analytical methods for determination of mycotoxins: A review. *Anal. Chim. Acta* **2009**, *632*, 168–180. [CrossRef]
10. Stroka, J.; Anklam, E. Development of a simplified densitometer for the determination of aflatoxins by thin-layer chromatography. *J. Chromatogr. A* **2000**, *904*, 263–268. [CrossRef]
11. Kolosova, A.Y.; Shim, W.B.; Yang, Z.Y.; Eremin, S.A.; Chung, D.H. Direct competitive ELISA based on a monoclonal antibody for detection of aflatoxin B_1 stabilization of ELISA kit components and application to grain samples. *Anal. Bioanal. Chem.* **2006**, *384*, 286–294. [CrossRef]
12. Khayoon, W.S.; Saad, B.; Yan, C.B.; Hashim, N.H.; Ali, A.S.M.; Salleh, M.I.; Salleh, B. Determination of aflatoxins in animal feeds by HPLC with multifunctional column clean-up. *Food Chem.* **2010**, *118*, 882–886. [CrossRef]
13. Kav, K.; Col, R.; Tekinsen, K.K. Detection of aflatoxin M_1 levels by ELISA in white-brined Urfa cheese consumed in Turkey. *Food Control* **2011**, *22*, 1883–1886. [CrossRef]
14. Selvaraj, J.N.; Zhou, L.; Wang, Y.; Zhao, Y.J.; Xing, F.G.; Dai, X.F.; Liu, Y. Mycotoxin detection-recent trends at global level. *Integr. Agric.* **2015**, *14*, 2265–2281. [CrossRef]
15. Abnous, K.; Danesh, N.M.; Alibolandi, M.; Ramezani, M.; Emrani, A.S.; Zolfaghari, R.; Taghdisi, S.M. A new amplified π-shape electrochemical aptasensor for ultrasensitive detection of aflatoxin B_1. *Biosens. Bioelectron.* **2017**, *94*, 374–379. [CrossRef]
16. Ahamadi, S.F.; Hojjatoleslamy, M.; Kiani, H.; Molavi, H. Monitoring of aflatoxin M_1 in milk using a novel electrochemical aptasensor based on reduced grapheme oxide and gold nanoparticles. *Food Chem.* **2021**, *373*, 131321. [CrossRef]
17. Shadjou, R.; Hasanzadeh, M.; Heidar-poor, M.; Shadjou, N. Electrochemical monitoring of aflatoxin M_1 in milk samples using silver nanoparticles dispersed on α-cyclodextrin-GQDs nanocomposite. *J. Mol. Recognit.* **2018**, *31*, e2699. [CrossRef]
18. Perez-Fernandez, B.; Escosura-Muniz, A.D.L. Electrochemical biosensors based on nanomaterials for aflatoxins detection: A review (2015–2021). *Anal. Chim. Acta* **2022**, *1212*, 339658. [CrossRef]
19. Gevaerd, A.; Banks, C.E.; Bergamini, M.F.; Marcolino-Junior, L.H. Nanomodified screen-printed electrode for direct determination of aflatoxin B1 in malted barley samples. *Sens. Actuator B Chem.* **2020**, *307*, 127547. [CrossRef]

20. Suzuki, Y.; Komatsu, H.; Ikeda, T.; Satio, N.; Araki, S.; Citterio, D.; Hisamoto, D.; Kitamura, Y.; Kubota, T.; Nakagawa, J.; et al. Design and Synthesis of Mg^{2+} Selective Fluoroionophores Based on a Coumarin Derivative and Application for Mg^{2+} Measurement in a Living Cell. *Anal. Chem.* **2002**, *74*, 1423–1428. [CrossRef]
21. Ciesienski, K.L.; Hyman, L.M.; Derisavifard, S.; Franz, K.J. Toward the detection of cellular copper(II) by a light-activated fluorescence increase. *Inorg. Chem.* **2010**, *49*, 6808–6810. [CrossRef]
22. Sheng, J.R.; Feng, F.; Qiang, Y.; Liang, F.G.; Sen, L.; Wei, F.H. A coumarin-derived fluorescence chemosensors selective for copper(II). *Anal. Lett.* **2008**, *41*, 2203–2213. [CrossRef]
23. Lim, N.C.; Brückner, C. DPA-substituted coumarins as chemosensors for zinc(II): Modulation of the chemosensory characteristics by variation of the position of the chelate on the coumarin. *Chem. Commun.* **2004**, *10*, 1094–1095. [CrossRef] [PubMed]
24. Kumar, R.; Gahlyan, P.; Yadav, N.; Bhandari, M.; Kakkar, R.; Manu, D.; Prasad, A.K. Bis-triazolylated-1,4-dihydropyridin—Highly selective hydrophilic fluorescent probe for detection of Fe^{3+}. *Dyes Pigm.* **2017**, *147*, 420–428. [CrossRef]
25. Xu, Y.K.; Liu, X.G.; Zhao, J.Y.; Wang, H.Y.; Liu, Z.; Yang, X.F.; Pei, M.S.; Zhang, G.Y. A new "ON–OFF–ON" fluorescent probe for sequential detection of Fe^{3+} and PPi based on 2-pyridin-2-ylethanamine and benzimidazo [2,1-a]benz[de]isoquinoline-7-one-12-carboxylic acid. *New J. Chem.* **2019**, *43*, 474–480. [CrossRef]
26. Wang, W.; Wu, J.; Liu, Q.L.; Gao, Y.; Liu, H.M.; Zhao, B. A highly selective coumarin-based chemosensor for the sequential detection of $Fe^{3+}$and pyrophosphate and its application in living cell imaging. *Tetrahedron Lett.* **2018**, *59*, 1860–1865. [CrossRef]
27. Zhao, B.; Liu, T.; Fang, Y.; Wang, L.Y.; Song, B.; Deng, Q.G. Two 'turn-off' Schiff base fluorescence sensors based on phenanthro [9,10-d]imidazole-coumarin derivatives for Fe^{3+} in aqueous solution. *Tetrahedron Lett.* **2016**, *57*, 4417–4423. [CrossRef]
28. Luna, D.M.N.; Avelino, K.Y.P.S.; Cordeiro, M.T.; Andrade, C.A.S.; Oliveira, M.D.L. Electrochemical immunosensor for dengue virus serotypes based on 4-mercaptobenzoic acid modified gold nanoparticles on self-assembled cysteine monolayers. Sens. *Actuators B Chem.* **2015**, *220*, 565–572. [CrossRef]
29. Jalalian, S.H.; Ramezani, M.; Danesh, N.M.; Alibolandi, M.; Abnous, K.; Taghdisi, S.M. A novel electrochemical aptasensor for detection of aflatoxin M1 based on target-induced immobilization of gold nanoparticles on the surface of electrode. *Biosens. Bioelectron.* **2018**, *117*, 487–492. [CrossRef]
30. Micheli, L.; Grecco, R.; Badea, M.; Moscone, D.; Palleschi, G. An electrochemical immunosensor for aflatoxin M1 determination in milk using screen-printed electrodes. *Biosens. Bioelectron.* **2006**, *21*, 588–596. [CrossRef]
31. Nguyen, B.H.; Tran, L.D.; Do, Q.P.; Nguyen, H.L.; Tran, N.H.; Nguyen, P.X. Label-free detection of aflatoxin M1 with electrochemical Fe_3O_4/polyaniline-based aptasensor. *Mater. Sci. Eng. C* **2013**, *33*, 2229–2234. [CrossRef]
32. Smolko, V.; Shurpik, D.; Porfireva, A.; Evtugyn, G.; Stoikov, I.; Hianik, T. Electrochemical Aptasensor Based on Poly(Neutral red) and Carboxylated Pillar [5]arene for Sensitive Determination of Aflatoxin M1. *Electroanalysis* **2018**, *30*, 486–496. [CrossRef]
33. Karczmarczyk, A.; Baeumner, A.J.; Feller, K.H. Rapid and sensitive inhibition-based assay for the electrochemical detection of Ochratoxin A and Aflatoxin M1 in red wine and milk. *Electrochim. Acta* **2017**, *243*, 82–89. [CrossRef]
34. Dinçkaya, E.; Kınık, Ö.; Sezgintürk, M.K.; Altuğ, Ç.; Akkoca, A. Development of an impedimetric aflatoxin M1 biosensor based on a DNA probe and gold nanoparticles. *Biosens. Bioelectron.* **2011**, *26*, 3806–3811. [CrossRef] [PubMed]
35. Patel-Sorrentino, N.; Mounier, S.; Benaim, J.Y. Excitation–emission fluorescence matrix to study pH influence on organic matter fluorescence in the Amazon basin rivers. *Water Res.* **2002**, *36*, 2571–2581. [CrossRef]
36. Hussain, E.; Li, Y.X.; Cheng, C.; Zhuo, H.P.; Shahzad, S.A.; Ali, S.; Ismail, M.; Qi, H.; Yu, C. Benzo[ghi]perylene and coronene as ratiometric fluorescence probes for the selective sensing of nitroaromatic explosives. *Talanta* **2020**, *207*, 120316. [CrossRef] [PubMed]

foods

MDPI

Article

Uncovering the Nutritive Profiles of Adult Male Chinese Mitten Crab (*E. sinensis*) Harvested from the Pond and Natural Water Area of Qin Lake Based on Metabolomics

Yuhui Ye, Yulong Wang, Pengyan Liu, Jian Chen and Cunzheng Zhang *

Jiangsu Key Laboratory for Food Quality and Safety-State Key Laboratory Cultivation Base of Ministry of Science and Technology, Institute of Food Safety and Nutrition, Jiangsu Academy of Agricultural Sciences, Nanjing 210014, China; yyh@jaas.ac.cn (Y.Y.); yllzzn@sina.com (Y.W.); 20210047@jaas.ac.cn (P.L.); chenjian@jaas.ac.cn (J.C.)
* Correspondence: zcz@jaas.ac.cn

Abstract: *E. sinensis*, normally harvested in October and November, is an economic aquatic product in China. Pond culture has been widely applied for the production of *E. sinensis*, wherein a stable food supply for crabs is provided. In order to improve the nutritional quality of *E. sinensis* products, this study evaluated the effect of the local pond culture on the nutritive profiles of *E. sinensis* and screened out the best harvest time for the nutrient-rich crabs, thereby guiding the local crab industry to improve its aquaculture mode and harvest strategy. The results indicated that pond culture enhanced the levels of protein, amino acids, and specific organic acid derivatives, and reduced the levels of peptides and polyunsaturated fatty acids (PUFAs). Compared with *E. sinensis* harvested in October, peptide levels were significantly increased, whereas sugar, phenolic acid, and nucleotide levels were decreased in those harvested in November. Overall, the study revealed that the nutritive profile of the pond-reared *E. sinensis* was significantly modulated by a high-protein diet, thus lacking the diversity of metabolites. Additionally, October could be more appropriate for harvesting *E. sinensis* than November.

Keywords: harvest time; metabolomic profiling; polyunsaturated fatty acid; pond culture

Citation: Ye, Y.; Wang, Y.; Liu, P.; Chen, J.; Zhang, C. Uncovering the Nutritive Profiles of Adult Male Chinese Mitten Crab (*E. sinensis*) Harvested from the Pond and Natural Water Area of Qin Lake Based on Metabolomics. *Foods* **2023**, *12*, 2178. https://doi.org/10.3390/foods12112178

Academic Editor: Edel Oddny Elvevoll

Received: 12 April 2023
Revised: 22 May 2023
Accepted: 24 May 2023
Published: 29 May 2023

1. Introduction

Chinese mitten crab (*E. sinensis*) is an important aquatic product native to the East Pacific coast of China and the Korean Peninsula [1]. Although *E. sinensis* has increasingly entered into Europe and America, China has accounted for the most of its production in the world, reaching 770,000 tons in 2020 [2–4]. Jiangsu province of China has the largest crab-producing water area, belonging to the Yangtze River, annually producing half of *E. sinensis* over the country [3].

Qin Lake is located in the north of Jiangsu province and flows as a tributary of Yangtze River. With a subtropical monsoon climate, the average temperature of Qin Lake is 15.3 °C, and the mean annual sunshine duration is 2075 h. Moreover, the lake water has a high dissolved oxygen content (5.5–8 mg/L) and relatively constant pH (7.2–8.6), providing an ideal environment to the local aquaculture [5]. The local people from Qin Lake harvest *E. sinensis* using a unique tool, 'Duan', which is a set of circle-shaped bamboo weirs located in the lake. Small and immature crabs normally pass around the weir circle; however, strong crabs are able to climb over the weir, thus falling down to the center of the weir circle and being captured afterwards [6]. Owing to this special harvest strategy and the ideal lake environment, *E. sinensis* harvested from Qin Lake may possess high nutritional quality.

Protein and fat are two of the major fundamental nutrients in crabs. Crab protein is considered nutritious, owing to the high level of essential amino acids, which not only contribute to the flavor but also exerted various biological effects after being absorbed and metabolized by the human gastrointestinal tract [7]. Fat is nutritionally significant in crustaceans, where SFA (saturated fatty acid), MUFA (monounsaturated fatty acid), and PUFA (polyunsaturated fatty acids) are the key biomarkers for human health. SFA was proven to elevate plasma cholesterol and could be involved in inflammation response [8,9]. However, MUFA and PUFA are able to reduce the level of low-density lipoprotein (LDL) and exhibit beneficial effects on the risk markers of cardiovascular disease, including inflammation, blood pressure, and vascular function [10,11].

Besides protein and fat, other delicate but important nutritional compounds have been uncovered using a metabolomics approach, which was an analytical profiling technique for measuring and comparing large numbers of metabolites present in cells, organs, tissues, or biofluids. As a high-throughput and high-sensitivity technique, metabolomics has been widely adopted recently for the characterization of aquatic products' nutritional quality. For example, the flesh quality of grass carp was evaluated by a UPLC-QTOF/MS-based metabolomics study [12]. Integrated metabolomic and gene expression analyses were conducted for analyzing the meat quality of large yellow croaker [13]. Therefore, metabolomics is an effective method for evaluating aquatic products' nutritional quality. Although the metabolomics data of *E. sinensis* has been collected recently [14], there has been no systematic metabolomics study comparing the nutritive profiles of the pond-reared and naturally produced *E. sinensis*. Normally, crabs reared in a pond are offered a stable food supply, but at the same time, they have limited space for living and reproducing. In contrast, naturally produced crabs are nourished by the lake's eco-system, whereas they should face the environmental changes independently. Hence, the way in which these different aquaculture modes affect the nutritive profiles of *E. sinensis* remains to be elaborated. Further, October and November have been regarded as the best season for crab harvesting in China [3], but no report is available that has investigated the delicate difference of the nutritional quality of crabs harvested between these two months.

Therefore, the present study aimed to decipher the nutritive profiles of *E. sinensis* harvested from the pond and natural water area of Qin Lake, as well as to compare the different nutritional compositions between *E. sinensis* harvested in October and November, thereby providing comprehensive metabolomics information for the aquaculture of *E. sinensis* and thus facilitating the production of *E. sinensis* with superior nutritional quality.

2. Materials and Methods

2.1. Sample Preparation and Sampling

The crabs were reared in a 0.8 ha water pond and fed a high-protein diet (91% of dry matter, 56% of crude protein, 9% of crude lipid, and 13% of ash) twice per day. However, the crabs from the natural water area of Qin Lake, China (32°31′11.47″–32°42′56.74″ N, 119°53′16.94″–120°16′14.52″ E), were mostly dependent on natural food, such as algae, zoobenthos, and planktons. Twenty-one male crabs (200.0 ± 10.0 g) were collected from the pond in October 2021 (P-M10) and November 2021 (P-M11), respectively. Twenty-one male crabs (200.0 ± 10.0 g) were collected from natural water area in October 2021 (N-M10) and November 2021 (N-M11), respectively. Then, the crabs were stored in a plastic foam box with a small amount of water and thereafter were transported to the laboratory within an hour. After that, the crabs were cleaned with water and sacrificed, and the crab meat, including breast, feet and claw meat, and crab cream, including the hepatopancreas and gonad, were picked by hand. All the samples were frozen at −20 °C before analyses. The 21 crabs of each group were divided into two parts: 3 crabs were chosen for the proximate composition, free amino acid, and fatty acid composition analyses, and the rest of the 18 crabs were analyzed by the metabolomics approach. The edible part of the crab, including the meat and cream, was mixed before analysis. Therefore, 18 edible parts of crabs were obtained, and they were further divided into 6 biological replicates by randomly mixing every 3 edible parts together.

2.2. *Proximate Composition Analysis*

The moisture content and protein content were analyzed on the basis of Association of Analytical Chemists Procedures [15]. For the ash content measurement, 0.5 g of the sample was added to the crucible and combusted at 550 ± 25 °C for 30 min. The ash was then cooled down and weighed. The ash content (%) was finally presented as the ratio of the ash weight to the sample weight. The total lipid was extracted with absolute ether by a Soxhlet apparatus (BSXT-02, Yuming Co., Ltd., Shanghai, China). A total of 1 g of the sample was added to the evaporating dish and removed the moisture via evaporation, and then we transferred the dried sample into a filter paper pocket. The paper pocket was then transferred to the Soxhlet apparatus, and absolute ether was added to extract the total lipid. The whole apparatus was run under a water bath for 6–8 h. After extraction, the total lipid was weighed, and the lipid content (%) was calculated by dividing by the original sample weight.

2.3. *Amino Acid Determination*

A total of 5 mL of 10% sulfosalicylic acid (304851-84-1, Sigma, Darmstadt, Germany) was added to the crushed crab meat or cream samples. Then, the solution was centrifuged at 12,000× *g* for 20 min. We collected the supernatant and added 1 mL of petroleum ether (101316-46-5, Sigma, Darmstadt, Germany). After 1 h, we removed the petroleum ether and obtained the solution in the bottom. Lastly, the solution was filtrated by a 0.22 μm syringe filter and loaded to an automatic amino acid analyzer (LA 8080, Hitachi, Japan) equipped with a Hitachi high-performance cation exchange column.

2.4. *Fatty Acid Composition*

A total of 20 mL of chloroform/methanol solution (*v*/*v*, 2:1, Sigma, Darmstadt, Germany) was added to the 1.0 g sample. After 2 h, we removed the chloroform by a rotatory evaporator (RE-52AA, Yarong Co., Ltd., Shanghai, China). Then, 3 mL of 0.5 mol/L sodium hydroxide methanol solution was added, and the mixture was heated to 50 °C for 20 min by a thermostatic water bath (HH-1, Yuexin Instrument Manufacturing, Co., Ltd., Changzhou, China). Next, methanolic boron trifluoride (3 mL, 13%, Sigma, Darmstadt, Germany) was added, and the mixture was heated again at 85 °C for 30 min. Finally, after cooling down to the room temperature, 1 mL of n-hexane was added, and the mixture was filtered through a 0.45 μm syringe filter. The solution was applied to an Agilent 6890 gas chromatograph equipped with an HP-5.5% phenyl methyl Si-loam capillary column (30.0 m × 25 mm, Agilent 19091J-413, Santa Clara, CA, USA).

2.5. *Wide Targeted Metabolomics Analysis*

A total of 20 mg of each replicate (6 biological replicates in each experimental group) was weighed and homogenized (30 HZ) for 20 s with a steel ball. The mixture was centrifuged (3000 rpm, 4 °C) for 30 s. Then, 400 μL of 70 % methanol was added, and the solution was shaken (1500 rpm) for 5 min and placed on ice for 15 min. After that, the solution was centrifuged at 12,000 rpm, 4 °C for 10 min, and 300 μL of the supernatant was obtained. Finally, the supernatant was centrifuged again (12,000 rpm, 4 °C) for 3 min, and the newly obtained supernatant was collected for LC-MS/MS analysis, where a LC-ESI-MS/MS system (UPLC, ExionLC AD; MS, QTRAP® System) was applied. The analytical conditions were as follows: UPLC column, Waters ACQUITY UPLC HSS T3 C18 (1.8 μm, 2.1 mm × 100 mm); column temperature, 40 °C; flow rate, 0.4 mL/min; injection volume, 5 μL; solvent system, water (0.1% formic acid): acetonitrile (0.1% formic acid); gradient program, 95:5 *v*/*v* at 0 min, 10:90 *v*/*v* at 10.0 min, 10:90 *v*/*v* at 11.0 min, 95:5 *v*/*v* at 11.1 min, 95:5 *v*/*v* at 14.0 min. Linear ion trap (LIT) and triple quadrupole (QQQ) scans were acquired on a triple quadrupole-linear ion trap mass spectrometer (QTRAP), QTRAP® LC-MS/MS System, equipped with an ESI Turbo Ion-Spray interface, operating in positive and negative ion modes and controlled by Analyst 1.6.3 software (Sciex). The ESI source operation parameters were as follows: source temperature, 500 °C; ion spray voltage (IS), 5500 V (positive) or −4500 V (negative); ion source gas I, gas II, and curtain gas were set at 50, 50, and 25 Psi, respectively; the collision gas was set as high. Instrument tuning and mass calibration were performed with 10 and 100 μmol/L polypropylene glycol solutions in QQQ and LIT modes, respectively.

2.6. Statistical Analysis

The significant differences among samples were analyzed by one-way ANOVA with SPSS statistics package software (version 25.0, Chicago, IL, USA), using Tukey's range test under a significant level of $p < 0.05$. In terms of MS data analysis, unsupervised PCA (principal component analysis) was performed by statistics function prcomp within R (www.r-project.org, accessed on July 2022). The data were unit variance scaled before unsupervised PCA. Variable importance of the projected (VIP) values and the permutation plots were extracted from orthogonal partial least squares discriminant analysis (OPLS-DA) result, which was generated using the R package MetaboAnalystR. The data was log transformed (log2) and mean centered before OPLS-DA. Differential metabolites were selected on the basis of VIP > 1, absolute Log_2FC (fold change: the ion intensity of the metabolite in experimental group/the ion intensity of the metabolite in control group) > 2, and $p < 0.05$.

3. Results

3.1. Proximate Composition of the E. sinensis

The proximate composition of the edible part of *E. sinensis* (meat and cream) is presented (Table S1). P-M11 showed the highest protein content in meat samples, followed by P-M10, N-M11, and N-M10. N-M11, P-M11, and P-M10 had significantly higher fat content compared to N-M10 ($p < 0.05$). For crab cream, the pond-reared crabs (P-M10 and P-M11) also had the most proteins, and crabs harvested in November possessed higher protein content compared to those harvested in October. The lowest fat content was observed in N-M11; however, no significant difference was found among the other groups ($p < 0.05$).

3.2. Amino Acid Determination

The pond-reared crabs had a significantly higher amount total amino acid in meat compared to crabs harvested in a natural water area ($p < 0.05$, Table 1).

Table 1. Free amino acid content of the meat and cream from *E. sinensis* (mg/g of wet weight).

	Meat				Cream			
	N-M10	**P-M10**	**N-M11**	**P-M11**	**N-M10**	**P-M10**	**N-M11**	**P-M11**
Asp [1]	N.D.	0.02 ± 0.00^{a}	0.01 ± 0.00^{a}	0.01 ± 0.00^{a}	1.08 ± 0.13^{C}	1.41 ± 0.07^{B}	1.72 ± 0.18^{A}	1.50 ± 0.11^{AB}
Thr	0.63 ± 0.07^{a}	0.59 ± 0.06^{ab}	N.D.	0.47 ± 0.05^{b}	2.49 ± 0.22^{A}	2.13 ± 0.16^{A}	2.58 ± 0.32^{A}	2.49 ± 0.33^{A}
Ser	0.18 ± 0.02^{b}	0.32 ± 0.07^{a}	0.15 ± 0.04^{b}	0.24 ± 0.03^{ab}	1.34 ± 0.12^{A}	1.41 ± 0.32^{A}	1.25 ± 0.23^{A}	0.98 ± 0.15^{B}
Glu	0.17 ± 0.01^{b}	0.32 ± 0.09^{a}	0.12 ± 0.02^{b}	0.31 ± 0.07^{a}	2.95 ± 0.15^{B}	2.91 ± 0.34^{B}	3.80 ± 0.29^{A}	3.88 ± 0.35^{A}
Gly	1.69 ± 0.31^{b}	3.59 ± 0.24^{a}	1.50 ± 0.24^{b}	3.80 ± 0.23^{a}	1.32 ± 0.16^{B}	1.09 ± 0.14^{B}	1.67 ± 0.18^{A}	1.89 ± 0.38^{A}
Ala	2.20 ± 0.07^{b}	4.87 ± 0.74^{a}	1.78 ± 0.34^{b}	5.07 ± 0.70^{a}	2.81 ± 0.36^{C}	2.72 ± 0.22^{C}	4.86 ± 0.12^{B}	5.50 ± 0.22^{A}
Cys	0.04 ± 0.01^{b}	0.08 ± 0.02^{a}	0.02 ± 0.00^{c}	0.07 ± 0.02^{a}	0.26 ± 0.03^{B}	0.32 ± 0.03^{A}	0.25 ± 0.01^{B}	0.28 ± 0.07^{AB}
Val	0.14 ± 0.03^{b}	0.27 ± 0.10^{a}	0.11 ± 0.04^{b}	0.29 ± 0.07^{a}	1.68 ± 0.18^{A}	1.92 ± 0.18^{A}	1.99 ± 0.47^{A}	2.04 ± 0.23^{A}
Met	0.14 ± 0.01^{bc}	0.31 ± 0.11^{a}	0.10 ± 0.03^{c}	0.23 ± 0.02^{ab}	0.83 ± 0.03^{A}	0.85 ± 0.12^{A}	0.81 ± 0.06^{A}	0.86 ± 0.04^{A}
Ile	0.07 ± 0.02^{bc}	0.15 ± 0.05^{a}	0.05 ± 0.01^{c}	0.13 ± 0.03^{ab}	1.23 ± 0.19^{A}	1.25 ± 0.35^{A}	1.49 ± 0.09^{A}	1.54 ± 0.22^{A}
Leu	0.16 ± 0.01^{ab}	0.29 ± 0.12^{a}	0.12 ± 0.02^{b}	0.22 ± 0.09^{ab}	2.36 ± 0.29^{A}	2.55 ± 0.44^{A}	2.70 ± 0.28^{A}	2.75 ± 0.12^{A}
Tyr	0.07 ± 0.02^{b}	0.27 ± 0.09^{a}	0.09 ± 0.02^{b}	0.14 ± 0.03^{b}	1.36 ± 0.29^{A}	1.32 ± 0.14^{A}	1.22 ± 0.25^{A}	1.16 ± 0.20^{A}
Phe	0.10 ± 0.04^{b}	0.29 ± 0.08^{a}	0.11 ± 0.03^{b}	0.23 ± 0.07^{a}	1.42 ± 0.05^{A}	1.57 ± 0.15^{A}	1.42 ± 0.15^{A}	1.45 ± 0.22^{A}
Lys	0.11 ± 0.02^{b}	0.46 ± 0.19^{a}	0.13 ± 0.01^{b}	0.51 ± 0.03^{a}	2.12 ± 0.27^{A}	2.34 ± 0.38^{A}	2.74 ± 0.24^{A}	2.09 ± 0.48^{A}
His	0.14 ± 0.03^{b}	0.31 ± 0.08^{a}	0.12 ± 0.03^{b}	0.25 ± 0.04^{a}	0.76 ± 0.06^{AB}	0.69 ± 0.17^{B}	0.89 ± 0.09^{A}	0.90 ± 0.07^{A}
Arg	2.53 ± 0.19^{b}	3.81 ± 0.18^{a}	1.86 ± 0.33^{c}	4.26 ± 0.17^{a}	2.67 ± 0.27^{A}	2.70 ± 0.29^{A}	2.99 ± 0.14^{A}	3.28 ± 0.52^{A}

Table 1. *Cont.*

	Meat				Cream			
	N-M10	P-M10	N-M11	P-M11	N-M10	P-M10	N-M11	P-M11
Pro	1.06 ± 0.06 b	1.19 ± 0.22 b	1.07 ± 0.05 b	2.32 ± 0.17 a	1.44 ± 0.11 B	1.50 ± 0.24 B	2.91 ± 0.40 A	2.43 ± 0.32 A
Essential AA	1.49 ± 0.09 c	2.67 ± 0.37 a	0.74 ± 0.13 d	2.33 ± 0.13 b	12.89 ± 0.45 C	13.30 ± 0.37 BC	14.62 ± 0.1 A	14.12 ± 0.57 AB
Total umami AA	0.17 ± 0.01 b	0.34 ± 0.09 a	0.13 ± 0.02 b	0.32 ± 0.07 a	4.03 ± 0.25 B	4.32 ± 0.39 B	5.52 ± 0.29 A	5.38 ± 0.24 A
Total sweet AA	5.76 ± 0.39 c	10.56 ± 0.80 b	4.50 ± 0.22 d	11.90 ± 0.49 a	9.40 ± 0.30 B	8.85 ± 0.22 B	13.27 ± 0.96 A	13.29 ± 1.10 A
Total bitter AA	3.39 ± 0.29 b	5.89 ± 0.53 a	2.60 ± 0.25 c	6.12 ± 0.28 a	13.07 ± 0.54 B	13.87 ± 0.44 B	15.03 ± 0.41 A	14.91 ± 0.32 A
Total AA	9.43 ± 0.31 c	17.14 ± 0.55 b	7.34 ± 0.04 d	18.55 ± 0.64 a	28.12 ± 0.15 B	28.68 ± 0.99 B	35.29 ± 0.67 A	35.02 ± 1.20 A

[1] Letters in lower case mark the significant difference of each line of the meat samples ($p < 0.05$); letters in upper case mark the significant difference of each line of the cream samples ($p < 0.05$). Sweet AA includes Thr, Ser, Gly, Ala, and Pro. Umami AA includes Asp and Glu. Bitter AA includes Val, Met, Ile, Leu, Phe, Lys, His, and Arg. N.D. = not detected.

Glu, Gly, Ala, Val, Phe, Lys, His, and Arg were the most significant amino acids accumulated in the meat of pond-reared crabs. The crabs harvested in November possessed a greater level of total amino acids in cream compared to those harvested in October ($p < 0.05$), which was mostly due to the increased levels of Glu, Gly, and Ala.

3.3. Fatty Acid Composition

For the crab meat, the pond-reared crabs had significantly higher MUFAs and lower SFAs and PUFAs than crabs from a natural water area ($p < 0.05$, Table 2).

Table 2. The fatty acid composition of the meat and cream from *E. sinensis* (% of total fatty acid).

	Meat				Cream			
	N-M10	P-M10	N-M11	P-M11	N-M10	P-M10	N-M11	P-M11
$C_{14:0}$	0.35 ± 0.06	0.49 ± 0.03	0.38 ± 0.02	0.62 ± 0.03	0.94 ± 0.02	1.28 ± 0.05	0.77 ± 0.06	1.23 ± 0.03
$C_{15:0}$	N.D.	0.32 ± 0.02	0.21 ± 0.01	0.28 ± 0.02	0.45 ± 0.03	0.73 ± 0.04	0.43 ± 0.02	0.56 ± 0.01
$C_{16:0}$	14.50 ± 0.46	15.13 ± 0.11	14.15 ± 0.15	14.04 ± 0.40	20.17 ± 0.36	21.83 ± 0.42	19.26 ± 0.18	19.49 ± 0.12
$C_{17:0}$	0.57 ± 0.08	0.66 ± 0.05	0.45 ± 0.02	0.54 ± 0.02	0.35 ± 0.09	0.48 ± 0.05	0.24 ± 0.01	0.35 ± 0.05
$C_{18:0}$	7.56 ± 0.35	7.29 ± 0.40	6.71 ± 0.25	6.40 ± 0.01	2.49 ± 0.30	2.92 ± 0.24	2.49 ± 0.13	2.60 ± 0.06
$C_{20:0}$	N.D.	N.D.	N.D.	N.D.	0.23 ± 0.02	N.D.	N.D.	0.21 ± 0.02
$C_{24:0}$	12.47 ± 0.47	9.98 ± 0.14	12.48 ± 0.16	11.13 ± 0.13	1.71 ± 0.20	1.64 ± 0.10	1.42 ± 0.12	1.54 ± 0.16
SFA [1]	35.45 ± 1.00 a	33.87 ± 0.43 bc	34.38 ± 0.55 b	33.01 ± 0.21 c	26.34 ± 0.26 B	28.88 ± 0.72 A	24.61 ± 0.42 C	25.98 ± 0.44 B
$C_{16:1\ n7}$	3.14 ± 0.17	3.20 ± 0.12	2.89 ± 0.02	2.79 ± 0.06	8.72 ± 0.15	7.81 ± 0.09	7.59 ± 0.10	7.71 ± 0.02
$C_{17:1\ n7}$	0.47 ± 0.06	0.51 ± 0.03	0.45 ± 0.06	0.44 ± 0.02	0.54 ± 0.05	0.50 ± 0.04	0.48 ± 0.02	0.52 ± 0.02
$C_{18:1\ n9}$	20.71 ± 0.51	22.74 ± 0.13	21.45 ± 0.11	21.75 ± 0.09	28.05 ± 0.17	27.45 ± 0.60	29.23 ± 0.16	29.06 ± 0.11
$C_{20:1\ n9}$	0.42 ± 0.08	0.51 ± 0.02	0.44 ± 0.01	1.58 ± 0.03	0.53 ± 0.05	0.74 ± 0.09	0.61 ± 0.02	0.92 ± 0.02
MUFA	24.74 ± 0.67 b	26.96 ± 0.05 a	25.24 ± 0.15 b	26.56 ± 0.21 a	37.84 ± 0.29 A	36.50 ± 0.53 B	37.92 ± 0.28 A	38.22 ± 0.13 A
$C_{18:2\ n6}$	17.27 ± 0.24	13.27 ± 0.22	18.11 ± 0.19	15.08 ± 0.09	25.43 ± 0.27	16.54 ± 0.43	26.47 ± 0.36	19.60 ± 0.19
$C_{18:3\ n3}$	1.66 ± 0.06	1.39 ± 0.06	1.34 ± 0.04	0.21 ± 0.02	2.39 ± 0.08	2.35 ± 0.19	1.70 ± 0.03	2.07 ± 0.01
$C_{20:2\ n6}$	1.31 ± 0.08	1.81 ± 0.11	1.37 ± 0.07	1.62 ± 0.10	0.90 ± 0.10	2.07 ± 0.29	1.09 ± 0.05	1.68 ± 0.10
$C_{20:3\ n3}$	0.35 ± 0.03	0.29 ± 0.33	0.34 ± 0.01	N.D.	0.31 ± 0.05	0.35 ± 0.48	0.32 ± 0.01	0.28 ± 0.03
$C_{20:4\ n6}$	4.62 ± 0.17	7.08 ± 0.23	4.56 ± 0.19	6.32 ± 0.17	1.08 ± 0.12	2.79 ± 0.20	1.11 ± 0.14	2.10 ± 0.09
$C_{22:6\ n3}$	6.30 ± 0.40	5.48 ± 0.26	6.45 ± 0.27	7.13 ± 0.20	0.93 ± 0.22	1.08 ± 0.06	0.89 ± 0.11	2.01 ± 0.19
PUFA	31.50 ± 0.64 a	29.31 ± 0.07 c	32.17 ± 0.37 a	30.36 ± 0.35 b	31.04 ± 0.22 A	25.18 ± 0.61 C	31.58 ± 0.63 A	27.74 ± 0.26 B
n3 PUFA	8.30 ± 0.46 a	7.15 ± 0.27 b	8.13 ± 0.27 a	7.34 ± 0.19 b	3.64 ± 0.22 B	3.78 ± 0.14 B	2.91 ± 0.11 C	4.36 ± 0.16 A
n6 PUFA	23.20 ± 0.23 b	22.16 ± 0.21 c	24.04 ± 0.11 a	23.02 ± 0.17 b	27.40 ± 0.28 B	21.40 ± 0.57 D	28.67 ± 0.54 A	23.39 ± 0.33 C
n3/n6	0.36 ± 0.02 a	0.32 ± 0.02 b	0.34 ± 0.01 ab	0.32 ± 0.01 b	0.13 ± 0.01 B	0.18 ± 0.01 A	0.10 ± 0.00 C	0.18 ± 0.01 A

[1] Letters in lower case mark the significant difference of the SFA, MUFA, and PUFA of the meat samples ($p < 0.05$); letters in upper case mark the significant difference of the SFA, MUFA, and PUFA of the cream samples ($p < 0.05$). N.D. = not detected.

Crabs harvested in October had significantly more SFAs and less MUFAs and PUFAs than those harvested in November ($p < 0.05$). For the cream samples, the levels of PUFAs were less in the crabs harvested in November when compared to those harvested in October ($p < 0.05$). Further, the pond-reared crabs had significantly more SFAs and less PUFAs compared to those from a natural water area ($p < 0.05$).

3.4. Overview of the Metabolic Profile of *E. sinensis*

A widely targeted metabolomic analysis was performed on the meat and cream (mixture) of *E. sinensis*. In total, 1920 metabolites were detected, including 680 amino acids and their metabolites; 158 organic acids and their metabolites; 156 heterocyclic compounds; 151 fatty acyls; 129 benzene and substituted derivatives; 117 nucleotides and their metabolites; 95 glycerophospholipids; 91 aldehyde, ketones, and esters; 84 alcohol and amines; 60 carboxylic acids and derivatives; 34 hormones and hormone-related compounds; 31 glycerolipids; and 46 undefined organic compounds and other minor components (less than 1% of total metabolites) (Figure 1A).

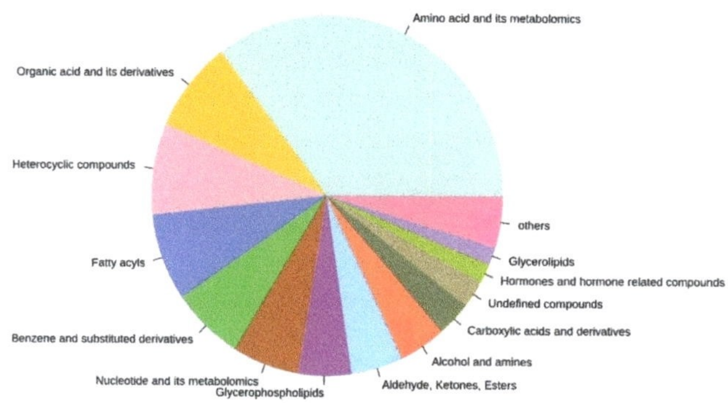

(A)

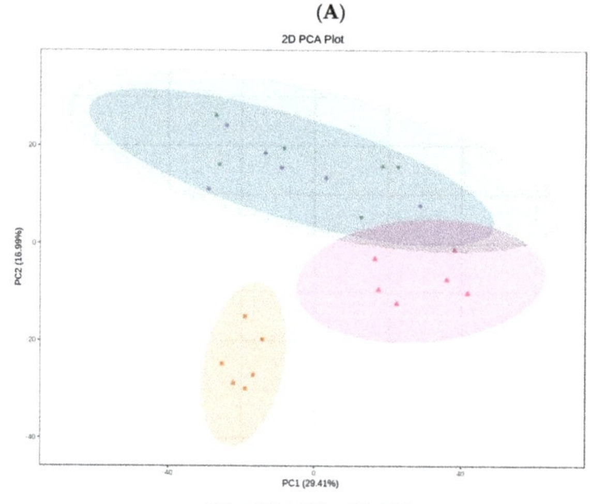

(B)

Figure 1. The overview of the metabolomic profile (**A**) and principal component analysis of *E. sinensis* harvested from the pond and natural water area of Qin Lake (**B**). Amino acid and its metabolomics: 35.42%; organic acid and its derivatives: 8.23%; heterocyclic compounds: 8.12%; fatty acyls: 7.86%; benzene and substituted derivatives: 6.72%; nucleotide and its metabolomics: 6.09%; glycerophospholipids: 4.95%; aldehyde, ketones, and esters: 4.74%; alcohol and amines: 4.38%; carboxylic acids and derivatives: 3.12%.

Unsupervised PCA was conducted to compare the differences of the metabolite profiles among all the groups, where six biological replicates were clustered together by different colors (Figure 1B). The first principal component (PC1) was with a 29.41% variance contribution value, and P-M10 and P-M11 were separated by line X = 0 as distinct groups. However, N-M10 and N-M11 were relatively similar, sharing large areas of overlap. On the basis of the second principal component (PC2), the line Y = 0 separated groups between crabs from the natural water area (N-M10, N-M11, Y > 0) and the pond (P-M10, P-M11, Y < 0). The result indicated that the pond culture significantly altered the metabolites of *E. sinensis*. In addition, the harvest time (maturity of crabs) remarkably affected the metabolites in pond-reared crabs, whereas metabolites in crabs from a natural water area were relatively similar between October and November.

3.5. *Differential Metabolites Analysis*

3.5.1. Effect of Pond Culture on the Metabolites of *E. sinensis*

The differential metabolites were analyzed according to pairwise comparisons, using the OPLS-DA model. The permutation tests of OPLS-DA models are shown in the Supplementary Materials (Figure S1), which reflects the robustness of the model. A further differential metabolites screening was applied, wherein metabolites with a fold change > 2 (upregulation) or <0.5 (downregulation), VIP > 1, and $p < 0.05$ were regarded as significant differential metabolites. In the present study, there were 275 upregulated and 174 downregulated metabolites in P-M10 compared with N-M10 (Figure 2A,C).

Six organic acid derivatives were among the top 10 upregulated metabolites, including stachydrine, methylsuccinic acid, and glutaric acid, whereas the top 10 downregulated metabolites mostly contain small peptides. In November, there were more downregulated metabolites (373) than upregulated metabolites (66) observed in P-M11 compared to N-M11 (Figure 2B,D). Similar to the results from October, peptides were the most downregulated metabolites (127 out of 373), followed by glycerophospholipids and heterocyclic compounds. Organic acid derivatives were also the most upregulated metabolites. This showed that the pond culture exhibited a similar effect on the nutritive profiles between *E. sinensis* harvested in October and November. In addition, the Venn diagram illustrates the common metabolites shared between P-M10 and P-M11, revealing the key metabolites modulated by pond breeding (Figure 2E).

3.5.2. Effect of Harvest Time (Maturity) on the Metabolites of *E. sinensis*

There were 165 differential metabolites identified in N-M11 compared to N-M10 (Figure 3A,C).

The most upregulated metabolites included 5,6-epoxy-eicosatrienoic acid (EET), N-stearamide, peptide HQTE, hypericin, and palmitoylethanolamide (PEA). Urocanic acid, trigonelline, and urea, followed by several sugars, including D-trehalose, lactose, lactulose, and maltose, were the most downregulated metabolites (Figure 4).

Regarding on the pond-reared crabs, the metabolites were significantly different between October and November, due to approximately one-third of the metabolites (674 out of 1908) being identified as differential metabolites according to our screening threshold (Figure 3B,D). Small peptides and carnitine (C18 and C14) were the most upregulated metabolites, whereas phenolic acids (8-paradol and coniferaldehyde) and nucleotides (2′-deoxyuridine and deoxyguanosine) were significantly downregulated (Figure 4). Moreover, 57 key metabolites affected by the maturity of crabs were identified (Figure 3E).

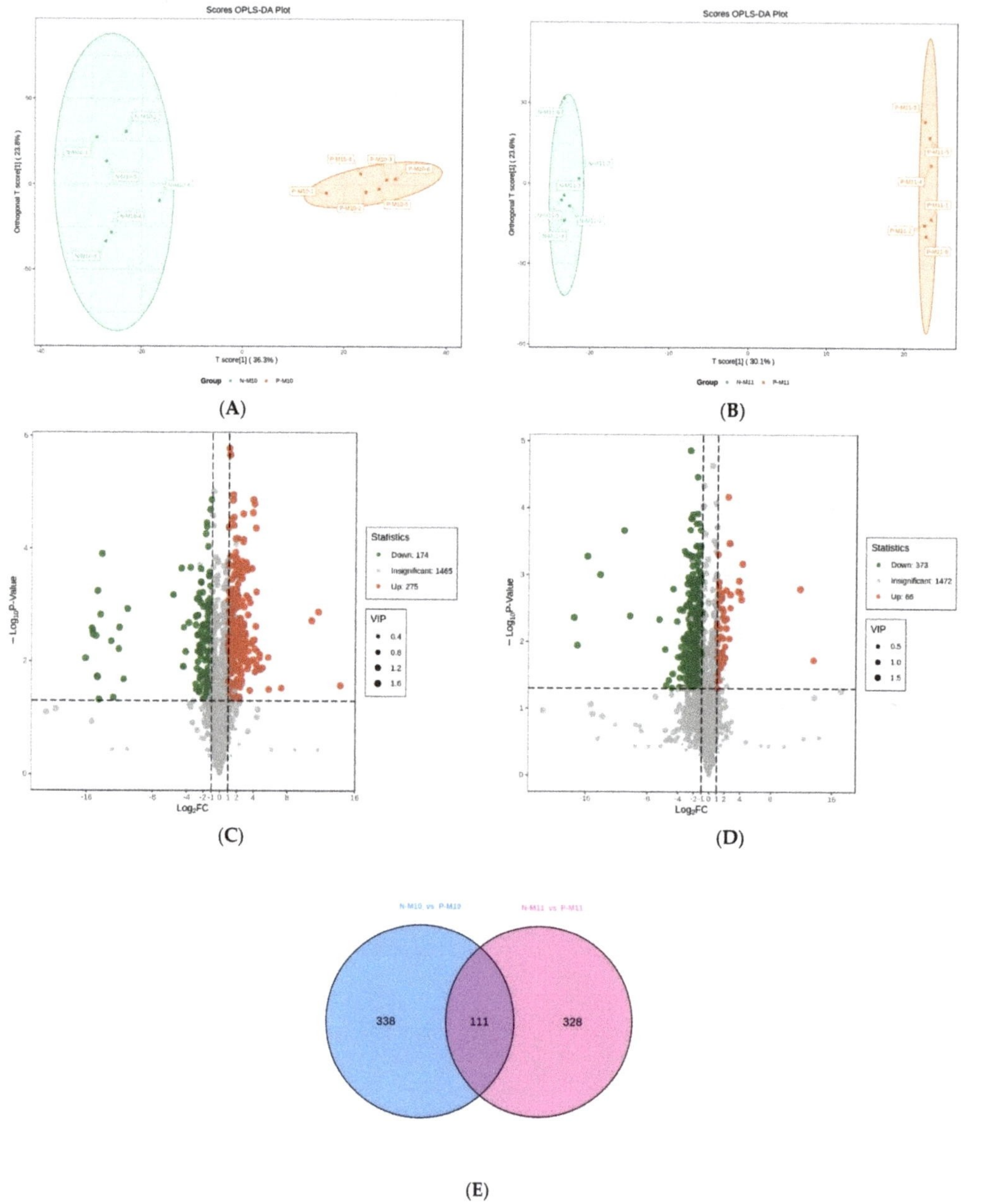

Figure 2. The score plots of OPLS-DA pairwise comparisons between N-M10 and P-M10 (**A**), N-M11 and P-M11 (**B**). Volcano plots of the differential metabolites between N-M10 and P-M10 (**C**), N-M11 and P-M11 (**D**). Venn diagram of differential metabolites (**E**). The T score of the score plot indicates the degree of variations between groups, and the orthogonal T score represents the degree of variations within the group. Percentages indicate how well a component explains the dataset. The volcano plot illustrates the distribution of the up-/downregulated differential metabolites. In the Venn diagram, each comparison group is illustrated as a circle, and the number in it indicates the number of differential metabolites. The number in the overlapped area of the circles indicates the number of the common differential metabolites among all the comparison groups.

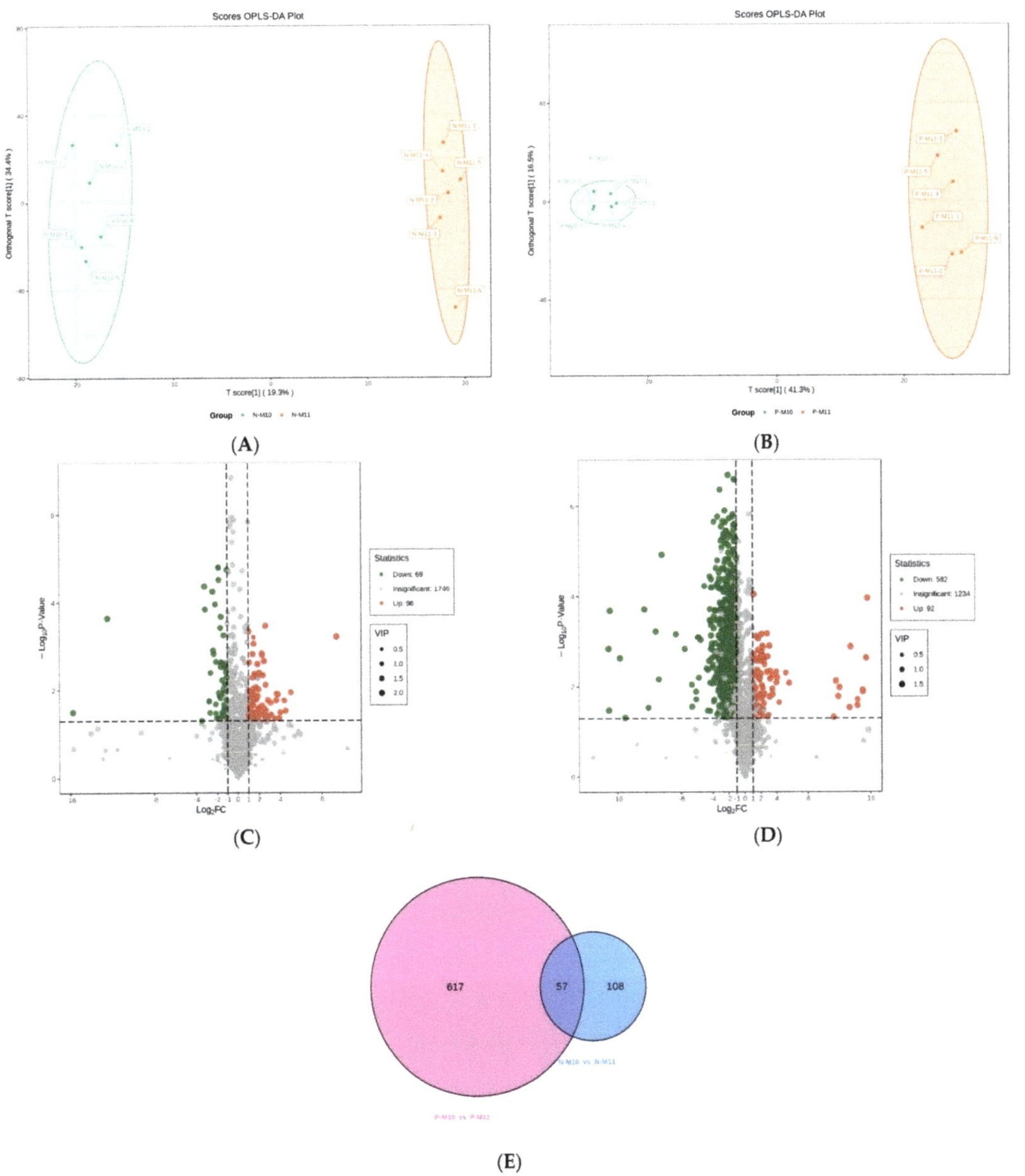

Figure 3. The score plots of OPLS-DA pairwise comparisons between N-M10 and N-M11 (**A**), P-M10 and P-M11 (**B**). Volcano plots of the differential metabolites between N-M10 and N-M11 (**C**), P-M10 and P-M11 (**D**). Venn diagram of differential metabolites (**E**). The T score of the score plot indicates the degree of variations between groups, and the orthogonal T score represents the degree of variations within the group. Percentages indicate how well a component explains the dataset. The volcano plot illustrates the distribution of the up-/downregulated differential metabolites. In the venn diagram, each comparison group is illustrated as a circle, and the number in it indicates the number of differential metabolites. The number in the overlapped area of the circles indicates the number of the common differential metabolites among all the comparison groups.

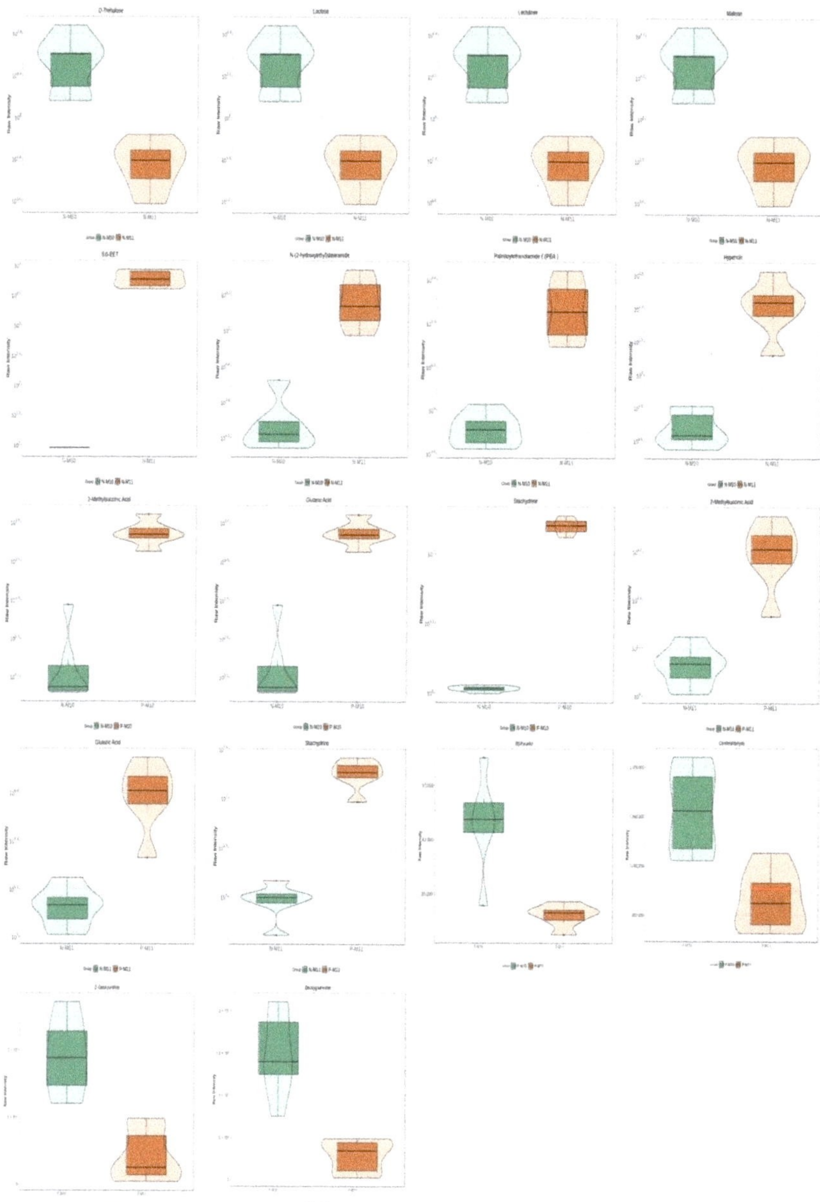

Figure 4. Representatives of the differential metabolites from *E. sinensis*. The box in the middle represents interquartile range. The 95% confidence interval is presented as a black line that penetrates through the box. The black line in the middle of the box is the median. The outer shape describes the density of distribution.

4. Discussion

The production of *E. sinensis* in China experienced a rapid increase, from 17,500 tons in 1993 to 77,000 tons in 2020, which brought in huge economic benefit, reaching a gross annual output value of more than 10 billion dollars [16]. Although a huge business success occurred for this large-scale crab industry, the nutritional quality of the products, which may be compromised for the yield, possessed great potential to be improved. The present study firstly focused on the aquaculture mode and therefore intended to figure out how the traditional pond culture at Qin Lake affected the nutritive profiles of *E. sinensis*. Secondly, the study elucidated the difference of nutritive profiles between crabs harvested in October and November, thereby screening out the best harvest time, in order to facilitate the production of *E. sinensis* with superior nutritional quality. As a metabolomics-based study, it should be noted that the crabs were transported to the laboratory freshly after harvesting without any freezing or drying steps. Therefore, the metabolites kept their best activities during the following extraction step.

The characterization of samples, regarding the proximate composition of the pond-reared and naturally produced crabs, was firstly determined. The pond culture showed a significant effect in terms of increasing the protein level of both crab meat and cream. Further, P-M10 and P-M11 contained significantly higher fat content than N-M11 in the crab cream ($p < 0.05$), which was in agreement with He et al. [17], where pond culture exhibited a significant growth-promoting effect on crabs, including enhancement of both fat and protein content. Pond culture was often applied as a traditional aquaculture mode, in order to ensure a stable growth of the aquatic products, thus facilitating a high yield and thereby creating economic values. However, people nowadays focused more on their health and began to consider whether the consumption of this unnaturally cultured aquatic products would lead to abnormal nutritional uptake or even health problems. Contrary to natural lake water, the water in the pond flowed less, and therefore may have contained less oxygen and fewer small organisms, although enough food for the crabs was supplemented. Several recent studies indicated that the strong crabs reared in ponds were not always associated with superior nutrition, suggesting that various kinds of metabolites, rather than protein and fat, also contributed to the crabs' nutritive profile and exhibition of high nutritional value [18–20]. This raised the need for a wide-targeted metabolomics study to decipher the subtle difference of the nutritive profiles between pond-reared and naturally produced crabs. Regarding the free amino acid, pond-reared crabs accumulated more Glu, Gly, Ala, Val, Phe, Lys, His, and Arg than naturally produced crabs. Additionally, the sum levels of essential amino acids in pond-reared crabs were significantly higher than those in naturally produced crabs ($p < 0.05$). A similar result was observed by He et al. [21], who found that almost all the amino acid levels were elevated in the pond-reared swimming crabs (Portunus trituberculatus). The amino acids in crabs not only provided flavor but also participated in amino acid metabolism after being absorbed by the human body, exerting numerous biological processes. For example, Glu/Gln and Asp/Asn were involved in nitrogen assimilation, synthesis of other amino acids and nitrogenous compounds, and carbon/nitrogen partitioning [22]. Lys was the major building blocks for growth factors and played important roles in cell division and wound-healing processes. Deficiency in Lys could cause impaired growth performance, immune response, and inflammation in grown-up grass carp [23]. Further, Lys and Met were the precursors for the synthesis of carnitine, which was able to convert fat into energy and suppress oxidative stress [24]. Additionally, regarded as a dietary essential amino acid and the precursor for various hormones, His showed great importance in kidney function, gastric secretion, the immune system, and neurotransmission [25]. Overall, the pond culture enriched these important amino acids in *E. sinensis*, owing to its stable supply of protein-rich food.

Fatty acids also played important roles in the energy metabolism and regulation of blood glucose [26]. However, in contrast to free amino acids, the pond-reared crabs had significantly less PUFAs than crabs from a natural water area, both in terms of the crab meat and cream ($p < 0.05$). Similar results were observed by Kong, Cai, Ye, Chen, Wu, Li, Chen, and Song [18], who indicated that the snails in the natural water area served as an extra source of PUFA for crabs, leading to an increase in PUFAs in the naturally produced crabs. Moreover, the naturally produced crabs had a larger space for living and traveling, thus being likely to develop more muscle, wherein the stearoyl-CoA desaturase 1 (SCD1) from crab muscle was able to catalyze SFA into PUFA [27]. PUFAs, including omega-3 and omega-

6 fatty acids, could maintain cell membrane fluidity and improve vascular endothelial cell function, as well as being important to the regulation of obesity, metabolic disease, cardiovascular and immune functions, and neuropsychiatric disorders [28]. However, pond culture led to decreased PUFAs in the present study, which may have been due to the deficiency of snails in the pond and the reduced muscle content in crabs.

The organic acid derivatives stachydrine, methylsuccinic acid, and glutaric acid were significantly upregulated in the pond-reared crabs compared to those cultured in a natural water area. Interestingly, stachydrine was a potent bioactive compound, having demonstrated various activities against tissue fibrosis, cardiovascular diseases, cancers, uterine disease, and inflammation [29]. However, methylsuccinic acid is a kind of urinary metabolite that has been regarded as one of the potential biomarkers for nephrotoxicity [30]. Glutaric acid could lead to apoptosis in immature oligodendrocytes and has been associated with glutaryl-CoA dehydrogenase deficiency, which is an inherited metabolic disease [31]. In addition, the levels of peptides were significantly downregulated in the pond-reared crabs, which could have been due to the amino acid metabolism, where peptides were consumed and free amino acids were produced.

Therefore, the sufficient food provided by ponds resulted in the accumulation of protein, free amino acids, and specific organic acids derivatives. However, it also led to the decreased release of peptides and PUFAs compared to naturally produced crabs. These results revealed that the pond culture significantly affected the nutritive profile of *E. sinensis* via the modulation of mostly protein, peptide, and amino acid levels. This could have been due to the food supplied by pond being mostly protein-based; however, naturally produced *E. sinensis* had access to various organisms in the lake, e.g., snails, zoobenthos, and planktons, leading to a more diverse nutritive profile, including elevated levels of PUFAs. Although snails, small fish, and animal-based freeze-dried powder could also be added to the pond, it not only raised the cost for crab rearing but also brought contaminants to the water, and therefore the local pond owners only used the cost-effective plant-based protein to rear *E. sinensis*. This implied that the pond culture was still relatively primitive at the Qin Lake area, although the environment there was highly suitable for crab rearing. The introduction of animal-based feed, combined with the improvement of a pond water circulation (inlet and drainage) system, growing aquatic plants with self-purification capacity, or reducing the culture density, could be ways to enhance the nutritional quality of the pond-reared crabs [16].

The nutritional quality of crabs was not only decided by the aquaculture mode but also affected by the harvest time when crabs with different maturities were collected. The local farmers at Qin Lake typically harvest crabs through the entire crab season, i.e., October, November, and possibly the beginning of December, thereby obtaining as many crabs as possible for selling. However, it was extremely important for the large-scale crab industry to focus on the best representative products instead of the yield, and therefore a comparison of the crab nutritive profiles between October and November was conducted.

In the present study, 5,6-EET significantly accumulated in the naturally produced crabs harvested in November, when compared to those harvested in October ($p < 0.05$). EETs were cytochrome-P450-epoxygenase-derived metabolites that play important biological roles in humans. Reports showed that they could not only improve the vasodilatation but also modulate insulin sensitivity, therefore being regarded as a potential therapeutic target for diabetes and ischemic cardiomyopathy [32,33]. Two amines, N-stearamide and PEA, were also remarkably upregulated in N-M11. However, they showed distinct bioactivities. N-stearamide was a potential biomarker for disordered fatty acid metabolism related to hepatic cirrhosis [34], whereas PEA was a ligand of peroxisome proliferator-activated receptors (PPAR-α) that exerted neuroprotective, anti-inflammatory, analgesic, and lipid-regulating effects [35,36]. Surprisingly, hypericin, a plant-originated bioactive compound that possesses anti-depressive, anti-neoplastic, anti-tumor, and anti-viral activities [37], was enriched in N-M11. This might provide evidence that the naturally produced crab consumes plant for survival in November. However, sugars, including D-trehalose, lactose, lactulose, and maltose, were significantly downregulated in N-M11. The consumption of carbohydrates provides crabs with enough energy to live through the cold weather in November.

Different to the naturally produced crabs, the pond-reared crabs had a stable energy supply during the entirety of November. This led to the significant increase in peptide levels in P-M11 compared to P-M10, as the protein-based food was largely digested. However, significant amounts of phenolic acids and nucleotides, which were not only important nutrients but also flavor compounds, were downregulated in P-M11. This was in line with studies that indicated that the total free amino acids, hydroxyeicosatrienoic acid, umami 5′-nucleotide compounds, and PUFAs were decreased in pond-reared crabs [18,20]. It was speculated that the strategy for a pond-reared crab to survive in November could be in consuming sufficient food and staying still in the pond, leading to a significant increase in peptide levels for energy provision. Overall, it could be concluded that October was more appropriate for harvesting *E. sinensis* than November, due to the elevated levels of sugars, phenolic acids, and nucleotides. Therefore, the local farmers could consider reducing the duration of the harvest time, instead of harvesting throughout the whole crab season, while finding a balance between the nutritional quality and the yield of crabs. In addition, controlling the fluctuation of water temperature was found to be crucial for the nutritional quality of *E. sinensis* in cold November.

5. Conclusions

To conclude, pond culture enhanced the levels of protein, free amino acids (Glu, Gly, Ala, Val, Phe, Lys, His, and Arg), and organic acid derivatives (stachydrine, methylsuccinic acid, and glutaric acid) and reduced the levels of peptides and PUFAs in *E. sinensis*. For pond-reared *E. sinensis* harvested in November, the levels of peptides were elevated, whereas phenolic acid (8-paradol and coniferaldehyde) and nucleotide (2′-deoxyuridine and deoxyguanosine) levels were declined, compared with those harvested in October. Above all, the metabolic profile of the pond-reared *E. sinensis* was significantly modulated by the high-protein diet, thus lacking the diversity of metabolites, especially PUFA levels. October could be more appropriate for harvesting *E. sinensis* in terms of crabs' nutritional quality. Therefore, our study facilitated the development of a high-end *E. sinensis* product by providing new understandings on the pond culture and the harvest strategy of crabs.

Supplementary Materials: The following supporting information can be downloaded at https://www.mdpi.com/article/10.3390/foods12112178/s1, Figure S1: The OPLS-permutation plot of the comparisons between N-M10 and P-M10 (A); N-M11 and P-M11 (B); N-M10 and N-M11 (C); and P-M10 and P-M11 (D). Table S1: The proximate composition of the meat and cream of *E. sinensis*.

Author Contributions: Conceptualization, Y.Y. and C.Z.; methodology, Y.Y.; validation, J.C.; formal analysis, Y.Y. and Y.W.; investigation, P.L. and J.C.; resources, C.Z.; data curation, Y.W.; writing—original draft preparation, Y.Y.; writing—review and editing, Y.Y. and C.Z.; visualization, Y.Y.; supervision, C.Z.; project administration, C.Z.; funding acquisition, C.Z. All authors have read and agreed to the published version of the manuscript.

Funding: This research and the APC were funded by National Key R&D Program of China, grant number 2022YFD2100605.

Data Availability Statement: The data presented in this study are available on request from the corresponding author.

Acknowledgments: Special thanks to all the collaborators in the National Key R&D Program of China (2022YFD2100605).

Conflicts of Interest: The authors declare no conflict of interest.

References

1. Xue, J.; Jiang, T.; Chen, X.; Liu, H.; Yang, J. Multi-mineral fingerprinting analysis of the Chinese mitten crab (*Eriocheir sinensis*) in Yangcheng Lake during the year-round culture period. *Food Chem.* **2022**, *390*, 133167. [CrossRef] [PubMed]
2. FAO. Fishery and Aquaculture Statistics. Global Aquaculture Production 1950–2019. Available online: https://www.fao.org/fishery/statistics-query/en/global_production (accessed on 23 March 2023).
3. MOAC. *China Fisheries Yearbook, 2021*; Ministry of Agriculture of China: Beijing, China, 2021.
4. Dittel, A.I.; Epifanio, C.E. Invasion biology of the Chinese mitten crab Eriochier sinensis: A brief review. *J. Exp. Mar. Biol. Ecol.* **2009**, *374*, 79–92. [CrossRef]
5. GIAP. Geographical Indication Agricultural Product: Qin Lake Bamboo Weir Crab. Available online: http://120.78.129.247/Home/Product/28879 (accessed on 23 March 2023).
6. QLTA. Qin Lake Tourist Attraction-Food Cultural 2-Qin Lake Bamboo Weir Crab. Available online: http://qinlake.com/en/show_04_03.html (accessed on 23 March 2023).
7. Chen, D.; Zhang, M.; Shrestha, S. Compositional characteristics and nutritional quality of Chinese mitten crab (*Eriocheir sinensis*). *Food Chem.* **2007**, *103*, 1343–1349. [CrossRef]
8. Hunter, J.E.; Zhang, J.; Kris-Etherton, P.M. Cardiovascular disease risk of dietary stearic acid compared with trans, other saturated, and unsaturated fatty acids: A systematic review. *Am. J. Clin. Nutr.* **2009**, *91*, 46–63. [CrossRef]
9. Nicholls, S.J.; Lundman, P.; Harmer, J.A.; Cutri, B.; Griffiths, K.A.; Rye, K.-A.; Barter, P.J.; Celermajer, D.S. Consumption of Saturated Fat Impairs the Anti-Inflammatory Properties of High-Density Lipoproteins and Endothelial Function. *J. Am. Coll. Cardiol.* **2006**, *48*, 715–720. [CrossRef] [PubMed]
10. Hooper, L.; Summerbell, C.D.; Thompson, R.L.; Sills, D.; Roberts, F.; Moore, H.J.; Smith, G.D. Effects of reducing or modifying dietary fat on CVD: A systematic review and meta-analysis of randomised controlled trials. *Proc. Nutr. Soc.* **2011**, *70*, E222. [CrossRef]
11. Mozaffarian, D.; Wu, J.H.Y. Omega-3 Fatty Acids and Cardiovascular Disease: Effects on Risk Factors, Molecular Pathways, and Clinical Events. *J. Am. Coll. Cardiol.* **2011**, *58*, 2047–2067. [CrossRef]
12. Du, H.; Lv, H.; Xu, Z.; Zhao, S.; Huang, T.; Manyande, A.; Xiong, S. The mechanism for improving the flesh quality of grass carp (*Ctenopharyngodon idella*) following the micro-flowing water treatment using a UPLC-QTOF/MS based metabolomics method. *Food Chem.* **2020**, *327*, 126777. [CrossRef] [PubMed]
13. Wang, J.; Jiang, H.; Alhamoud, Y.; Chen, Y.; Zhuang, J.; Liu, T.; Cai, L.; Shen, W.; Wu, X.; Zheng, W.; et al. Integrated metabolomic and gene expression analyses to study the effects of glycerol monolaurate on flesh quality in large yellow croaker (*Larimichthys crocea*). *Food Chem.* **2022**, *367*, 130749. [CrossRef]
14. Zhang, L.; Tao, N.-P.; Wu, X.; Wang, X. Metabolomics of the hepatopancreas in Chinese mitten crabs (*Eriocheir sinensis*). *Food Res. Int.* **2022**, *152*, 110914. [CrossRef]
15. AOAC. *Official Methods of Analysis of Official Analytical Chemists International*, 16th ed.; AOAC (Association of Official Analytical Chemists): Arlington, VA, USA, 1995.
16. Cheng, Y.; Wu, X.; Li, J. Chinese mitten Crab Culture: Current Status and Recent Progress Towards Sustainable Development. In *Aquaculture in China*; John Wiley & Sons Ltd: Hoboken, NJ, USA, 2018; pp. 197–217.
17. He, J.; Wu, X.; Li, J.; Huang, Q.; Huang, Z.; Cheng, Y. Comparison of the culture performance and profitability of wild-caught and captive pond-reared Chinese mitten crab (*Eriocheir sinensis*) juveniles reared in grow-out ponds: Implications for seed selection and genetic selection programs. *Aquaculture* **2014**, *434*, 48–56. [CrossRef]
18. Kong, L.; Cai, C.; Ye, Y.; Chen, D.; Wu, P.; Li, E.; Chen, L.; Song, L. Comparison of non-volatile compounds and sensory characteristics of Chinese mitten crabs (*Eriocheir sinensis*) reared in lakes and ponds: Potential environmental factors. *Aquaculture* **2012**, *364–365*, 96–102. [CrossRef]
19. Long, X.; Pan, J.; Wade, N.M.; Sun, Y.; Liu, Y.; Cheng, Y.; Wu, X. Effects of three feeding modes on the metabolism, antioxidant capacity, and metabolome of the adult male Chinese mitten crab *Eriocheir sinensis*. *Aquac. Int.* **2022**, *30*, 1101–1119. [CrossRef]
20. Chen, X.; Chen, H.; Liu, Q.; Ni, K.; Wang, C. High Plasticity of the Gut Microbiome and Muscle Metabolome of Chinese Mitten Crab (*Eriocheir sinensis*) in Diverse Environments. *J. Microbiol. Biotechnol.* **2021**, *31*, 240–249. [CrossRef] [PubMed]
21. He, J.; Xuan, F.; Shi, H.; Xie, J.; Wang, W.; Wang, G.; Xu, W. Comparison of nutritional quality of three edible tissues of the wild-caught and pond-reared swimming crab (*Portunus trituberculatus*) females. *LWT* **2017**, *75*, 624–630. [CrossRef]
22. Blanco, A.; Blanco, G. Chapter 16—Amino Acid Metabolism. In *Medical Biochemistry*; Blanco, A., Blanco, G., Eds.; Academic Press: Boston, MA, USA, 2017; pp. 367–399.
23. Hu, Y.; Feng, L.; Jiang, W.; Wu, P.; Liu, Y.; Kuang, S.; Tang, L.; Zhou, X. Lysine deficiency impaired growth performance and immune response and aggravated inflammatory response of the skin, spleen and head kidney in grown-up grass carp (*Ctenopharyngodon idella*). *Anim. Nutr.* **2021**, *7*, 556–568. [CrossRef]
24. Pekala, J.; Patkowska-Sokoła, B.; Bodkowski, R.; Jamroz, D.; Nowakowski, P.; Lochyński, S.; Librowski, T. L-carnitine-metabolic functions and meaning in humans life. *Curr. Drug Metab.* **2011**, *12*, 667–678. [CrossRef]
25. Brosnan, M.E.; Brosnan, J.T. Histidine Metabolism and Function. *J. Nutr.* **2020**, *150*, 2570s–2575s. [CrossRef]
26. Turner, N.; Cooney, G.J.; Kraegen, E.W.; Bruce, C.R. Fatty acid metabolism, energy expenditure and insulin resistance in muscle. *J. Endocrinol.* **2014**, *220*, T61–T79. [CrossRef]

27. Rogowski, M.P.; Flowers, M.T.; Stamatikos, A.D.; Ntambi, J.M.; Paton, C.M. SCD1 activity in muscle increases triglyceride PUFA content, exercise capacity, and PPARδ expression in mice. *J. Lipid Res.* **2013**, *54*, 2636–2646. [CrossRef]
28. Kumar, A.; Butt, N.A.; Levenson, A.S. Chapter 39—Natural Epigenetic-Modifying Molecules in Medical Therapy. In *Medical Epigenetics*; Tollefsbol, T.O., Ed.; Academic Press: Boston, MA, USA, 2016; pp. 747–798.
29. Cheng, F.; Zhou, Y.; Wang, M.; Guo, C.; Cao, Z.; Zhang, R.; Peng, C. A review of pharmacological and pharmacokinetic properties of stachydrine. *Pharmacol. Res.* **2020**, *155*, 104755. [CrossRef]
30. Gowthaman, N.S.K.; Lim, H.N.; Gopi, S.; Amalraj, A. 17—Identification of toxicology biomarker and evaluation of toxicity of natural products by metabolomic applications. In *Inflammation and Natural Products*; Gopi, S., Amalraj, A., Kunnumakkara, A., Thomas, S., Eds.; Academic Press: Boston, MA, USA, 2021; pp. 407–436.
31. Gerstner, B.; Gratopp, A.; Marcinkowski, M.; Sifringer, M.; Obladen, M.; Bührer, C. Glutaric Acid and Its Metabolites Cause Apoptosis in Immature Oligodendrocytes: A Novel Mechanism of White Matter Degeneration in Glutaryl-CoA Dehydrogenase Deficiency. *Pediatr. Res.* **2005**, *57*, 771–776. [CrossRef]
32. Romashko, M.; Schragenheim, J.; Abraham, N.G.; McClung, J.A. Epoxyeicosatrienoic Acid as Therapy for Diabetic and Ischemic Cardiomyopathy. *Trends Pharmacol. Sci.* **2016**, *37*, 945–962. [CrossRef] [PubMed]
33. Sisignano, M.; Park, C.-K.; Angioni, C.; Zhang, D.D.; von Hehn, C.; Cobos, E.J.; Ghasemlou, N.; Xu, Z.-Z.; Kumaran, V.; Lu, R.; et al. 5,6-EET Is Released upon Neuronal Activity and Induces Mechanical Pain Hypersensitivity via TRPA1 on Central Afferent Terminals. *J. Neurosci.* **2012**, *32*, 6364. [CrossRef] [PubMed]
34. Lian, J.S.; Liu, W.; Hao, S.R.; Guo, Y.Z.; Huang, H.J.; Chen, D.Y.; Xie, Q.; Pan, X.P.; Xu, W.; Yuan, W.X.; et al. A serum metabonomic study on the difference between alcohol- and HBV-induced liver cirrhosis by ultraperformance liquid chromatography coupled to mass spectrometry plus quadrupole time-of-flight mass spectrometry. *Chin. Med. J.* **2011**, *124*, 1367–1373. [CrossRef] [PubMed]
35. Keppel Hesselink, J.M. Chapter 4—Palmitoylethanolamid and Other Lipid Autacoids Against Neuroinflammation, Pain, and Spasms in Multiple Sclerosis. In *Nutrition and Lifestyle in Neurological Autoimmune Diseases*; Watson, R.R., Killgore, W.D.S., Eds.; Academic Press: Boston, MA, USA, 2017; pp. 29–37.
36. McIlwain, M.E.; Minassian, A.; Perry, W. Chapter 9—The Cannabinoid System in Nicotine Dependence and Withdrawal. In *Negative Affective States and Cognitive Impairments in Nicotine Dependence*; Hall, F.S., Young, J.W., Der-Avakian, A., Eds.; Academic Press: San Diego, CA, USA, 2017; pp. 153–176.
37. Kubin, A.; Wierrani, F.; Burner, U.; Alth, G.; Grünberger, W. Hypericin—The facts about a controversial agent. *Curr. Pharm. Des.* **2005**, *11*, 233–253. [CrossRef]

foods

MDPI

Article

Microwave-Induced Rapid Shape Change of 4D Printed Vegetable-Based Food

Xiaohuan Chen [1,2], Min Zhang [1,3,*] and Tiantian Tang [1,2]

1 State Key Laboratory of Food Science and Resources, School of Food Science and Technology, Jiangnan University, Wuxi 214122, China; 6200112010@stu.jiangnan.edu.cn (X.C.); 7210112048@stu.jiangnan.edu.cn (T.T.)
2 Jiangsu Province International Joint Laboratory on Fresh Food Smart Processing and Quality Monitoring, Jiangnan University, Wuxi 214122, China
3 China General Chamber of Commerce Key Laboratory on Fresh Food Processing & Preservation, Jiangnan University, Wuxi 214122, China
* Correspondence: min@jiangnan.edu.cn; Tel./Fax: +86-(0)510-85807976

Abstract: Microwave heating acts as an environmental stimulus factor to induce rapid shape changes in 4D-printed stereoscopic models over time. The influence of microwave power and model structure on the shape change behavior was explored, and the applicability of the deformed method to other vegetable-based gels was verified. The results described that the G′, G″, η, and proportion of bound water of yam gels increased with the increase in yam powder content, and the yam gel with 40% content had the best printing effect. The IR thermal maps showed the microwaves first gathered in the designed gully region caused the swelling phenomenon, which induced the printed sample to undergo a bird-inspired "spreading of wings" process within 30 s. Increasing the microwave power and microwave heating time were able to increase the bending angles and dehydration rates of the printed samples, thus improving the deformed degree and deformed speed. Different model base thicknesses (4, 6, 8, and 10 mm) also had significant effects on the shape change of the printed structures. The efficiency of the shape changes of 4D-printed structures under microwave induction can be judged by studying the dielectric properties of the materials. In addition, the deformed behaviors of other vegetable gels (pumpkin and spinach) verified the applicability of the 4D deformed method. This study aimed to create 4D-printed food with personalized and rapid shape change behavior, providing a basis for the application scenarios of 4D-printed food.

Keywords: 4D printing; shape change; microwave induction; vegetable-based food; temperature distribution

Citation: Chen, X.; Zhang, M.; Tang, T. Microwave-Induced Rapid Shape Change of 4D Printed Vegetable-Based Food. *Foods* **2023**, *12*, 2158. https://doi.org/10.3390/foods12112158

Academic Editor: Amalia Conte

Received: 17 April 2023
Revised: 29 April 2023
Accepted: 23 May 2023
Published: 26 May 2023

1. Introduction

Three-dimension (3D) printing technology, first proposed in 1984, is a development method derived from rapid prototyping technology, which is a type of additive manufacturing technology [1,2]. In recent years, with the continuous understanding of model structures, people are no longer satisfied with the static shapes of 3D-printed food structures. The demand for changes in the dynamic appearances and structural properties of food has promoted the emergence and development of four-dimensional (4D) food printing technology [3,4]. Four-dimensional printing technology enables three-dimensionally printed food to undergo predictable dynamic changes in appearance and structural properties (shape, color, flavor, etc.) over time when it is stimulated correctly in the appropriate environment (temperature, pH, light, moisture content, electromagnetic radiation, etc.), thus allowing for transformation from 3D to 4D [5–7]. Therefore, to smoothly print a 4D model, 3 factors need to be met: a stimulus–response material with specific properties, an environmental stimulus that can trigger the state change of the 3D-printed model, and the time demanded for the change in the state of the model [8,9]. 4D-printed food can enhance the interactions between food and consumers, thereby enriching the eating experiences of consumers.

There are two main strategies for studying the shape change achieved by 4D food printing at present: the use of double-layer materials and the design of special model structures. In materials science, the use of two materials with different expansion and shrinkage rates in response to specific stimuli is one of the most common ways to achieve shape changes of materials [10]. This approach has also been used to produce food products. Wang et al. [11] took advantage of the difference in water absorption capacity between gelatin and ethyl cellulose to design a gelatin/ethyl cellulose double-layer structure that could lead to shape change under the stimulation of hydration induction, and on this basis, invented deformable pasta. Chen et al. [12] made use of the different responses of pumpkin and printing paper to temperature to allow pumpkin products combined with printing paper to change shape. There are many ways to induce shape changes by designing special model structures. For instance, ideal deformable models can be obtained by designing filling structures of 3D printed models. Tao et al. [13] changed the product shape by controlling the depth, direction, and number of grooves in the model to control the changes between the dehydration and expansion rates of the material. He et al. [4] achieved shape change behavior of the models under microwave induction by designing filling rates and filling angles of the printed purple potato gel models. The results described that the filling rates and filling angles of the printed models were related to the deformation directions and degrees. However, these studies mainly focused on the shape transformation from planar structures (two-dimension, 2D) to stereoscopic structures (3D), and the amount of time required for shape change was often very long (a few min to more than ten min). On this basis, Shi et al. [14] used the different microwave heating efficiencies of purple potato puree and oil gel to achieve shape changes of the printed stereoscopic structures. Guo et al. [15] realized the directional shape change from 3D structure to 3D structure by using the swelling behavior of the printed food at hot-spots. However, there are still few studies on the rapidly deformed behavior of 4D-printed food.

Microwave is a kind of electromagnetic wave which has the advantages of selective heating, fast heating speed, and short drying time. Microwave heating has important applications in the food industry and in home cooking [16]. During microwave heating, the microwave can penetrate the inside of the material to produce heat so as to allow for volume heating from the inside to the outside [17]. At present, microwave heating technology has been successfully applied to the environment-induced change process of 4D food printing [4,15]. Chen et al. [18] used a microwave to stimulate the decomposition of baking soda in lotus root powder gel to increase the pH of the gel system, thereby inducing the color change process of the lotus root powder gel containing curcumin. Guo et al. [19] used soybean protein extracts and oats as food inks to achieve the shape change of a 4D-printed structure under microwave stimulation.

Vegetables contain a great number of vitamins, minerals, and cellulose, which the human body requires [20]. The daily intake of the vegetable can help to improve the body's immunity and prevent chronic diseases [21]. The application of vegetables in food printing has a profound influence on the production of personalized nutritious foods. Yams are a kind of food with high nutritional value. Studies have found that yams have antioxidant, anticancer, hypoglycemic, and lipid-lowering functions [22]. Pumpkin and spinach, as common vegetables, have good nutritional function and printing performance, and can also be used as good raw materials for 4D food printing [12].

In this study, microwave heating acted as the environmental stimulus factor for 4D printing to investigate the rapid shape changes of 3D-printed stereoscopic models. The influences of yam powder content on the material properties and printing properties of the mixed gels were studied. The influences of microwave power and model design structures on the deformed behaviors of 3D printed components were investigated, and pumpkin gel and spinach gel were used for the validation tests of microwave-induced shape change.

This study aimed to create 4D-printed food with personalized and rapid shape change behavior, providing a basis for the application scenarios of 4D-printed food.

2. Materials and Methods

2.1. Materials

Yam powder containing 7.6% water, 73.5% carbohydrate, 4.7% protein, and 3.7% crude fat was obtained from Duomai Food Co., Ltd. (Zaozhuang, China). Pumpkin powder containing 7.8% water, 61.2% carbohydrate, 6.4% protein, and 2.6% crude fat was collected from Baofeng Biotechnology Co., Ltd. (Bozhou, China). Spinach powder containing 8.1% water, 55.6% carbohydrate, 19.5% protein, and 2.7% crude fat was collected from Lvshuai Food Co., Ltd. (Xinghua, China). Low acyl gellan gum (food grade) was obtained from Fufeng Biotechnology Co., Ltd. (Urumqi, China).

2.2. Preparation of Printing Materials

Yam gel was obtained by mixing the yam powder with deionized water, and the contents of yam powder were 25%, 30%, 35%, 40%, and 45% (*w*/*w*), respectively. The yam gels were gelatinized in a water bath at 75 ± 0.2 °C for 0.5 h. During gelatinization, the container mouth was wrapped with plastic wrap to avoid water loss. According to preliminary test results, 0.5% (*w*/*w*) gellan gum was added to the gelatinized yam gels and mixed evenly to form smooth pastes without any clumps. The deformation model put forward higher requirements for the printability and post-processing abilities of materials. Adding an appropriate amount of gellan gum can enhance the support and stability of the printed samples in the shape change process. Finally, the yam gels were placed in a normal-temperature environment for 30 min to stabilize them for the next study.

Pumpkin gel and spinach gel were prepared by the same preparation process as yam gel to verify the applicability of the 4D deformed method. The selection of printing ink was mainly based on rheological properties, moisture properties, and printing performance. The formula selection of the yam gel will be described later. By the same formula selection method, the contents of pumpkin powder and spinach powder in the pumpkin gel and spinach gel were determined to be 33% and 36% (*w*/*w*), respectively.

2.3. Determination of Rheological Properties

A rotating rheometer (DHR-3, TA Instruments, New Castle, DE, USA) was used to analyze the rheological properties of mixed gels with different yam powder contents, referring to the experimental method provided by Yang et al. [23]. The diameter of the experimental parallel plate was 40 mm, and the measurement gap and temperature were 1000 μm and 25 °C, respectively. Firstly, the storage modulus (G′) and loss modulus (G″) of the printing material were determined at angular frequencies ranging from 1 to 100 rad/s by an oscillation test under constant 0.1% strain (within the linear viscoelastic region). A flow scanning test was then carried out at shear rates ranging from 0.1 to 100 s^{-1} to determine the apparent viscosity (η).

2.4. LF-NMR Analysis

A low-field nuclear magnetic resonance (LF-NMR) analyzer (MicroMR20-030V-I, Niumag, Shanghai, China) was applied to analyze the moisture statuses and distributions of the mixed gels with different yam powder contents by referring to the experimental method described by Ezeanaka et al. [24]. Approximately 4.00 g of the material was wrapped using plastic wrap and then placed into a nuclear magnetic tube (Ø30 × 200 mm), which was placed into the instrument for detection. The experimental parameters were set as sampling frequency (SW) = 100 KHz, scanning number (NS) = 4, echo time (TE) = 0.400 ms, echo number (NECH) = 1000, and waiting time (TW) = 2500 ms. After the test, the transverse relaxation time (T_2) of the printing material was obtained by inverting the test results.

2.5. Model Design and 3D Printing Process

A cone model (Ø30 mm × 40 mm) was built using Rhinocero 6.0 software (Robert McNeel&Associates, Settle, WA, USA) to examine the printability of the mixed gels with different yam powder contents. A bird model with two wings (40 mm base length, 15 mm base width, and 18 mm wing height) was designed for deformed testing as shown in Figure 1A, with the model base thickness set at 4, 6, 8, and 10 mm, respectively.

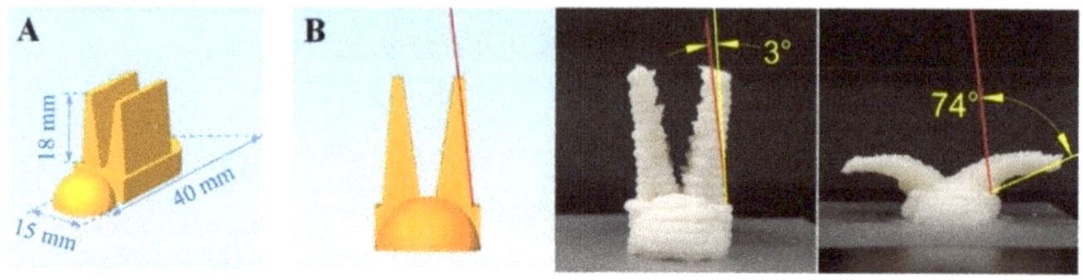

Figure 1. Schematic diagrams of the model's dimensions (**A**) and the method for determining the bending angles (**B**) of the printed samples.

The 3D printing process was performed using a printer (Foodbot-Mf, Shiyin, Hangzhou, China) at 25 °C. The printing material was first filled into the printing tube, which was then inserted into the material barrel of the printer for printing. The printer and computer were connected through a USB port, and the printer was controlled by the computer for printing. The model data in STL format were transformed into machine codes using Simplify 3D software, and the X-Y-Z printing paths were automatically generated. Through the preliminary tests, the printing parameters were optimized and set as follows: 1.2 mm layer height and nozzle diameter, 22 mm/s printing speed, and 1.9 mm Z-value.

2.6. Microwave Heating and Determination of Temperature Distribution

The printed samples were placed into a microwave oven (M1-L213B, Midea Group Co., Ltd., Foshan, China) with the microwave power set to low fire (70 W), medium fire (350 W), and high fire (700 W). The microwave oven used in this study was a traditional microwave oven, the output power of which was a constant value (700 W). The amount of power was controlled by adjusting the on/off time of microwave emission at different gear positions. According to the preliminary experimental results, the microwave heating times were set as 0, 5, 10, 15, 30, 45, and 60 s, and the samples were immediately taken out for weighing and determination after being heated for the set amount of time.

An infrared thermal imager (IRI-4010, Infrared Integrated Systems Ltd., Northampton, UK) was used to measure the temperature distribution of the microwave-induced samples, referring to the experimental method described by Tang et al. [25]. Only the tested sample was controlled in the measurement background, the focal length was adjusted for measurement, and photos were taken to record the infrared (IR) thermal map. IRISYS 4000 Series Imager software was used to process the map.

2.7. Measurement of Bending Angle

Photographs were taken to record the images and shape change behaviors of the samples at different heating times. To quantify the degree of shape change, AutoCAD 2019 software (Autodesk Inc., San Francisco, CA, USA) was purchased to measure the bending angles of 3D-printed samples after microwave heating treatment by referring to the experimental method of Liu et al. [26], as shown in Figure 1B. Both wings of each printed model were measured.

2.8. Determination of Dielectric Property

A vector network analyzer (E5062A, Agilent Technologies, Santa Clara, CA, USA) was used to determine the dielectric properties of the printing materials by referring to the experimental method provided by Guo et al. [19]. Firstly, the equipment was calibrated by measuring the air and deionized water. The range of the frequency was set as 2.0 to 3.0 GHz, and the measuring probe contacted the printed samples for testing as far as possible to avoid the existence of air between the measuring probe and the sample, in order to ensure that the dielectric constant (ε') and loss factor (ε'') reached the accuracy of the measurement.

2.9. Statistical Analysis

The experimental data were analyzed using SPSS 25.0 statistical software (IBM, Chicago, IL, USA). Duncan's test was used for significance analysis at 95% confidence intervals, with $p < 0.05$ indicating obvious differences between the samples. Charts were made using AutoCAD 2019 software and Origin 2018 software (OriginLab Corporation, Northampton, MA, USA). All experiments were performed at least three times in parallel, and the resulting values were represented as mean value $\pm$ standard deviation.

3. Results and Discussion

3.1. Rheological Properties of Printing Materials

The rheological properties of food materials have important effects on the printing performance of 4D-printed samples and the post-processing abilities in the microwave-induced process [27,28]. Figure 2 shows the rheological characteristics of mixed gels with different yam powder contents. The G' and G'' values of all yam gels presented frequency dependence and content dependence; the modulus values increased as the angular frequency and the yam powder content increased. G' was related to the mechanical strength of the printed materials and their supporting performance after printing [29,30]. The increase in G' was conducive to the molding of the printed samples and the maintenance of the structures before inducing the shape change [31,32].

The η and shear-thinning characteristics of the printing materials were obtained to evaluate the extrusion properties of the materials during extruding 3D printing [31]. As shown in Figure 2C, the η values of all printing materials reduced with the addition of the shear rate, showing shear-thinning of the gel, which was propitious to the extrusion of materials from the printing nozzle [28,33]. The shear-thinning property of yam gels was possibly related to the structure and entanglement of polysaccharides' molecular chains, such as starch and gellan gum in the system and their intermolecular interaction [34]. At lower shear rates, the molecule chains of the polysaccharides were entangled with each other in an irregular manner to create a gel network structure [35]. When the shear rate was increased, the internal network structure and the intermolecular forces were damaged and weakened, which led to decreases in the flow resistance and the η value.

Furthermore, the increase in yam powder content induced an increase in the η value of the material. The elevated yam powder content meant that more starch molecules expanded and split to create a yam puree with higher viscosity in the gelatinization process, and the gel structure of the material was more closely combined, which was also conducive to the supporting ability and printing accuracy of the printed samples after deposition [27]. However, the increase in η put forward higher requirements for the extrusion characteristics of the materials, and the larger the η was, the more difficult the extrusion of the material became, which was also confirmed in the subsequent printing process.

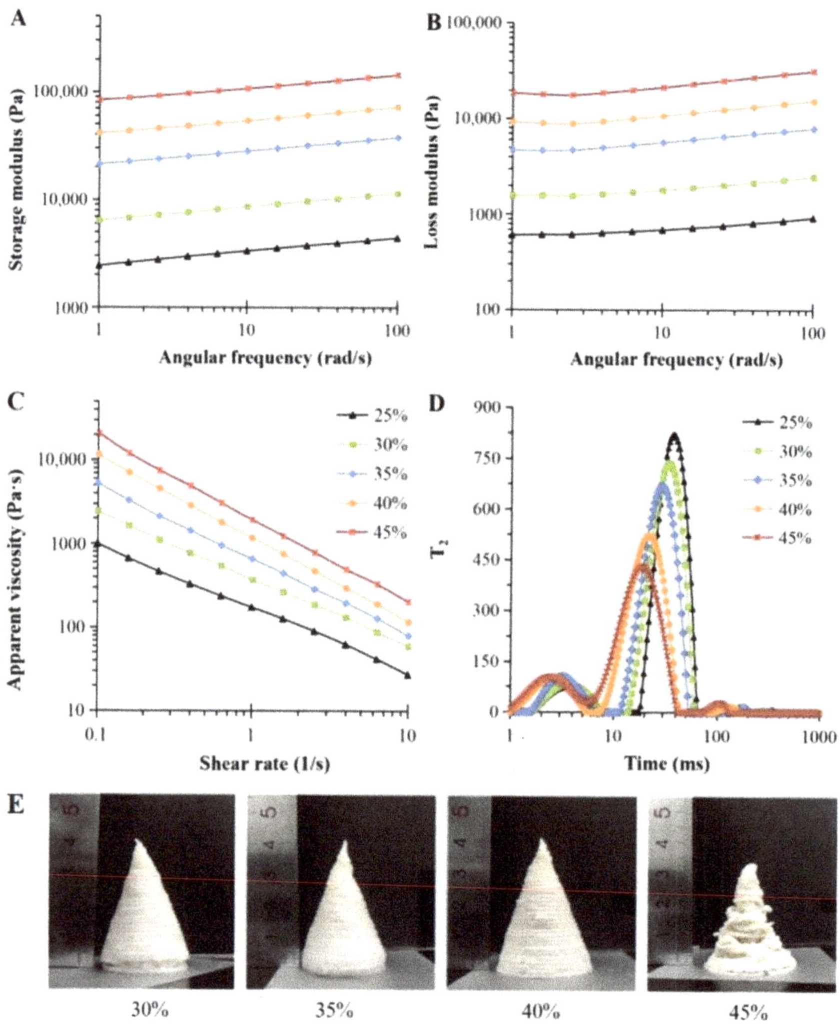

Figure 2. Storage modulus (**A**), loss modulus (**B**), apparent viscosity (**C**), LF-NMR profiles (**D**), and printed results (**E**) of yam gels with different yam powder contents.

3.2. Moisture Characteristics of Printing Materials

LF-NMR spectroscopy revealed the moisture state and moisture distribution of the mixed gels with different yam powder contents (Figure 2D), which correlated with the rheological, mechanical, and printing characteristics of the materials [36]. Different peak areas (A_2) were obtained to represent the proportions of moisture in different states [37]. Three peaks—T_{21}, T_{22}, and T_{23}—were found in LF-NMR spectroscopy of all of the yam gels, corresponding to the peak areas of A_{21}, A_{22}, and A_{23}, respectively (Table 1). In general, the peak center of T_{21} was located within 1–10 ms, representing the water closely bound with macromolecules in the gel matrix, with low degrees of freedom. T_{22} with a peak center of 10–100 ms indicated semi-bound water, which had a certain fluidity. T_{23} with a peak center of 100–1000 ms represented free water, which was move freely; and this type of water showed easy migration and evaporation [32]. As shown in Figure 2D, with the increase in yam powder content, the peak centers of all T_2 peaks shifted to the left, while the T_{22} peaks decreased significantly. The freedom of the water in the materials was reduced, and the difficulty of water migration and dehydration evaporation was increased, but the supporting performances of the materials were improved. Table 1 shows the significant effects of yam powder content on the T_2 and A_2 values of the mixed gel ($p < 0.05$). T_2 was related to the freedom and fluidity of the water molecules. The higher the value of T_2, the higher the freedom and fluidity of the material [25]. The T_{21}, T_{22}, and T_{23} values of the yam gel were reduced with the increase in yam powder content, meaning that the freedom and fluidity of the water in the material decreased, which was consistent with the left shift result of the T_2 peak center. As the yam powder content increased, the A_{21} value gradually increased, while the A_{22} value gradually decreased, showing that part of the water changed from semi-bound to bound. This meant that the water, starch, and gellan gum in the system were combined more closely, and the mechanical strength of the material was enhanced along with the increase in yam powder content [35]. These results were identical to the rheological analysis results of the mixed gels, which were conducive to the printing molding and shape maintenance before the induced shape change of the printed models.

Table 1. LF-NMR analysis of different yam gels.

	T_{21} (ms)	T_{22} (ms)	T_{23} (ms)	A_{21} (%)	A_{22} (%)	A_{23} (%)
25%	7.58 ± 0.56^{a}	48.30 ± 0.00^{a}	138.07 ± 21.29^{a}	10.79 ± 1.63^{e}	95.90 ± 1.19^{a}	0.32 ± 0.03^{d}
30%	8.25 ± 0.66^{a}	47.46 ± 1.21^{a}	143.23 ± 14.46^{a}	12.21 ± 1.16^{d}	88.97 ± 1.63^{ab}	0.24 ± 0.03^{d}
35%	6.91 ± 1.54^{a}	43.19 ± 0.77^{b}	102.64 ± 13.74^{b}	13.92 ± 1.05^{c}	87.47 ± 1.17^{b}	0.95 ± 0.07^{b}
40%	4.87 ± 0.46^{b}	35.70 ± 0.71^{c}	89.93 ± 5.56^{bc}	16.00 ± 0.21^{b}	83.05 ± 0.26^{c}	0.50 ± 0.07^{c}
45%	4.60 ± 0.03^{b}	36.43 ± 0.00^{c}	74.83 ± 9.92^{c}	19.18 ± 0.08^{a}	79.19 ± 0.05^{d}	1.23 ± 0.14^{a}

Different letters in the same column indicate significant differences between values ($p < 0.05$).

3.3. *Printing Performance of Materials*

Four-dimensional food printing technology has put forward higher requirements for the printability and printing precision of materials [38]. To further evaluate the 3D printability of different yam gels, a cone structure was printed for analysis. Figure 2E shows the effects of different yam powder contents on the printing results of the mixed gels. The cone model could be successfully printed with the mixed gels containing 30–40% yam powder content, and the height of the model was close to the designed height of 40 mm. However, with the reduction in yam powder content, the printing accuracy of the yam gel decreased, as did the supporting performance and stability. The model printed by the mixed gel with 30% content of yam powder showed obvious retraction at the bottom, because the reduction of yam powder content reduced the G', G'', η, and the proportion of bound water in the system, which weakened the mechanical strength and enhanced the fluidity of the gel. Therefore, due to the weak self-supporting performance, the mixed gel with low yam powder content was unable to support its structure after printing, and was prone to collapse to a certain extent. The mixed gel with 25% yam powder content lacked sufficient mechanical strength, making difficult to print. On the contrary, although the mixed gel with 45% yam powder content had high G' and G'' values, its excessive η also limited the extrusion characteristics of the material. The extrusion of the material was difficult during 3D printing, and there was a serious line fracture phenomenon. The ideal printing material should have the appropriate mechanical strength to improve the printing precision and maintain the stability of the model structure, but at the same time, the η and mechanical strength should not be too high to avoid encountering difficulty during material extrusion [5]. According to the analysis results of the material properties and printing characteristics, the mixed gel with 40% yam powder content was selected as the best printing material in terms of the shape change of 4D-printed food.

3.4. Shape Change of the Stereoscopic Model Induced by Microwave

Through the active design of the 4D-printed model, a bionic model of a bird with two wings was designed. The connection between the two wings formed a gully structure with a certain angle, which was propitious to the aggregation of microwaves and the formation of hot-spot regions [39]. Microwave heating can make the water molecules inside the food evaporate by heating, resulting in a difference in the environmental pressure between the inside and outside, causing the swelling of the material [40,41]. The gully structure effectively aggregated microwaves and generated a local high temperature in this region, as a hot-spot. The microwave induced the swelling phenomenon of the hot-spot region, which pushed the wings of the bird model to bend outwards and achieve the ideal deformation effect of "spreading of wings" [15]. Figure 3A shows the shape change behaviors of the printed samples at 700 W of microwave power. After microwave heating treatment, the shape of the sample changed and the wings of the model opened outwards. With the extension of the microwave heating time, the shape change degree of the printed structure increased. The two wings of the bird model opened outwards and bent, lowering in height and exhibiting the expected "spreading of wings" behavior. However, when the microwave heating time exceeded 45 s, yellow lines began to appear at the gully part of the printed sample, showing unexpected results. This was caused by the non-uniformity of microwave heating [42].

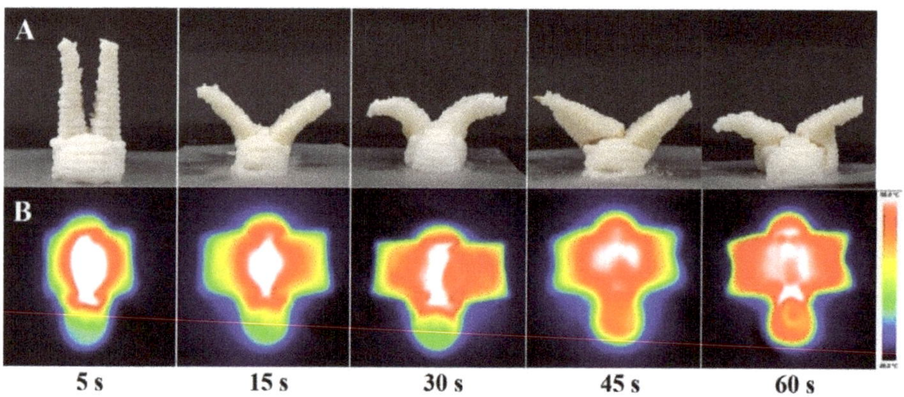

Figure 3. Shape change photographs (**A**) and IR thermal maps (**B**) of the printed samples at 700 W of microwave power (temperature distribution between 20 °C and 80 °C).

The IR thermal maps of the printed samples at 700 W of microwave power are shown in Figure 3B. Because of the uneven distribution of the microwaves, they first concentrated in the gully region between the wings of the designed model. The heat of the gully part was greater than that of the other parts, inducing local swelling and pushing the wings to spread out [42]. This result agreed with the expected model design and hot-spot setting. Guo et al. [15] achieved the shape change effect of "flowering" by using the printing method of local swelling induced through non-uniform microwaves. Moreover, with the extension of the microwave heating time, the area range of the hot-spot rose in temperature, inducing a more obvious swelling phenomenon and resulting in a greater shape change. However, when the microwave heating time reached 45 s, the hot-spot area spread to the head area of the bird model, which caused the head gully region to overheat and led to the appearance of yellow lines.

3.5. Effect of Microwave Power on Shape Change

Figure 4 shows the influence of microwave power on the deformation behaviors of the printed structures, which were deformed at different angles under different microwave power levels and microwave heating times. To quantify the shape change behavior, the bending angle was used to represent the degree of shape change (Figure 4C). Due to the ink-swelling effect of food materials, the diameters of the material lines deposited in the printed samples were greater than those in the design models, making the initial bending angles of the printed structures negative [43]. During 0–30 s of microwave heating time, the bending angles of the samples increased as the microwave power and heating time increased, demonstrating power dependence and time dependence. When the heating time reached 30 s and the microwave power was 70, 350, or 700 W, the bending angles of the model wings reached 59.14°, 63.56°, and 71.64°, respectively. When the microwave heating time reached 45 s, the printed models demonstrated unexpected and uncontrollable deformation behavior. This was due to the large amount of materials deposited in the bottom of the model, which concentrated the microwaves and swelled during the microwave treatment. When the microwave heating time was 45 s or 60 s, excessive swelling occurred in the bottom of the models and affected the bending change of the upper wings, as shown in Figure 4A.

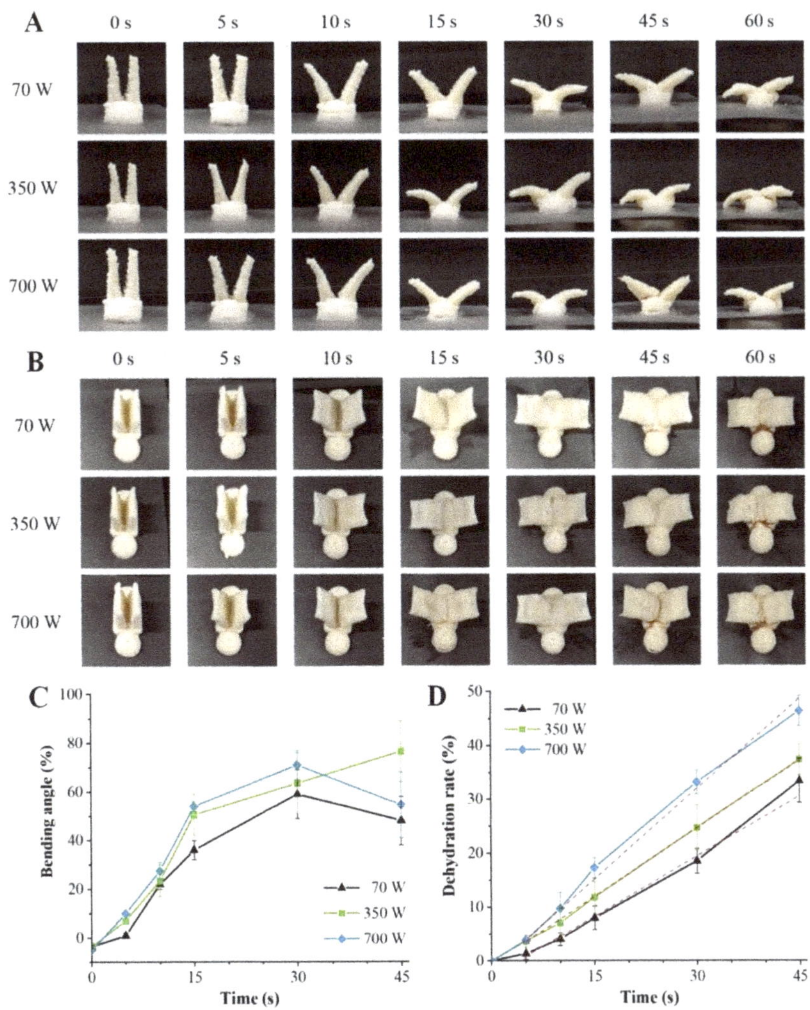

Figure 4. Shape change photographs in front view (**A**) and lateral view (**B**) of the printed samples, and the changes in the bending angle (**C**) and dehydration rate (**D**) at different microwave power levels.

The microwave heating process is also a drying and dehydration process [4]. Figure 4D shows the influences of microwave power and microwave heating time on the dehydration rates of the printed food. The dehydration rates of the printed food raised with the increase in microwave heating time, in a time-dependent manner. The linear fitting of the dehydration rate–microwave heating time curves at 70, 350, and 700 W of microwave power showed strong correlations with the fitting values (R^2) of 0.994, 0.999, and 0.989, respectively. Microwave heating mainly caused high-frequency reciprocating motion or oscillation of the polar substances in the materials through the electromagnetic field, as well as the rotation of dipoles (such as water molecules). The polar molecules or dipoles then showed mutual friction or collision, thus consuming electric energy, transforming into heat energy, and causing water to evaporate [16,44]. With the extension of heating time, water molecules absorbed more microwave energy, which caused increases in friction heat and temperature, thus increasing the dehydration rates. Microwave power also showed an obvious influence on the dehydration rates of the printed food ($p < 0.05$), and the dehydration rates of the printed samples were raised with the increase in microwave power. With 45 s of microwave heating time, the maximum dehydration rates of the printed samples at 70, 350, and 700 W of microwave power were 33.44%, 37.41%, and 46.40%, respectively.

3.6. Effect of Model Structure on Shape Change

The pictures depicting the shape changes of the printed structures with different model base thicknesses under 700 W of microwave power are shown in Figure 5A–C, which display the influence of the model base thickness on the deformed angle of the printed sample. During 0–30 s of microwave heating time, the bending angles of the printed structures showed an increasing trend with time evolution. Within 0–15 s, the bending angles of the printed structures rose as the model base thickness rose. The increase in the base thickness provided the model base with a larger volume, which then absorbed more microwave energy to induce a greater swelling phenomenon during microwave heating and to further increase the degree of shape change [15].

However, when the microwave heating time was 15 s or 30 s, there was no distinct difference in the deformed angles of the printed samples with base thicknesses of 4, 6, and 8 mm ($p > 0.05$), while the bending angle of the sample with a base thickness of 10 mm was significantly larger than that of other samples, showing a greater deformed degree. As shown in Figure 5B, when the microwave heating time reached 30 s or 45 s, the material deposited between the two wings part of the printed sample with a thickness of 10 mm showed visible fractures, which was related to the overheating temperature in the gully region. At this time, the bending angle of the wing part of the printed sample without traction from the model base was not only affected by the driving force produced by the swelling of the gully part, but also caused to collapse downward by the gravity effect of the wing part itself, showing a greater bending angle. Moreover, the bending angles of the printed structures with different base thicknesses increased or decreased at 45 s, at which time the bending angle of the printed structure was positively correlated with the base thickness. The maximum angles of the models with 4, 6, 8, and 10 mm of base thickness were 74.85°, 71.64°, 70.58°, and 106.42° within 45 s of microwave heating time. As the base thickness was raised from 4 mm to 10 mm, the deformed angle increased by 1.42 times.

Figure 5D shows the influence of model base thickness on the dehydration rate of the printed samples. The dehydration rates of the printed food were positively correlated with microwave heating time, and the R^2 of the dehydration rate–microwave heating time curves of the samples with thicknesses of 4, 6, 8, and 10 mm were 0.991, 0.994, 0.998, and 0.999, respectively. The base thicknesses (6, 8, and 10 mm) also had a certain influence on the dehydration rates of the printed food samples, which reduced with the increase in the base thickness. This was because microwave heating was volumetric, and the microwaves directly entered the interiors of the substances and interacted with each other. The temperature gradients of the material subjected to microwave were high on the inside and low on the outside [17]. The volume of the printed model with a larger base thickness was greater, and the discharge rate of the internal water vapor was slowed down by the material. The exception was that in the early stage of microwave heating (0–10 s), the printed sample with a thickness of 4 mm showed a low dehydration rate due to its weak microwave absorption ability. After 10 s, the dehydration rate of the printed sample was accelerated, and yellow lines appeared at 30 s, which was the earliest compared with the other samples (Figure 5).

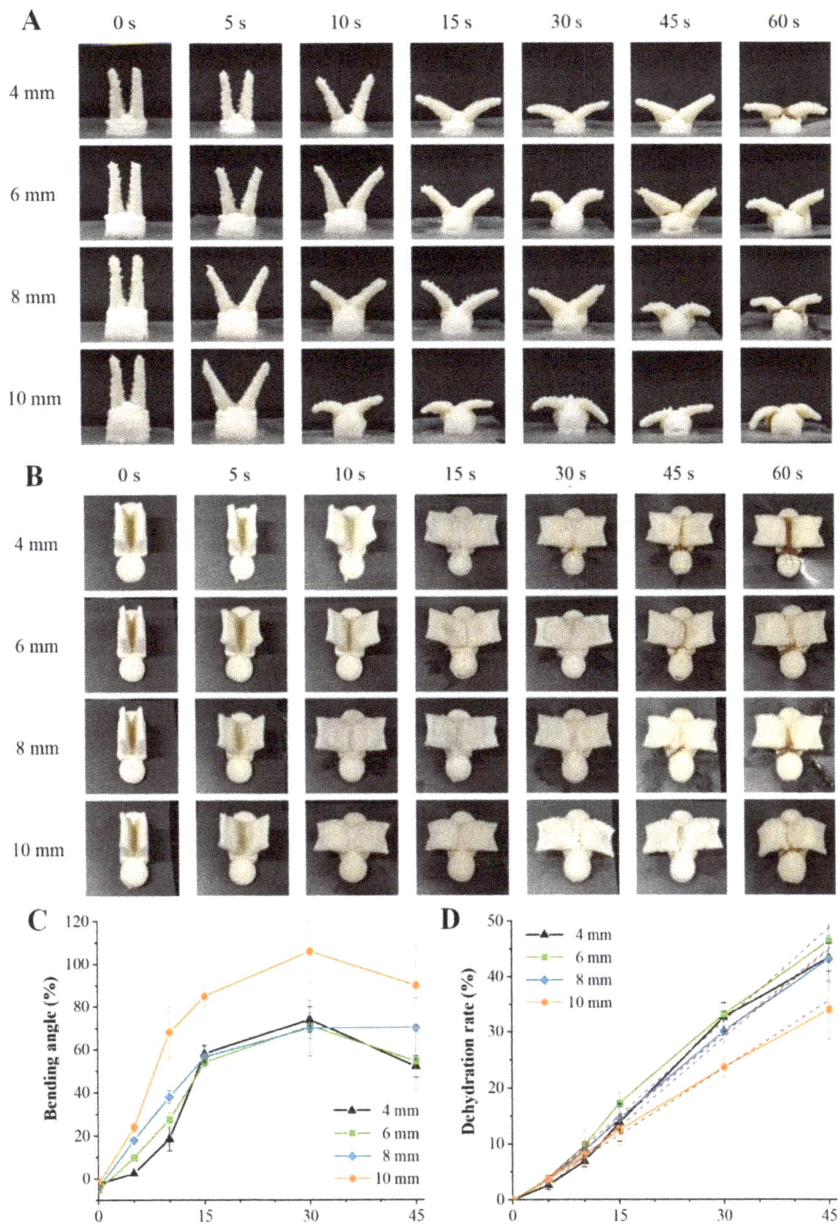

Figure 5. Shape change photographs in front view (**A**) and lateral view (**B**) of the printed samples, and the changes in bending angle (**C**) and dehydration rate (**D**) with different model structures.

3.7. Shape Change Behavior of Other Vegetable Gels

Figure 6A shows the shape change behaviors of the printed samples prepared with pumpkin gel and spinach gel at 700 W of microwave power. Under the action of the microwave, the wings of the bird models opened outwards along with time evolution and the "spreading of wings" behavior of the bird model was simulated, which verified the feasibility and applicability of the microwave-induced shape change method of the 4D stereoscopic model. The bending angle of the printed model rose over time. Compared with the deformed behavior of the sample printed with yam gel, shown in Figure 3, the magnitude of the deformed efficiency of different vegetable gels was pumpkin gel > yam gel > spinach gel. The times to reach the maximum controllable and ideal bending angles of the samples printed with pumpkin gel, yam gel, and spinach gel were 15 s, 30 s, and 30 s, respectively, and the corresponding maximum bending angles were 76.17°, 71.64°, and 41.47°, respectively. This was related to the dielectric properties of the vegetable gels.

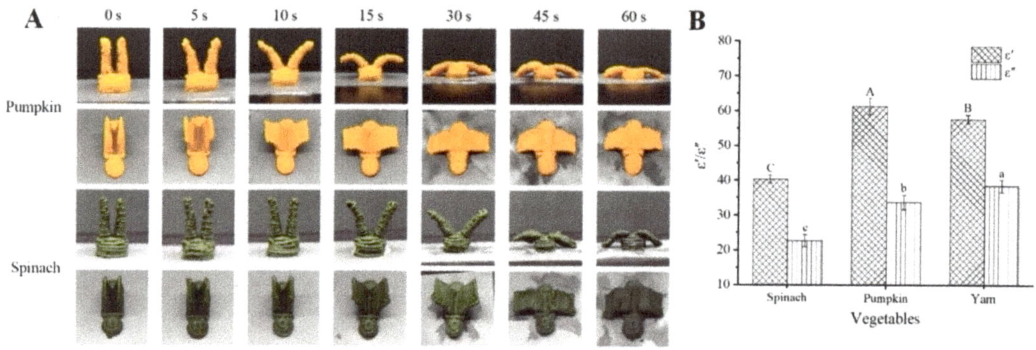

Figure 6. Shape change photographs (**A**) and dielectric properties (**B**) of different vegetable materials. Different letters indicate significant differences between values ($p < 0.05$).

In the process of microwave heating, the difference in materials directly affect the microwave heating efficiency, which can be essentially determined by the dielectric property of the food materials. The dielectric property is the response of bound charges in the molecules to the applied electromagnetic field [45]. The dielectric properties ε' and ε'' determined the interactions between the electromagnetic field and the materials, representing the capacity of the materials to store electromagnetic energy and convert it into thermal energy, respectively [16]. The dielectric properties of the three vegetable gels are shown in Figure 6B. The ε' values, from high to low, were arranged in the following order: pumpkin gel (61.33) > yam gel (57.67) > spinach gel (40.33). The ε'' values, from high to low, were in the order of yam gel (38.33) > pumpkin gel (33.67) > spinach gel (22.67). The differences in the dielectric properties of the materials were mainly related to the type of materials and their material characteristics (bulk density, water content, etc.) [45]. As a kind of polar molecule, water molecules moved under the action of the microwaves and produced a friction effect, which increased the internal energy of the materials and achieved the effect of rapid heating [14]. The ε'' indicates the capacity of the food material to absorb microwaves, and the greater the ε'' value, the faster the heating rate [35]. Compared with spinach gel, the yam and pumpkin gels exhibited stronger dielectric properties and higher microwave absorption properties, corresponding to higher microwave heating efficiency and 4D change efficiency. The results demonstrated that the deformed efficiency of the 4D-printed samples under microwave induction can be predicted by studying the dielectric properties of the materials.

4. Conclusions

The yam powder content affected the printing effect of the mixed gel by influencing its material characteristics. The increase in yam powder content optimized the rheological and moisture characteristics of the materials, but also increased the difficulty of extrusion while improving the printing accuracy. Gel mixed with 40% yam powder content was used as 4D printing ink. A bird model with two wings was designed, and the shape change behavior and IR thermal map of the printed models demonstrated the success of the method, inducing a shape change by forming local hot-spots at the pre-designed gully regions. Increasing the microwave power and heating time increased the bending angles and dehydration rates of the printed samples, and improve the degree and speed of the shape change. However, when the microwave heating time exceeded 45 s, the hot-spot area expanded greatly, causing unexpected yellow lines and affecting the deformed behavior of the printed sample. The model's base thickness also had an obvious influence on the deformed degrees and dehydration rates of the printed samples. Moreover, the microwave-induced 4D shape change method was also applicable to other vegetable-based gels, and the dielectric properties of the materials affected the microwave heating effect, which could be used to predict the efficiency of the 4D change. In this study, rapid and controllable shape changes of printed food within 30 s were achieved through microwave induction so as to enrich the application scenarios of 4D-printed food.

Author Contributions: X.C. completed the data analysis, graphic drawing, and writing of the manuscript text; M.Z. reviewed, supervised, and revised the manuscript; and T.T. revised the manuscript. All authors have read and agreed to the published version of the manuscript.

Funding: This research was funded by the National Key R&D Program of China [No. 2022YFF1101300], the National Natural Science Foundation Program of China [No. 31872902], and the Jiangsu Province Key Laboratory Project of Advanced Food Manufacturing Equipment and Technology [No. FMZ202003].

Data Availability Statement: The data are available from the corresponding author upon suitable request.

Conflicts of Interest: There are no conflict of interest to declare.

References

1. Ford, S.; Despeisse, M. Additive manufacturing and sustainability: An exploratory study of the advantages and challenges. *J. Clean. Prod.* **2016**, *137*, 1573–1587. [CrossRef]
2. Hull, C. Apparatus for Production of Three-Dimensional Objects by Stereolithography. U.S. Patent US4575330A, 8 August 1984.
3. Ghazal, A.F.; Zhang, M.; Liu, Z. Spontaneous color change of 3D printed healthy food product over time after printing as a novel application for 4D food printing. *Food Bioprocess Technol.* **2019**, *12*, 1627–1645. [CrossRef]
4. He, C.; Zhang, M.; Devahastin, S. Investigation on spontaneous shape change of 4D printed starch-based purees from purple sweet potatoes as induced by microwave dehydration. *ACS Appl. Mater. Interfaces* **2020**, *12*, 37896–37905. [CrossRef]
5. Teng, X.; Zhang, M.; Mujumdar, A.S. 4D printing: Recent advances and proposals in the food sector. *Trends Food Sci. Technol.* **2021**, *110*, 349–363. [CrossRef]
6. Khoo, Z.X.; Teoh, J.E.M.; Liu, Y.; Chua, C.K.; Yang, S.; An, J.; Leong, K.F.; Yeong, W.Y. 3D printing of smart materials: A review on recent progresses in 4D printing. *Virtual Phys. Prototyp.* **2015**, *10*, 103–122. [CrossRef]
7. Rastogi, P.; Kandasubramanian, B. Breakthrough in the printing tactics for stimuli-responsive materials: 4D printing. *Chem. Eng. J.* **2019**, *366*, 264–304. [CrossRef]
8. Gao, B.; Yang, Q.; Zhao, X.; Jin, G.; Ma, Y.; Xu, F. 4D bioprinting for biomedical applications. *Trends Biotechnol.* **2016**, *34*, 746–756. [CrossRef]
9. Momeni, F.; Hassani, N.; Liu, X.; Ni, J. A review of 4D printing. *Mater. Des.* **2017**, *122*, 42–79. [CrossRef]
10. Tao, Y.; Gu, J.; An, B.; Cheng, T.; Chen, X.A.; Zhang, X.; Zhao, W.; Do, Y.; Zhang, T.; Yao, L. Demonstrating thermorph: Democratizing 4d printing of self-folding materials and interfaces. In Proceedings of the CHI EA '18: Extended Abstracts of the 2018 CHI Conference on Human Factors in Computing Systems, Montreal, QC, Canada, 21–26 April 2018; Volume 405, pp. 1–4. [CrossRef]
11. Wang, W.; Yao, L.; Zhang, T. Transformative appetite: Shape-changing food transforms from 2D to 3D by water interaction through cooking. In Proceedings of the CHI '17: 2017 CHI Conference on Human Factors in Computing Systems, Denver, CO, USA, 6–11 May 2017; pp. 6123–6132. [CrossRef]

12. Chen, F.; Zhang, M.; Liu, Z.; Bhandari, B. 4D deformation based on double-layer structure of the pumpkin/paper. *Food Struct.* **2021**, *27*, 100168. [CrossRef]
13. Tao, Y.; Do, Y.; Yang, H.; Lee, Y.; Wang, G.; Mondoa, C.; Cui, J.; Wang, W.; Yao, L. Morphlour: Personalized four-based morphing food induced by dehydration or hydration method. In Proceedings of the UIST '19: 32nd Annual ACM Symposium on User Interface Software and Technology, New Orleans, LA, USA, 20–23 October 2019; pp. 329–340.
14. Shi, Y.; Zhang, M.; Phuhongsung, P. Microwave-induced spontaneous deformation of purple potato puree and oleogel in 4D printing. *J. Food Eng.* **2022**, *313*, 110757. [CrossRef]
15. Guo, C.; Zhang, M.; Bhandari, B. Investigation on simultaneous change of deformation, color and aroma of 4D printed starch-based pastes from fruit and vegetable as induced by microwave. *Food Res. Int.* **2022**, *157*, 111214. [CrossRef] [PubMed]
16. Kumar, C.; Karim, M.A. Microwave-convective drying of food materials: A critical review. *Crit. Rev. Food Sci. Nutr.* **2019**, *59*, 379–394. [CrossRef] [PubMed]
17. Chandrasekaran, S.; Ramanathan, S.; Basak, T. Microwave food processing—A review. *Food Res. Int.* **2013**, *52*, 243–261. [CrossRef]
18. Chen, C.; Zhang, M.; Guo, C.; Chen, H. 4D printing of lotus root powder gel: Color change induced by microwave. *Innov. Food Sci. Emerg.* **2021**, *68*, 102605. [CrossRef]
19. Guo, J.; Zhang, M.; Li, J.; Fang, Z. Using soy protein isolate to improve the deformation properties of 4D-Printed oat flour butterfly. *Food Bioprocess Technol.* **2023**, *16*, 1165–1176. [CrossRef]
20. Watzl, B.; Stehle, P.; Boeing, H.; Kroke, A.; Oberritter, H.; Ellinger, S.; Schulze, M.; Bechthold, A.; Bub, A.; Muller, M.J.; et al. Critical review: Vegetables and fruit in the prevention of chronic diseases. *Eur. J. Clin. Nutr.* **2012**, *51*, 637. [CrossRef]
21. Larson, N.I.; Dianne, R.; Harnack, L.J.; Wall, M.M.; Story, M.T.; Eisenberg, M.E. Fruit and vegetable intake correlates during the transition to young adulthood. *Am. J. Prev. Med.* **2008**, *35*, 33–37. [CrossRef]
22. Rosida, A.E.E.; Djajati, S. Characterization of soyghurt synbiotic drink from soymilk and purple yam extract. *J. Phys. Conf. Ser.* **2020**, *1569*, 032014. [CrossRef]
23. Yang, F.; Zhang, M.; Wang, W. 3D printing physical properties of mochi as affected by different compositions. *J. Food Sci. Biotechnol.* **2020**, *39*, 71–75. Available online: https://ss.zhizhen.com/detail_38502727e7500f26bd22f7bc9ed20ad3fafda09420a450431921b0a3ea255101fc1cf1fbb4666ae6a028f5216aca1f418c296cca829f7db313a7e2e6324330dd99961bedc10b5d6efe0811fc55552d49?&apistrclassfy=0_18_17,0_18_19. (accessed on 3 November 2022).
24. Ezeanaka, M.C.; Zhang, M.; Chen, K.; Wang, Y. Influence of microwave blanching and modified atmosphere packaging on physiochemical properties and quality of carrot used. *J. Food Sci. Biotechnol.* **2022**, *41*, 46–55. Available online: https://ss.zhizhen.com/detail_38502727e7500f26ea4998eb1cc7d2bbbe842c48ab19bdd91921b0a3ea255101fc1cf1fbb4666ae62f4796aa74119a677f57c8396b6f5f62e7b5ac980794a20803a1689cd533c47016ec98d68539181d?&apistrclassfy=0_18_19. (accessed on 12 January 2023).
25. Tang, T.; Zhang, M.; Mujumdar, A.S.; Teng, X. 3D printed white radish/potato gel with microcapsules: Color/flavor change induced by microwave-infrared heating. *Food Res. Int.* **2022**, *158*, 111496. [CrossRef] [PubMed]
26. Liu, Z.; He, C.; Guo, C.; Chen, F.; Bhandari, B.; Zhang, M. Dehydration-triggered shape transformation of 4D printed edible gel structure affected by material property and heating mechanism. *Food Hydrocoll.* **2021**, *115*, 106608. [CrossRef]
27. Feng, C.; Zhang, M.; Bhandari, B. Materials properties of printable edible inks and printing parameters optimization during 3D printing: A review. *Crit. Rev. Food Sci. Nutr.* **2018**, *59*, 3074–3081. [CrossRef] [PubMed]
28. Yang, F.; Guo, C.; Zhang, M.; Bhandari, B.; Liu, Y. Improving 3D printing process of lemon juice gel based on fluid flow numerical simulation. *LWT* **2019**, *102*, 89–99. [CrossRef]
29. Liu, Z.; Zhang, M.; Bhandari, B. Effect of gums on the rheological, microstructural and extrusion printing characteristics of mashed potatoes. *Int. J. Biol. Macromol.* **2018**, *117*, 1179–1187. [CrossRef] [PubMed]
30. Liu, Z.; Zhang, M.; Yang, C. Dual extrusion 3D printing of mashed potatoes/strawberry juice gel. *LWT* **2018**, *96*, 589–596. [CrossRef]
31. Zhu, S.; Stieger, M.A.; van der Goot, A.J.; Schutyser, M.A.I. Extrusion-based 3D printing of food pastes: Correlating rheological properties with printing behaviour. *Innov. Food Sci. Emerg.* **2019**, *58*, 102214. [CrossRef]
32. Yang, F.; Zhang, M.; Bhandari, B.; Liu, Y. Investigation on lemon juice gel as food material for 3D printing and optimization of printing parameters. *LWT* **2018**, *87*, 67–76. [CrossRef]
33. Liu, Z.; Bhandari, B.; Prakash, S.; Mantihal, S.; Zhang, M. Linking rheology and printability of a multicomponent gel system of carrageenan-xanthan-starch in extrusion based additive manufacturing. *Food Hydrocoll.* **2019**, *87*, 413–424. [CrossRef]
34. Ai, Y.; Jane, J.-L. Gelatinization and rheological properties of starch. *Starke* **2015**, *67*, 213–224. [CrossRef]
35. He, C.; Zhang, M.; Devahastin, S. Microwave-induced deformation behaviors of 4D printed starch-based food products as affected by edible salt and butter content. *Innov. Food Sci. Emerg.* **2021**, *70*, 102699. [CrossRef]
36. Liu, Z.; Zhang, M.; Ye, Y. Indirect prediction of 3D printability of mashed potatoes based on LF-NMR measurements. *J. Food Eng.* **2020**, *287*, 110137. [CrossRef]
37. Zhang, Q.; Li, W.; Li, H.; Chen, X.; Jiang, M.; Dong, M. Low-field nuclear magnetic resonance for online determination of water content during sausage fermentation. *J. Food Eng.* **2017**, *212*, 291–297. [CrossRef]
38. Guo, C.; Zhang, M.; Devahastin, S. Color/aroma changes of 3D-Printed buckwheat dough with yellow flesh peach as triggered by microwave heating of gelatin-gum Arabic complex coacervates. *Food Hydrocoll.* **2021**, *112*, 106358. [CrossRef]

39. Ferrari-John, R.S.; Katrib, J.; Palade, P.; Batchelor, A.; Dodds, C.; Kingman, S. A tool for predicting heating uniformity in industrial radio frequency processing. *Food Bioprocess Technol.* **2016**, *9*, 1865–1873. [CrossRef]
40. Pompe, R.; Briesen, H.; Datta, A.K. Understanding puffing in a domestic microwave oven. *J. Food Process. Eng.* **2020**, *43*, 13429. [CrossRef]
41. Zheng, X.; Chen, K.; Su, X.; Liu, H.; Xu, H.; Wang, X.; Han, L. Analysis of process of microwave puffing blueberry snacks. *J. Northeast Agric. Univ. (Engl. Ed.)* **2018**, *25*, 62–69.
42. Fan, H.; Zhang, M.; Liu, Z.; Ye, Y. Effect of microwave-salt synergetic pre-treatment on the 3D printing performance of SPI-strawberry ink system. *LWT* **2020**, *122*, 109004. [CrossRef]
43. Al-Muslimawi, A.; Tamaddon-Jahromi, H.R.; Webster, M.F. Simulation of viscoelastic and viscoelastoplastic die-swell flows. *J. Non-Newton. Fluid Mech.* **2013**, *191*, 45–56. [CrossRef]
44. Zhang, Y.; Zeng, H.; Huang, Z.; Wang, Y.; Zheng, B. Drying characteristics and kinetics of anoectochilus roxburghii by microwave vacuum drying. *J. Food Process. Pres.* **2014**, *38*, 2223–2231. [CrossRef]
45. Torrealba-Meléndez, R.; Sosa-Morales, M.E.; Olvera-Cervantes, J.L.; Corona-Chávez, A. Dielectric properties of beans at different temperatures and moisture content in the microwave range. *Int. J. Food Prop.* **2016**, *19*, 564–577. [CrossRef]

foods

MDPI

Article

Three-Dimensional Appearance and Physicochemical Properties of *Pleurotus eryngii* under Different Drying Methods

Jun-Wen Bai [1], Yu-Chi Wang [1], Jian-Rong Cai [1], Lu Zhang [1], Yi Dai [1], Xiao-Yu Tian [1,*] and Hong-Wei Xiao [2,*]

[1] School of Food and Biological Engineering, Jiangsu University, Zhenjiang 212013, China; bjw@ujs.edu.cn (J.-W.B.)
[2] College of Engineering, China Agricultural University, Beijing 100083, China
* Correspondence: tianxy@ujs.edu.cn (X.-Y.T.); xhwcaugxy@163.com (H.-W.X.)

Abstract: This study investigated the effects of different drying methods on the drying characteristics, three-dimensional (3D) appearance, color, total polysaccharide content (TPC), antioxidant activity, and microstructure of *Pleurotus eryngii* slices. The drying methods included hot air drying (HAD), infrared drying (ID), and microwave drying (MD). The results showed that the drying method and conditions significantly influenced the drying time, with MD having a significant advantage in reducing the drying time. The 3D appearance of *P. eryngii* slices was evaluated based on shrinkage and roughness as quantitative indexes, and the best appearance was obtained by hot air drying at 55 and 65 °C. HAD and ID at lower drying temperatures obtained better color, TPC, and antioxidant activity, but MD significantly damaged the color and nutritional quality of *P. eryngii*. The microstructure of dried *P. eryngii* slices was observed using scanning electron microscopy, and the results showed that drying methods and conditions had an obvious effect on the microstructure of *P. eryngii* slices. Scattered mycelia were clearly observed in *P. eryngii* samples dried by HAD and ID at lower drying temperatures, while high drying temperatures led to the cross-linking and aggregation of mycelia. This study offers scientific and technical support for choosing appropriate drying methods to achieve a desirable appearance and quality of dried *P. eryngii*.

Keywords: three-dimensional appearance; drying methods; quality; microstructure; *Pleurotus eryngii*

Citation: Bai, J.-W.; Wang, Y.-C.; Cai, J.-R.; Zhang, L.; Dai, Y.; Tian, X.-Y.; Xiao, H.-W. Three-Dimensional Appearance and Physicochemical Properties of *Pleurotus eryngii* under Different Drying Methods. *Foods* **2023**, *12*, 1999. https://doi.org/10.3390/foods12101999

Academic Editor: Alessandra Fratianni

Received: 22 March 2023
Revised: 9 May 2023
Accepted: 12 May 2023
Published: 15 May 2023

1. Introduction

Pleurotus eryngii, also known as the king oyster mushroom, is called xingbaogu in China. As a member of the *Pleurotus species*, it is a valuable edible fungus that is rich in bioactive compounds, such as polysaccharides, proteins, and free amino acids. It also possesses numerous physiological functionalities, such as antimicrobial, antioxidant, hypolipidemic, hypoglycemic, and immune regulatory activities [1]. However, one of the challenges associated with using *P. eryngii* is its high moisture content, which can reach up to 90% on a wet basis. This makes fresh *P. eryngii* sensitive to bacterial growth and shortens its shelf life if not stored properly [2]. Therefore, there is an urgent need to implement processing techniques for *P. eryngii* in order to increase its shelf life and enhance its value-added potential throughout the industrial chain.

Drying is one of the most common processing methods for edible fungi [3,4]. The objective of drying is to reduce the moisture content, eliminate microorganisms, increase the soluble solids content, inhibit enzyme activity, extend the storage period, and enhance the value of commodities [5,6]. The most widely used drying method is hot air drying (HAD) due to its low cost, easy operation, and simple control [7,8]. In comparison, infrared drying (ID) can significantly shorten drying time and improve product quality, but its effectiveness can be influenced by infrared wavelengths, food surface porosity, and composition [9]. Although microwave drying (MD) is a fast and effective alternative technique, it has some drawbacks, such as a high initial cost, uneven heating, limited microwave penetration, and quality loss [10]. Appropriate drying methods and conditions can minimize nutrient and color degradation during the drying process of *P. eryngii*, reduce energy consumption, and ensure product quality.

The appearance of fruits and vegetables changes visibly during the drying process due to their porous tissue structure, resulting in shrinkage, curling, roughness, and collapse. With the advancement of computer vision (CV) technology, researchers use it to measure and analyze the changes in the appearance dimensions of materials during drying. Onwude et al. investigated the shrinkage of sweet potatoes using a combined digital and backscattering imaging system and found that shrinkage was affected by the product thickness, drying temperature, and drying time [11]. Pei et al. applied a camera to obtain the surface area and shrinkage of ginger slices, important quality parameters [12]. Li et al. observed the phenomenon of surface wrinkling of shiitake mushrooms during the drying process and obtained the ratio of the wrinkled surface area by region segmentation [13]. However, a single camera can only capture data on the projected area of the sample's surface and cannot effectively measure changes in material thickness. To address this issue, Sampson et al. constructed a top and side dual-view computer vision system to measure the volume of apple slices [14]. Nonetheless, even with multiple cameras, the data obtained are still based on the analysis of two-dimensional (2D) projected images.

To obtain the three-dimensional (3D) appearance and shape changes of materials during the drying process, Cai et al. used the Kinect sensor to acquire depth and color images of potato slices and calculated the shrinkage and curling degree of the slices [15]. However, the accuracy limitations of the Kinect sensor [16] restrict its ability to distinguish changes within a range of 2 mm in practical applications. Bai et al. employed a binocular snapshot sensor to obtain point cloud images with a measurement accuracy of 0.035 mm [17]. They also calculated the shrinkage and height standard deviation (HSD) to evaluate the three-dimensional (3D) deformation, such as surface curl and flatness, of apple slices. Regarding edible fungi, particularly *P. eryngii*, the measurement and quantifiable representation indices for their 3D appearance changes are still limited in research.

This study aimed to investigate the effects of different drying methods and conditions on the drying characteristics, 3D appearance, physicochemical properties, and microstructure of dried *P. eryngii* slices. Furthermore, the study provides quantitative indexes for the analysis and evaluation of 3D appearance and offers technical support for selecting suitable drying methods for *P. eryngii*.

2. Materials and Methods

2.1. Samples and Chemicals

Fresh *P. eryngii* samples were purchased from a supermarket near Jiangsu University in Zhenjiang City, with an average wet basis moisture content of about 90.0 ± 1.0%. All the samples were stored in a refrigerator at 4 ± 1 °C and 90% relative humidity prior to the experiments. The analytical grade 2,2-diphenyl-1-picrylhydrazyl (DPPH), ethanol, concentrated sulfuric acid, phenol, and anhydrous glucose were purchased from Sinopharm Chemical Reagent Co., Ltd. (Shanghai, China).

2.2. Drying Experiments

Fresh *P. eryngii* samples were cut into circular slices approximately 5 mm in thickness with a diameter of 4.5 cm. The weight of raw samples was kept at 50.0 ± 0.5 g for all runs. The safe moisture content was set at 10% wet basis, according to the GB7096-2014 "National Food Safety Standard for Edible Fungi and Their Products". The *P. eryngii* samples were periodically taken from the drying chamber for weighing and 3D point cloud collection until the samples reached the final moisture content during drying. All the drying experiments were conducted in triplicate.

Hot air drying (HAD): *P. eryngii* slices were dried under different temperatures of 55, 65, 75, and 85 °C. The air velocity was 3 m/s, and the relative humidity was 20%. The samples were laid flat on trays in a hot air dryer, which was self-developed by Jiangsu University [18].

Infrared drying (ID): Samples were dried under different temperatures of 60, 65, 75, and 85 °C. The infrared equipment was a Hauswirt electric infrared oven (I7, Hauswirt, China). No airflow was used in the infrared oven, and the distance between the samples and the infrared source was approximately 15 cm.

Microwave drying (MD): Drying experiments were carried out in a domestic digital microwave oven (P70d2otl, Galanz, China), with a maximum power output capacity of 700 W at 2450 MHz. The microwave oven has a capability to operate at four different microwave output powers (119, 157, 196, and 280 W), with a measurement accuracy of ±10 W. Processing time and microwave output power were adjusted using the digital control on the microwave oven. The *P. eryngii* slices were placed in a single layer on a rotating glass plate in the oven to ensure uniform heating.

2.3. Drying Characteristics

The initial moisture content of *P. eryngii* slices was measured using a quick moisture analyzer (DHS-16, Lichen instrument technology Co., Ltd., Shanghai, China). The moisture ratio was calculated using Equations (1) and (2) [19,20];

$$MR = \frac{M_t - M_e}{M_0 - M_e} \tag{1}$$

where M_0 is the initial dry basis moisture content; M_t is the dry basis moisture content at the drying time; MR is the moisture ratio; M_e is the equilibrium moisture content. As in the equilibrium moisture content, M_e is much smaller than M_t and M_0, it can generally be ignored [21]. Therefore, the calculation of MR can be simplified as

$$MR = \frac{M_t}{M_0} \tag{2}$$

2.4. Point Cloud Data Acquisition

A 3D point cloud acquisition platform developed by our research group [17,18] was used in this study. The platform uses a binocular snapshot Gocator3210 sensor (LMI, Canada) with a detection accuracy of 0.035 mm. Before data collection, the datum plane was calibrated, and then point cloud images of the *P. eryngii* slices were successively captured during the drying process.

2.5. Three-Dimensional Appearance Evaluation Index

The 3D point cloud processing software Cloud Compare was used to process the acquired images, including background removal, point cloud filtering, surface reconstruction, and surface smoothing [19]. After the point cloud processing, the images were subjected to feature extraction, and two quantitative indexes (shrinkage and roughness) were selected to evaluate the 3D appearance of the *P. eryngii* slices during drying.

2.5.1. Shrinkage

The surface model was composed of tens of thousands of triangles. First, the distance between two points was calculated using Euclid's formula to obtain the three side lengths of each triangle. For example, the distance between the point p_1 (x_1, y_1, z_1) and p_2 (x_2, y_2, z_2) can be calculated using Equation (3). Then, the area of each triangle was calculated using Helen's formula, as shown in Equation (4), and the sum of the area of all triangles was calculated, which was the surface area. The shrinkage of the *P. eryngii* slices during drying can be calculated by the change in surface area at different drying time points using Equation (5). The specific formulas are as follows:

$$d_{p_1p_2} = \sqrt{(x_1 - x_2)^2 + (y_1 - y_2)^2 + (z_1 - z_2)^2} \tag{3}$$

$$S_{ABC} = \sqrt{p(p - d_{AB})(p - d_{AC})(p - d_{BC})} \quad (4)$$

$$Shrinkage = \frac{S_0 - S_t}{S_0} \quad (5)$$

where d_{p1p2} is the distance between the two points of p_1 and p_2, S_{ABC} is the area of the triangle ABC, and p is half of the perimeter of the triangle ABC. S_0 is the surface area of the sample before drying, and S_t is the surface area of the sample during drying.

2.5.2. Roughness

During the drying process, the surface of the material becomes convex and uneven, so roughness was selected to evaluate the smoothness of the material surface. The principle of calculation is as follows: for each point in the point cloud data, the radius of the input parameter R is used to find the adjacent points in this range and fit the least-squares plane. The distance d of the point to the least-squares plane is calculated as the roughness. If the least-squares plane cannot be fitted to a point in the point cloud data, the roughness of the point is defined as null. According to the experimental results, the best input radius was set to 2 mm.

2.6. Color Measurement

The color of the dried samples was measured using a colorimeter (SC-10, Shenzhen 3nh Technology Co., Ltd., Shenzhen, China). The color coordinates included L^* (lightness, 0~100), a^* (redness, −60~60), and b^* (yellowness, −60~60). Fresh *P. eryngii* slices were used as the control, and the total color difference (ΔE) was calculated according to Equation (6) [22]. The color parameters of each sample were measured 9 times and averaged:

$$\Delta E = \sqrt{(L^* - L_0^*)^2 + (a^* - a_0^*)^2 + (b^* - b_0^*)^2} \quad (6)$$

where L_0^*, a_0^*, b_0^* are the color parameters of fresh *P. eryngii* slices, and L^*, a^*, b^* are the color parameters of dried *P. eryngii* slices.

2.7. Determination of Total Polysaccharide Content (TPC)

To determine the crude polysaccharide content, we followed the NY/T 1676-2008 method for determining the crude polysaccharide content in edible fungi. The total polysaccharide content of the extraction was measured using the phenol–sulfuric acid method, as described by Su et al. [23] and Zhu et al. [24]. The crude polysaccharides from *P. eryngii* slices were extracted using ultrasonic-assisted extraction (YM-040S, Fangao Microelectronics Co., Ltd., Shenzhen, China). The results were measured using dry mass, and all units were expressed as mg/g. Each experiment was repeated three times.

2.8. Determination of DPPH Radical Scavenging Activity

The sample was homogenized with 1:40 (*w*/*v*) ethanol to obtain the extract, which was used for scavenging DPPH free radical activity. After filtering, the mixture was centrifuged (4000 rpm, 10 min), and then the supernatant was collected. The following measurements were conducted following the method described by Liu et al. [25]. A suitable volume of the *P. eryngii* slices extract was mixed with 2 mL of DPPH ethanol solution and placed in the dark at 37 °C for 30 min. The absorbance of the reaction solution was measured at 517 nm, and a blank control was established using ethanol instead of the sample solution in the DPPH ethanol mixture. The DPPH radical scavenging activity of *P. eryngii* slices was calculated using Equation (7) as described by Bechlin et al. [26]. Each group of samples was repeated three times.

$$DPPH\ radical\ scavenging\ activity = (1 - \frac{A}{A_0}) \times 100\% \quad (7)$$

where A is the absorbance value of dried samples, and A_0 is the absorbance value of the control.

2.9. Microstructure Analysis

Scanning electron microscopy (S-3400N, Hitachi Co., Ltd., Tokyo, Japan), at an accelerating voltage of 15 kV vacuum, was used to examine the microstructure of the dried *P. eryngii* slices. Each sample was cut into blocks with a cross-section of 5 mm × 5 mm, and microstructure images of 200× and 1000× were obtained after sample preparation and gold spraying.

2.10. Statistical Analysis

All experiments were conducted in triplicate, and data were presented in the form of mean ± standard deviation values. The experimental data of the drying process were analyzed using origin 2018 and Cloud Compare software. The results were evaluated using one-way analysis of variance (ANOVA) followed by Duncan's new multiple range test (SPSS 25.0 software). Differences were considered significant at $p < 0.05$.

3. Results and Discussions

3.1. Drying Curves

Figure 1 shows the moisture ratio (MR) curves versus drying time of *P. eryngii* slices under different drying conditions. From Figure 1a, it can be seen that the time required to reduce the moisture from the initial moisture content to the desired moisture content was approximately 220, 180, 140, and 160 min, respectively, for hot air temperatures of 55, 65, 75, and 85 °C. It is worth noting that the drying time of *P. eryngii* slices at 85 °C is about 20 min longer than that at 75 °C. This could be attributed to the high drying temperature accelerating the drying of the surface, causing surface hardening, which hindered the transfer of moisture inside the samples [27]. Demirel and Turhan also reported that the effective moisture diffusivity of banana slices at 70 °C was lower than at 60 °C due to starch gelatinization and surface hardening [28].

From Figure 1b, it can be observed that the drying time of *P. eryngii* slices dried by ID was approximately 280, 260, 210, and 140 min at drying temperatures of 60, 65, 75, and 85 °C, respectively. The drying time decreased by 50% with the ID temperature increasing from 60 to 85 °C. Several studies have shown that ID can reduce the drying time compared to HAD. For example, Cao et al. [29] reported that the drying time of ID was reduced by 38.27%, 34.62%, and 30.56% compared to HAD under the corresponding temperatures of 60, 70, and 80 °C, respectively. However, the results of this study showed the opposite. Nowak et al. reported that the drying time of ID was dependent on emitter–sample distance, air velocity, air temperature, and infrared radiation intensity [30,31]. The difference in results from Cao et al. could be due to the absence of airflow in the ID equipment and the relatively long distance from irradiation.

For microwave drying (Figure 1c), the drying time was about 56, 36, 20, and 12 min for 119, 157, 196, and 280 W, respectively. It is evident that MW resulted in a considerable reduction in drying time compared to HAD and ID. Albanese et al. also reported that using MW for apricots reduced the drying time by more than 25% when compared to HAD [32]. Therefore, MW has significant advantages in reducing the drying time of *P. eryngii* slices. However, the appearance and quality characteristics of the product under different drying methods still require further investigation.

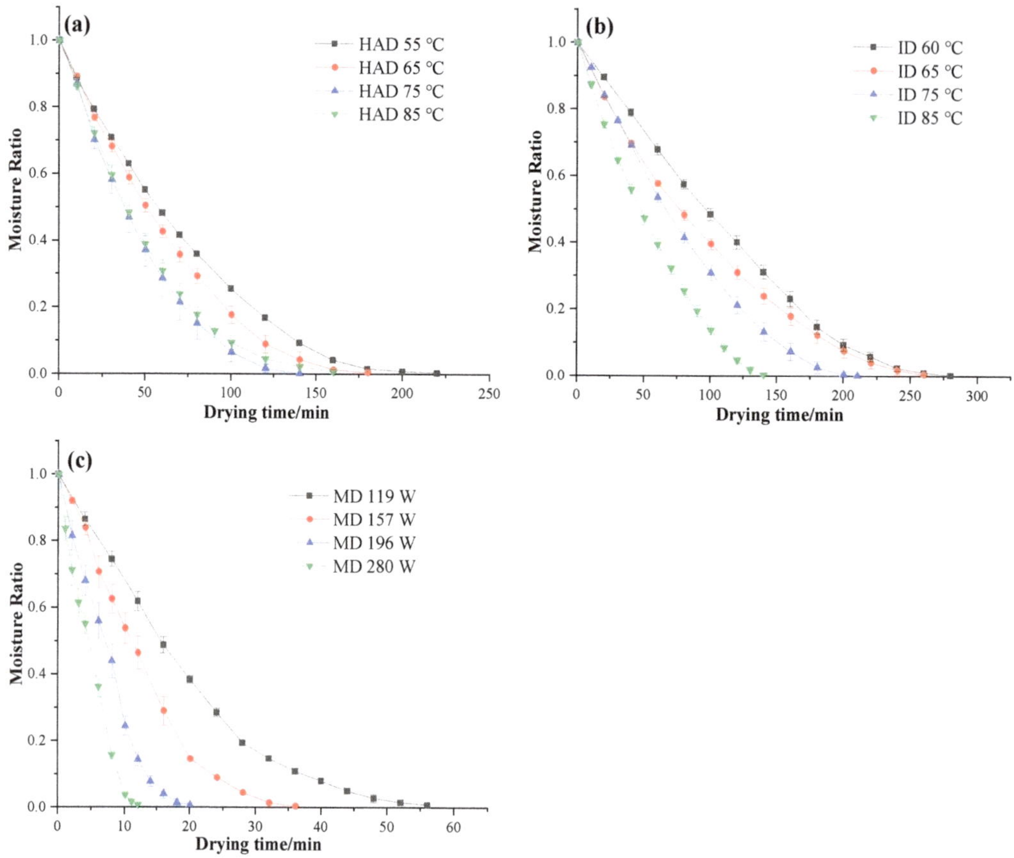

Figure 1. Drying curves of *P. eryngii* slices under different drying conditions. (**a**) Hot air drying, (**b**) infrared drying, (**c**) microwave drying.

3.2. Three-Dimensional Appearance Characterization

The time-varying appearance images of a *P. eryngii* slice at 75 °C under HAD are displayed in Figure 2I. In each column, the three images from top to bottom represent a color physical image, a height-distributed image, and a 3D reconstructed image. The height distribution of the *P. eryngii* slice in Figure 2 is depicted using pseudo-color images, where the color from blue to red indicates that the height value of the pixels on the material changed from small to large. The 3D reconstructed images, created by the greedy projection algorithm, accurately reproduce the 3D appearance changes of the *P. eryngii* slice during drying [33]. According to the height-distributed images, the thickness of the *P. eryngii* slices was relatively uniform at the beginning of drying, and the only morphological change was a decrease in thickness. After 30 min of drying, the central part of the sample began to thin and develop a concave shape. After 120 min, the edges of the sample curled significantly, and the surface became rough with an increase in shrinkage.

The appearance images of the dried *P. eryngii* slices under different drying methods are shown in Figure 2II. It was found that the appearance of the *P. eryngii* slices had significant curling, shrinkage, and browning after the drying process. Moreover, the appearance of the dried *P. eryngii* slices varied greatly under different methods and conditions. The appearance of microwave-dried *P. eryngii* slices changed most seriously, with the color and shape differing greatly from the initial state.

Figure 2(IIa–IId) shows the appearance of *P. eryngii* slices under HAD at 55, 65, 75, and 85 °C. The figure illustrates that at lower temperatures (55 °C and 65 °C), *P. eryngii* slices dried by HAD had a relatively flat appearance. However, when the drying temperature was increased, the surface of *P. eryngii* slices underwent significant browning, severe curling, and shrinkage. In addition, brown spots and bumps also appeared on the surface of the samples. The appearance of *P. eryngii* slices under ID at 60, 65, 75, and 85 °C is shown in Figure 2(IIe–IIh). The images show that as the drying temperature increased, the samples with ID did not shrink uniformly from all directions towards the center but rather sank downwards in the middle and curled upwards around the edges. Due to uneven heating during MD, some parts of the *P. eryngii* slices were burnt, resulting in pits or bumps in the middle, along with a burning smell (Figure 2(IIi–III)). Additionally, the appearance of the samples shrank significantly after microwave drying, and the surface exhibited obvious browning. To address the challenge of controlling material temperatures during the later stages of MD, a common practice involves utilizing microwave predrying as a fast initial step, followed by traditional hot air convection drying. This combined method is effective in achieving a balanced drying rate while maintaining product quality [34]. The corresponding 3D reconstructions and height distribution images vividly show the appearance and color changes of *P. eryngii* slices under different drying methods.

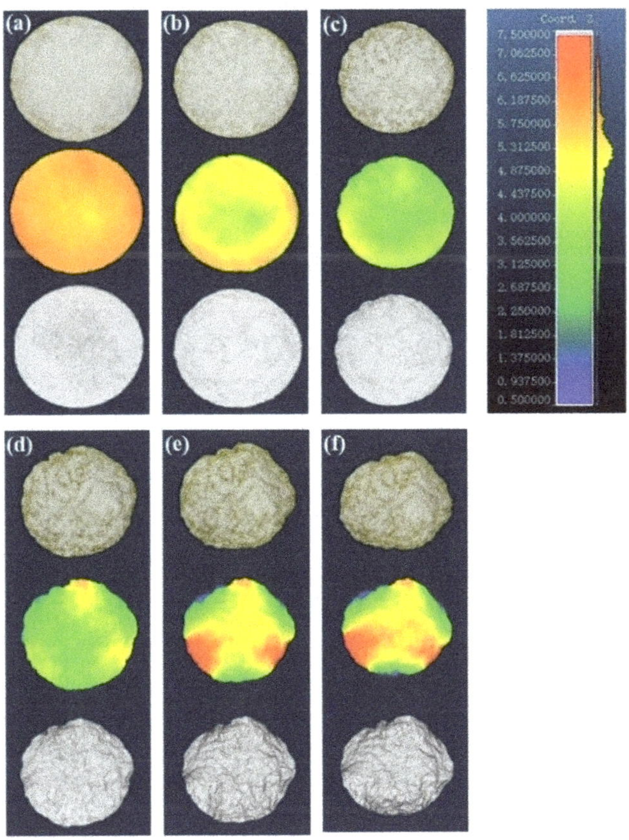

(I)

Figure 2. *Cont.*

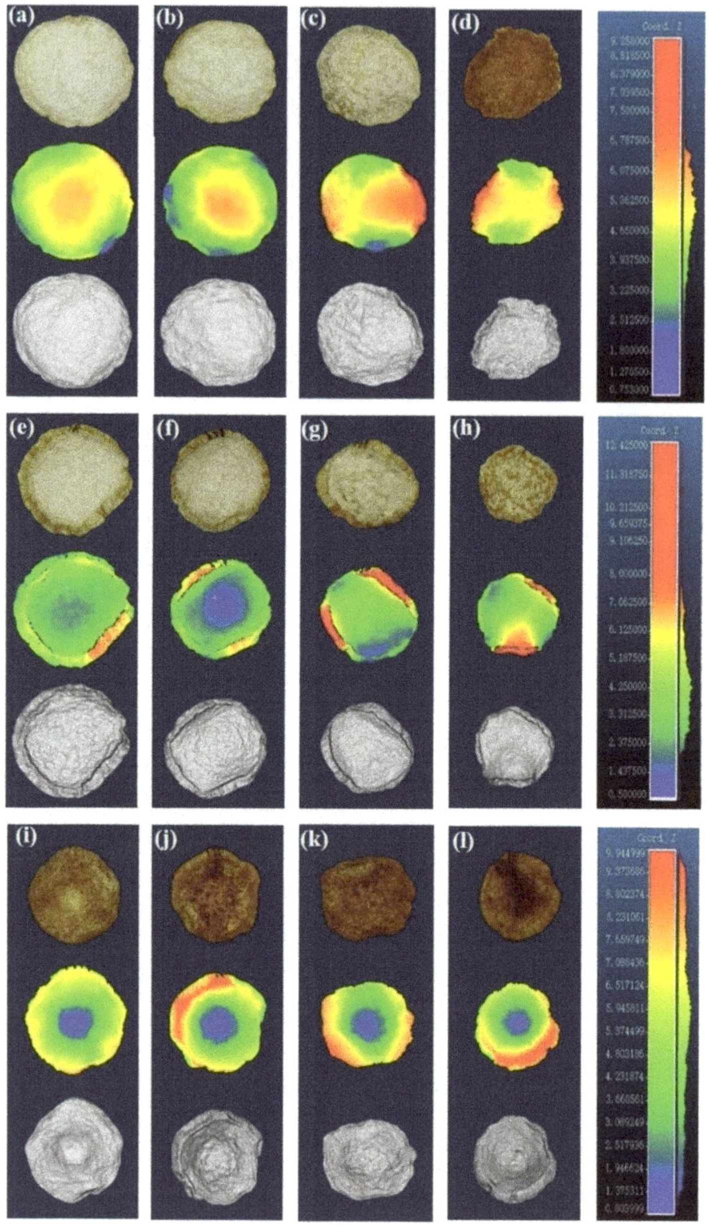

(II)

Figure 2. (**I**) Time-varying appearance images of one *P. eryngii* slice during drying. (**Ia–If**) represent *P. eryngii* slices dried at 75 °C, 20% RH, 3 m/s for 0, 30, 60, 90, 120, 140 min under hot air drying, respectively. The three images from top to bottom in each column represent color physical images, height-distributed images, and 3D reconstructed images. (**II**) Three-dimensional appearance images of dried *P. eryngii* slices under different drying methods. (**IIa–IId**) Hot air drying at 55, 65, 75, and 85 °C. (**IIe–IIh**) Infrared drying at 60, 65, 75, and 85 °C. (**IIi–IIl**) Microwaves drying at 119, 157, 196, 280 W. The three images from top to bottom in each column represent color physical images, height-distributed images, and 3D reconstructed images of *P. eryngii* slices.

3.3. Shrinkage

Figure 3 depicts the shrinkage curves of *P. eryngii* under different drying conditions. The curves indicate that the shrinkage of *P. eryngii* slices occurs mainly during the early drying stage and gradually slows down in the later drying stage. Previous research showed that shrinkage at the initial stage of drying is approximately equal to the volume of moisture lost. In the middle and late stages, as the "skeleton" becomes fixed, the shrinkage slows down [35]. Figure 3a shows that the shrinkage of *P. eryngii* samples dried by HAD were 0.3625, 0.3718, 0.5372, and 0.7647 at drying temperatures of 55, 65, 75, and 85 °C, respectively. As the drying temperature increased, the shrinkage rate of *P. eryngii* significantly increased, indicating that high temperatures exacerbate the shrinkage of dried samples. This phenomenon was caused by accelerated water removal under high temperatures, which led to more stress in the matrix of the food material. Wang et al. observed that the HAD process on jujube slices caused serious volume shrinkage due to the high temperatures involved [36]. Udomkun et al. also found that an increase in drying temperature from 50 to 80 °C led to an increase in papaya shrinkage at the same final moisture content [37].

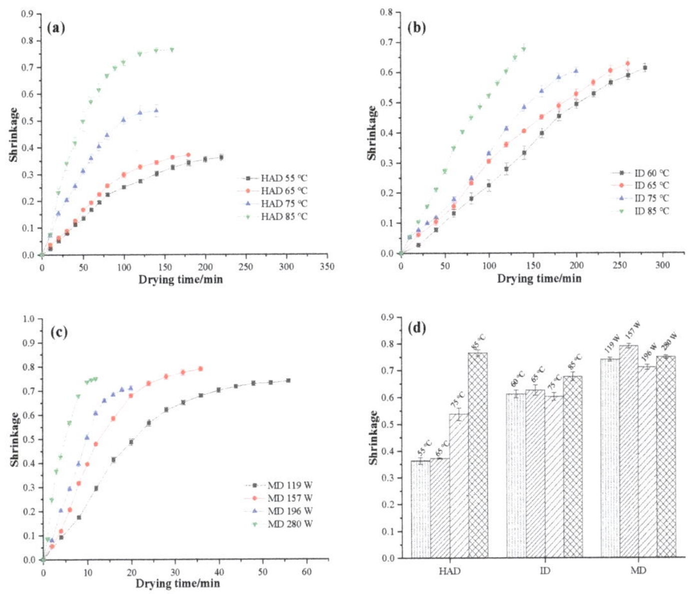

Figure 3. Shrinkage of *P. eryngii* slices during drying under different drying conditions, such as (**a**) hot air drying, (**b**) infrared drying, and (**c**) microwave drying, (**d**) the shrinkage of *P. eryngii* at the end point of drying.

As shown in Figure 3b, the shrinkage of ID *P. eryngii* samples was 0.6122, 0.6264, 0.6024, and 0.6767 at drying temperatures of 60, 65, 75, and 85 °C, respectively. The shrinkage rate increased with increasing drying time. Shrinkage at 75 °C was significantly lower ($p < 0.05$) than that at 60 and 65 °C. This may be because the higher temperatures during case hardening hindered further shrinkage. It is worth noting that, compared to HAD dried samples, the shrinkage of ID samples was more obvious. Kaveh et al. also found that green beans dried by infrared drying had greater shrinkage than those dried by hot air drying [38]. The shrinkage of *P. eryngii* samples dried by MD ranged from 0.6712 to 0.7760. Compared to the other two drying methods, the shrinkage of *P. eryngii* by MD was more severe. This may be due to the fact that microwaves can directly heat the interior of the sample, which may increase its internal temperature and produce water vapor, providing an additional driving force for water migration, thus increasing the drying rate and accelerating shrinkage [39].

Rokhbin and Azadbakht found that any rise in microwave power (from 90 to 900 W) led to an intensification of orange slice shrinkage [40]. However, in this experiment, the shrinkage rate did not increase continuously with the increase in microwave power. According to the report by Mahiuddin et al., tissue structure is an important factor affecting contraction behavior [41], so the difference between the obtained data and previous studies might be due to the difference in the tissue structure of different plant species.

3.4. Roughness

The effects of drying conditions on the roughness of *P. eryngii* are shown in Figure 4. It is evident that the surface roughness increases with the extension of drying time. Moreover, the roughness of dried *P. eryngii* also increases with an increase in the HAD temperature from 55 °C to 85 °C (Figure 4a). This phenomenon can be attributed to the uneven distribution of moisture and the surface stress formed by the heat effect, resulting in an uneven appearance. Furthermore, polysaccharides tend to depolymerize into small molecules of polysaccharides or monosaccharides under higher temperatures, increasing their hygroscopicity and leading to mycelium aggregation, which can cause an uneven surface texture.

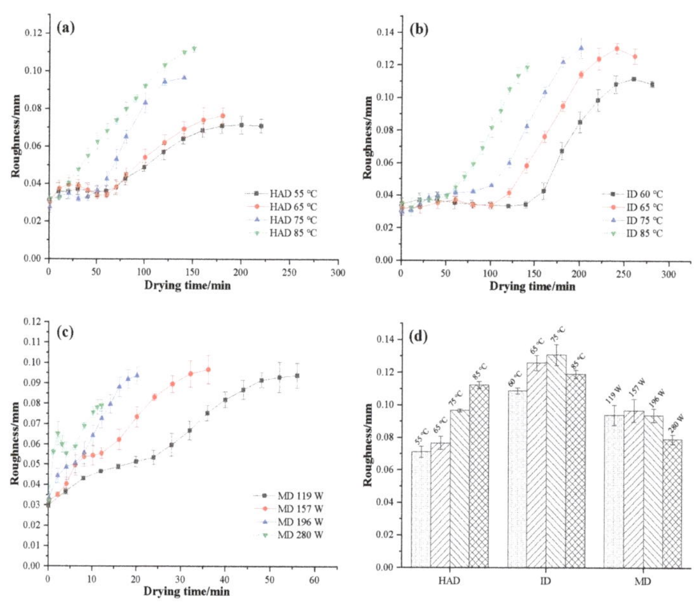

Figure 4. Roughness of *P. eryngii* slices during drying under different drying conditions, such as (**a**) hot air drying, (**b**) infrared drying, and (**c**) microwave drying, (**d**) the roughness of *P. eryngii* at the end point of drying.

During infrared drying, the roughness of *P. eryngii* increased gradually over the drying time, showing a similar trend as that of hot air drying. However, the roughness of *P. eryngii* dried by ID was noticeably higher than that of samples dried by HAD. Yang et al. noted that the uneven distribution of water diffusion, as well as the collapse of the cellular structure on the surface, ultimately resulted in uneven structures, which may be attributed to the limited depth of infrared radiation [42].

After microwave drying, the roughness values of *P. eryngii* slices at the end point of drying ranged from 0.0782 to 0.0947. The roughness of samples dried by MD was lower than that of samples dried by ID, but higher than those dried at 75 °C and 85 °C using HAD. Considering all three drying methods and conditions, the roughness of *P. eryngii* was lowest when dried with low-temperature HAD, resulting in the smoothest surface of the dried samples. It was reported that the larger difference between the vapor pressure of water on the surface of the mushrooms and the bulk air partial pressure of water vapor resulted in a larger driving force for drying, leading to more rapid dehydration, which promoted wrinkling [13]. Under low-temperature drying conditions, the driving force for the drying of *P. eryngii* is relatively weak. This facilitates the release of stress caused by the uneven distribution of moisture on the surface of the material, thereby reducing wrinkling and shrinkage on its surface.

3.5. Color

Color is one of the most important sensory attributes of edible mushrooms, as it influences the customer's perception and purchasing power. The color values of fresh and dried *P. eryngii* slices using different drying methods are presented in Table 1. The L^* values of dried samples were significantly ($p < 0.05$) lower, while the values of a^* and b^* were significantly higher ($p < 0.05$) than the values of the fresh *P. eryngii* slices. Similar results were also reported in shiitake mushrooms [43]. It is generally believed that the observer can detect the color difference clearly when the ΔE is greater than 5 [44]. The ΔE of the dried *P. eryngii* slices obtained by different drying methods were higher than 24, indicating that the drying process significantly changed the color of *P. eryngii* slices.

Table 1. Physicochemical properties of P. eryngii slices under different drying conditions.

Drying Condition		L^*	a^*	b^*	ΔE	TPC (mg/g)	DPPH Radical Scavenging Activity (%)
Fresh	——	85.62 ± 0.33^{a}	3.95 ± 0.19^{h}	12.85 ± 0.39^{f}	——	48.37 ± 0.75^{a}	68.87 ± 1.03^{a}
HAD	55 °C	76.15 ± 4.87^{b}	15.63 ± 0.72^{f}	31.74 ± 2.27^{c}	24.40 ± 1.15^{g}	$30.70 \pm 0.32^{b,c}$	$43.91 \pm 2.04^{b,c}$
	65 °C	$71.46 \pm 5.86^{c,d}$	14.68 ± 0.66^{g}	28.54 ± 2.04^{d}	24.71 ± 2.23^{g}	30.88 ± 0.15^{b}	40.30 ± 2.08^{d}
	75 °C	$66.21 \pm 4.53^{e,f}$	16.20 ± 0.59^{f}	31.74 ± 1.52^{c}	29.99 ± 2.76^{f}	$30.57 \pm 0.08^{b,c}$	32.80 ± 0.15^{e}
	85 °C	50.19 ± 1.81^{h}	$21.83 \pm 0.80^{b,c}$	38.34 ± 2.00^{b}	47.25 ± 0.31^{c}	$30.50 \pm 0.1^{b,c}$	32.69 ± 0.65^{e}
ID	60 °C	$68.89 \pm 1.97^{d,e}$	16.18 ± 0.29^{f}	32.01 ± 0.66^{c}	28.29 ± 0.95^{f}	$30.55 \pm 0.13^{b,c}$	42.24 ± 0.70^{d}
	65 °C	$66.08 \pm 1.22^{e,f}$	16.06 ± 0.52^{f}	31.16 ± 1.07^{c}	29.40 ± 1.51^{f}	$30.63 \pm 0.20^{b,c}$	$43.87 \pm 0.71^{b,c}$
	75 °C	$74.18 \pm 3.09^{b,c}$	16.05 ± 0.22^{f}	$32..30 \pm 0.63^{c}$	25.77 ± 1.21^{g}	$30.32 \pm 0.3^{b,c}$	45.21 ± 0.75^{b}
	85 °C	63.92 ± 1.09^{f}	19.97 ± 0.53^{e}	38.46 ± 1.41^{b}	37.20 ± 1.68^{e}	$30.00 \pm 0.94^{b,c,d}$	46.55 ± 0.24^{b}
MD	119 W	57.26 ± 2.90^{g}	$22.73 \pm 0.52^{a,b}$	42.42 ± 0.95^{a}	45.13 ± 1.93^{d}	$29.90 \pm 0.92^{b,c,d}$	40.25 ± 3.11^{c}
	157 W	37.87 ± 2.65^{i}	$20.61 \pm 2.15^{d,e}$	25.69 ± 3.65^{e}	52.41 ± 1.04^{a}	$29.95 \pm 0.90^{b,c,d}$	39.20 ± 2.30^{d}
	196 W	47.09 ± 4.23^{h}	23.22 ± 1.37^{a}	36.58 ± 4.27^{b}	49.56 ± 1.14^{b}	$29.32 \pm 0.07^{c,d}$	32.29 ± 1.95^{e}
	280 W	40.77 ± 3.63^{i}	$21.39 \pm 1.23^{c,d}$	28.97 ± 3.29^{d}	$50.94 \pm 2.32^{a,b}$	28.82 ± 0.50^{d}	25.89 ± 0.90^{f}

Note: The data are presented as the mean ± standard deviation for three replicates. Different letters (a, b, c, d, e, f, g, h, and i) in the same column for each parameter indicate significant differences.

By observing Figure 2(II) and Table 1, it can be seen that the color of the sample is better at lower drying temperatures. This can be explained by the fact that enzymatic browning reactions and nonenzymatic browning reactions were significantly attenuated at lower temperatures [42]. Gouzi et al. [45] found that the polyphenol oxidase (PPO) activities of the mushroom *Agaricus bisporus* were still retained by approximately 15% at 60 °C after 30 min of heat treatment. Although the drying temperature is higher than the temperature of PPO enzyme inactivation, the internal temperature of the material is usually less than the drying temperature in most of the drying time due to the effect of moisture evaporation and heat absorption [46]. It was also reported that both nonenzymatic browning and enzymatic browning take place during the drying process of button mushrooms, but the effect of enzymatic browning is more significant [47]. MD-dried samples presented significantly lower L^* values than the HAD and ID samples. Meanwhile, MD also particularly influenced the a^* and b^* values, resulting in the highest value of ΔE (50.94). This might be due to the phenomenon of coke spot caused by the excessive heating of materials in the MD process. Therefore, using color as the evaluation index, it can be concluded that the HAD and ID methods produce acceptable dried *P. eryngii*, whereas the MD method is not recommended.

3.6. TPC

The TPC of fresh and dried *P. eryngii* slices under different conditions is listed in Table 1. The TPC of all dried samples was lower than that of the fresh samples (48.37 ± 0.75 mg/g). This can be attributed to the thermal decomposition of polysaccharides and the Maillard reaction during heat treatment. By comparing the TPC of dried *P. eryngii* slices under three drying methods, it was found that the retention of total polysaccharides by HAD and ID had little difference. However, MD-dried *P. eryngii* slices had the lowest retention rate of total polysaccharides. This result may be due to the limited availability of water at the final stages of the MD process, which can cause the temperature of the material to easily rise to a level that accelerates the degradation of polysaccharides [48,49]. The TPC of *P. eryngii* slices decreased with increasing microwave power, and the lowest TPC was 28.82 ± 0.50 mg/g.

3.7. DPPH Radical Scavenging Activity

The values of DPPH radical scavenging activity of *P. eryngii* slices under different drying methods and conditions are shown in Table 1. According to the DPPH assay, the antioxidant activity of the fresh sample was significantly stronger than that of the dried samples ($p < 0.05$). Similarly, Gasecka et al. found that DPPH scavenging activity in fresh samples was significantly higher than that in dried samples [50]. Normally, the antioxidant activity of plant materials depends on their content of bioactive components. Drying is a thermal process that results in a decrease in the content of heat-sensitive substances with antioxidant activity.

For the HAD and MD processes, the DPPH radical scavenging activity of *P. eryngii* significantly decreased with increasing drying temperatures and microwave power. Different from the above two drying methods, the DPPH radical scavenging activity of *P. eryngii* slices dried by ID increased with the increase in drying temperature. Infrared drying at 85 °C obtained the highest antioxidant activity (46.55%), while microwave drying at 280 W showed the lowest antioxidant activity (25.89%). In contrast to our study, Ren et al. found that the antioxidant activity of *Coprinus comatus* dried by microwave drying was superior to that of hot air and infrared drying [51]. This difference in results may be attributed to the significantly higher microwave power used in Ren et al.'s study, which was set at 700 W, compared to the range of drying conditions used in our study. This high microwave power setting may have greatly reduced the drying time, thereby preserving more antioxidant compounds in the dried samples. The discrepancy in microwave power used in the two studies may have led to the difference in the antioxidant activity observed in the dried mushroom products.

Overall, the differences in the DPPH radical scavenging activity of dried samples under different drying methods and conditions may be due to the effects of these factors on the preservation of antioxidants in the samples. The retention of the antioxidants in the dried samples depends on both the processing temperature and drying time. Higher drying temperatures may not help to preserve the antioxidant compounds, but high temperatures can significantly shorten the drying time, thus reducing the exposure time of materials to oxygen and thermal effects. Wu et al. also discovered that reducing the drying time improved the DPPH radical scavenging activity of the dried samples [52]. The antioxidant activity of dried samples is determined by the combination of drying methods and conditions; therefore, selecting the appropriate drying methods and conditions is crucial for improving the antioxidant activity of dried products.

3.8. Microstructure

The microstructures of dried *P. eryngii* slices under different drying conditions (Figure 5) were observed using SEM. The images showed that the drying methods and conditions had an obvious effect on the tissue structure of *P. eryngii* slices. When the samples were dried by HAD at 65 °C (Figure 5a,b) and ID at 65 °C (Figure 5e,f), their loose porous structure was clearly visible, and the individual mycelium could be easily distinguished. When the samples were dried by microwave (Figure 5i–l), hot air at 85 °C (Figure 5c,d), and infrared at 85 °C (Figure 5g,h), the mycelium was completely gathered and cross-linked. This could be due to the depolymerization of mycelium cell wall polysaccharides under high temperatures, which could damage the structure of mycelium. Additionally, the Maillard reaction may also cause the loss of polysaccharides and the cross-link reaction with proteins [51,53]. By matching Figure 5 with the appearance images in Figure 2, it can be concluded that the appearance changes of *P. eryngii* were related to the microstructure.

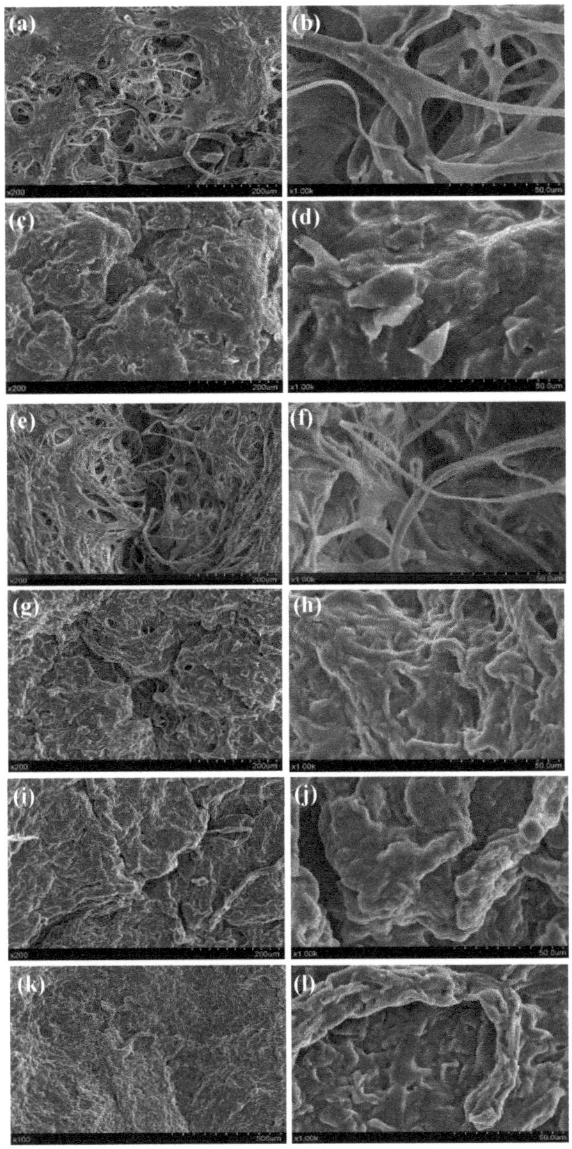

Figure 5. Microstructure of dried *P. eryngii* slices under different drying conditions at different magnifications. (**a**,**b**) Hot air drying at 65 °C; (**c**,**d**) hot air drying at 85 °C; (**e**,**f**) infrared drying at 65 °C; (**g**,**h**) infrared drying at 85 °C; (**i**,**j**) microwave drying at 157 W; (**k**,**l**) microwave drying at 280 W.

4. Conclusions

The effects of different drying methods and conditions on the drying characteristics, 3D appearance, physicochemical properties, and microstructure of *P. eryngii* slices were investigated. The drying method and conditions significantly influenced the drying time of *P. eryngii*, and microwave drying had a significant advantage in reducing drying time. The results showed that the shrinkage and roughness of *P. eryngii* slices gradually increased during drying, which was significantly related to the drying method and conditions. The best appearance of *P. eryngii* was obtained by hot air drying at 55 and 65 °C. The color, total polysaccharide content, antioxidant activity, and microstructure of dried *P. eryngii* slices were also analyzed. HAD and ID at lower drying temperatures achieved better product quality, but MD significantly damaged the color and nutritional quality of *P. eryngii*, so MD cannot be recommended. The microstructures of dried *P. eryngii* slices were observed using SEM. The observation of microstructures is helpful to understand the appearance and quality changes of *P. eryngii* during drying.

Author Contributions: Conceptualization, H.-W.X. and X.-Y.T.; data curation, Y.-C.W. and L.Z.; formal analysis, Y.-C.W. and Y.D.; funding acquisition, J.-W.B. and J.-R.C.; investigation, Y.-C.W. and Y.D.; methodology, J.-W.B., X.-Y.T. and Y.-C.W.; project administration, X.-Y.T.; supervision, J.-W.B. and X.-Y.T.; visualization, Y.-C.W. and L.Z.; writing—original draft, J.-W.B. and Y.-C.W.; writing—review and editing, J.-W.B., Y.-C.W. and H.-W.X. All authors have read and agreed to the published version of the manuscript.

Funding: This research is supported by the National Natural Science Foundation of China (31601578) and the China Agriculture Research System of MOF and MARA (CARS26).

Data Availability Statement: The data that support the findings of this study are available from the corresponding author upon reasonable request.

Conflicts of Interest: The authors declare no conflict of interest.

References

1. Li, X.; Feng, T.; Zhou, F.; Zhou, S.; Liu, Y.; Li, W.; Ye, R.; Yang, Y. Effects of drying methods on the tasty compounds of Pleurotus eryngii. *Food Chem.* **2015**, *166*, 358–364. [CrossRef] [PubMed]
2. Wang, Q.; Li, S.; Han, X.; Ni, Y.; Zhao, D.; Hao, J. Quality evaluation and drying kinetics of shitake mushrooms dried by hot air, infrared and intermittent microwave–assisted drying methods. *LWT* **2019**, *107*, 236–242. [CrossRef]
3. Piskov, S.; Timchenko, L.; Grimm, W.; Rzhepakovsky, I.; Avanesyan, S.; Sizonenko, M.; Kurchenko, V. Effects of Various Drying Methods on Some Physico-Chemical Properties and the Antioxidant Profile and ACE Inhibition Activity of Oyster Mushrooms (*Pleurotus Ostreatus*). *Foods* **2020**, *9*, 160. [CrossRef] [PubMed]
4. Chen, D.; Qin, L.; Geng, Y.; Kong, Q.; Wang, S.; Lin, S. The Aroma Fingerprints and Discrimination Analysis of Shiitake Mushrooms from Three Different Drying Conditions by GC-IMS, GC-MS and DSA. *Foods* **2021**, *10*, 2991. [CrossRef]
5. Xiao, H.; Mujumdar, A.S. Importance of drying in support of human welfare. *Dry. Technol.* **2020**, *38*, 1542–1543. [CrossRef]
6. Qin, L.; Gao, J.; Xue, J.; Chen, D.; Lin, S.; Dong, X.; Zhu, B. Changes in Aroma Profile of Shiitake Mushroom (*Lentinus edodes*) during Different Stages of Hot Air Drying. *Foods* **2020**, *9*, 444. [CrossRef]
7. Motevali, A.; Minaei, S.; Khoshtaghaza, M.H.; Amirnejat, H. Comparison of energy consumption and specific energy requirements of different methods for drying mushroom slices. *Energy* **2011**, *36*, 6433–6441. [CrossRef]
8. Bai, J.; Dai, Y.; Wang, Y.; Cai, J.; Zhang, L.; Tian, X. Potato Slices Drying: Pretreatment Affects the Three-Dimensional Appearance and Quality Attributes. *Agriculture* **2022**, *12*, 1841. [CrossRef]
9. Pawar, S.B.; Pratape, V.M. Fundamentals of Infrared Heating and Its Application in Drying of Food Materials: A Review. *J. Food Process Eng.* **2017**, *40*, e123082017. [CrossRef]
10. Zielinska, M.; Michalska, A. Microwave-assisted drying of blueberry (*Vaccinium corymbosum* L.) fruits: Drying kinetics, polyphenols, anthocyanins, antioxidant capacity, colour and texture. *Food Chem.* **2016**, *212*, 671–680. [CrossRef]
11. Onwude, D.I.; Hashim, N.; Abdan, K.; Janius, R.; Chen, G. The potential of computer vision, optical backscattering parameters and artificial neural network modelling in monitoring the shrinkage of sweet potato (*Ipomoea batatas* L.) during drying. *J. Sci. Food Agric.* **2018**, *98*, 1310–1324. [CrossRef] [PubMed]
12. Pei, Y.; Li, Z.; Ling, C.; Jiang, L.; Wu, X.; Song, C.; Li, J.; Song, F.; Xu, W. An improvement of far-infrared drying for ginger slices with computer vision and fuzzy logic control. *J. Food Meas. Charact.* **2022**, *16*, 3815–3831. [CrossRef]
13. Li, X.; Liu, Y.; Gao, Z.; Xie, Y.; Wang, H. Computer vision online measurement of shiitake mushroom (*Lentinus edodes*) surface wrinkling and shrinkage during hot air drying with humidity control. *J. Food Eng.* **2021**, *292*, 110253. [CrossRef]

14. Sampson, D.J.; Chang, Y.K.; Rupasinghe, H.P.V.; Zaman, Q.U. A dual-view computer-vision system for volume and image texture analysis in multiple apple slices drying. *J. Food Eng.* **2014**, *127*, 49–57. [CrossRef]
15. Cai, J.R.; Lu, Y.; Bai, J.W.; Sun, L.; Xiao, H.W. Three-dimensional imaging of morphological changes of potato slices during drying. *Trans. Chin. Soc. Agric. Eng.* **2019**, *35*, 278–284.
16. Wasenmüller, O.; Stricker, D. Comparison of Kinect V1 and V2 Depth Images in Terms of Accuracy and Precision. *Comput. Vis.-ACCV 2016 Workshops* **2017**, *10117*, 34–45. [CrossRef]
17. Bai, J.; Zhang, L.; Aheto, J.H.; Cai, J.; Wang, Y.; Sun, L.; Tian, X. Effects of different pretreatment methods on drying kinetics, three-dimensional deformation, quality characteristics and microstructure of dried apple slices. *Innov. Food Sci. Emerg. Technol.* **2023**, *83*, 103216. [CrossRef]
18. Sun, L.; Zhang, P.; Zheng, X.; Cai, J.; Bai, J. Three-dimensional morphological changes of potato slices during the drying process. *Curr. Res. Food Sci.* **2021**, *4*, 910–916. [CrossRef] [PubMed]
19. Xiao, A.; Ding, C. Effect of Electrohydrodynamic (EHD) on Drying Kinetics and Quality Characteristics of Shiitake Mushroom. *Foods* **2022**, *11*, 1303. [CrossRef] [PubMed]
20. Liu, Z.; Zielinska, M.; Yang, X.; Yu, X.; Chen, C.; Wang, H.; Wang, J.; Pan, Z.; Xiao, H. Moisturizing strategy for enhanced convective drying of mushroom slices. *Renew. Energy* **2021**, *172*, 728–739. [CrossRef]
21. Loureiro, A.D.C.; Souza, F.D.C.D.; Sanches, E.A.; Bezerra, J.D.A.; Lamarão, C.V.; Rodrigues, S.; Fernandes, F.A.N.; Campelo, P.H. Cold plasma technique as a pretreatment for drying fruits: Evaluation of the excitation frequency on drying process and bioactive compounds. *Food Res. Int.* **2021**, *147*, 110462. [CrossRef]
22. Liu, Z.; Xie, L.; Zielinska, M.; Pan, Z.; Deng, L.; Zhang, J.; Gao, L.; Wang, S.; Zheng, Z.; Xiao, H. Improvement of drying efficiency and quality attributes of blueberries using innovative far-infrared radiation heating assisted pulsed vacuum drying (FIR-PVD). *Innov. Food Sci. Emerg. Technol.* **2022**, *77*, 102948. [CrossRef]
23. Su, D.; Lv, W.; Wang, Y.; Wang, L.; Li, D. Influence of microwave hot-air flow rolling dry-blanching on microstructure, water migration and quality of pleurotus eryngii during hot-air drying. *Food Control* **2020**, *114*, 107228. [CrossRef]
24. Zhu, L.; Lu, Y.; Sun, Z.; Han, J.; Tan, Z. The application of an aqueous two-phase system combined with ultrasonic cell disruption extraction and HPLC in the simultaneous separation and analysis of solanine and Solanum nigrum polysaccharide from Solanum nigrum unripe fruit. *Food Chem.* **2020**, *304*, 125383. [CrossRef] [PubMed]
25. Liu, D.; Guo, Y.; Wu, P.; Wang, Y.; Kwaku Golly, M.; Ma, H. The necessity of walnut proteolysis based on evaluation after in vitro simulated digestion: ACE inhibition and DPPH radical-scavenging activities. *Food Chem.* **2020**, *311*, 125960. [CrossRef]
26. Bechlin, T.R.; Granella, S.J.; Christ, D.; Coelho, S.R.M.; Paz, C.H.D.O. Effects of ozone application and hot-air drying on orange peel: Moisture diffusion, oil yield, and antioxidant activity. *Food Bioprod. Process.* **2020**, *123*, 80–89. [CrossRef]
27. Moon, J.H.; Kim, M.J.; Chung, D.H.; Pan, C.; Yoon, W.B. Drying Characteristics of Sea Cucumber (*Stichopus japonicas Selenka*) Using Far Infrared Radiation Drying and Hot Air Drying. *J. Food Process Preserv.* **2014**, *38*, 1534–1546. [CrossRef]
28. Demirel, D.; Turhan, M. Air-drying behavior of Dwarf Cavendish and Gros Michel banana slices. *J. Food Eng.* **2003**, *59*, 1–11. [CrossRef]
29. Cao, Z.; Zhou, L.; Bi, J.; Yi, J.; Chen, Q.; Wu, X.; Zheng, J.; Li, S. Effect of different drying technologies on drying characteristics and quality of red pepper (*Capsicum frutescens* L.): A comparative study. *J. Sci. Food Agric.* **2016**, *96*, 3596–3603. [CrossRef]
30. Nowak, D.; Lewicki, P.P. Infrared drying of apple slices. *Innov. Food Sci. Emerg. Technol.* **2004**, *5*, 353–360. [CrossRef]
31. Nozad, M.; Khojastehpour, M.; Tabasizadeh, M.; Azizi, M.; Miraei Ashtiani, S.; Salarikia, A. Characterization of hot-air drying and infrared drying of spearmint (*Mentha spicata* L.) leaves. *J. Food Meas. Charact.* **2016**, *10*, 466–473. [CrossRef]
32. Albanese, D.; Cinquanta, L.; Cuccurullo, G.; Di Matteo, M. Effects of microwave and hot-air drying methods on colour, β-carotene and radical scavenging activity of apricots. *Int. J. Food Sci. Technol.* **2013**, *48*, 1327–1333. [CrossRef]
33. Thayyil, S.B.; Yadav, S.K.; Polthier, K.; Muthuganapathy, R. Local Delaunay-based high fidelity surface reconstruction from 3D point sets. *Comput. Aided Geom. Des.* **2021**, *86*, 101973. [CrossRef]
34. Qin, J.; Wang, Z.; Wang, X.; Shi, W. Effects of microwave time on quality of grass carp fillets processed through microwave combined with hot-air drying. *Food Sci. Nutr.* **2020**, *8*, 4159–4171. [CrossRef] [PubMed]
35. Wang, J.; Law, C.; Nema, P.K.; Zhao, J.; Liu, Z.; Deng, L.; Gao, Z.; Xiao, H. Pulsed vacuum drying enhances drying kinetics and quality of lemon slices. *J. Food Eng.* **2018**, *224*, 129–138. [CrossRef]
36. Wang, X.; Gao, Y.; Zhao, Y.; Li, X.; Fan, J.; Wang, L. Effect of different drying methods on the quality and microstructure of fresh jujube crisp slices. *J. Food Process Preserv.* **2021**, *45*, e15162. [CrossRef]
37. Udomkun, P.; Nagle, M.; Argyropoulos, D.; Mahayothee, B.; Müller, J. Multi-sensor approach to improve optical monitoring of papaya shrinkage during drying. *J. Food Eng.* **2016**, *189*, 82–89. [CrossRef]
38. Kaveh, M.; Abbaspour Gilandeh, Y.; Fatemi, H.; Chen, G. Impact of different drying methods on the drying time, energy, and quality of green peas. *J. Food Process Preserv.* **2021**, *45*, e15503. [CrossRef]
39. Chen, A.; Achkar, G.E.; Liu, B.; Bennacer, R. Experimental study on moisture kinetics and microstructure evolution in apples during high power microwave drying process. *J. Food Eng.* **2021**, *292*, 110362. [CrossRef]
40. Rokhbin, A.; Azadbakht, M. The shrinkage of orange slices during microwave drying and ohmic pretreatment. *J. Food Process Preserv.* **2021**, *45*, e154002021. [CrossRef]
41. Mahiuddin, M.; Khan, M.I.H.; Kumar, C.; Rahman, M.M.; Karim, M.A. Shrinkage of Food Materials During Drying: Current Status and Challenges. *Compr. Rev. Food Sci. Food Saf.* **2018**, *17*, 1113–1126. [CrossRef] [PubMed]

42. Yang, R.; Li, Q.; Hu, Q. Physicochemical properties, microstructures, nutritional components, and free amino acids of Pleurotus eryngii as affected by different drying methods. *Sci. Rep.* **2020**, *10*, 121. [CrossRef] [PubMed]
43. Tian, Y.; Zhao, Y.; Huang, J.; Zeng, H.; Zheng, B. Effects of different drying methods on the product quality and volatile compounds of whole shiitake mushrooms. *Food Chem.* **2016**, *197*, 714–722. [CrossRef] [PubMed]
44. Xiao, H.; Law, C.; Sun, D.; Gao, Z. Color Change Kinetics of American Ginseng (*Panax quinquefolium*) Slices during Air Impingement Drying. *Dry. Technol.* **2014**, *32*, 418–427. [CrossRef]
45. Gouzi, H.; Depagne, C.; Coradin, T. Kinetics and Thermodynamics of the Thermal Inactivation of Polyphenol Oxidase in an Aqueous Extract fromAgaricus bisporus. *J. Agric. Food Chem.* **2012**, *60*, 500–506. [CrossRef] [PubMed]
46. Ju, H.; Zhang, Q.; Mujumdar, A.S.; Fang, X.; Xiao, H.; Gao, Z. Hot-air Drying Kinetics of Yam Slices under Step Change in Relative Humidity. *Int. J. Food Eng.* **2016**, *12*, 783–792. [CrossRef]
47. Duan, X.; Liu, W.C.; Ren, G.Y.; Liu, L.L.; Liu, Y.H. Browning behavior of button mushrooms during microwave freeze-drying. *Dry. Technol.* **2015**, *34*, 1373–1379. [CrossRef]
48. Xu, L.; Fang, X.; Wu, W.; Chen, H.; Mu, H.; Gao, H. Effects of high-temperature pre-drying on the quality of air-dried shiitake mushrooms (*Lentinula edodes*). *Food Chem.* **2019**, *285*, 406–413. [CrossRef]
49. Zhang, M.; Tang, J.; Mujumdar, A.S.; Wang, S. Trends in microwave-related drying of fruits and vegetables. *Trends Food Sci. Technol.* **2006**, *17*, 524–534. [CrossRef]
50. Gąsecka, M.; Siwulski, M.; Magdziak, Z.; Budzyńska, S.; Stuper-Szablewska, K.; Niedzielski, P.; Mleczek, M. The effect of drying temperature on bioactive compounds and antioxidant activity of *Leccinum scabrum* (Bull.) Gray and *Hericium erinaceus* (Bull.) Pers. *J. Food Sci. Technol.* **2020**, *57*, 513–525. [CrossRef]
51. Ren, S.; Zheng, E.; Zhao, T.; Hu, S.; Yang, H. Evaluation of physicochemical properties, equivalent umami concentration and antioxidant activity of *Coprinus comatus* prepared by different drying methods. *LWT* **2022**, *162*, 113479. [CrossRef]
52. Wu, B.; Guo, X.; Guo, Y.; Ma, H.; Zhou, C. Enhancing jackfruit infrared drying by combining ultrasound treatments: Effect on drying characteristics, quality properties and microstructure. *Food Chem.* **2021**, *358*, 129845. [CrossRef] [PubMed]
53. Zivanovic, S.; Buescher, R.; Kim, S.K. Mushroom Texture, Cell Wall Composition, Color, and Ultrastructure as Affected by pH and Temperature. *J. Food Sci.* **2003**, *68*, 1860–1865. [CrossRef]

foods

MDPI

Article

Artificial Neural Network Modeling and Genetic Algorithm Multiobjective Optimization of Process of Drying-Assisted Walnut Breaking

Taoqing Yang [1,2,3], Xia Zheng [1,2,3,*], Sriram K. Vidyarthi [4], Hongwei Xiao [5], Xuedong Yao [1,2,3], Yican Li [1,2,3], Yongzhen Zang [1,2,3] and Jikai Zhang [1,2,3]

1 College of Mechanical and Electrical Engineering, Shihezi University, Shihezi 832003, China; 20212009002@stu.shzu.edu.cn (T.Y.)
2 Key Laboratory of Northwest Agricultural Equipment, Ministry of Agriculture and Rural Affairs, Shihezi 832003, China
3 Key Laboratory of Modern Agricultural Machinery Corps, Shihezi 832003, China
4 Department of Biological and Agricultural Engineering, University of California, One Shields Avenue, Davis, CA 95616, USA
5 College of Engineering, China Agricultural University, Beijing 100080, China
* Correspondence: zx_mac@shzu.edu.cn

Citation: Yang, T.; Zheng, X.; Vidyarthi, S.K.; Xiao, H.; Yao, X.; Li, Y.; Zang, Y.; Zhang, J. Artificial Neural Network Modeling and Genetic Algorithm Multiobjective Optimization of Process of Drying-Assisted Walnut Breaking. *Foods* **2023**, *12*, 1897. https://doi.org/10.3390/foods12091897

Academic Editor: Ramón F. Moreira

Received: 22 March 2023
Revised: 15 April 2023
Accepted: 4 May 2023
Published: 5 May 2023

Abstract: This study combined an artificial neural network (ANN) with a genetic algorithm (GA) to obtain the model and optimal process parameters of drying-assisted walnut breaking. Walnuts were dried at different IR temperatures (40 °C, 45 °C, 50 °C, and 55 °C) and air velocities (1, 2, 3, and 4 m/s) to different moisture contents (10%, 15%, 20%, and 25%) by using air-impingement technology. Subsequently, the dried walnuts were broken in different loading directions (sutural, longitudinal, and vertical). The drying time (DT), specific energy consumption (SEC), high kernel rate (HR), whole kernel rate (WR), and shell-breaking rate (SR) were determined as response variables. An ANN optimized by a GA was applied to simulate the influence of IR temperature, air velocity, moisture content, and loading direction on the five response variables, from which the objective functions of DT, SEC, HR, WR, and SR were developed. A GA was applied for the simultaneous maximization of HR, WR, and SR and minimization of DT and SEC to determine the optimized process parameters. The ANN model had a satisfactory prediction ability, with the coefficients of determination of 0.996, 0.998, 0.990, 0.991, and 0.993 for DT, SEC, HR, WR, and SR, respectively. The optimized process parameters were found to be 54.9 °C of IR temperature, 3.66 m/s of air velocity, 10.9% of moisture content, and vertical loading direction. The model combining an ANN and a GA was proven to be an effective method for predicting and optimizing the process parameters of walnut breaking. The predicted values under optimized process parameters fitted the experimental data well, with a low relative error value of 2.51–3.96%. This study can help improve the quality of walnut breaking, processing efficiency, and energy conservation. The ANN modeling and GA multiobjective optimization method developed in this study provide references for the process optimization of walnut and other similar commodities.

Keywords: walnut; shell breaking; drying; artificial neural network; genetic algorithm; multi-objective optimization

1. Introduction

Walnut is the second largest nut in the world and has high edible and medicinal value. Breaking the shell to take the kernel is the premise of deep walnut processing. The shell is hard, and the gap between the shell and the kernel is small, leading to difficulty in breaking the shell to procure the kernel. Therefore, new walnut breaking processes must be explored.

The moisture content of walnuts has a considerable effect on their shell breaking characteristics [1]. At present, the research on the walnut drying process focuses on the drying kinetics and physicochemical qualities of walnuts [2–5], but few studies have linked the drying process with walnut breaking. In addition, adopting the proper loading direction can also improve the processing efficiency and quality of walnut breaking [6]. If drying is used as a pretreatment to assist in walnut breaking, a new walnut breaking process can be obtained through the selection of an appropriate loading direction. In addition, drying technology can ensure the food security of walnuts by reducing their moisture content [7].

Air-impingement technology ejects pressurized hot air through a nozzle and removes water through the impact and heating of hot air on a material. It is characterized by a high drying rate and high heat transfer coefficient. The application of this technology in walnut drying can increase the drying rate and reduce energy consumption.

The primary goal of walnut breaking is shell breaking and protecting the kernel. The shell-breaking rate is used to evaluate the difficulty of shell breaking, and it is the ratio of the number of cracked walnuts to the total number of walnuts in walnut breaking. High kernel rate is the ratio of the weight of the walnut kernel that is not completely crushed to the total weight of the walnut kernel. Whole kernel rate is the ratio of the weight of the whole walnut kernel to the total weight of the walnut kernel. High kernel and whole kernel rates are used to evaluate the integrity of the walnut kernel. At a drying temperature of 43 °C, walnut drying in an industrial hot air dryer requires a long drying time (more than 24 h) and consumes a large amount of energy (1284.6 MJ for natural gas and 85.1 MJ for electricity consumption per ton of dried walnuts) [8]. Therefore, the drying time and energy consumption must be minimised whilst ensuring the quality of walnut breaking. The optimization of the walnut breaking process depends on several parameters, including the shell-breaking rate (SR), high kernel rate (HR), whole kernel rate (WR), drying time (DT) and specific energy consumption (SEC).

The relationship between the process parameters of drying-assisted walnut breaking (including air velocity, temperature, moisture content, and loading direction) and the process evaluation index is highly nonlinear. Researchers have developed mathematical models to describe nonlinear models, such as theoretical, semitheoretical and empirical models. However, these models can accurately predict experimental data only under highly specific conditions, and no general equations are available to describe complete models [9]. Artificial intelligence technologies can model nonlinear systems where the relationships between variables are unknown [10]. An artificial neural network (ANN) is a kind of artificial intelligence technology that can simulate the behaviour of highly nonlinear systems dynamically [11]. ANNs have been successfully applied to the modeling and prediction of engineering problems, especially in areas where mathematical modeling methods fail [12]. ANN models have also been used in the moisture content prediction of pumpkin slice drying [13], prediction of the moisture ratio and color parameters of ginkgo biloba seed drying [14], prediction of the energy and exergy of mushroom slice drying [15] and moisture content prediction of carrot drying [16]. However, no study has examined its application in walnut drying and breaking. The model formula of an ANN is expressed by weights and thresholds. The essence of the ANN training process is the optimization of weights and thresholds. The initial weights and thresholds of the ANN are generated randomly. When the initial weights and thresholds are unreasonable, the convergence speed of the neural network is slow, and the training result is a local optimal value rather than a global optimal value. The genetic algorithm (GA) first preliminarily optimizes the initial weights and thresholds of the ANN, and then the neural network uses the optimized weights and thresholds for training to accelerate the convergence speed of the network and obtain the global optimal value.

The optimization of the walnut breaking process is multiobjective optimization (MOO), which is a method of finding the optimal target within the range of constraints based on the model. A GA is an optimization technique used to obtain the optimal value of a complex objective function by simulating biological evolutionary processes. Winiczenko, R applies a GA to the MOO during the apple cube drying process [17]. Raj, GVSB. used a GA to optimize the microwave vacuum drying process of dragon fruit slices [18]. MOO problems involve multiple objective functions, and these functions restrict one another. Therefore, obtaining a global optimal solution is difficult. Usually, MOO problems tend to obtain a set of optimal solutions called Pareto optimal solutions. A Pareto optimal solution is not dominated by other solutions in the solution space, and the improvement of one of the objectives requires the sacrifice of others [17]. The set of Pareto optimal solutions is called the Pareto front. The major goal of MOO is to find the Pareto front [19]. Obtaining the global optimal solution in the Pareto front in accordance with the specific optimization purpose is easy [20].

ANN modeling can be combined with GA MOO, which provides objective functions for MOO, and the genetic algorithm finds the optimal solution under the given constraint. This model is an effective method for predicting and optimizing any complex process parameters [21]. It has been applied in other fields [22,23].

In this study, drying was used as a pretreatment for walnut breaking for the first time, and the quality of walnut breaking and energy efficiency were utilised to evaluate the shell breaking process. An ANN and GA were applied to the modeling and optimization of the walnut breaking process for the first time. The objectives of this study are to: (1) analyse the effects of IR temperature (T), air velocity (V), moisture content (MC) and loading direction (D) on DT, SEC, HR, WR and SR; (2) develop an ANN nonlinear model with T, V, MC, and D as input variables, and DT, SEC, HR, WR, and SR as output variables; and (3) optimize the process parameters of drying-assisted walnut breaking by using a GA and minimize DT and SEC whilst improving the quality of walnut breaking. This study will provides an optimal process for drying-assisted walnut breaking.

2. Materials and Methods

2.1. Materials

Walnuts (cultivar Xinjiang Wen-185) of uniform size produced in Aksu, Xinjiang were selected and stored in a freezer (temperature: 5 °C) for use. The average weight, diameter, initial moisture content, and thickness of the walnuts were 16.52 ± 0.2 g, 36.43 ± 0.1 mm, 30.12 ± 0.65%, and 1.1 ± 0.1 mm, respectively.

2.2. Processing Equipment

Drying experiments were conducted using an air-impingement technology (Taizhou Senttech Infrared Technology Co., Ltd., Taizhou, China; temperature accuracy of ±0.1 °C, and power range of 0–2 kW). A principal diagram of the technology is shown in Figure 1. The air velocity at the nozzle was measured using an anemometer (AT816, SMART SENSOR, Thincol, Guangzhou, China; accuracy of ±0.1 m/s). Prior to the experiment, the technology was run for 30 min to stabilize the equipment. The walnut drying characteristics were used to predict when the walnuts reached the target moisture content. An electronic balance (BSM-5200.2, Shanghai Zhuojing Electronic Technology Co., Ltd., Shanghai, China; accuracy of 0.01 g) was adopted to weigh the walnuts and judge whether the walnuts reached the target moisture content. Drying was stopped when the walnuts reached the target moisture content, and continued otherwise. The drying experiments were performed at IR temperatures of 40 °C, 45 °C, 50 °C, and 55 °C; air velocities of 1, 2, 3, and 4 m/s, and moisture contents of 10%, 15%, 20%, and 25%. The dried walnuts were broken in three loading directions: sutural, longitudinal, and vertical.

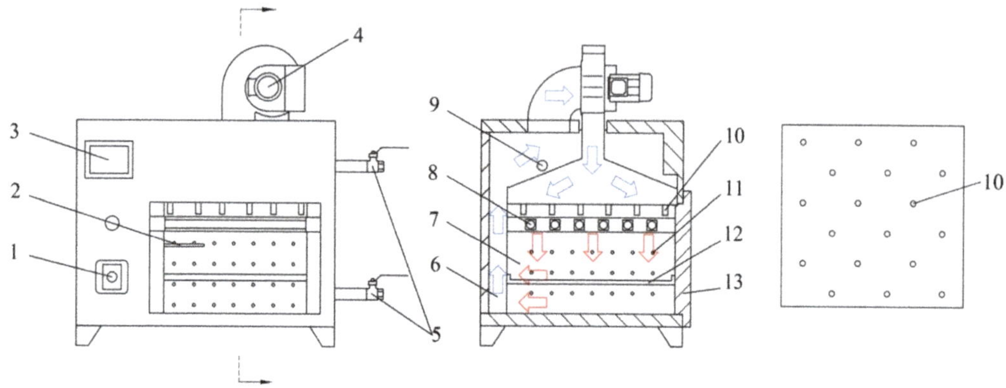

Figure 1. The principal diagram of the air-impingement technology. (1) air velocity adjustment knob; (2) Temperature transducer; (3) temperature control touch panel; (4) centrifugal fan; (5) wet discharge valve; (6) drying outer chamber; (7) drying inner chamber; (8) infrared heating tube; (9) weephole; (10) air nozzle; (11) air outlet; (12) drying tray; (13) door.

2.3. *Drying Experiments*

The drying process parameters included IR temperature, air velocity and moisture content. IR temperature refers to the temperature of the hot air ejected from the nozzle, and air velocity refers to the velocity of the hot air ejected from the nozzle. The inner diameter of the nozzle was 10 mm, and the number of nozzles was 18. The distribution of the nozzles is shown in Figure 1. Moisture content refers to the moisture content (dry basis) of the walnut at the end of drying.

The IR temperature range was set to be 40–55 °C to ensure drying efficiency and product quality [5]. The maximum air velocity of the equipment was 4 m/s. To sufficiently study the influence of air velocity on walnut breaking, the range of air velocity was set to 1–4 m/s. To prevent production reduction caused by walnuts with a moisture content below the safe storage moisture content of 8%, the minimum moisture content of the walnuts was set to 10%. The range of moisture content was 10–25% [1]. The drying experiments were conducted at IR temperatures of 40 °C, 45 °C, 50 °C, and 55 °C; air velocities of 1, 2, 3, and 4 m/s, and moisture contents of 10%, 15%, 20%, and 25%. The dried walnuts were broken in three loading directions: sutural, longitudinal and vertical. A total of 192 experimental groups ($4 \times 4 \times 4 \times 3$) were established.

2.4. *Loading Direction(L)*

Breaking experiments were performed using a DF-9000 dynamic and static universal material testing machine, and the loading rate was set to 10 mm/min. The walnuts' loading direction is shown in Figure 2.

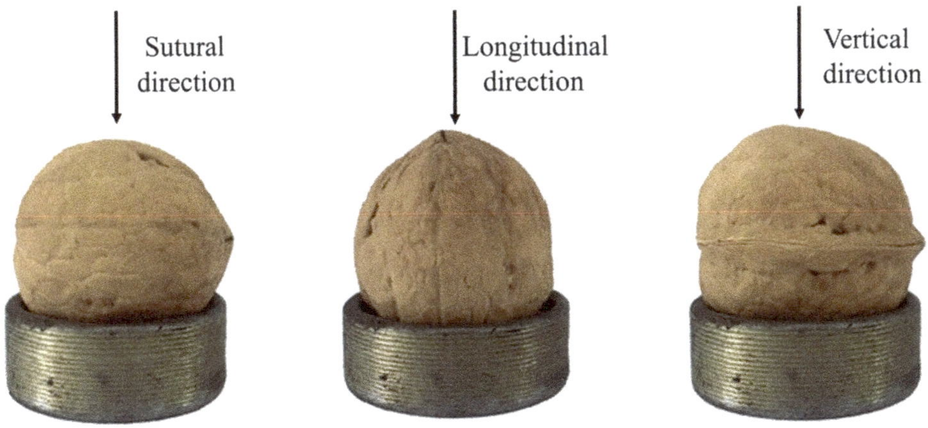

Figure 2. Walnuts' loading direction.

2.5. High Kernel Rate (HR), Whole Kernel Rate (WR) and Shell-Breaking Rate (SR)

The integrity of walnut kernels was evaluated with the high kernel and whole kernel rates. The walnut kernels were divided into Classes A, B, C, and D according to their integrity. Whole walnut kernels belonged to Class A, half walnut kernels belonged to Class B, quarter walnut kernels were classified as Class C and the rest of the crushed walnut kernels belonged to Class D.

High kernel rate was calculated using Equation (1), and whole kernel rate was calculated with Equation (2).

$$HR = \frac{M_A + M_B + M_C}{M_A + M_B + M_C + M_D} \times 100\% \quad (1)$$

$$WR = \frac{M_A}{M_A + M_B + M_C + M_D} \times 100\% \quad (2)$$

The difficulty of shell breaking was evaluated by the shell-breaking rate (*SR*). *SR* is the ratio of the number of walnuts that are cracked by more than 3/4 to the total number of walnuts in the experiment. SR was calculated with Equation (3).

$$SR = \frac{L}{N} \times 100\% \quad (3)$$

where L is the number of walnuts that are cracked by more than 3/4; N is the total number of walnuts, M_A is the weight of Class A walnut kernels, M_B is the weight of class B walnut kernels; M_C is the weight of Class C walnut kernels and M_D is the weight of Class D walnut kernels.

2.6. Drying Time and Specific Energy Consumption

The walnuts were dried in a hot-air dryer at 105 °C for 24 h to measure the dry weight, W_d [24]. The moisture content of the walnuts at time T was calculated by Equation (4) [25]. The time required for the walnuts to dry to the target moisture content was the drying time.

$$M_t = \frac{W_T - W_d}{W_d} \times 100\% \quad (4)$$

The energy required to dry 1 kg of walnuts was defined as the specific energy consumption, which was calculated with Equation (5).

$$SEC = \frac{E}{m_{walnut}} \quad (5)$$

where E is the electrical energy consumed during drying and m_{walnut} is the total walnut mass during drying.

2.7. Shell Kernel Clearance (SKC)

Shell kernel clearance refers to the distance between the walnut shell and the walnut kernel, and it was calculated using Equation (6) [26].

$$SKC = \frac{A + B + C - a - b - c - 6d}{6} \quad (6)$$

where A is the maximum longitudinal dimension of the inner wall of the walnut shell (mm), B is the maximum vertical dimension of the inner wall of the walnut shell, (mm), C is the maximum sutural dimension of the inner wall of the walnut shell (mm), a is the maximum longitudinal dimension of the walnut kernel (mm), b is the maximum vertical dimension of walnut kernel, (mm), c is the maximum sutural dimension of the walnut kernel (mm) and d is the thickness of the walnut shell (mm).

2.8. Walnut Shell Hardness

Hardness is one of the important factors that affect walnut breaking. A DF-9000 dynamic and static universal material testing machine was used for the walnut breaking experiment, and the loading rate was set to 10 mm/min. Hardness H was calculated using Equation (7) [27].

$$H = \frac{F_r}{D_r} \quad (7)$$

where F_r is the maximum breaking force and D_r is the deformation when the maximum breaking force is reached. D_r can be obtained from the DF-9000 dynamic and static universal material testing machine.

2.9. Artificial Neural Network (ANN) Model

Back propagation ANN (BP-ANN) is a feed-forward network-trained model in accordance with the error-back propagation algorithm, and it is one of the most widely used ANN models. The model of the process of drying-assisted walnut breaking was established using BP-ANN. The topology of the ANN model consisted of three layers (input layer, hidden layer and output layer) and two transfer functions between the three layers, as shown in Figure 3. The input layer had four factors: IR temperature (T), air velocity (V), moisture content (MC), and loading direction (D). The output layer had five factors: drying time (DT), specific energy consumption (SEC), high kernel rate (HR), whole kernel rate (WR), and shell-breaking rate (SR). The range of the number of neurons in the hidden layer was determined with the Equations (8)–(11). The optimal number of neurons in the hidden layer was obtained by trial and error. Nonlinear transfer functions, including tansig sigmoidal and logsig sigmoidal functions, were used in the input and hidden layers, and a linear transfer function (Pureline) was employed in the hidden layers and output layers. The topology of neural networks influences two of the most important evaluation criteria of neural network training: generalization and training time [28]. Improper topology of the neural network causes many redundancies, which makes the neural network fall into local optimization and considerably prolongs the training time [29]. Therefore, determining the optimal neural network topology is important.

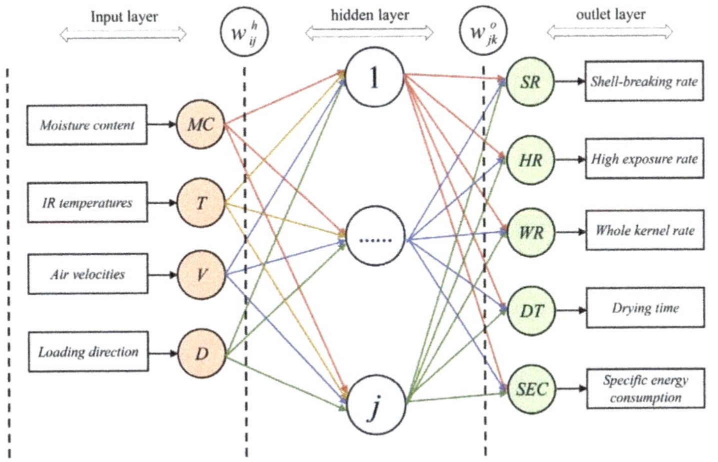

Figure 3. Topology of ANN model.

The range of the number of neurons in the hidden layer was determined using the following equations [30]:

$$j < n - 1 \tag{8}$$

$$j < \sqrt{m + n} + a \tag{9}$$

$$j = \log_2 n \tag{10}$$

$$j = 2n + 1 \tag{11}$$

where j is the number of neurons in the hidden layer, n is the number of factors in the input layer, m is the number of factors in the output layer and a is a constant between 0 and 10.

The union of Equations (8)–(11) was determined as the range of the number of neurons in the hidden layer to ensure that the optimal number of neurons in the hidden layer could be achieved. Generally, the number of neurons in the hidden layer should be more than one; then, the range is as follows: $1 < j < 13$. To obtain the variation in neural network performance beyond the range, the range of the number of neurons in the hidden layer was determined as $1 < j \leq 13$ in this study.

The coefficient of determination (R^2) and root mean square error (*RMSE*) were calculated to determine the optimal number of neurons in the hidden layer and the transfer function in the input and hidden layers by using Equations (12) and (13), respectively [31,32].

$$RMSE = \sqrt{\frac{\sum_{i=1}^{N} \left(y_{act,i} - y_{pre,i}\right)^2}{N}} \tag{12}$$

where $y_{act,i}$ is the actual value and $y_{pre,i}$ is the predicted value of the sample. In general, the smaller the *RMSE* value is, the higher the model accuracy is.

$$R^2 = 1 - \frac{\sum_{i=0}^{N} (y_i - \hat{y}_i)^2}{\sum_{i=0}^{N} (y_i - \overline{y_i})^2} \tag{13}$$

where y_i is the actual value of the sample, $\hat{y}$ is the predicted value of the sample, $\overline{y}$ is the average of the actual values, and N is the number of test samples. The larger the R^2 value is, the better the predictive ability of the model is.

After determining the optimal topology structure of the neural network, the neural network was trained. A total of 192 groups of experimental data ($4 \times 4 \times 4 \times 3$) were established. One group consisted of four input variables and five output variables. To ensure the effectiveness of neural network training and the accuracy of the test data, 80% of the experimental data (154 groups) were randomly selected as training data, and 20% (38 groups) were adopted as test data [33]. The algorithm flow of the neural network is shown in the Figure 4.

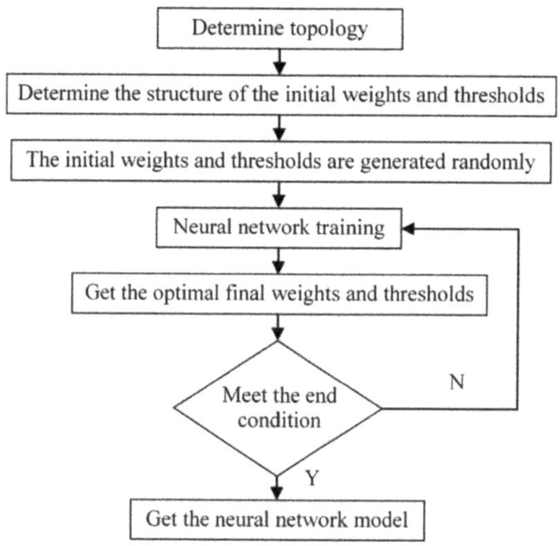

Figure 4. Flow of ANN algorithm.

The weights and thresholds of the neural network were obtained after the training was completed, and the neural network model was established. The neural network model had four parts: the input variables were normalized; the weight, threshold, and transfer function of the hidden layer were inputted into Equations (15) and (16) to obtain the output of the hidden layer; the weight, threshold, and transfer function of the output layer were inputted into Equation (17) to obtain the output of the output layer and the output of the output layer was reversely normalized to obtain the output variables.

The input variables were normalized, and the normalized range was (−1, 1).

$$x_i^g = \frac{2(x_i - x_{i,\min})}{x_{i,\max} - x_{i,\min}} - 1 \quad (14)$$

where x_i^g (i = 1, 2, 3, 4) is the normalized input variable, x_i is the input variable; x_1 is T, x_2 is V, x_3 is MC, x_4 is D, $x_{i,\min}$ is the minimum of the training data for the input variable x_i and $x_{i,\max}$ is the maximum of the training data for input variable x_i.

The output in the hidden layer is:

$$h_j = \log\operatorname{sig}\left(\sum w_{ij}^h x_i^g + b_j^h\right) \quad (15)$$

$$\log\operatorname{sig}(x) = \frac{1}{1 + \exp(-x)} \quad (16)$$

where h_j (j = 1, 2, 3, 4, 5, 6, 7, 8, 9, 10, 11, 12) refers to the outputs of each neuron in the hidden layer, w_{ij}^h is the weight between the input layer and the hidden layer, and b_j^h is the threshold in the hidden layer. The values of weights w^h_{ij} and threshold b^h_j are used in Equation (15).

The output in the output layer is:

$$y^g_k = purelin(\sum w^o_{jk}h_j + b^o_k) = \sum w^o_{jk}h_j + b^o_k \tag{17}$$

where y^g_k (k = 1, 2, 3, 4, and 5) refers to the outputs of each neuron in the output layer, w^o_{jk} is the weight between the hidden layer and the output layer, and b^o_k is the threshold in the output layer. The values of weights w^o_{jk} and threshold b^o_k are used in Equation (17).

Reverse normalization of the output in the layer output was implemented as follows:

$$DT = \frac{(y_{1,\max} - y_{1,\min})(y^g_1 + 1)}{2} + y_{1,\min} \tag{18}$$

$$SEC = \frac{(y_{2,\max} - y_{2,\min})(y^g_2 + 1)}{2} + y_{2,\min} \tag{19}$$

$$HR = \frac{(y_{3,\max} - y_{3,\min})(y^g_3 + 1)}{2} + y_{3,\min} \tag{20}$$

$$WR = \frac{(y_{4,\max} - y_{4,\min})(y^g_4 + 1)}{2} + y_{4,\min} \tag{21}$$

$$SR = \frac{(y_{5,\max} - y_{5,\min})(y^g_5 + 1)}{2} + y_{5,\min} \tag{22}$$

where $y_{k,\min}$ is the minimum of the training data for output variable y_k and $y_{k,\max}$ is the maximum of the training data for the output variable y_k.

2.10. Optimization of Artificial Neural Network Using Genetic Algorithm

The neural network is sensitive to the initial weights and thresholds. Therefore, the training results are greatly affected by the initial weights and thresholds, and they easily fall into the local minimum [34]. GA is a global optimization method based on the principle of biological evolution, namely, survival of the fittest [35]. By optimizing the initial weights and thresholds, GA can move the neural network training process from the local optimal domain to the global optimal domain [36]. Therefore, for the walnut breaking process model, the initial weights and thresholds of the ANN can be optimized by the GA first, followed by ANN training with the optimized initial weights and thresholds, resulting in global optimal weights and thresholds. The parameters of the GA are shown in Table 1.

Table 1. Parameters of genetic algorithm.

Population Type	Double Vector
variable range	(−3, 3)
Population size	200
Number of generations	1000
Crossover rate	90%
Mutation rate	10%

The GA assisted by ANN modeling was divided Into three parts. First, by analyzing R^2 and RMSE, the optimal topology of the neural network was selected, and then the structure of the initial weights and thresholds was obtained. Second, the weights and thresholds of the neural network were adopted as the optimization object, the error of the neural network model was used as the fitness function, and the initial weights and thresholds were optimized using the GA. Lastly, the optimized weights and thresholds were used for neural network training, and the global optimal weights and thresholds were obtained. The algorithm flow is shown in Figure 5.

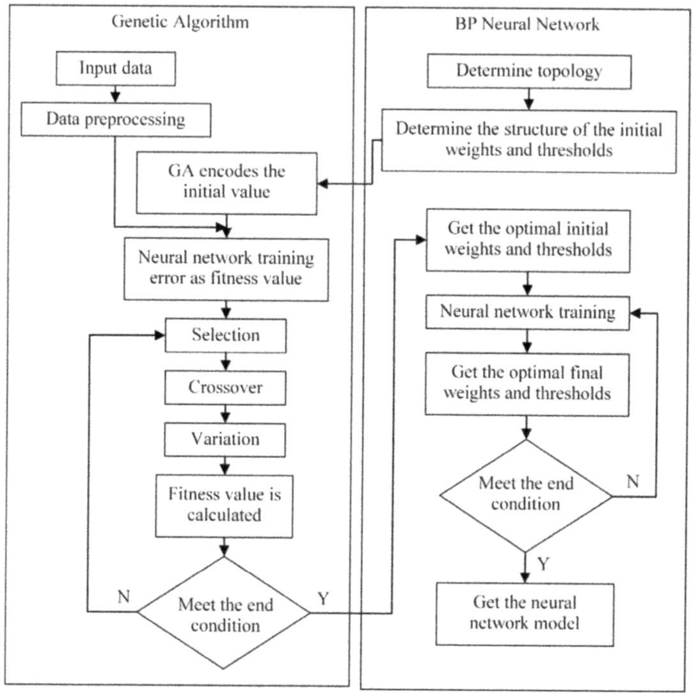

Figure 5. Flow of GA-ANN algorithm.

2.11. *Multiobjective Optimization*

The MOO of the walnut breaking process was applied to reduce the energy consumption and drying time while improving the quality of walnut breaking. The optimization goal was to minimize DT and SEC and maximize HR, WR, and SR. The upper and lower limits of the input variables were the upper and lower limits of the experimental conditions in the walnut breaking experiment, respectively, as shown in Equation (23). In terms of the loading direction, 1 was the sutural direction, 2 was the Longitudinal direction, and 3 was the Vertical direction.

Five objective functions and four constraints were entered in MATLAB, and the GAMULTIOBJ function was used to obtain the Pareto optimal solution. The parameters of the GAMULTIOBJ function are listed in Table 2.

$$objectives = \begin{cases} Min\ DT(T, V, MC, D) \\ Min\ SEC(T, V, MC, D) \\ Max\ HR(T, V, MC, D) \\ Max\ WR(T, V, MC, D) \\ Max\ SR(T, V, MC, D) \\ 40 \leq T \leq 55\ ^{\circ}C \\ 1 \leq V \leq 4m{\cdot}x^{-1} \\ 10 \leq MC \leq 25\% \\ D = (1, 2, 3) \end{cases} \quad (23)$$

Table 2. The parameters of GAMULTIOBJ function.

Population Type	Double Vector
Pareto front population fraction	0.3
Population size	120
Number of generations	1000
Crossover rate	90%
Mutation rate	10%

2.12. Statistical Analysis

The experimental data were processed with Microsoft Excel 2019, and drawing was performed with Origin 2019b. ANN modeling and GA MOO were implemented in MATLAB R2021a.

3. Results and Discussion

3.1. Drying Time and Specific Energy Consumption

The effect of IR temperature on DT and SEC under a constant air velocity of 3 m/s and moisture content of 15% is shown in Figure 6a. DT was greatly influenced ($p < 0.01$) by the IR temperature. DT decreased with increasing IR temperature due to increments in heat and moisture transfer rates and thermal radiation intensity [37]. With increasing IR temperature, DT decreased considerably, but SEC did not decrease rapidly. The increase in IR temperature led to an increase in equipment power, and thus increased the equipment's energy consumption [38]. Figure 6b shows the effect of air velocity on DT and SEC under a constant IR temperature of 50 °C and moisture content of 15%. DT decreased considerably with increasing air velocity due to the accelerated discharge of water vapor from the drying chamber caused by an increase in the heat convection rate [39]. SEC decreased considerably with increased air velocity. First, the increase in air velocity reduced the drying time, which refers to the time for the equipment to operate. Second, fan power consumed a small portion of the total power of the equipment, and the increase in air velocity did not greatly improve the total power of the equipment. Thus, the energy consumption did not increase much. As shown in Figure 6c, DT and SEC decreased greatly with increasing moisture content under an IR temperature of 50 °C and air velocity of 3 m/s. Notably, the DT and SEC of the sample with 15% moisture content decreased by 45.11% and 47.79%, respectively, compared with that of the sample with 10% moisture content. This result was obtained because the drying rate decreased substantially at the last stage of drying, resulting in a large increase in drying time [40]. The energy consumption of equipment was directly related to the length of drying time, so SEC decreased rapidly with increasing moisture content [20]. Among the three drying parameters, moisture content had the greatest influence on drying time and SEC. Thus, increasing the moisture content is the best approach for minimizing drying time and SEC, provided that the quality of walnut breaking is not reduced.

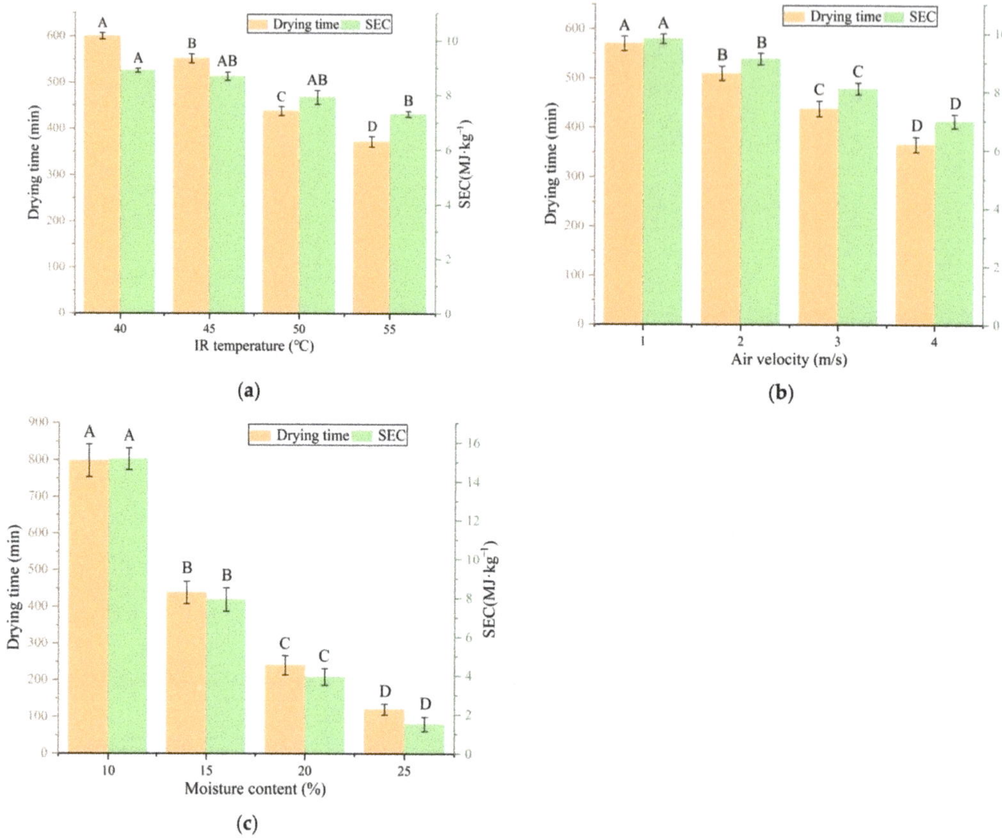

Figure 6. Effects of T, V and MC on DT and SEC. (**a**) Air velocity was 3 m/s and moisture content was 15%. (**b**) IR temperature was 50 °C and moisture content was 15%. (**c**) IR temperature was 50 °C and air velocity was 3 m/s. Note: different letters of columns of same color indicate significant differences between the mean values ($p < 0.01$).

3.2. High Kernel Rate and Whole Kernel Rate

As shown in Figure 7a, at a constant air velocity of 3 m/s, moisture content of 15% and longitudinal loading direction, IR temperature had no notable influence on the high kernel and whole kernel rates. No remarkable difference in shell kernel clearance was observed among the four groups. The effect of air velocity on the high kernel rate was not substantial at a constant IR temperature of 50 °C, moisture content of 15% and in the longitudinal loading direction, as shown in Figure 7b. With the increase in air velocity, the whole kernel rates of Groups 3 and 4 decreased compared with those of Groups 1 and 2. The whole kernel rate is related to shell kernel clearance, and it can only be guaranteed when the deformation of shell breaking is smaller than the shell kernel clearance [26]. Therefore, the decrease in whole kernel rate may be due to the decrease in shell kernel clearance. Figure 7c shows that the high kernel rate and whole kernel rate decreased significantly with an increase in the moisture content from 10% to 25% at a fixed IR temperature of 50 °C, an air velocity of 3 m/s, and longitudinal loading direction. Wang Jiannan et al. had the same finding [1]. As presented in Figure 7d, under a constant IR temperature of 50 °C, air velocity of 3 m/s and moisture content of 15%, the high kernel and whole kernel rates of sutural direction were the highest, followed by those of the longitudinal and vertical directions in turn. Under the same drying conditions, the shell kernel clearance of the three groups was almost undifferentiated. "Shell kernel clearance" refers to overall shell kernel clearance and is the mean value of shell kernel clearance in three directions of the walnut. However, the shell kernel clearance in different directions obviously differed, which resulted in a great difference in the high kernel and the whole kernel rates of different loading directions. In conclusion, moisture content had the greatest influence on the high kernel and whole kernel rates, and reducing the moisture content helped improve the high kernel and whole kernel rates. However, this result is contradictory to the minimization of drying time and SEC. The optimization of the walnut breaking process needs to achieve balance between energy efficiency and quality of walnut breaking.

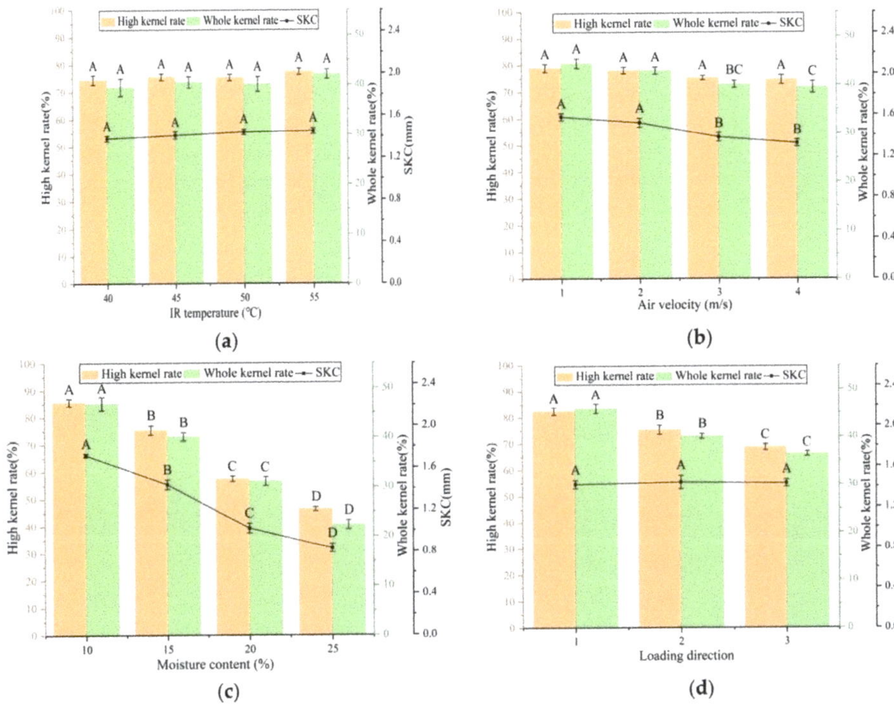

Figure 7. Effects of T, V, MC and DT on HR and WR. (**a**) Air velocity was 3 m/s, moisture content was 15% and loading direction was Longitudinal. (**b**) IR temperature was 50 °C, moisture content was 15% and loading direction was Longitudinal. (**c**) IR temperature was 50 °C, air velocity was 3 m/s and loading direction was Longitudinal. (**d**) IR temperature was 50 °C, air velocity was 3 m/s and moisture content was 15%. Note: different letters of columns of the same color indicate significant differences between the mean values ($p < 0.01$).

3.3. *Shell-Breaking Rate*

Figure 8a shows the effects of different IR temperatures on the shell-breaking rate when the air velocity, moisture content and loading direction were 3 m/s, 15%, and longitudinal, respectively. The shell-breaking rate and hardness of Groups 3 and 4 increased compared with those of Groups 1 and 2 when the IR temperature continued to increase to 55 °C. The water loss rate of the walnut shells increased with the increasing IR temperature, which accelerated the change in porosity and rapidly increased the density of walnut shells, leading to an increase in walnut shell hardness [41]. With high walnut shell hardness, high breaking force would need be exerted under the same deformation, and walnut shells are more likely to be broken. Feizollah Shahbazi also believed that the shell-breaking rate was related to the hardness of walnut shell [42]. As shown in Figure 8b, at a constant IR temperature of 50 °C, moisture content of 15%, and longitudinal loading direction, air velocity had no notable influence on the shell-breaking rate. No remarkable difference in the hardness was observed among the four groups of experiments. As shown in Figure 8c, at a constant IR temperature of 50 °C, air velocity of 3 m/s, and longitudinal loading direction, the shell-breaking rate decreased significantly with increasing moisture content. The increase in walnut moisture content reduced the walnut shell hardness, indicating a small shell-breaking force was required to achieve the same deformation. Feizollah Shahbazi found that the decrease in the shell-breaking rate is due to the softening of walnuts at higher moisture contents [42]. At the same time, as the moisture content of walnuts decreases, the hygrothermal stress in the drying process causes the shell to produce microcracks and other damage, and the shell becomes prone to fracture and breakage [43]. Figure 8d show that the shell-breaking rate was the highest in the vertical direction when the IR temperature was fixed at 50 °C, the air velocity was 3 m/s, and the moisture content was 15%. The thickness and structure of the walnut shells in different directions varied, resulting in different shell-breaking rates in different loading directions. Liu Kui et al. also found that walnuts are likely to be broken in the vertical direction [44]. In conclusion, the influence law of moisture content on the shell-breaking rate is similar to that of the high kernel and whole kernel rates. In addition, the shell-breaking rate can be optimized by selecting an appropriate IR temperature and loading direction.

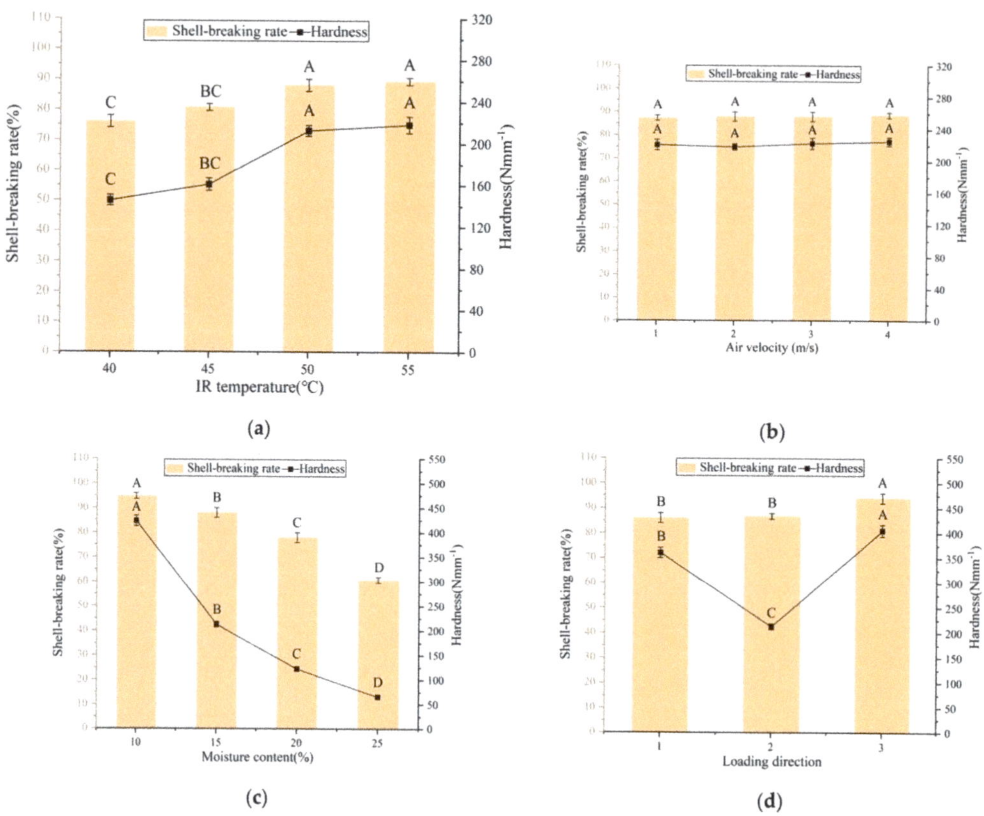

Figure 8. Effects of T, V, MC and DT on SR. (**a**) Air velocity was 3 m/s, moisture content was 15% and loading direction was Longitudinal. (**b**) IR temperature was 50 °C, moisture content was 15% and loading direction was Longitudinal. (**c**) IR temperature was 50 °C, air velocity was 3 m/s and loading direction was Longitudinal. (**d**) IR temperature was 50 °C, air velocity was 3 m/s and moisture content was 15%. Note: different letters of columns of the same color indicate significant differences between the mean values ($p < 0.01$).

3.4. Construction of Artificial Neural Network Model

The construction of the ANN model proceeded through three steps. The first step was to determine the optimal topology of the ANN. The second step was to optimize the initial weight and threshold of the ANN by using a GA. In the third step, the ANN used the optimized weights and thresholds for network training.

To obtain the optimal topology of the neural network, the transfer function ('transig' sigmoidal or 'logsig' sigmoidal) between the input layer and the hidden layer and the number of neurons in the hidden layer (2–13) needed to be determined. Nonoptimized weights and thresholds for network training can cause a large deviation in R^2 and RMSE under the same topology structure. Therefore, the results of the same topology structure were adopted as the mean value of 10 simulations. The simulation results of different ANN topologies is shown in Table 3.

Table 3. Simulation results of different ANN topologies (shaded group is the optimal topology; bold font indicates the optimal solution).

Group	Transfer Function of the Hidden Layer	Number of Neurons in the Hidden Layer	DT		SEC		HR		WR		SR	
			R^2	*RMSE*	R^2	*RMSE*	R^2	*RMSE*	R^2	*RMSE*	R^2	*RMSE*
1		2	0.6101	87	0.9170	0.7809	0.7262	3.6053	0.8348	2.4655	0.7695	3.4578
2		3	0.7330	70	0.9430	0.6930	0.8490	2.7626	0.9029	1.9311	0.8680	2.9473
3		4	0.8053	62	0.9290	0.5373	0.8953	2.2088	0.9284	1.6161	0.9034	2.6510
4		5	0.8455	50	0.9113	0.5288	0.9198	2.2477	0.9377	1.5535	0.9231	2.5961
5		6	0.8688	42	0.9172	0.4820	0.9331	2.0987	0.9411	1.3893	0.9231	2.8564
6	Tansig	7	0.8858	38	0.9234	0.4868	0.9430	1.9751	0.9474	1.3514	0.9245	2.5263
7		8	0.8965	36	0.9204	0.4876	0.9509	1.9454	0.9522	1.3170	0.9246	2.6684
8		9	0.9030	37	0.9261	0.4557	0.9561	1.8920	0.9560	1.2482	0.9214	2.3708
9		10	0.9113	36	0.9256	0.4839	0.9589	1.9481	0.9556	1.3308	0.9207	2.4248
10		11	0.9176	34	0.9276	0.4794	0.9619	1.7221	0.9570	**1.1531**	0.9158	**2.3483**
11		12	0.9207	37	0.9296	0.4642	0.9633	1.7937	0.9573	1.2995	0.898	2.4721
12		13	0.9018	38	0.9248	0.4782	0.9610	1.8238	0.9568	1.3210	0.9103	2.5131
13		2	0.8142	68	0.9584	0.8237	0.8740	4.2739	0.8135	3.1225	0.8150	4.1104
14		3	0.8468	58	0.9197	0.8994	0.8915	3.0264	0.8413	2.4166	0.8564	3.9061
15		4	0.8762	48	0.9328	0.5449	0.9127	2.4617	0.8734	1.9124	0.8643	3.3415
16		5	0.9010	44	0.9385	0.5140	0.9338	2.1232	0.8980	1.6816	0.8876	3.0413
17		6	0.9170	43	0.9478	0.4458	0.9457	2.0833	0.9198	1.6132	0.9120	3.0922
18	Logsig	7	0.9271	36	0.9539	0.4363	0.9600	1.8385	0.9384	1.4448	0.9320	2.9298
19		8	0.9338	38	0.9582	0.3943	0.9620	2.0749	0.9411	1.4663	0.9359	2.8583
20		9	0.9363	36	0.9620	0.3795	0.9682	1.6762	0.9519	1.2627	0.9409	2.6781
21		10	0.9412	**30**	0.9650	0.4118	0.9720	1.7408	0.9604	1.3396	0.9468	2.7705
22		11	0.9459	34	0.9665	**0.3551**	0.9760	1.6523	0.9651	1.2436	0.9490	2.7766
23		12	**0.9500**	33	**0.9682**	0.3631	**0.9786**	**1.6521**	**0.9695**	1.1869	**0.9555**	2.5246
24		13	0.9432	35	0.9641	0.3728	0.9731	1.7239	0.9543	1.2165	0.9461	2.6987
	Types of activation functions				$\text{Tansig}(x) = \frac{2}{1+\exp(-2x)}$		$\text{logsig}(x) = \frac{1}{1+\exp(-x)}$					

The highest R^2 values of DT, SEC, HR, WR, and SR were achieved in Group 23 ('logsig' sigmoidal + 12 neurons). The minimum RMSEs of DT, SEC, HR, WR, and SR were achieved in Groups 21, 22, 23, 10 and 10, respectively. The subminimum RMSEs of DT, SEC and WR appeared in the Group 23, with a deviation of 10%, 2.25% and 2.93% from the minimum RMSE, respectively. The RMSE of SR in Group 23 differed by 7.51% from the minimum RMSE of SR. In summary, the topology of Group 23 ('logsig' sigmoidal + 12 neurons) was the optimal topology.

The results showed that different topologies had a great influence on R^2 and RMSE. The 'logsig' sigmoidal function between the input layer and hidden layer was much better than the 'transig' sigmoidal function. The larger the number of neurons in the hidden layer was, the better the prediction ability of the model was. However, having too many neurons can lead to overfitting, resulting in reduced model accuracy. Therefore, the number of neurons in the hidden layer was set as 12 based on the optimal topology and GA to optimize the initial weights and thresholds of the neural network. The parameters of the GA are shown in Table 1.

The neural network was trained with the optimized weights and thresholds. A comparison of the results of the BP neural network model and the GA neural network model (GA-ANN) is shown in Table 4.

Table 4. Comparison results between BP model and GA-ANN model.

Algorithm	DT		SEC		HR		WR		SR	
	R^2	*RMSE*	R^2	*RMSE*	R^2	*RMSE*	R^2	*RMSE*	R^2	*RMSE*
BP	0.950	33	0.968	0.363	0.979	1.652	0.970	1.187	0.956	2.525
GA-ANN	0.996	22.176	0.998	0.279	0.990	1.471	0.991	0.926	0.993	1.250
Error (%)		32.8		23.14		10.96		21.99		50.5

In the GA-ANN model, the R^2 values of DT, SEC, HR, WR and SR were all above 0.990; RMSE decreased by 32.8%, 23.14%, 10.96%, 21.99%, and 50.5% compared with the BP model, respectively. The GA-ANN model had a remarkable optimization effect, indicating that the GA-ANN model removed the local optimal value and obtained the global optimal value.

The experimental data and predicted values of the GA-ANN model are shown in Figure 9. The GA-ANN model demonstrated a sufficient prediction ability and can be further applied to the MOO of the process of drying-assisted walnut breaking.

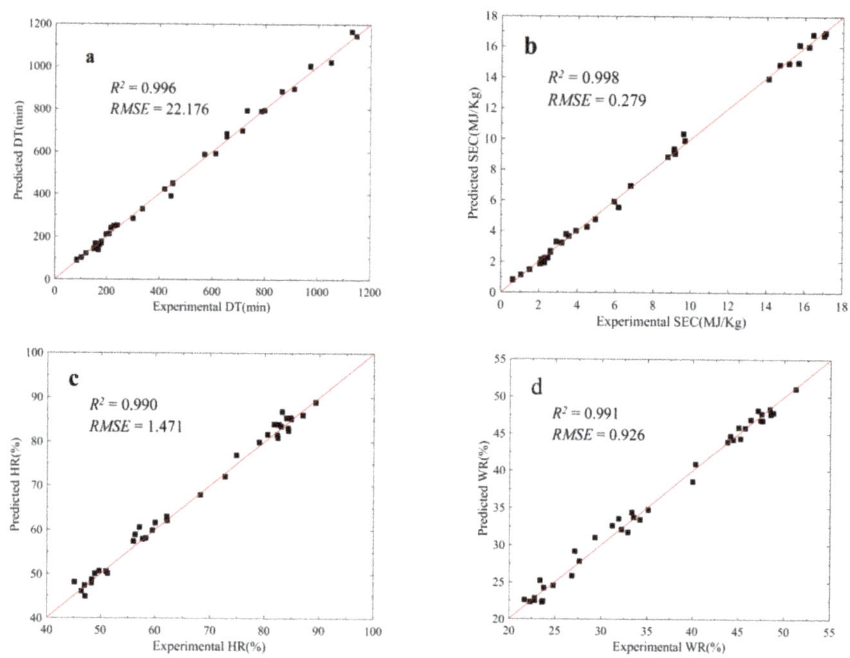

Figure 9. *Cont.*

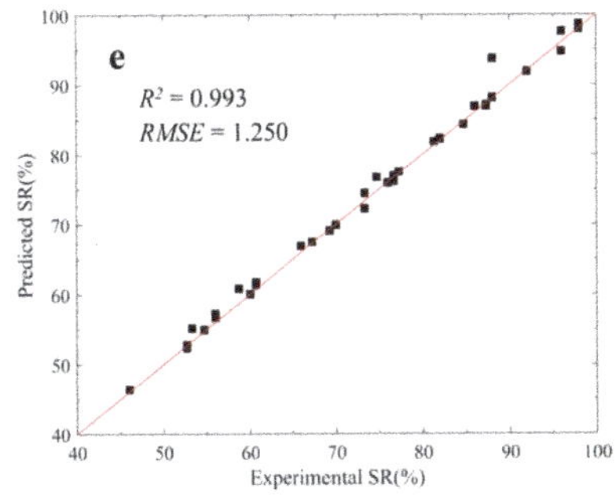

Figure 9. Experimental data and predicted values of the GA-ANN model: (**a**) drying time; (**b**) specific energy consumption; (**c**) high kernel rate; (**d**) whole kernel rate, and (**e**) shell-breaking rate.

3.5. Multi-Objective Optimization

The Pareto optimal set, which included 30 optimal solutions, was obtained using the GAMULTIOBJ function, as shown in Table 5. The selection of specific optimal solutions depended on the primary purpose of the process. For example, Group 1 was selected as the optimal process for the highest HR of 89%. Group 3 was a slightly worse process because HR and SR decreased by 0.2% and 3.6%, respectively, and DT increased by 12.9%. Moreover, it had a similar SEC of 16.78 MJ/kg relative to Group 1, but WR increased by 0.6%. In addition, the lowest DT of 87.3 min (90.9% lower than that of Group 1) and the lowest SEC of 0.81 MJ/kg (95.2% lower than that of Group 1) were obtained in Group 30. Under this processing condition, HR was 46.1% (48.2% lower than that of Group 1), WR was 22.4% (56% lower than that of Group 1) and the SR was 61.8% (33.4% lower than that of Group 1). Given that the primary purpose of MOO was to ensure the quality of walnut breaking, Group 30 was not be selected as the optimal solution. Group 7 had the highest SR of 99.8% (7.5% higher than that of Group 1), HR of 81.7% that of (8.2% lower than that of Group 1), and WR of 43.7% (14.1% lower than that of Group 1). DT decreased by 39.0%, and SEC decreased by 23.3%, relative to Group 1. The reason was that Group 7 adopted the highest possible values of IR temperature and air velocity to reduce the drying time and energy consumption. In addition, the drying rate of walnuts at the last stage of drying was extremely low. Group 7 slightly increased the moisture content, which considerably reduced the drying time and energy consumption. Group 7 reduced the drying time and energy consumption as much as possible without notably reducing the quality of walnut breaking. In conclusion, Group 7's parameters (T = 54.9, V = 3.66, MC = 10.9, vertical loading direction) were selected as the optimal parameters for the process of drying-assisted walnut breaking.

Verification experiments were conducted under the optimal parameters to verify the optimization results. As shown in Table 6, the errors between the predicted values and the experimental data were 3.56%, 3.88%, 2.51%, 3.32%, and 3.96% for DT, SEC, HR, WR, and SR, respectively, indicating good accuracy. The ANN model optimized by the GA could accurately predict the process parameters of drying-assisted walnut breaking, and the MOO results of the GA were effective.

Table 5. Pareto optimal set (shaded groups are the optimum sets in the Pareto front).

Pareto ID	T (°C)	V (m/s)	MC	D	DT (min)	SEC (MJ/kg)	HR	WR	SR
1	51.5	1.04	10.0	2	958.4	16.77	89.0	50.9	92.8
2	53.3	1.20	10.1	2	888.5	16.49	88.9	50.4	93.6
3	47.0	1.01	10.0	2	1082.2	16.78	88.8	51.2	89.5
4	44.7	1.50	11.1	2	974.0	14.77	86.2	48.4	86.8
5	52.5	3.59	23.2	2	137.9	1.84	48.8	24.9	68.2
6	43.1	1.02	10.0	2	1193.4	16.78	88.7	51.2	85.3
7	54.9	3.66	10.9	3	585.1	12.86	81.7	43.7	99.8
8	54.9	3.98	20.3	3	181.4	2.78	58.6	32.1	84.8
9	54.7	3.97	21.3	2	157.2	2.33	54.1	28.8	75.7
10	49.6	1.01	10.0	2	1012.0	16.79	88.9	51.0	91.6
11	53.4	3.83	17.4	2	264.8	4.70	69.2	38.3	85.3
12	54.7	3.85	23.9	2	109.6	1.29	47.8	23.8	66.1
13	54.2	3.25	25.0	3	103.9	1.08	56.8	29.3	69.3
14	52.5	2.27	16.0	2	407.9	7.45	75.9	41.6	87.3
15	54.9	3.13	17.0	2	299.7	5.65	72.3	39.5	87.0
16	50.6	2.16	10.2	2	847.7	15.48	87.5	49.1	92.9
17	47.7	1.97	12.8	2	696.3	11.68	82.5	45.4	88.3
18	54.9	3.82	24.2	2	102.8	1.15	47.3	23.3	64.6
19	52.5	2.92	12.9	2	535.3	10.54	82.1	44.9	91.9
20	43.6	1.39	14.2	2	729.8	10.64	80.2	44.0	82.1
21	49.8	3.38	12.4	2	573.7	10.85	82.0	44.6	91.3
22	43.6	1.02	10.1	2	1169.3	16.65	88.6	51.0	85.8
23	47.3	1.75	10.9	2	901.0	14.80	86.4	48.5	89.7
24	50.4	3.71	17.2	2	299.8	5.11	69.5	38.4	84.0
25	52.9	3.65	20.4	3	200.6	3.06	58.6	32.2	83.5
26	52.5	2.74	16.6	2	353.5	6.39	73.3	40.3	86.3
27	52.7	2.03	15.7	2	436.6	8.03	77.2	42.3	87.9
28	55.0	3.98	23.3	2	116.8	1.45	48.7	24.7	68.5
29	54.9	3.93	23.0	2	123.3	1.60	49.4	25.2	69.6
30	55.0	3.99	25.0	2	87.3	0.81	46.1	22.4	61.8

Table 6. Prediction and validation results of five response variables under the optimal parameters.

Results	Operating Conditions				Response Variables				
	T (°C)	V (m/s)	MC	D	DT (min)	SEC (MJ/kg)	HR	WR	SR
Prediction	54.9	3.66	10.9	3	585.1	12.86	81.7	43.7	99.8
validation	54.9	3.66	10.9	3	565	12.38	83.8	45.2	96
Error (%)					3.56	3.88	2.51	3.32	3.96

4. Conclusions

A process of drying-assisted walnut breaking was developed in this study. An ANN model optimized by a GA (GA-ANN) was established to simulate the effects of input variables (T, V, MC, and D) on output variables (DT, SEC, HR, WR, and SR). The GA-ANN model demonstrated sufficient prediction ability, with coefficients of determination of 0.996, 0.998, 0.990, 0.991, and 0.993 for DT, SEC, HR, WR, and SR, respectively. RMSE decreased by 32.8%, 23.14%, 10.96%, 21.99%, and 50.5% compared with those in the ANN model. On the basis of the GA-ANN model, the GA was applied to MOO of the walnut breaking process with the aim of minimizing DT and SEC and maximizing HR, WR and SR. The optimal process parameters were determined to be T = 54.9, V = 3.66, MC = 10.9, and vertical loading direction, which led to DT = 585.1 min, SEC = 12.86 MJ/kg, HR = 81.7%, WR = 43.7%, and SR = 99.8%. The model that combines ANN and GA was proven to be effective in predicting and optimizing the process parameters of walnut breaking. The predicted values under the optimized process parameters fitted the experimental data well, with low relative error values of 3.56%, 3.88%, 2.51%, 3.32%, and 3.96% for DT, SEC, HR, WR, and SR, respectively. The process of drying-assisted walnut breaking considerably improved the quality of walnut breaking and reduced the energy consumption and drying time. The ANN modeling and the method of GA MOO developed in this study could be applied to other similar commodities.

Author Contributions: Conceptualization, T.Y. and X.Z.; methodology, T.Y.; experiments, T.Y.; software, T.Y.; formal analysis, T.Y. and X.Z.; investigation, T.Y., X.Z.; validation, T.Y., X.Z., S.K.V., H.X., X.Y., Y.L., Y.Z. and J.Z.; writing—original draft preparation, T.Y.; writing—review and editing, T.Y., X.Z., S.K.V., H.X. and X.Y.; supervision, X.Z.; project administration, X.Z. All authors have read and agreed to the published version of the manuscript.

Funding: This research was funded by the project from Xinjiang Corps Industrial and High-tech Science and Technology Research and Achievement Transformation Program (No. 2016AB003); Shihezi University Achievement Transformation and Technology Promotion Project (No. CGZH201808); National Natural Science Foundation of China (Grant No. 31960488).

Institutional Review Board Statement: Not applicable.

Informed Consent Statement: Not applicable.

Data Availability Statement: The data presented in this study are available on request from the corresponding author.

Conflicts of Interest: The authors declare no conflict of interest.

References

1. Wang, J.; Liu, M.; Wu, H.; Peng, J.; Peng, B.; Yang, Y.; Cao, M.; Wei, H.; Xie, H. Design and Key Parameter Optimization of Conic Roller Shelling Device Based on Walnut Moisture-Regulating Treatments. *Agriculture* **2022**, *12*, 561. [CrossRef]
2. Jin, W.G.; Pei, J.J.; Wang, S.Q.; Chen, X.H.; Gao, R.C.; Tan, M.Q. Effect of continuous and intermittent drying on water mobility of fresh walnuts (*Juglans regia* L.): A LF-NMR study. *Dry. Technol.* **2020**, *40*, 254–264. [CrossRef]
3. Pakrah, S.; Rahemi, M.; Nabipour, A.; Zahedzadeh, F.; Kakavand, F.; Vahdati, K. Sensory and nutritional attributes of Persian walnut kernel influenced by maturity stage, drying method, and cultivar. *J. Food Process. Preserv.* **2021**, *45*, e15513. [CrossRef]
4. Chen, C.; Venkitasamy, C.; Zhang, W.P.; Deng, L.Z.; Meng, X.Y.; Pan, Z.L. Effect of step-down temperature drying on energy consumption and product quality of walnuts. *J. Food Eng.* **2020**, *285*, 110105. [CrossRef]
5. Zhou, X.; Gao, H.Y.; Mitcham, E.J.; Wang, S.J. Comparative analyses of three dehydration methods on drying characteristics and oil quality of in-shell walnuts. *Dry. Technol.* **2018**, *36*, 477–490. [CrossRef]
6. Bao, X.L.; Chen, B.Y.; Dai, P.; Li, Y.S.; Mao, J.C. Construction and verification of spherical thin shell model for revealing walnut shell crack initiation and expansion mechanism. *Agriculture* **2022**, *12*, 1446. [CrossRef]
7. Deng, L.Z.; Xiong, C.H.; Sutar, P.P.; Mujumdar, A.S.; Pei, Y.P.; Yang, X.H.; Ji, X.W.; Zhang, Q.; Xiao, H.W. An emerging pretreatment technology for reducing postharvest loss of vegetables—A case study of red pepper (*Capsicum annuum* L.) drying. *Dry. Technol.* **2022**, *40*, 1620–1628. [CrossRef]
8. Atungulu, G.G.; Teh, H.E.; Wang, T.; Fu, R.; Wang, X.; Khir, R.; Pan, Z. Infrared pre-drying and dry-dehulling of walnuts for improved processing efficiency and product quality. *Appl. Eng. Agric.* **2013**, *29*, 961–971.
9. Martinez-Martinez, V.; Gomez-Gil, J.; Stombaugh, T.S.; Montross, M.D.; Aguiar, J.M. Moisture content prediction in the switchgrass (*Panicum virgatum*) drying process using artificial neural networks. *Dry. Technol.* **2015**, *33*, 1708–1719. [CrossRef]
10. Liu, Z.L.; Nan, F.; Zheng, X.; Zielinska, M.; Duan, X.; Deng, L.Z.; Wang, J.; Wu, W.; Gao, Z.J.; Xiao, H.W. Color prediction of mushroom slices during drying using Bayesian extreme learning machine. *Dry. Technol.* **2020**, *38*, 1869–1881. [CrossRef]
11. Liu, Z.L.; Wei, Z.Y.; Vidyarthi, S.K.; Pan, Z.L.; Zielinska, M.; Deng, L.Z.; Wang, Q.H.; Wei, Q.; Xiao, H.W. Pulsed vacuum drying of kiwifruit slices and drying process optimization based on artificial neural network. *Dry. Technol.* **2020**, *39*, 405–417. [CrossRef]
12. Shojaeefard, M.H.; Etghani, M.M.; Akbari, M.; Khalkhali, A.; Ghobadian, B. Artificial neural networks based prediction of performance and exhaust emissions in direct injection engine using castor oil biodiesel-diesel blends. *J. Renew. Sustain. Energy* **2013**, *4*, 063130. [CrossRef]
13. Bai, J.; Zhou, C.; Cai, J.; Xiao, H.; Gao, Z.; Ma, H. Vacuum pulse drying characteristics and moisture content prediction of pumpkin slices. *Trans. Chin. Soc. Agric. Eng.* **2017**, *33*, 290–297.

14. Bai, J.W.; Xiao, H.W.; Ma, H.L.; Zhou, C.S. Artificial neural network modeling of drying kinetics and color changes of *Ginkgo biloba* seeds during microwave drying process. *J. Food Qual.* **2018**, *2018*, 3278595. [CrossRef]
15. Liu, Z.L.; Bai, J.W.; Wang, S.X.; Meng, J.S.; Wang, H.; Yu, X.L.; Gao, Z.J.; Xiao, H.W. Prediction of energy and exergy of mushroom slices drying in hot air impingement dryer by artificial neural network. *Dry. Technol.* **2020**, *38*, 1959–1970. [CrossRef]
16. Sun, Q.; Zhang, M.; Mujumdar, A.S.; Yang, P.Q. Combined LF-NMR and Artificial Intelligence for Continuous Real-Time Monitoring of Carrot in Microwave Vacuum Drying. *Food Bioprocess Technol.* **2019**, *12*, 551–562. [CrossRef]
17. Winiczenko, R.; Górnicki, K.; Kaleta, A.; Martynenko, A.; Janaszek-Mańkowska, M.; Trajer, J. Multi-objective optimization of convective drying of apple cubes. *Comput. Electron. Agric.* **2018**, *145*, 341–348. [CrossRef]
18. Raj, G.V.S.B.; Dash, K.K. Microwave vacuum drying of dragon fruit slice: Artificial neural network modelling, genetic algorithm optimization, and kinetics study. *Comput. Electron. Agric.* **2020**, *178*, 105814.
19. Shojaeefard, M.H.; Behnagh, R.A.; Akbari, M.; Givi, M.K.B.; Farhani, F. Modelling and Pareto optimization of mechanical properties of friction stir welded AA7075/AA5083 butt joints using neural network and particle swarm algorithm. *Mater. Des.* **2013**, *44*, 190–198. [CrossRef]
20. Zhang, W.; Wang, K.; Chen, C. Artificial Neural Network Assisted Multiobjective Optimization of Postharvest Blanching and Drying of Blueberries. *Foods* **2022**, *11*, 3347. [CrossRef]
21. Rajendra, M.; Jena, P.C.; Raheman, H. Prediction of optimized pretreatment process parameters for biodiesel production using ANN and GA. *Fuel* **2009**, *88*, 868–875. [CrossRef]
22. Asadi, P.; Aliha, M.R.M.; Akbari, M.; Imani, D.M.; Berto, F. Multivariate optimization of mechanical and microstructural properties of welded joints by FSW method. *Eng. Fail. Anal.* **2022**, *140*, 106528. [CrossRef]
23. Akbari, M.; Asadi, P.; Besharati Givi, M.K.; Khodabandehlouie, G. Artificial neural network and optimization. *Adv. Frict. -Stir Weld. Process.* **2014**, *13*, 543–599.
24. Chen, C.; Venkitasamy, C.; Pan, Z.L. Effective moisture diffusivity and drying simulation of walnuts under hot air. *Int. J. Heat Mass Transf.* **2020**, *150*, 119283. [CrossRef]
25. Chen, C.; Upadhyaya, S.; Khir, R.; Pan, Z.L. Simulation of walnut drying under hot air heating using a nonequilibrium multiphase transfer model. *Dry. Technol.* **2022**, *40*, 987–1001. [CrossRef]
26. Zheng, X.; Zhang, E.; Kan, Z.; Zhang, H.; Li, H.; Chou, W. Improving cracking characteristics and kernel percentage of walnut by optimal position of cutting on shell. *Trans. Chin. Soc. Agric. Eng.* **2018**, *34*, 300–308.
27. Gulsoy, E.; Kus, E.; Altikat, S. Determination of physico-mechanical properties of some domestic and foreign walnut (*Juglans regia* L.) varieties. *Acta Sci. Pol.-Hortorum Cultus* **2019**, *18*, 67–74. [CrossRef]
28. Blanco, A.; Delgado, M.; Pegalajar, M.C. A genetic algorithm to obtain the optimal recurrent neural network. *Int. J. Approx. Reason.* **2000**, *23*, 67–83. [CrossRef]
29. Chai, Z.H.; Yang, X.; Liu, Z.L.; Lei, Y.L.; Zheng, W.H.; Ji, M.Y.; Zhao, J.F. Correlation Analysis-Based Neural Network Self-Organizing Genetic Evolutionary Algorithm. *IEEE Access* **2019**, *7*, 135099–135117. [CrossRef]
30. Martin, T.H.; Howard, B.D. *Neural Network Design*, 2nd ed.; China Machine Press: Beijing, China, 2002; pp. 11–15.
31. Kalathingal, M.S.H.; Basak, S.; Mitra, J. Artificial neural network modeling and genetic algorithm optimization of process parameters in fluidized bed drying of green tea leaves. *J. Food Process Eng.* **2020**, *43*, e13128. [CrossRef]
32. Huang, X.; Li, Y.B.; Zhou, X.; Wang, J.; Zhang, Q.; Yang, X.H.; Zhu, L.C.; Geng, Z.H. Prediction of Apple Slices Drying Kinetic during Infrared-Assisted-Hot Air Drying by Deep Neural Networks. *Foods* **2022**, *11*, 3486. [CrossRef] [PubMed]
33. Zhu, K.; Ren, G.; Duan, X.; Qiu, C.; Li, L.; Chu, Q.; Yu, Z. Backward Propagation (BP) Neural Network-Based Prediction of Moisture Ratio of Fresh In-shell Peanut during. *Food Sci.* **2022**, *43*, 9–18.
34. Zhang, F.Q.; Wu, S.Y.; Wang, Y.O.; Xiong, R.; Ding, G.Y.; Mei, P.; Liu, L.Y. Application of Quantum Genetic Optimization of LVQ Neural Network in Smart City Traffic Network Prediction. *IEEE Access* **2020**, *8*, 104555–104564. [CrossRef]
35. Ozcelik, B.; Oktem, H.; Kurtaran, H. Optimum surface roughness in end milling Inconel 718 by coupling neural network model and genetic algorithm. *Int. J. Adv. Manuf. Technol.* **2005**, *27*, 234–241. [CrossRef]
36. Yuan, P.; Mao, J.L.; Xiang, F.H.; Liu, L.; Zhang, M.X. Fault Diagnosis of Power System Based on Improved Genetic Optimized BP-NN. *Int. Conf. Eng. Technol. Appl. (ICETA)* **2015**, *22*, 01050. [CrossRef]
37. Jeevarathinam, G.; Pandiselvam, R.; Pandiarajan, T.; Preetha, P.; Krishnakumar, T.; Balakrishnan, M.; Thirupathi, V.; Ganapathy, S.; Amirtham, D. Design, development, and drying kinetics of infrared-Assisted hot air dryer for turmeric slices. *J. Food Process Eng.* **2022**, *45*, e13876. [CrossRef]
38. Vimercati, W.C.; Araujo, C.D.; Macedo, L.L.; Maradini, A.M.; Saraiva, S.H.; Teixeira, L.J.Q. Influence of drying temperature on drying kinetics, energy consumption, bioactive compounds and cooking quality of pasta enriched with spinach. *J. Food Process Eng.* **2020**, *43*, e13571. [CrossRef]
39. Liu, Z.L.; Bai, J.W.; Yang, W.X.; Wang, J.; Deng, L.Z.; Yu, X.L.; Zheng, Z.A.; Gao, Z.J.; Xiao, H.W. Effect of high-humidity hot air impingement blanching (HHAIB) and drying parameters on drying characteristics and quality of broccoli florets. *Dry. Technol.* **2019**, *37*, 1251–1264. [CrossRef]
40. Chang, A.T.; Zheng, X.; Xiao, H.W.; Yao, X.D.; Liu, D.C.; Li, X.Y.; Li, Y.C. Short—and Medium-Wave Infrared Drying of Cantaloupe (*Cucumis melon* L.) Slices: Drying Kinetics and Process Parameter Optimization. *Processes* **2022**, *10*, 114. [CrossRef]
41. Lu, Y.; Ren, G.; Duan, X.; Zhang, L.; Ling, Z. Moisture Migration Properties and Quality Changes of Fresh In-Shell Peanuts during Hot Air Drying. *Food Sci.* **2020**, *41*, 86–92.

42. Shahbazi, F. Effects of Moisture Content and Impact Energy on the Cracking Characteristics of Walnuts. *Int. J. Food Eng.* **2014**, *10*, 149–156. [CrossRef]
43. Man, X. Optimization of the Process Parameters of Hot-Air and Microwave-Vacuum Synergistic Drying for Walnut Shell Breaking. Master's Thesis, Tarim University, Alaer, China, 2021.
44. Liu, K.; Guo, W.; Zhu, Z. Study on various factors compression test of Xinfeng walnut. *Food Mach.* **2020**, *36*, 124–128.

foods

MDPI

Article

Evaluation of Temperature Uniformity in a Middle-Refrigerated Truck Loaded with Pig Carcasses

Hongwu Bai [1,2], Guanghong Zhou [1,*] and Xianjin Liu [2]

1 Key Laboratory of Meat Processing and Quality Control, Ministry of Education, College of Food Science and Technology, Nanjing Agricultural University, Nanjing 210095, China; hongwu.bai@jaas.ac.cn
2 Institute of Food Safety and Nutrition, Jiangsu Academy of Agricultural Sciences, Nanjing 210014, China; jaasliu@163.com
* Correspondence: ghzhou@njau.edu.cn; Tel.:+86-25-84395754

Abstract: In this study, we constructed a calculation model to determine the internal temperature field distribution in a medium-sized refrigeration truck with the dimensions of 4.1 m × 2.2 m × 2.2 m. Wind speed, air temperature, and carcass temperature were designated as the initial conditions. The k-ε model of computational fluid dynamics was used to simulate different wind speeds and ventilation duct settings on the carriage. Additionally, under specific boundary conditions, the speed of the air outlet, the types of ventilation ducts, and the carcass loads were all varied to determine the uniformity of the temperature field. The results showed that, when the air outlet speed was 5 m/s, the temperature field in the refrigerated truck was relatively more uniform. The simulated results were in good agreement with the measured results. The average absolute error was 0.35 °C, and the average relative error was 9.23%.

Keywords: refrigerated truck; computational fluid dynamics; carcass traceability; temperature field

1. Introduction

Temperature is a key factor affecting the rapid propagation of microorganisms. If the temperature of a food storage area is not correct, then food spoilage will increase [1–3]. Refrigerated trucks are an important means of land transportation. During transportation, the internal temperature field within a refrigerated truck is affected by the external environment, internal structures, and the stacking methods used [4,5]. Many scholars have used computational fluid dynamics (CFD) to study refrigerated trucks, and the results of their work have helped to overcome the shortcomings of traditional theoretical analysis and enable simpler, more effective computation [6,7]. Scholars have also conducted numerical simulations of transport-compartment temperature fields based on airflow velocity, stacking patterns, and airflow organization [8]. Several temperature and humidity sensors in a refrigerated truck are usually managed at certain periods to simulate the corresponding temperature and humidity curves [4,9,10]. The analysis of turbulent flow with complex heat transfer in a refrigerated body was conducted by Kayansayan et al. [11]. Their study investigated the impact of the shape factors of the body by examining two types of cold air outlets, three types of aspect ratios, and the Reynolds number of the wind speed. The study concluded that the optimum cold-air speed value could be determined depending on the shape of the refrigerated body. However, to develop accurate modeling, experimental work should accompany these methods. Margeirsson et al. [12] developed a 3D heat transfer model that predicted changes in product temperature for supercooled cod fillets packaged in loaded EPS boxes. The model was validated by comparing with experimental results but was found to be slightly different from the actual results. In another study, Hoang [13] presented a refrigerated vehicle heat transfer model based on computational fluid dynamics (CFD) to predict the temperature of cargo. The CFD modeling considered various phenomena, such as the respiration heat of the product and infiltration of outside

Citation: Bai, H.; Zhou, G.; Liu, X. Evaluation of Temperature Uniformity in a Middle-Refrigerated Truck Loaded with Pig Carcasses. *Foods* **2023**, *12*, 1837. https://doi.org/10.3390/foods12091837

Academic Editor: Begoña Panea

Received: 29 March 2023
Revised: 22 April 2023
Accepted: 26 April 2023
Published: 28 April 2023

air due to the opening of the door. The results obtained reflect the influence of the thermal characteristics of the refrigerator compartment on the airflow structure, but they cannot truly reflect the convection heat transfer in the transport environment [14,15]. In the present study, 36 temperature and humidity nodes were installed in a passenger compartment to establish a wireless environmental detection system for refrigerated trucks and record the truck temperature in real time. The internal temperature field of the refrigerated truck was numerically simulated using Fluent 17 with the k-ε turbulence model. The distribution of the temperature field inside the truck was calculated under different wind speeds and ventilation modes. Finally, a temperature-based assessment model was ultimately developed, which could monitor the quality variations of pig carcasses during their transportation via refrigerated trucks [16].

2. Materials and Methods

2.1. Test Materials

In this work, we studied the temperature field of pig carcasses in a middle-refrigerated truck, because this type of truck is suitable for medium- and short-distance cold-chain transportation in big cities due to its low cost and widespread use. The middle-refrigerated truck had external dimensions of 4.3 m × 2.4 m × 2.4 m (length × width × height), an internal clearance of 4.1 m × 2.2 m × 2.2 m, a body thickness of 105 mm, and a covering made of Fiberglass reinforced plastic (FRP) with a thickness of 2.5 mm. The "sandwich" structure was a 100 mm polyurethane foam board with double doors at the back of the cabin, and the air outlet was placed directly in front of the carriage.

A refrigerated truck is not always fully loaded. Hence, it was necessary to compare the standard temperature fields with the same volume of pig carcass areas, to produce a credible evaluation model. The volumes of the standard temperature fields were equal. The carcass-volume ratio refers to the proportion between the volume of the carcass area being evaluated within the refrigerated compartment and the volume of the maximum allowable cargo area. An upper limit to the actual cargo volume of the refrigerated truck was specified and the carriage was not filled entirely, to ensure that airflow circulated in the interior part of the refrigerated truck. Figure 1 presents a schematic illustration of a large pig carcass area in a carriage with a standard temperature field. The selected area of the standard temperature field was used as the reference area, and the entire pig carcass area was reduced to a voided cube (Figure 1b). A constant distance between the pig carcass area and the inner wall of the carriage was assumed, and the height of the pig carcass area was unchanged. The distance from the top plate was termed h_1 and the distance from the bottom plate was termed h_2.

$$x_1 = x_2 = y_1 = y_2 \tag{1}$$

$$V_x = (x_0 - x_1 - \Delta x_2) \times (y_0 - y_1 - y_2) \times (h_0 - h_1 - h_2) \tag{2}$$

(a) (b)

Figure 1. Structure of a middle-refrigerated truck and area occupied by pig carcasses. (**a**) An empty compartment. (**b**) A compartment with pig carcasses.

If x_1, x_2, y_1, and y_2 are 0, then

$$V_x = V_0 = x_0 y_0 h_0 \tag{3}$$

$$\eta = \frac{V_x}{V_0} \tag{4}$$

These formulas express the maximum volume of the standard temperature field, where η is the standard temperature field volume ratio. The length, width, and height (h_0) of the compartment were all taken into consideration.

2.2. Test Conditions

The main unit (Testo 480) of a multi-functional environmental tester (Detu Instrument International Trading Co., Ltd., Shanghai, China) was used for the study. The probe parameters were as follows:

Flexible wind-speed probe with a range of 0–20 m/s and a measurement accuracy of ±0.03 m/s;
Surface-temperature probe with a range from −60 °C to 300 °C and a measuring accuracy of ±0.5 °C;
Immersion-type PT100 temperature probe with a measuring range of −100 °C to 400 °C and a measuring accuracy of ±0.15 °C; and an air-humidity probe with a range of 0–100% RH and a measurement accuracy of ±1.8% RH.

An Internet-of-Things temperature and humidity node (Nanjing Huiming Instrument Co., Ltd., Nanjing, China) was used with the following parameters:

Humidity measurement range of 0–100% RH;
Humidity accuracy of ±1.8% RH with humidity resolution of ±0.05% RH;
Temperature measurement range from −40 °C to 123.8 °C;
Temperature accuracy of ±0.3 °C and temperature resolution of ±0.01 °C.

2.3. Test Methods

Many factors affect the temperature fields of refrigerated compartments. The main considerations in this study were the external temperature, the refrigerated vehicle ventilation mode, and the temperature of the air outlet in the compartment. These three factors, taken together, completely determine the standard temperature fields. Before each test, the temperature and humidity nodes were calibrated with a high-precision temperature and humidity sensor to ensure testing accuracy and consequently reduce testing error. In addition, the refrigerator truck was turned on for pre-cooling. During the test, the humidity node acquisition interval was set to 30 s, and the starting temperature of the vehicle compartment was determined. Wind speed has a transient instability value, so a wind-speed probe was affixed to the cold air outlet. The average wind speed under the maximum cooling capacity was used as the initial value. Before and after the testing, nine temperature and humidity nodes were arranged, at specific positions of 0.5, 2.0, and 4.0 m from the outlet location. Each test chain comprised four nodes suspended at a distance from the top of the carriage (i.e., 0.5, 1.0, 1.5, and 2.0 m). A temperature and humidity node was arranged on the outer wall of the carriage to measure outdoor temperature. The gateway, router, and GPRS transmission module were all located in the cab.

3. Models and Assumptions

3.1. Model Assumptions

To facilitate subsequent calculations, the following points were assumed for the model:

1. The air in the compartment is a transparent radiation medium according to Boussinesq's hypothesis.

The gas in the truck is a Newtonian fluid, and the pig carcass is considered to be a porous medium.

The effect of temperature changes on the physical parameters of air and swine fever can be ignored.

Any influence of pipes, rails, and wires in the refrigerated truck on the airflow can be ignored.

The walls of the truck body are insulated and the truck body is fully sealed, so that there is no leakage phenomenon.

The effects of moisture loss and latent heat of evaporation on the body temperatures of pig carcasses can be ignored.

The issue of heat dissipation in pig carcasses is not considered.

3.2. Basic Equations

In fluid heat transfer analysis, researchers often regard the mass conservation equation, momentum conservation equation, and energy conservation equation as the basic control equations [17]. The three basic equations of CFD can be referred to simply as follows:

mass conservation equation:

$$\frac{\partial \rho}{\partial t} + \frac{\partial}{\partial x_i}(\rho u_i) = S_m \quad (5)$$

conservation of momentum:

$$\frac{\partial}{\partial t}(\rho u_i) + \frac{\partial}{\partial x_j}(\rho u_i u_j) = \frac{\partial p}{\partial x_i} + \frac{\partial \tau_{ij}}{\partial x_j} + \rho g_i + F_i \quad (6)$$

energy conservation theorem:

$$\frac{\partial}{\partial t}(\rho E) + \nabla \cdot \left(\vec{v}(\rho E + \rho)\right) = \nabla \cdot \left(k_{eff} \nabla T - \sum_j h_j \vec{J}_j + \overline{\tau}_{eff} \cdot \vec{v}\right) + S_h \quad (7)$$

subsequently,

$$\tau_{ij} = \left[\mu \left(\frac{\partial u_i}{\partial x_j} + \frac{\partial u_j}{\partial x_i}\right)\right] - \frac{2}{3}\mu \frac{\partial u_l}{\partial x_l}\delta_{ij} \quad (8)$$

$$\mathrm{E} = h + \frac{p}{\rho} + \frac{v^2}{2} \quad (9)$$

$$\mathrm{h} = \sum_j Y_j h_j \quad (10)$$

$$h_j = \int_{T_{ref}}^{T} c_{p,j} dt \quad (11)$$

$$\overline{\tau}_{eff} = \mu_{eff}\left(\frac{\partial u_j}{\partial x_i} + \frac{\partial u_i}{\partial x_j}\right) - \frac{2}{3}\mu_{eff}\frac{\partial u_i}{\partial x_i}\delta_{ij} \quad (12)$$

where

S_m—source item;
k_{eff}—effective thermal conductivity;
E—total energy;
τ_{ij}—stress tensor;
Y_j—mass fraction of component;
T_{ref}—298.15 K;
τ_{eff}—partial stress equations;
x—flow field direction;
t—time;
ρ—fluid density;
μ—velocity vector;

p—static pressure;
h—height;
S—fluid quality;
J—fluid heat;
T—force;
F—surface force.

3.3. Thermal Conduction and Heat Convection

When the refrigerated truck is running on the road, the heat transfer of its compartment enclosure structure continues to change [18]. At the same time, the goods in the refrigerated truck are coupled to the heat transfer. Convection, conduction, and radiation all affect the temperature field, and the entire refrigerated truck is in a state of dynamic thermal equilibrium. The heat conduction equation of the inner and outer walls of the compartment can be expressed as follows:

$$q_c = \frac{t_o - t_i}{\frac{\delta_i}{\lambda_i} + \frac{\delta_m}{\lambda_m} + \frac{\delta_o}{\lambda_o}} \quad (13)$$

where

t_0—temperature of the outer wall of the refrigerated compartment;
t_i—temperature of the wall surface of the refrigerated compartment;
δ_i—thickness of the FRP outside the refrigerator compartment;
δ_m—thickness of the polyurethane insulation board;
δ_0—thickness of the FRP in the refrigerated compartments;
λ_i —λ_0—thermal conductivity of the FRP of the refrigerated truck;
λ_m—thermal conductivity of polyurethane.

To investigate the refrigerated compartments during movement, we dynamically calculated the heat flow on the surfaces of the refrigerated trucks. The external heat flow mainly comes from solar radiant heat and the convection heat transfer between the inner and outer walls of the cabin. Because fresh pig carcasses are the object of this study, there is no respiratory heat in the animal tissue. The average heat flux is defined as the amount of heat that flows into the interior of a compartment per unit of time, per unit area of the wall that separates the compartment from the outside environment. Internal and external convection heat transfer must also be considered in addition to vehicle body heat transfer. The average heat flux density q_h of the refrigerated truck compartment wall can thus be expressed as:

$$q_h = \frac{t_o - t_i}{\frac{1}{h_o} + \frac{\delta_m}{\lambda_m} + \frac{1}{h_i}} \quad (14)$$

where

h_o—convection heat transfer coefficient of the outer wall of the refrigerated compartment and
h_i—convection heat transfer coefficient of the inner wall of the refrigerated compartment.

The other parameters are the same as those in Equation (13). When the outer wall of the refrigerated truck is considered, the total heat transfer is, therefore:

$$Q_c = q_c \times A_0 \quad (15)$$

where

A_0—total area of the outer wall of the refrigerated compartment.

During transportation, the thermal load of the external thermal environment (i.e., heat transfer through the wall) accounts for more than 80% of the total heat load in the refrigerated truck. If the external temperature of the compartment is higher than the internal temperature, the heat-insulation layer of the compartment can only achieve a small amount of heat insulation. Within the unit area, when the difference between internal and external temperature increases, the heat from the external wall to the interior of the refrigerated compartment initially increases. Then, the evaporator inside the compartment fan spreads this heat to all parts of the compartment to form an unstable temperature field. Finally, the heat that enters the compartment exits through the truck's cooling system. By using such means, a low-temperature environment can be maintained inside the compartment.

3.4. Temperature Field Uniformity Index

Refrigerated trucks in operation are not always fully loaded, and different cargoes may require different stacking methods. Therefore, the temperature field where a cargo is located should not be compared with the standard temperature field of the entire truck, but with the same volume as the stacking area [4]. The standard temperature field of the refrigerated truck is called the measured area, which is equal to the volume of this same area. In this study, the ratio of the cargo area volume to the maximum allowable cargo area volume is called the refrigerated truck volume. To ensure that airflow circulates in the interior, the actual capacity of the refrigerated truck is given an upper limit (i.e., it cannot fill the entire carriage). In other words, the maximum allowable cargo volume of the refrigerated truck is the same as the maximum reference volume of the standard temperature field.

$$\mu_0 = \sum_{i=1}^{n=4} \propto_i \eta^{i-1} \ (\mu_0 = \propto_0 + \propto_1 \eta + \propto_2 \eta^2 + \propto_3 \eta^3) \tag{16}$$

$$\sigma_0 = \sum_{i=1}^{n=4} \beta_i \eta^{i-1} \ (\sigma_0 = \beta_0 + \beta_1 \eta + \beta_2 \eta^2 + \beta_3 \eta^3) \tag{17}$$

$$R_0 = \sum_{i=1}^{n=4} \gamma \eta^{i-1} \ (R_0 = \gamma_0 + \gamma_1 \eta + \gamma_2 \eta^2 + \gamma_3 \eta^3) \tag{18}$$

$$\eta = \frac{V_x}{V_0} \tag{19}$$

$$\mu_1 = \sum_{i=1}^{n} {t_i}/{n} \tag{20}$$

$$\sigma_1 = \sqrt{\sum_{i=1}^{n} (t_i - \mu)^2 / n} \Big/ \mu \tag{21}$$

$$R_1 = \text{Max}(t) - \min(t) \tag{22}$$

$$\psi_1 = (\mu - t_0)/(\mu_0 - t_0) \tag{23}$$

$$\psi_2 = {\sigma}/{\sigma_0} \tag{24}$$

$$\psi_3 = {R}/{R_0} \tag{25}$$

$$\xi = \psi_1 + \psi_2 + \psi_3 \tag{26}$$

where

μ_0—average temperature of the standard temperature field;
ρ_0—temperature non-uniform coefficient of the standard temperature field;
R_0—temperature of the standard temperature field;

μ_1—average temperature of the temperature field to be measured;
ρ_1—temperature non-uniform coefficient of the temperature field to be measured;
R_1_temperature of the temperature field to be measured;
η—volumetric rate of the actual cargo carried with the standard temperature field in the refrigerated truck;
α_i—coefficient of the average temperature equation of the standard temperature field;
β_i—coefficient of the temperature non-uniform coefficient equation of the standard temperature field;
γ_i_coefficient of the temperature extreme equation of the standard temperature field;
t_0—outlet temperature;
ψ_1_average deviation from the temperature difference ratio;
ψ_2—temperature non-uniform coefficient ratio;
ψ_3—temperature range ratio;
ξ—temperature field uniformity index.

The basic indicators (Formulas 16–18) used to evaluate the temperature field in the refrigerated truck are the average temperature of the temperature field, the temperature non-uniform coefficient, and the temperature range. The average deviation (Formulas 23–25) can be used to compare the temperature field with the standard temperature field and reveal any advantages or disadvantages. Formula 26 expresses the comprehensive inhomogeneity coefficient in the refrigerator truck.

3.5. Boundary Conditions

1. The bottom of the box is set as the adiabatic boundary condition;
2. The air outlet adopts the speed boundary conditions;
3. The returning-air inlets on both sides of the box are simplified using free-flow boundary conditions;
4. The relevant physical parameters and initial boundary conditions used in the calculation model are shown in Table 1.

Table 1. Physical parameters and initial boundary conditions.

Parameter	Initial Value	Unit	Instructions
Density of FRP	2100	kg/m^3	Merchant
Thickness of FRP	0.0025	m	Merchant
Density of polyurethane foam plate	40	kg/m^3	FRT properties
Thickness of polyurethane foam plate	0.1	m	FRT properties
Specific heat capacity of polyurethane foam plate	871	J/kg·K	FRT properties
Thermal conductivity of polyurethane foam plate	0.020	W/m·k	FRT properties
Space size of simulation	$4.1 \times 2.2 \times 2.2$	m^3	Clearance size
Vent speed	1–7	m/s	Control
Vent temperature	0	°C	Constant
Initial temperature in the body of the box	20	°C	Multipoint
External temperature	35	°C	Temperature
Air density	1.293	kg/m^3	Air properties
Air-specific heat capacity	1006.43	J/kg·K	Air properties
Air thermal conductivity	0.0242	W/m·k	Air properties
Air kinetic viscosity	1.7894×10^{-5}	kg/m·s	Air properties

3.6. Simulation Conditions

The simulation was performed on a DELL R720 PC. The hardware parameters are as follows:

Xeon 6 core E5-2630 × 2 2.6 GHz; 6 × 16 G 1333 MHz; 3 × 300 G 3.5 15 K; RAID5; PERC H310; single 750 W; and a silent fan.

The k-ε model is used for calculations. The discrete solver based on pressure has a first-order upwind scheme for momentum, energy, turbulent kinetic energy, diffusion rate, and a SIMPLE algorithm for pressure–velocity coupling. The equations for turbulent kinetic energy transport are derived entirely, while the equations for dissipation rate are obtained through a combination of physical reasoning and mathematical simulations based on similar primitive equations. The model assumes that the flow is completely turbulent and that molecular viscosity can be ignored [19].

The turbulent kinetic energy k and dissipation rate ε in the standard k-ε model [20] are calculated as follows:

$$\rho\frac{Dk}{Dt} = \frac{\partial}{\partial x_i}[(\mu + \frac{\mu_t}{\sigma_k})\frac{\partial k}{\partial x_i}] + G_k + G_b - \rho\varepsilon - Y_M \tag{27}$$

$$\rho\frac{D\varepsilon}{Dt} = \frac{\partial}{\partial x_i}\left[\left(\mu + \frac{\mu_t}{\sigma_k}\right)\frac{\partial \varepsilon}{\partial x_i}\right] + C_{1\varepsilon}\frac{\varepsilon}{k}(G_k + C_{3\varepsilon}G_b) - C_{2\varepsilon}\rho\frac{\varepsilon^2}{k} \tag{28}$$

where

G_k—turbulent kinetic energy due to the average velocity gradient;
G_b—a parameter used to generate turbulent kinetic energy due to buoyancy effects;
Y_M—effect of turbulent pulsation expansion on total dissipation rate;
μ_t—turbulent viscosity coefficient;
D_t—total derivative of time;
D_ε—the turbulent Prandtl numbers of ε;
D_k—the turbulent Prandtl numbers of k;
C—constant.

$$\mu_t = \rho C_\mu \frac{k^2}{\varepsilon} \tag{29}$$

In Fluent software, $C_{1\varepsilon}$ = 1.44, $C_{1\varepsilon}$ = 1.92, C_μ = 0.09, σ_k = 1.0, and σ_ε = 1.3. These default values are determined experimentally in basic turbulent flows, including commonly encountered shear flows, such as boundary layers, mixed layers, and jets, and attenuated isotropic mesh turbulence [21,22].

4. Results and Analysis

4.1. Flow Field Status Judgment

The laminar and turbulent flow states are based on the Reynolds number, which is determined in Equation (30). When the fluid sweeps the plate, the critical Reynolds number is 5×10^5. If the actual Reynolds number is less than the critical Reynolds number [23], then the flow state is laminar. If the actual Reynolds number is greater than the critical Reynolds number, then the flow state is turbulent.

$$\mathrm{Re} = \frac{\sigma \mathrm{VL}}{\mu} \tag{30}$$

where

Re—Reynolds number;
ρ—air density;
V—airflow velocity at the outlet (m/s);
L—flow distance of the airflow (m);
μ—kinematic viscosity of the air (m^2/s).

The air density at 0 °C is 1.293 kg/m^3, the air kinematic viscosity is 13.28×10^{-6} m^2/s [24], and the flow distance of the airstream is 1.5 m. Using Formula (31), the critical velocity in the empty truck is, therefore, calculated to be 0.125 m/s, as follows:

$$\mathrm{V} = \frac{\mathrm{Re} \times \mu}{\sigma \times \mathrm{L}} = \frac{50{,}000 \times 13.28 \times 10^{-6}}{1.293 \times 4.1} = 0.125 \tag{31}$$

The flow field along the four walls has a direct impact on the temperature field inside the truck. The turbulent airflow near the front, rear, left, or right walls can lead to an uneven temperature field near the inner wall of the cabin. On the other hand, laminar airflow near the walls can create a stable laminar boundary layer near the inner wall surface, resulting in an improved steady-state effect of the mainstream flow field. Figure 2 illustrates this phenomenon.

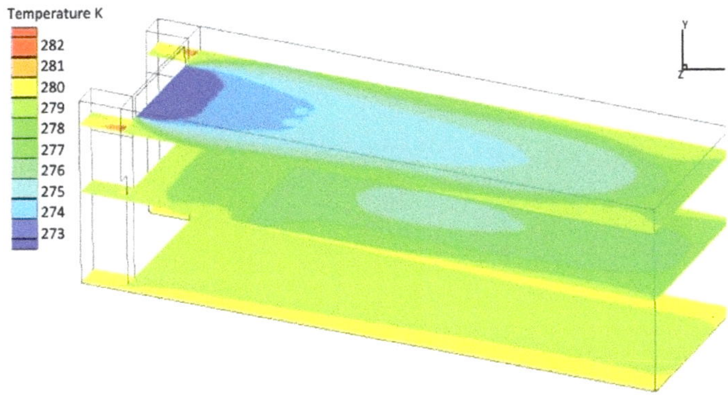

Figure 2. The standard temperature contour in a refrigerated truck with low-speed air supply.

4.2. *Influence of Wind Speed on Temperature Field*

The simulated operating conditions are as follows: external temperature 35.0 °C; truck speed 20.8 m/s (at approximately 75.0 km/h); outlet velocity 1.0 m/s; and temperature 0 °C. The result becomes stable and steady after 12 min of iterative calculation. From the central axis (z = 1.1 m) to the end of the refrigerated truck, the air outlet velocity gradually declines from 1 m/s to 0.1 m/s, and the temperature rises from 2 °C to 10 °C. As shown in Figure 3, the average temperature is high, and most of the areas are red. When the wind speed is gradually increased to 3 m/s, the average temperature tends to fall, and only half of the area is red. When wind speed rises to 5 m/s, the overall temperature becomes uniform (i.e., the blue region dominates), and the average temperature is close to the outlet temperature. An increase in wind speed to 7 m/s can further improve the homogeneity of the temperature inside the compartment. However, it is worth noting that high-wind velocity can cause carcasses to become darker and drier, ultimately resulting in a lower commodity value. Due to the potential negative impact of high-wind speeds on carcass quality, a wind speed of 5 m/s may be considered the optimal choice, even though higher wind speeds can result in a more uniform temperature distribution within the refrigerated truck.

4.3. *Effect of Ventilation Ducts on Temperature Field*

Three air-supply ducts are installed on the top of the refrigerated truck joining with the air outlet. The main air-supply duct leads directly to the rear of the truck. The secondary air-supply duct leads to the middle part of the carriage. This design optimizes the airflow direction of the air supply, in order to improve the temperature uniformity of the refrigerated truck. When v = 1 m/s, the main area of the original truck appears mostly in red, which means that the high-temperature zone dominates. When the upper ventilation duct is added, the cold air is directly delivered to the middle and tail, and the efficiency of the air-conditioning system appears to improve. The simulated result appears in light red. For the second working condition, 16 U-shaped air ducts are installed on the floor of the refrigerated truck directly; these are called lower ventilation ducts. Under this condition, the turbulent flow at the floor is enhanced and the bottom temperature zone becomes greener. The third working condition involves the simultaneous operation of the upper and lower ventilation ducts. Under this condition, the temperature field becomes uniform over the same simulation time, and the average temperature approximates the target temperature more ideally, which is shown in Figure 4.

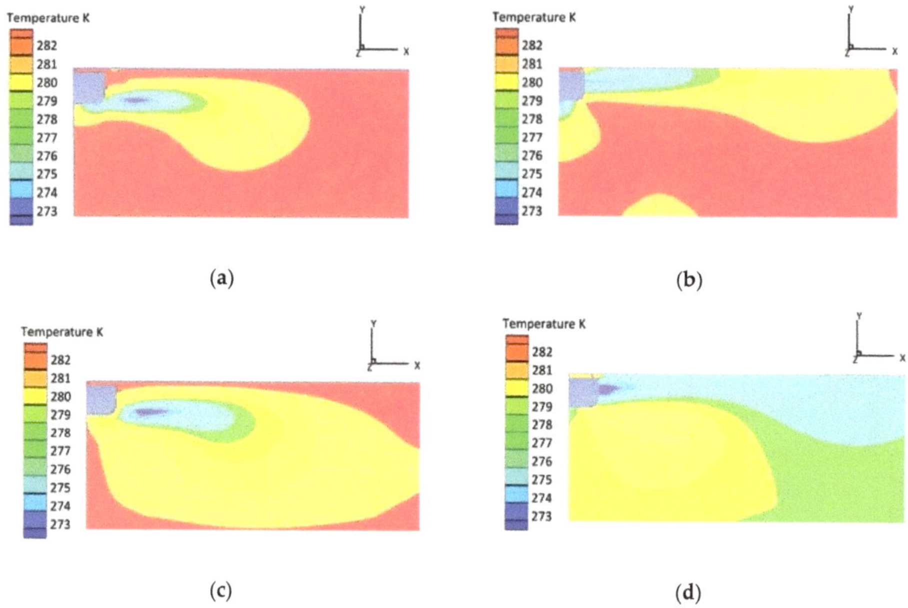

Figure 3. Temperature contours with different wind speeds (t = 0.1 h, z = 1.1 m): (**a**) wind speed (v = 1 m/s); (**b**) wind speed (v = 3 m/s); (**c**) wind speed (v = 5 m/s); and (**d**) wind speed (v = 7 m/s).

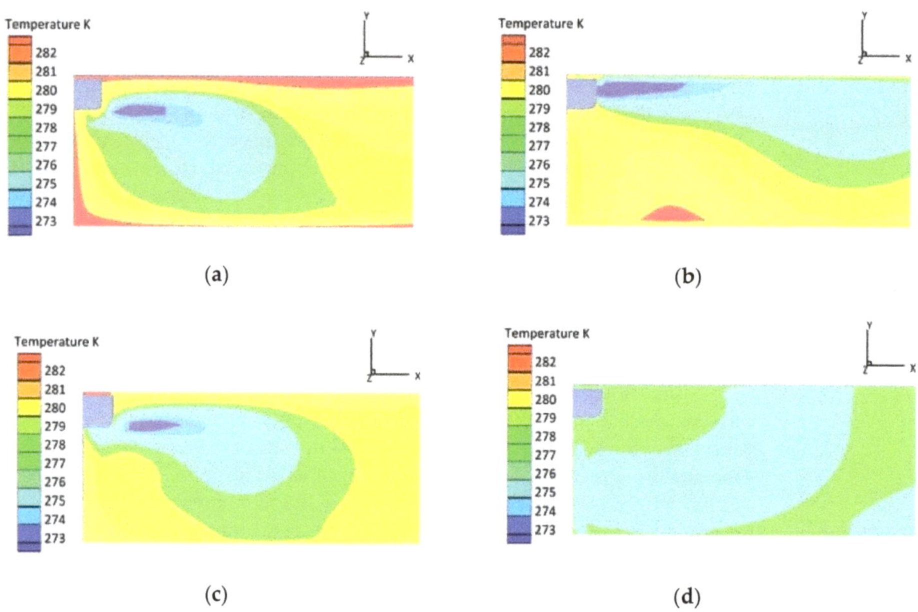

Figure 4. Temperature contours with different ventilation ducts: (**a**) original refrigerated truck (v = 1 m/s, h = 0.1); (**b**) with added upper duct (v = 1 m/s, h = 0.1); (**c**) with added lower duct (v = 1 m/s, h = 0.1); and (**d**) with added upper and lower ducts (v = 1 m/s, h = 0.1).

4.4. Evaluation of Temperature Field Uniformity

In this study, we numerically simulated four kinds of operating conditions: an original refrigerated truck; a truck with an upper duct; a truck with a lower duct; and a truck with upper and lower ducts. An evaluation model of a standard temperature field in a middle-refrigerated truck loaded with carcasses was established using experimental and simulated data. According to Equation (4), the volumetric rate of the experimental carriage could be calculated as follows:

$$\eta = \frac{V_x}{V_0} = \frac{3.5 \times 2.0 \times 2.0}{4.1 \times 2.2 \times 2.2} = 0.706 \tag{32}$$

η—Volume ratio of refrigerated vehicle load;
V_x—Volume of pig carcasses loaded in a refrigerated truck;
V_0—headroom volume of refrigerated truck.

The relevant data were reviewed and the following variables were obtained: the coefficient α is the average temperature equation of the standard temperature field; the coefficient β is the coefficient of non-uniformity coefficient; and the coefficient γ is the temperature range difference equation. Based on the results obtained from Equations (16) to (18), the inconsistent coefficients of the measured temperature field (Equation (20)), average temperature (Equation (21)), and temperature difference (Equation (22)) were calculated. The other required values were derived by using the average temperature ratio, the non-uniform coefficient ratio, and the temperature range ratio of Equations (23)–(25). As shown in Table 2, a smaller numerical value indicates a closer approximation to the ideal standard temperature field for the truck.

Table 2. Evaluation index of the temperature field to be measured.

Pattern	Average Deviation from the Temperature Difference Ratio (ψ_1)	Inhomogeneity Coefficient Ratio (ψ_2)	Maximum Difference Ratio (ψ_3)	Temperature Field Uniformity Index (ξ)
Original state	1.74	7.49	0.89	10.12
Upper ventilation ducts	1.54	6.44	0.76	8.74
Lower ventilation ducts	1.69	5.38	0.54	7.61
Upper and lower ventilation ducts	1.38	4.39	0.36	6.13

4.5. Test Verification

Figure 5 shows a comparison of the simulation results with real-world data, as measured using 36 temperature sensors. It can be seen that the simulated temperature values are consistent with the experimental values, with an average absolute error of 0.35 °C and an average relative error of 9.23%. For the truck in the original state, the average absolute error is 0.27 °C and the average relative error is 11.98%. For the truck with an upper ventilation duct, the average absolute error is 0.20 °C and the average relative error is 6.18%. For the truck with a lower ventilation duct, the average absolute error is 0.56 °C and the average relative error is 11.98%. For the truck with both upper and lower ventilation ducts, the average absolute error is 0.45 °C and the average relative error is 8.25%.

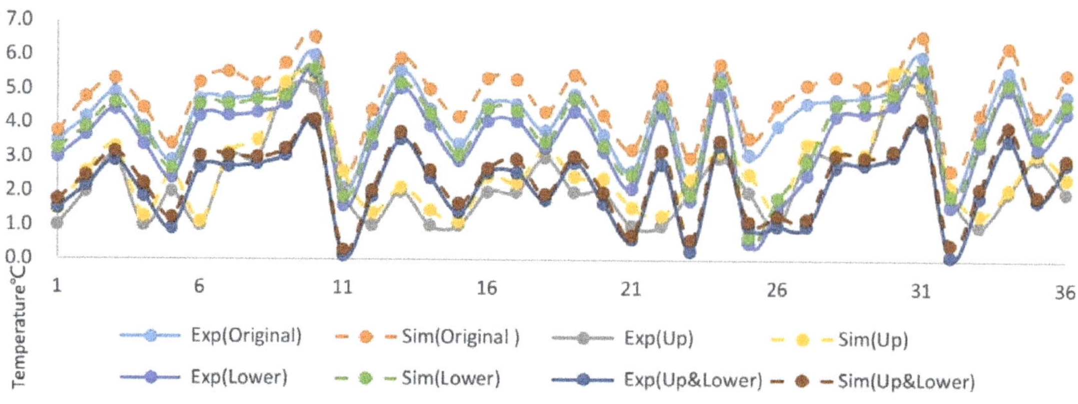

Figure 5. Experimental and simulated temperature values for different operating conditions. Exp(Original): data measured on the original unmodified refrigerated truck; Exp(Lower): data measured on a refrigerated truck added lower ventilation ducts; Sim(Original): the simulated values on the original unmodified refrigerated truck; Sim(Lower): the simulated values on a refrigerated truck added lower ventilation ducts; Exp(Up): data measured on a refrigerated truck added up ventilation ducts; Exp(Up and Lower): data measured on a refrigerated truck added both up and lower ventilation ducts; Sim(Up): the simulated values on the refrigerated truck added up ventilation ducts; Sim(Up and Lower): the simulated values on the refrigerated truck added both up and lower ventilation ducts.

5. Conclusions

For this study, a typical middle-refrigerated truck was used, with a clearance of 4.1 m × 2.2 m × 2.2 m. Firstly, using experimental and simulated data, we established an evaluation model for a standard temperature field in a middle-refrigerated truck loaded with pig carcasses. We found that the greater the air-supply speed, the shorter the average temperature stabilization time. Secondly, we established a temperature field evaluation system for refrigerated trucks loaded with pig carcasses. To evaluate the performance of the refrigerated truck's temperature field, parameters such as the average temperature, temperature range, and non-uniform coefficient were considered. By assigning appropriate weight coefficients to these parameters, the average deviation temperature ratio, temperature non-uniform coefficient ratio, and temperature range ratio were combined to produce a uniformity index for the temperature field inside the refrigerated truck. Finally, we studied how the use of upper and/or lower ventilation ducts affected the temperature field in the compartment of the refrigerated truck. The study found that incorporating three upper ventilation ducts and sixteen lower ventilation ducts led to an improvement in the uniformity of the temperature field inside the refrigerated truck. The results showed that the condition with 16 lower ventilation ducts was more effective, possibly because it allowed for a sufficient airflow return path in a middle-refrigerated truck that was almost fully loaded with pig carcasses.

Author Contributions: Conceptualization, H.B.; formal analysis, G.Z.; investigation, X.L.; data curation, H.B.; writing—original draft preparation, H.B.; writing—review and editing, G.Z.; funding acquisition, G.Z. All authors have read and agreed to the published version of the manuscript.

Funding: This work was funded by the Key Technologies Research and Development Program (2022YFD2100602), and the Jiangsu Provincial Key Research and Development Program (BE2020383).

Institutional Review Board Statement: Not applicable.

Informed Consent Statement: Not applicable.

Data Availability Statement: The datasets generated for this study are available on request to the corresponding author.

Acknowledgments: Chunbao Li and Huhu Wang were acknowledged for help in sampling.

Conflicts of Interest: The authors declare no conflict of interest.

References

1. Dave, D.; Ghaly, A.E. Meat spoilage mechanisms and preservation techniques: A critical review. *Am. J. Agric. Biol. Sci.* **2011**, *6*, 486–510.
2. Göransson, M.; Jevinger, Å.; Nilsson, J. Shelf-life variations in pallet unit loads during perishable food supply chain distribution. *Food Control* **2018**, *84*, 552–560. [CrossRef]
3. Grijspeerdt, K.; Mortier, L.; De Block, J.; Van Renterghem, R. Applications of modelling to optimise ultra high temperature milk heat exchangers with respect to fouling. *Food Control* **2004**, *15*, 117–130. [CrossRef]
4. Wu, X.; Chang, Z.; Yuan, P.; Lu, Y.; Ma, Q.; Yin, X. The optimization and effect of back panel structure on the performance of refrigerated display cabinet. *Food Control* **2014**, *40*, 278–285. [CrossRef]
5. Hsiao, H.-I.; Huang, K.-L. Time-temperature transparency in the cold chain. *Food Control* **2016**, *64*, 181–188. [CrossRef]
6. Norton, T.; Sun, D.-W.; Grant, J.; Fallon, R.; Dodd, V. Applications of computational fluid dynamics (CFD) in the modelling and design of ventilation systems in the agricultural industry: A review. *Bioresour. Technol.* **2007**, *98*, 2386–2414. [CrossRef]
7. Laguerre, O.; Hoang, M.; Osswald, V.; Flick, D. Experimental study of heat transfer and air flow in a refrigerated display cabinet. *J. Food Eng.* **2012**, *113*, 310–321. [CrossRef]
8. Norton, T.; Sun, D.-W. Computational fluid dynamics (CFD)—An effective and efficient design and analysis tool for the food industry: A review. *Trends Food Sci. Technol.* **2006**, *17*, 600–620. [CrossRef]
9. Akdemir, S.; Ozturk, S.; Edis, F.O.; Bal, E.; Singh, U.; Singh, A.; Parvez, S.; Sivasubramaniam, A. CFD modelling of two different cold stores ambient factors. *IERI Procedia* **2013**, *5*, 28–40. [CrossRef]
10. Wang, J.; Yue, H. Food safety pre-warning system based on data mining for a sustainable food supply chain. *Food Control* **2017**, *73*, 223–229. [CrossRef]
11. Kayansayan, N.; Alptekin, E.; Ezan, M.A. Thermal analysis of airflow inside a refrigerated container. *Int. J. Refrig.* **2017**, *84*, 76–91. [CrossRef]
12. Margeirsson, B.; Palsson, H.; Gospavic, R.; Popov, V.; Jónsson, M.Þ.; Arason, S. Numerical modelling of temperature fluctuations of chilled and superchilled cod fillets packaged in expanded polystyrene boxes stored on pallets under dynamic temperature conditions. *J. Food Eng.* **2012**, *113*, 87–99. [CrossRef]
13. Hoang, M.; Laguerre, O.; Moureh, J.; Flick, D. Heat transfer modelling in a ventilated cavity loaded with food product: Application to a refrigerated vehicle. *J. Food Eng.* **2012**, *113*, 389–398. [CrossRef]
14. Chaomuang, N.; Flick, D.; Laguerre, O. Experimental and numerical investigation of the performance of retail refrigerated display cabinets. *Trends Food Sci. Technol.* **2017**, *70*, 95–104. [CrossRef]
15. So, J.-H.; Joe, S.-Y.; Hwang, S.-H.; Jun, S.; Lee, S.-H.J. Analysis of the Temperature Distribution in a Refrigerated Truck Body Depending on the Box Loading Patterns. *Foods* **2021**, *10*, 2560. [CrossRef] [PubMed]
16. Jia, X.; Nian, Y.; Zhao, D.; Wu, J.; Li, C. Changes of the Microbiota Composition on the Surface of Pig Carcasses during Chilling and Its Associations with Alterations in Chiller's Temperature and Air Humidity. *Foods* **2021**, *10*, 2195. [CrossRef] [PubMed]
17. Ambaw, A.; Delele, M.; Defraeye, T.; Ho, Q.T.; Opara, L.; Nicolai, B.; Verboven, P. The use of CFD to characterize and design post-harvest storage facilities: Past, present and future. *Comput. Electron. Agric.* **2013**, *93*, 184–194. [CrossRef]
18. Currle, J. Numerical simulation of the flow in a passenger compartment and evaluation of the thermal comfort of the occupants. *SAE Trans.* **1997**, 806–816.
19. Singh, U.; Singh, A.; Parvez, S.; Sivasubramaniam, A. CFD-Based Operational Thermal Efficiency Improvement of a Production Data Center. *SustainIT* **2010**.
20. Kuffi, K.D.; Defraeye, T.; Nicolai, B.M.; De Smet, S.; Geeraerd, A.; Verboven, P. CFD modeling of industrial cooling of large beef carcasses. *Int. J. Refrig.* **2016**, *69*, 324–339. [CrossRef]
21. Launder, B.; Spalding, D. The Numerical Computation of Turbulent Flows. In *Numerical Prediction of Flow, Heat Transfer, Turbulence and Combustion*; Pergamon: Amsterdam, The Netherlands, 1983; pp. 96–116.
22. Troshko, A.; Hassan, Y. A two-equation turbulence model of turbulent bubbly flows. *Int. J. Multiph. Flow* **2001**, *27*, 1965–2000. [CrossRef]
23. Lamrini, B.; Della Valle, G.; Trelea, I.C.; Perrot, N.; Trystram, G. A new method for dynamic modelling of bread dough kneading based on artificial neural network. *Food Control* **2012**, *26*, 512–524. [CrossRef]
24. Wallhäußer, E.; Hussein, M.; Becker, T. Detection methods of fouling in heat exchangers in the food industry. *Food Control* **2012**, *27*, 1–10. [CrossRef]

foods

MDPI

Article

Effect of Apple Polyphenols on the Antioxidant Activity and Structure of Three-Dimensional Printed Processed Cheese

Yiqiu Deng [1], Guangsheng Zhao [2], Kewei Cheng [3], Chuanchuan Shi [4] and Gongnian Xiao [1,*]

[1] Key Laboratory of Agricultural Products Chemical and Biological Processing Technology, Zhejiang University of Science and Technology, Hangzhou 310023, China; 212003817007@zust.edu.cn
[2] Hangzhou New Hope Shuangfeng Dairy Products Co., Ltd., Hangzhou 311100, China
[3] Hangzhou Institute for Food and Drug Control, Hangzhou 310017, China; jhckw@163.com
[4] Panda Dairy Group Co., Ltd., Wenzhou 325400, China; shichuanchuan1105@126.com
* Correspondence: xiaogongnian@126.com; Tel./Fax: +86-571-85070370

Abstract: Additives can influence the processability and quality of three-dimensional (3D)-printed foods. Herein, the effects of apple polyphenols on the antioxidant activity and structure of 3D-printed processed cheese were investigated. The antioxidant activities of processed cheese samples with different contents of apple polyphenols (0%, 0.4%, 0.8%, 1.2%, or 1.6%) were evaluated using 2,2′-azinobis-(3-ethylbenzothiazoline-6-sulfonic acid) (ABTS) and 2,2-di(4-*tert*-octylphenyl)-1-picrylhydrazyl (DPPH) assays. In addition, the rheological properties and structural characteristics of the processed cheeses were investigated using rheometry, Fourier transform infrared spectroscopy, and fluorescence spectroscopy. Then, the final printed products were analyzed for comparative molding effects and dimensional characteristics. it was found that apple polyphenols can significantly improve the antioxidant activity of processed cheese. When the amount of apple polyphenols added was 0.8%, the 3D shaping effect was optimal with a porosity rate of 4.1%. Apple polyphenols can be used as a good antioxidant additive, and the moderate addition of apple polyphenols can effectively improve the antioxidant and structural stability of 3D-printed processed cheese.

Keywords: apple polyphenols; three-dimensional printing; processed cheese; antioxidant activity; structural stability

Citation: Deng, Y.; Zhao, G.; Cheng, K.; Shi, C.; Xiao, G. Effect of Apple Polyphenols on the Antioxidant Activity and Structure of Three-Dimensional Printed Processed Cheese. *Foods* **2023**, *12*, 1731. https://doi.org/10.3390/foods12081731

Academic Editors: Min Zhang and Hong-Wei Xiao

Received: 18 March 2023
Revised: 9 April 2023
Accepted: 20 April 2023
Published: 21 April 2023

1. Introduction

Three-dimensional (3D)-printed food, which can be digitally designed, has an important role in the field of food processing and production [1]. The increasing demand for functional foods has greatly contributed to new 3D printing applications in the food industry [2]. Hot-extrusion printing, which is currently the most widely used 3D printing technology in the food sector, has good compatibility with traditional food materials [3]. Compared with traditional food manufacturing processes, 3D printing technology can meet the demand for personalized customized functional food by preparing food ink with functional active ingredients or additives added to the raw materials [4].

Cheese, as a nutritious dairy product, can be roughly divided into two categories: natural cheese and processed cheese. Processed cheese is a homogeneous and easy-to-store dairy product made by heating and stirring different natural cheeses with added emulsifying salts and other dairy or non-dairy ingredients [5]. Compared with natural cheese, processed cheese is easier to store, more delicious, and can effectively improve the functional and structural characteristics of raw milk cheese by adding food additives [6]. With the continuous growth of cheese consumption in China and in response to the demand for healthy functional food from consumers, the combination of processed cheese and 3D printing technology can provide new ideas for expanding new and high-value functional processed cheese products.

Polyphenols are secondary metabolites of plants and have advantages such as antioxidant and antibacterial properties, which can endow food with good functional characteristics [7]. Among them, apple polyphenols are a common polyphenolic compound with strong antioxidant properties. Studies have shown that the antioxidant properties of apple polyphenols can effectively improve the function of the liver and intestines and prevent liver and digestive system diseases. For example, Huang et al. [8] added apple polyphenols to pig feed, which activated the Nrf2/Keap1 signaling pathway, improved the antioxidant capacity of the pig intestine, and effectively improved the mechanical and immune barriers of the intestine. This can provide effective references for the development of high-value functional 3D-printed processed cheese.

Related research reports that apple polyphenols can also form complex gels with protein macromolecules, but their structure may be affected by apple polyphenols, thereby affecting the properties and functions of the gel [9]. For example, Zhou et al. [10] found that apple polyphenols have a significant impact on the gel properties of whey protein isolate, and moderate addition of apple polyphenols can effectively improve its structural characteristics. It is worth noting that the gel properties of food are also an important factor affecting the 3D printing process. Therefore, while fully utilizing the antioxidant advantages of apple polyphenols, it is also necessary to further understand the influence of apple polyphenols on the gel structure of food and its application value.

Based on 3D printing technology and using Anjia cream cheese as the main raw material, this experiment intends to use apple polyphenols to improve processed cheese and develop high-antioxidant 3D-printed processed cheese, exploring the effects of apple polyphenols on the antioxidant and structural stability of 3D-printed processed cheese, and providing effective references for the production and application of functional 3D-printed processed cheese.

2. Materials and Methods

2.1. Reagents

Apple polyphenols (75% purity; food grade) were obtained from Shanghai Yuanye Biological Co., Ltd. Anjia cream cheese (15% protein, 57% fat, 1% carbohydrate, and 14% sodium; food grade) was obtained from Shanghai Ever Natural Trading Co., Ltd. (Shanghai, China), TG (food grade) was obtained from Hefei Bomei Biotechnology Co., Ltd. (Hefei, China), Emulsifying salts (sodium polyphosphate and sodium pyrophosphate; food grade) were obtained from Zhejiang Noyi Biotechnology Co., Ltd. (Haining, China), Phosphate buffer solution (PBS) was obtained from Fly Clean Biotechnology Co., Ltd. (Shanghai, China), Ethanol and KBr were obtained from Shanghai Lingfeng Chemical Reagent Co., Ltd. (Shanghai, China), 2,2′-azinobis-(3-ethylbenzothiazoline-6-sulfonic acid) (ABTS) and 2,2-di(4-tert-octylphenyl)-1-picrylhydrazyl (DPPH) were obtained from Sangon Biotech Co., Ltd. (Shanghai, China).

2.2. Sample Preparation

Anjia cream cheese (50 g) was mixed with 1.5% emulsifying salts (sodium pyrophosphate and sodium polyphosphate, 1:1, v/v) and 25% distilled water in a constant temperature water bath at 80 °C for 20 min. The mixture was placed under a high-speed homogenizer and homogenized intermittently at 6500 rpm for 2 min until uniform and non-grainy in appearance. After cooling to room temperature, 0.4% TG was added and then different amounts of apple polyphenols were added (0%, 0.4%, 0.8%, 1.2%, or 1.6%). The mixture was placed in a constant temperature water bath at 55 °C for 30 min, sterilized at 85 °C for 30 min, cooled, refrigerated at 4 °C for 24 h, and then used for 3D printing and sample analysis.

2.3. Determination of Antioxidant Activity

2.3.1. Preparation of Sample Solution

pH 7.4, 10 mmol/L phosphate buffer solution was prepared and used as a solvent to prepare 1 mg/mL of processed cheese stock solution. This was refrigerated at 4 °C and used as the sample solution for subsequent determination of antioxidant capacity.

2.3.2. 2,2-Di(4-tert-octylphenyl)-1-picrylhydrazyl (DPPH) Assay

First, a 0.06 mmol/L DPPH solution was prepared. A blank was prepared by mixing 4 mL of the DPPH solution with 0.2 mL of ethanol. The initial absorbance of the blank (A_0) was measured at 515 nm. The test solution was prepared by mixing 4 mL of the DPPH solution with 0.2 mL of sample. After incubating at room temperature and protecting from light for 1 h, the absorbance of the test solution (A_p) was measured at 515 nm. A control solution was prepared by mixing 4 mL of ethanol with 0.2 mL of sample, and the absorbance of this solution (A_c) was measured at 515 nm [11]. Parallel experiments were performed three times. The DPPH free radical scavenging rate was calculated as follows:

$$\text{DPPH free radical scavenging rate (\%)} = (A_0 - A_p - A_c)/A_0 \times 100\% \tag{1}$$

2.3.3. 2,2′-Azinobis-(3-ethylbenzothiazoline-6-sulfonic Acid) (ABTS) Assay

A 7 mmol/L ABTS free radical solution and a 2.45 mmol/L potassium persulfate solution were prepared and mixed in equal volumes. The mixture was placed in the dark for 12 h and then diluted with anhydrous ethanol to obtain an ABTS working solution with an absorbance of 0.70 ± 0.02 at 732 nm. A 1.0 mL sample solution of processed cheese with different apple polyphenol contents of 1.0 mg/mL was mixed with 3.9 mL of the ABTS working solution and reacted for 6 min at room temperature. Anhydrous ethanol was used as a blank instead of the sample solution, and the absorbance of the reaction sample was measured at 734 nm [11]. Parallel experiments were performed three times. The ABTS free radical scavenging rate was calculated as follows:

$$\text{ABTS free radical scavenging rate (\%)} = (1 - A_1/A_0) \times 100 \tag{2}$$

where A_1 is the absorbance value of the processed cheese sample after reacting with ABTS, and A_0 is the absorbance value of anhydrous ethanol instead of the sample solution after reacting with ABTS.

2.4. Determination of Rheological Characteristics

Rheological characterization was performed according to the method of Huang et al. [12] with slight modifications. The dynamic rheological properties were evaluated using a rheometer (DHR-2, TA Instruments, New Castle, DE, USA) to determine the effect of different apple polyphenol contents on the rheological characteristics of the cheese samples. A 40 mm diameter plate was used with a gap of 1 mm. The samples were placed on the platform for 1 min to warm up to 25 °C and then subjected to dynamic frequency scanning at angular frequencies of 0.1–100 rad/s. In addition, to evaluate the printability of the processed cheese samples, apparent viscosity profiles were collected. The apparent viscosity was recorded as the shear rate was scanned in the range of 0.1–100 s^{-1} at 25 °C.

2.5. Determination of Structural Properties

2.5.1. Fourier Transform Infrared (FT-IR) Spectroscopy

Processed cheese samples with different amounts of apple polyphenols were placed in a freezer at −80 °C for 24 h. After freeze-drying, the processed cheese samples were ground into powder and mixed with KBr (1:100, v/v). FT-IR spectra were recorded at room temperature using an FT-IR spectrometer (V70 IR Spectrometer, Bruker, Mannheim, Germany) [13].

2.5.2. Endogenous Tryptophan Fluorescence

The fluorescence spectra of the processed cheese samples with different apple polyphenol contents were measured using a fluorescence spectrometer (F4500, Hitachi, Tokyo, Japan) according to the method of Geng et al. [14]. The samples were diluted to 0.1 mg/mL using 0.5 mol/L phosphate buffer solution (PBS) and then placed in a cuvette. Fluorescence spectra were recorded in the range of 280–400 nm using an excitation wavelength of 295 nm, an excitation spacing of 5 nm, an emission spacing of 5 nm, and a data acquisition rate of 500 nm/min.

2.5.3. Scanning Electron Microscopy

A scanning electron microscope (SU1510, Hitachi, Japan) was used to observe the microstructure of the processed cheese samples with different apple polyphenol contents. The samples were freeze-dried at −80 °C for 24 h, fixed on a copper plate, and then sprayed with gold. Scanning electron micrographs were obtained at a magnification of 1000× and an accelerating voltage of 15 kV [15].

2.6. Measurement of 3D Printing Molding Effects

To evaluate the 3D printing and molding performance of the processed cheese samples, a food 3D printer (FOODBOT-S2, Hangzhou Shiyin Technology Co., Ltd., Hangzhou, China) was used and a geometric model of a horse was selected from the food 3D printer gallery. The model volume was (48 mm × 28 mm × 3 mm). The 3D printer parameters were as follows: nozzle diameter, 0.8 mm; nozzle temperature, 25 °C; fill rate, 10%; and print rate, standard. The size features of the printed product were measured using Vernier calipers, and the porosity of samples was determined using Image J software. Parallel experiments were performed three times [16].

2.7. Color Analysis

The L*, a*, and b* values of the finished product were measured using a spectrophotometer to analyze its color. The L* value represents brightness, with higher values indicating a whiter sample; the a* value represents the red-green degree, with higher values indicating a more red sample; the b* value represents the yellow-blue degree, with higher values indicating a deeper yellow color in the sample [16].

2.8. Sensory Evaluation

According to the method of Lao et al. [17], sensory evaluation was conducted on cheese samples with apple polyphenols added. Sensory evaluation indicators were divided into: color (0–3 points), appearance (0–3 points), flavor (0–4 points), and a total of 10 points. If the cumulative score exceeds 7 points, the overall acceptability of the product is considered high, and the average score of all people is used as the final score of the product.

2.9. Statistical Analysis

The statistical analysis was performed using SPSS software (version 26.0, IBMCorp., Armonk, NY, USA). The data were collated and plotted using Origin 9.1. software.

3. Results and Discussion

3.1. DPPH and ABTS Free Radical Scavenging Rates

The antioxidant activities of the processed cheese samples with added apple polyphenols were evaluated by both DPPH and ABTS methods. As shown in Figure 1, the DPPH free radical scavenging rate increased gradually with the addition of apple polyphenols. The DPPH free radical scavenging rate was positively correlated to the polyphenol content, and the highest scavenging rate of 71.9% was achieved at an apple polyphenol content of 1.6%. When the content of apple polyphenols is greater than 0.4%, the DHHP free radical scavenging rate is above 70%. Thus, apple polyphenols can effectively improve the antioxidant activity of processed cheese. The free radical scavenging activity of polyphenols is attributable to their free phenolic hydroxyl groups, which have a strong ability to provide protons that can reduce highly oxidative free radicals and terminate free radical chain reactions [18].

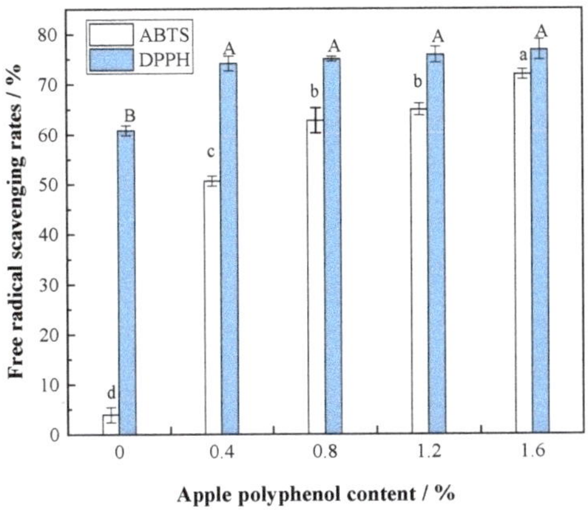

Figure 1. Effect of apple polyphenol addition on the free radical scavenging rates (%) of 2,2-di(4-*tert*-octylphenyl)-1-picrylhydrazyl (DPPH) and 2,2′-azinobis-(3-ethylbenzothiazoline-6-sulfonic acid) (ABTS) in processed cheese. Different letters in the figure indicate significant differences between values ($p < 0.05$).

The ABTS free radical scavenging rates of the processed cheese samples were also positively correlated to the amount of added apple polyphenols (Figure 1). The differences in the radical scavenging rates of ABTS and DPPH are due to differences in solubility. As ABTS is water-soluble, it is more suitable for the determination of water-soluble substances, whereas DPPH is soluble in alcohols and is thus more suitable for the determination of alcohol-soluble substances [19]. The highest ABTS free radical scavenging rate of 76.8% was achieved at an apple polyphenol content of 1.6%. When the content of apple polyphenols is greater than 0.4%, the ABTS free radical scavenging rate is above 60%. These results confirm that the addition of apple polyphenols can effectively improve the antioxidant activity of processed cheese.

3.2. Rheological Analysis

The rheological characteristics of the processed cheese samples with added apple polyphenols were investigated. As shown in Figure 2A, the apparent viscosity of all the processed cheese samples gradually decreased with increasing shear rate. Thus, irrespective of the apple polyphenol content, the processed cheese samples exhibited shear thinning, which indicates that the gelation properties are suitable for food 3D printing [20]. This behavior is due to the main phenolic substances in apple polyphenols, including chlorogenic acid and proanthocyanidins, which bind to proteins through hydrogen bonding and hydrophobic interactions to form aggregates [21]. As the content of apple polyphenols increased, the apparent viscosity gradually decreased, which may be due to the interactions between polyphenols and proteins reducing the viscosity of the processed cheese gel system. Notably, the viscosity of a material has an important influence on its 3D printing properties.

Figure 2B shows the energy storage modulus (G′), which is indicative of the elasticity of the protein gel network structure, whereas Figure 2C shows the loss modulus (G″, also known as the viscosity modulus), which provides a measure of gel viscosity [22]. As shown in Figure 2B, when the apple polyphenol content increases from 0% to 1.6%, the G′ values of the processed cheese samples first increased and then decreased as the apple polyphenol content increased, with the maximum value observed at an apple polyphenol content of 0.8%. Thus, within a certain content range, the addition of apple polyphenols promoted the formation of a protein gel network structure and improved the elasticity of processed cheese, but excess polyphenols caused the elasticity of the protein gel system to decrease. As shown in Figure 2C, the G″ values of the processed cheese samples tended to decrease as the apple polyphenol content increased. However, when the apple polyphenol content was 0.8%, 1.2%, and 1.6%, the G′ values were always greater than the G″ values, indicating that apple polyphenols promote the formation of a cross-linked protein structure that provides enhanced elasticity.

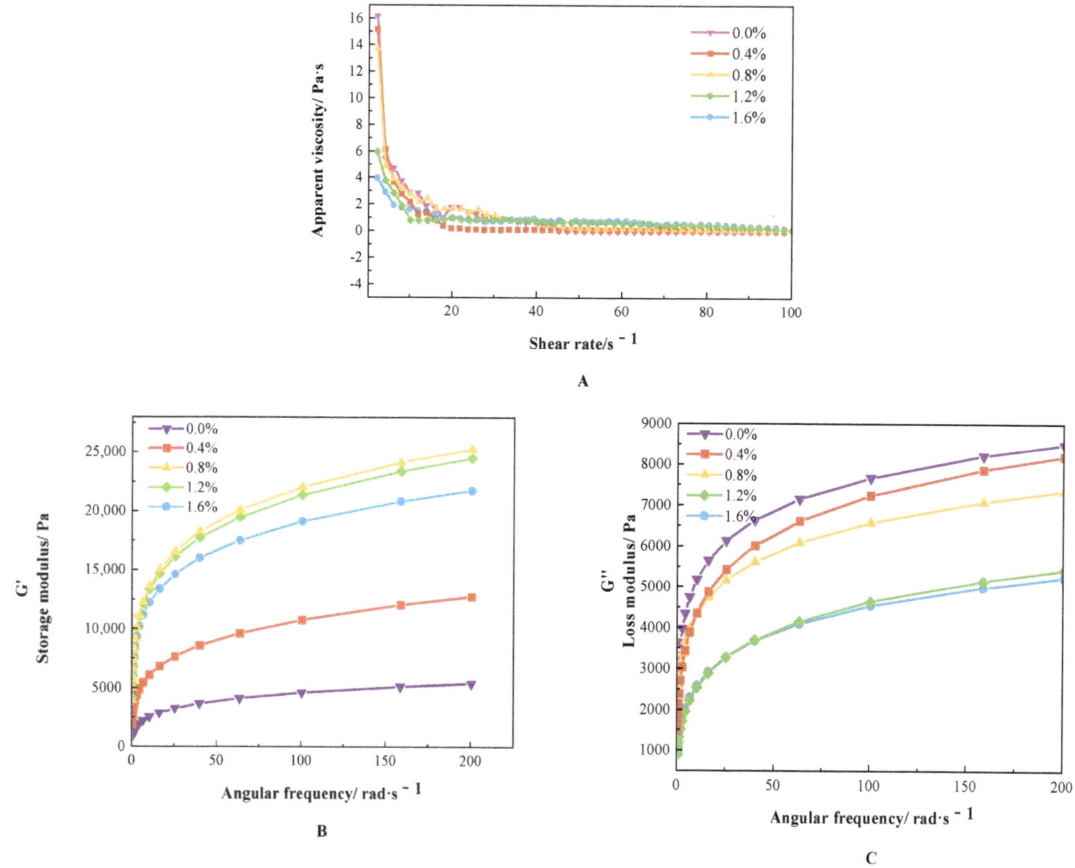

Figure 2. (**A**) Effect of added apple polyphenols on the apparent viscosity curves of processed cheese; (**B**,**C**) Effect of added apple polyphenols on the dynamic rheological properties of processed cheese: (**B**) storage modulus (G′) and (**C**) loss modulus (G″).

3.3. Structural Analysis

3.3.1. Scanning Electron Microscopy

The addition of apple polyphenols significantly affected the microstructure of processed cheese (Figure 3). The gel network structure of processed cheese without added apple polyphenols was loose with obvious surface folds and cavities (Figure 3A). With the addition of apple polyphenols, the protein network structure gradually became tighter. Compared with the sample without apple polyphenols (Figure 3A), the samples with added apple polyphenols had flatter and denser structures with some protein cross-linking (Figure 3B–E). Previous studies on protein–polyphenol interactions found that the presence of apple polyphenols enhances the tight binding of protein gels [23], suggesting that the added polyphenols act as a protein cross-linker and thus promote the formation of a cheese gel structure with enhanced mechanical strength.

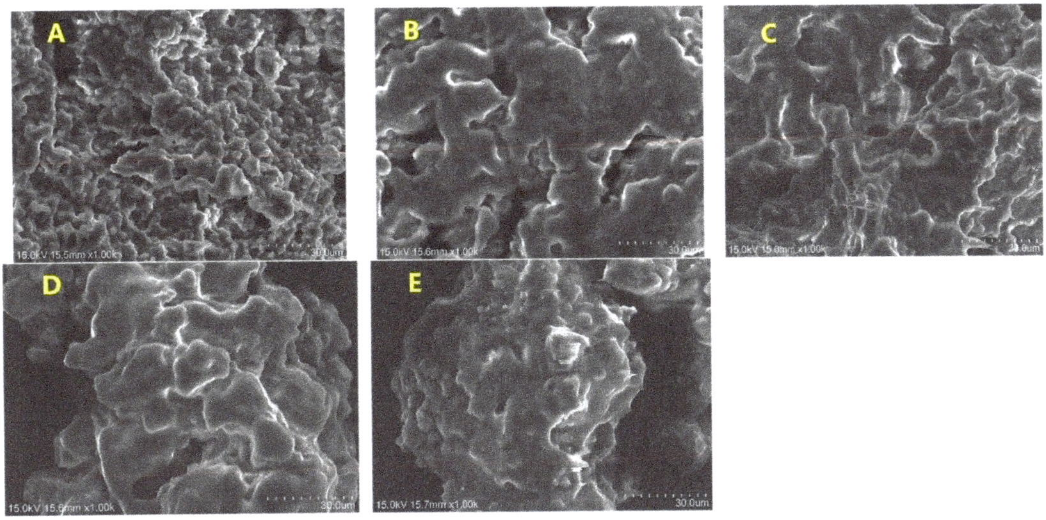

Figure 3. Scanning electron micrographs of processed cheese samples with (**A**) 0.0%, (**B**) 0.4%, (**C**) 0.8%, (**D**) 1.2%, and (**E**) 1.6% apple polyphenols.

3.3.2. FT-IR Spectroscopy

The effect of apple polyphenol addition on the structural characteristics of processed cheese was further investigated using FT-IR spectroscopy. As shown in Figure 4A, the FT-IR spectra did not change significantly after the addition of apple polyphenols, indicating that no new functional groups were generated and that the polyphenols were bound to the proteins through non-covalent bonds. The intensities of various characteristic peaks varied depending on the apple polyphenol content, possibly due to changes in the secondary structure of the protein [24]. One of the most useful spectral bands for the analysis of protein secondary structure and conformation is the amide I band (1600–1700 cm^{-1}) due to its sensitivity to hydrogen bonding patterns and dipole–dipole interactions [25,26]. Peaks corresponding to various secondary structures, including β-folds (1611–1640 cm^{-1}), random curls (1642–1650 cm^{-1}), α-helices (1654–1662 cm^{-1}), and β-turns (1665–1693 cm^{-1}), are located within the spectral range of the amide I band. Therefore, to clarify the effect of apple polyphenols on the secondary structure of processed cheese proteins, PeakFit software was used to calculate the relative percentages of various secondary structures based on the integrated areas of their peaks.

The changes in the gelation properties and strength of the processed cheese samples were related to the changes in the relative content of each protein secondary structure. A higher α-helix content results in a more rigid and tight protein structure, whereas a higher β-fold content results in a more ordered and stable protein structure [27]. As shown in Figure 4B, at apple polyphenol contents of 0.4% and 0.8%, the random curl content decreased, but the α-helix and β-fold contents increased to some extent, indicating a protective effect on the protein gel structure [28], which is consistent with the scanning electron microscopy results (Figure 3A–E). Upon further increasing the apple polyphenol content (>0.8%), the α-helix content gradually decreased, which may be due to excess polyphenols hindering the formation of hydrogen bonds in the protein gel network while promoting protein unfolding and the conversion of α-helices to β-turns [29]. Therefore, the addition of an appropriate amount of apple polyphenols to processed cheese could lead to improved structural stability.

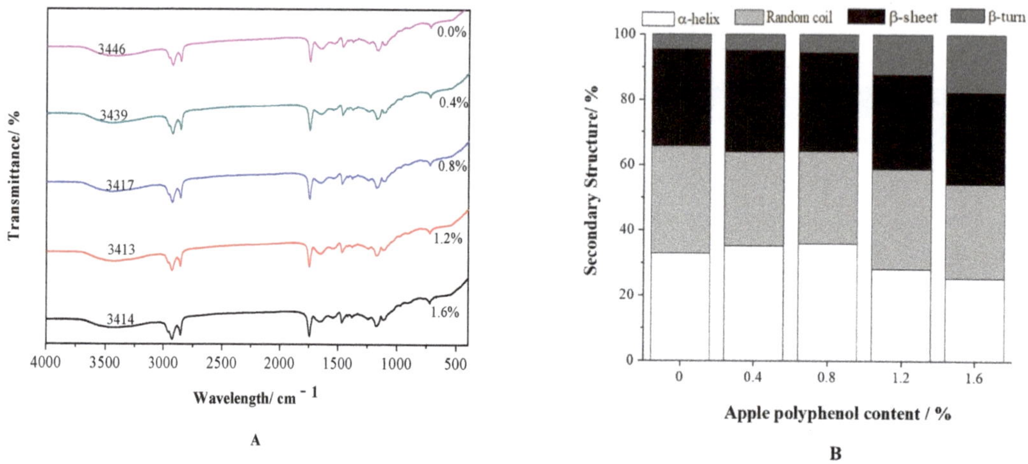

Figure 4. (**A**) Fourier transform infrared (FT-IR) spectra of processed cheese samples with added apple polyphenols; (**B**) Effect of apple polyphenol addition on the protein secondary structure of processed cheese.

3.3.3. Fluorescence Spectroscopy

Endogenous tryptophan fluorescence is often used to reveal protein conformation. When a protein is in a folded state, tryptophan residues are present in hydrophobic structures within the protein, resulting in relatively high fluorescence intensity. In contrast, when a protein is partially or completely unfolded, the tryptophan residues are more exposed to the protein surface, resulting in decreased fluorescence intensity [30]. As shown in Figure 5, the fluorescence of the proteins in the processed cheese samples group was gradually quenched as the amount of added apple polyphenols increased. This behavior indicates that apple polyphenols induce protein unfolding in processed cheese, thus exposing more tryptophan residues and reducing the fluorescence intensity, which is consistent with FT-IR spectroscopy results.

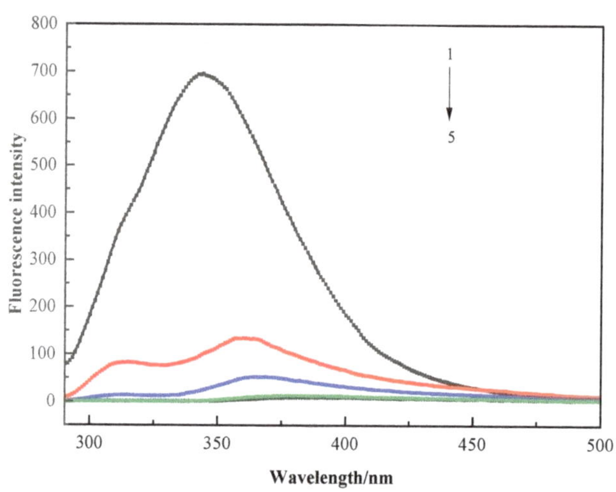

Figure 5. Effect of apple polyphenol addition on the fluorescence spectra of processed cheese. Labels 1 to 5 indicate apple polyphenol content of 0% to 1.6%.

3.4. D Printing Performance and Colour Analysis

To assess the 3D printing performance of the processed cheese samples with added apple polyphenols, 3D printing was carried out using a horse model. As shown in Table 1, after the addition of apple polyphenols, the best 3D printing effect was achieved at an apple polyphenol content of 0.8%, in which the porosity was 4.1%. In this case, the sample was well extruded. This sample exhibited closely stacked layers, no obvious collapse, a smooth surface, and good deformation resistance with no obvious deviations from the blank group sample. At higher apple polyphenol contents (1.2% and 1.6%), the effect of polyphenols on the protein gel network of processed cheese could lead to a decrease in the viscosity of the material, resulting in a poor forming effect and a rougher shape; in these cases, the porosity was 10.6% and11.1%, respectively. These samples showed gaps, indicating that the ductility was poor when the filaments were extruded, and broken lines appeared, causing the upper layers to be affected by the lower layers, which is consistent with the rheological properties of the samples.

Table 1. Size characteristics and molding effects of 3D-printed processed cheese with different apple polyphenol content.

Addition (%)	Image	Length (mm)	Width (mm)	Height (mm)	Porosity (%)
0.0		48.13 ± 0.01 ^{c}	27.96 ± 0.05 ^{b}	2.95 ± 0.01 ^{a}	4.16 ± 0.06 ^{b}
0.4		48.22 ± 0.03 ^{a}	28.33 ± 0.21 ^{b}	2.95 ± 0.02 ^{a}	4.23 ± 0.08 ^{b}
0.8		48.12 ± 0.02 ^{c}	27.93 ± 0.06 ^{b}	2.95 ± 0.03 ^{a}	4.1 ± 0.10 ^{b}
1.2		48.15 ± 0.05 ^{b}	29.67 ± 0.13 ^{a}	2.92 ± 0.03 ^{a}	10.6 ± 0.46 ^{a}
1.6		48.17 ± 0.08 ^{b}	30.33 ± 0.58 ^{a}	2.93 ± 0.02 ^{a}	11.1 ± 0.77 ^{a}

Different letters in the same line indicate significant differences between values ($p < 0.05$).

The effect of apple polyphenols on the color of the 3D-printed processed cheese samples is shown in Table 2. The addition of apple polyphenols had a significant effect on the sample color ($p < 0.05$). Specifically, apple polyphenol addition resulted in lower L* and b* values but higher a* values, indicating that the samples became less bright, darker, and more yellow in color, and these changes were positively correlated to the polyphenol content.

Table 2. Color analysis of 3D-printed processed cheese with different apple polyphenol content.

Addition (%)	L*	a*	b*
0.0	84.69 ± 0.03 [a]	−1.12 ± 0.01 [e]	9.56 ± 0.01 [b]
0.4	62.03 ± 0.02 [b]	−0.08 ± 0.02 [d]	12.89 ± 0.01 [a]
0.8	55.03 ± 0.01 [c]	0.21 ± 0.01 [c]	8.71 ± 0.01 [d]
1.2	52.02 ± 0.01 [d]	0.51 ± 0.01 [b]	8.52 ± 0.02 [e]
1.6	50.81 ± 0.01 [e]	1.35 ± 0.01 [a]	8.84 ± 0.00 [c]

Different letters in the same line indicate significant differences between values ($p < 0.05$). The L* value is the brightness, the a* value is the redness and greenness, and the b* value is the yellowness and blueness.

3.5. Sensory Evaluation

Figure 6 shows the effect of different contents of apple polyphenols on the sensory quality of 3D-printed processed cheese products. According to the product's color score, the appearance score of the product with 1.6% apple polyphenol content is the lowest, while the color score of the product with 0.4% apple polyphenol content is the highest, which is consistent with the results of the color analysis. According to the appearance score, the product with 0.8% apple polyphenol content has the highest appearance score, while the product with 1.6% apple polyphenol content has the lowest appearance score, which is consistent with the results of the size characteristics analysis. According to the flavor score, the product with 0.8% apple polyphenol content has the highest flavor score, but there is no significant difference in flavor compared to the product with 0.4% apple polyphenol content. The product with 1.6% apple polyphenol content has the lowest flavor score; this is because the excessive addition of apple polyphenols will present a certain bitterness. The overall scores of the products with 0.4% and 0.8% apple polyphenol contents exceeded the critical value of consumer acceptance (7 points). However, because the appearance and flavor scores of the product with 0.8% apple polyphenol content are higher than those of the product with 0.4% apple polyphenol content, the product with 0.8% apple polyphenol content has a higher consumer acceptance.

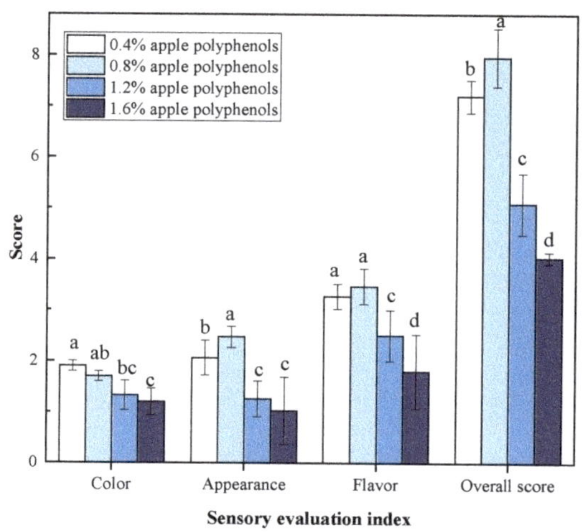

Figure 6. Effect of different apple polyphenol contents on the sensory evaluation of 3D-printed processed cheese samples. Different letters in the figure indicate significant differences between values ($p < 0.05$).

4. Conclusions

This study applied apple polyphenols to the 3D printing of processed cheese, and explored the effects of different apple polyphenol addition amounts on the antioxidant and structural stability of 3D-printed processed cheese. The addition of apple polyphenols can effectively improve the antioxidant capacity of 3D-printed processed cheese. When the content of apple polyphenols is greater than 0.4%, the ABTS free radical scavenging rate is above 60%, and the DHHP free radical scavenging rate is above 70%. At the same time, polyphenol molecules change the secondary structure of the protein in processed cheese by binding to it, and excessive addition of apple polyphenols will cause the viscosity of the processed cheese to continue to decrease and the gel properties to deteriorate, affecting the printing effect of the product. When the amount of added apple polyphenols is 0.8%, the 3D printing effect is optimal with a porosity rate of 4.1%. Overall, when the content of apple polyphenols is 0.8%, the resulting processed cheese can have both good antioxidant capacity and structural stability for 3D printing. This provides a new way for personalized antioxidant functional food.

Author Contributions: Methodology, Y.D.; investigation, C.S.; data curation, G.Z.; formal analysis, K.C.; writing—original draft preparation, Y.D. and K.C.; writing—review and editing, G.Z. and C.S.; funding acquisition, C.S. and G.X. All authors have read and agreed to the published version of the manuscript.

Funding: This study was funded by the Major Science and Technology Projects of Zhejiang Province (2020C02041), Zhejiang Province New Products Manufacture Projects (2022D60SA3B12123), Research on a new technology for personalised reconstituted cheese (2022GG27), and the Chick Eagle project of the Marketing Bureau (CY2022345).

Data Availability Statement: No new data were created or analyzed in this study. Data sharing in not applicable to this article.

Conflicts of Interest: The author declares no conflict of interest.

References

1. Enfield, R.; Pandya, J.; Lu, J.; McClements, D.; Kinchla, A. The Future of 3D Food Printing: Opportunities for Space Applications. *Crit. Rev. Food Sci. Nutr.* **2022**, *20*, 7299. [CrossRef] [PubMed]
2. Liu, Z.; Bhandari, B.; Zhang, M. Incorporation of probiotics (*Bifidobacterium animalis* subsp. *Lactis*) into 3D printed mashed potatoes: Effects of variables on the viability. *Food Res. Int.* **2020**, *128*, 108795. [CrossRef]
3. An, Y.; Guo, C.; Zhang, M.; Zhong, Z. Investigation on characteristics of 3D printing using *Nostoc sphaeroides* biomass. *J. Sci. Food Agric.* **2019**, *99*, 639–646. [CrossRef] [PubMed]
4. Wang, L.; Zhang, M.; Bhandari, B.; Yang, C. Investigation on fish surimi gel as promising food material for 3D printing. *J. Food Eng.* **2018**, *220*, 101–108. [CrossRef]
5. Johnson, M.E.; Kapoor, R.; McMahon, D.J.; McCoy, D.R.; Narasimmon, R.G. Reduction of Sodium and Fat Levels in Natural and Processed Cheeses: Scientific and Technological Aspects. *Compr. Rev. Food Sci. Food Saf.* **2009**, *8*, 252–268. [CrossRef]
6. Talbot-Walsh, G.; Kannar, D.; Selomulya, C. A review on technological parameters and recent advances in the fortification of processed cheese. *Trends Food Sci. Tech.* **2018**, *81*, 193–202. [CrossRef]
7. Lu, Y.; Du, Y.; Qin, X.; Wu, H.; Huang, Y.; Cheng, Y. Comprehensive evaluation of effective polyphenols in apple leaves and their combinatory antioxidant and neuroprotective activities. *Ind. Crops Prod.* **2019**, *129*, 242–252. [CrossRef]
8. Huang, T.; Che, Q.; Chen, X.; Chen, D.; Yu, B.; He, J.; Chen, H.; Yan, H.; Zheng, P.; Luo, Y. Apple Polyphenols Improve Intestinal Antioxidant Capacity and Barrier Function by Activating the Nrf2/Keap1 Signaling Pathway in a Pig Model. *J. Agric. Food Chem.* **2022**, *70*, 7576–7585. [CrossRef]
9. Li, H.; Pan, Y.; Lan, Y.; Yang, Z.; Rao, J.; Chen, B. Molecular interaction mechanism and structure–activity relationships of protein–polyphenol complexes revealed by side-directed spin labeling-electron paramagnetic resonance (SDSL-EPR) spectroscopy. *Food Chem.* **2023**, *402*, 134354. [CrossRef]
10. Zhou, Q.; Zhang, H.; Zhao, Y.; Zhao, M.; Wang, S. Effect of apple polyphenols on the gel properties of whey protein isolate. *Food Hydrocol.* **2019**, *90*, 433–440. [CrossRef]
11. Gülçin, I.; Huyut, Z.; Elmastas, M.; Aboul-Enein, H.Y. Radical scavenging and antioxidant activity of tannic acid. *Arab. J. Chem.* **2010**, *3*, 43–53. [CrossRef]
12. Huang, Q.; Huang, X.; Liu, L.; Wang, G.; Song, H.; Geng, F.; Luo, P. Effect of nano eggshell calcium on the structure, physicochemical, and gel properties of threadfin bream (*Nemipterus virgatus*) actomyosin. *LWT Food Sci. Technol.* **2021**, *150*, 112047. [CrossRef]
13. Mandalari, G.; Tomaino, A.; Arcoraci, T.; Martorana, M.; Turco, V.; Cacciola, F.; Rich, G.T.; Bisignano, C.; Saija, A.; Dugo, P.; et al. Characterization of polyphenols, lipids and dietary fibre from almond skins (*Amygdalus communis* L.). *J. Food Compos. Anal.* **2010**, *23*, 166–174. [CrossRef]

14. Geng, F.; Xie, Y.; Wang, Y.; Wang, J. Depolymerization of chicken egg yolk granules induced by high-intensity ultrasound. *Food Chem.* **2021**, *354*, 129580. [CrossRef]
15. Wang, J.; Liu, X.; Li, S.; Ye, H.; Luo, W.; Huang, Q.; Geng, F. Ovomucin may be the key protein involved in the early formation of egg-white thermal gel. *Food Chem.* **2022**, *366*, 130596. [CrossRef]
16. Yang, F.; Zhang, M.; Liu, Y. Effect of Post-Treatment Microwave Vacuum Drying on the Quality of 3D-Printed Mango Juice Gel. *Dry Technol.* **2019**, *37*, 1757–1765. [CrossRef]
17. Lao, Y.; Zhang, M.; Devahastin, S.; Ye, Y. Effect of Combined Infrared Freeze Drying and Microwave Vacuum Drying on Quality of Kale Yoghurt Melts. *Drying Technol.* **2020**, *38*, 621–633. [CrossRef]
18. Staroszczyk, H.; Kusznierewicz, B.; Malinowska-Pańczyk, E.; Sinkiewicz, I.; Gottfried, K.; Kołodziejska, I. Fish gelatin films containing aqueous extracts from phenolic-rich fruit pomace. *LWT Food Sci. Technol.* **2020**, *117*, 108613. [CrossRef]
19. Riaz, A.; Lei, S.; Akhtar, H.M.S.; Wan, P.; Chen, D.; Jabbar, S.; Abid, M.; Hashim, M.M.; Zeng, X. Preparation and characterization of chitosan-based antimicrobial active food packaging film incorporated with apple peel polyphenols. *Int. J. Biol. Macromol.* **2018**, *114*, 547–555. [CrossRef]
20. Chen, H.; Xie, F.; Chen, L.; Zheng, B. Effect of rheological properties of potato, rice and corn starches on their hot-extrusion 3D printing behaviors. *J. Food Eng.* **2019**, *244*, 150–158. [CrossRef]
21. Millet, M.; Poupard, P.; Guilois-Dubois, S.; Poiraud, A.; Fanuel, M.; Rogniaux, H.; Guyot, S. Heat-unstable apple pathogenesis-related proteins alone or interacting with polyphenols contribute to haze formation in clear apple juice. *Food Chem.* **2020**, *309*, 125636. [CrossRef] [PubMed]
22. Kim, T.; Lee, M.; Yong, H.; Jang, H.; Jung, S.; Choi, Y. Impacts of fat types and myofibrillar protein on the rheological properties and thermal stability of meat emulsion systems. *Food Chem.* **2021**, *346*, 128930. [CrossRef]
23. Zhang, L.; Li, X.; Janaswamy, S.; Chen, L.; Chi, C. Further insights into the evolution of starch assembly during retrogradation using SAXS. *Int. J. Biol. Macromol.* **2020**, *154*, 521–527. [CrossRef] [PubMed]
24. Xu, Y.; Han, M.; Huang, M.; Xu, X. Enhanced heat stability and antioxidant activity of myofibrillar protein-dextran conjugate by the covalent adduction of polyphenols. *Food Chem.* **2021**, *352*, 129376. [CrossRef]
25. Anvari, M.; Chung, D. Dynamic rheological and structural characterization of fish gelatin–gum Arabic coacervate gels cross-linked by tannic acid. *Food Hydrocoll.* **2016**, *60*, 516–524. [CrossRef]
26. Jia, J.; Gao, X.; Hao, M.; Tang, L. Comparison of binding interaction between β-lactoglobulin and three common polyphenols using multi-spectroscopy and modeling methods. *Food Chem.* **2017**, *228*, 143–151. [CrossRef]
27. You, G.; Niu, G.; Long, H.; Zhang, C.; Liu, X. Elucidation of interactions between gelatin aggregates and hsian-tsao gum in aqueous solutions. *Food Chem.* **2020**, *319*, 126532. [CrossRef] [PubMed]
28. Yang, Y.; Liu, G.; Wang, H. Investigation of the mechanism of conformational alteration in ovalbumin as induced by glycation with different monoses through conventional spectrometry and liquid chromatography high-resolution mass spectrometry. *J. Agric. Food Chem.* **2019**, *67*, 3096–3105. [CrossRef] [PubMed]
29. Lin, D.; Xiao, L.; Wen, Y.; Qin, W.; Wu, D.; Chen, H.; Zhang, Q.; Zhang, Q. Comparison of apple polyphenol-gelatin binary complex and apple polyphenol-gelatin-pectin ternary complex: Antioxidant and structural characterization. *LWT Food Sci. Technol.* **2021**, *148*, 111740. [CrossRef]
30. Gan, X.; Li, H.; Wang, Z.; Emara, A.; Zhang, D.; He, Z. Does protein oxidation affect proteolysis in low sodium Chinese traditional bacon processing. *Meat Sci.* **2019**, *150*, 14–22. [CrossRef]

foods

MDPI

Article

Effect of Vanillin on the Anaesthesia of Crucian Carp: Effects on Physiological and Biochemical Indices, Pathology, and Volatile Aroma Components

Lexia Jiang [1,2], Jiaming Tang [1,2], Baosheng Huang [1,3,4,*], Changfeng Zhang [1,3], Peihong Jiang [1,3] and Dongjie Chen [1,3]

1 Key Laboratory of Agricultural Products Storage, Transportation and Preservation Technology of Shandong Province, Shandong Institute of Commerce and Technology, Jinan 250103, China; 13977695205@163.com (L.J.); tang19980503@163.com (J.T.); zcf202@163.com (C.Z.); jiangpeihongjph@163.com (P.J.); dongjie613@163.com (D.C.)
2 College of Food Science and Technology, Guangdong Ocean University, Zhanjiang 524088, China
3 National Agricultural Products Modern Logistics Engineering Technology Research Center, Jinan 250103, China
4 Shandong Guonong Logistics Technology Co., Ltd., Jinan 250103, China
* Correspondence: huangbs@stu.hubu.edu.cn

Abstract: The anaesthetic effect of vanillin on crucian carp was investigated using different concentrations of vanillin, with a nonvanillin control. The effective concentration range of vanillin anaesthesia was determined from the behavioural characteristics of crucian carp during the anaesthesia onset and recovery phases. Physiological and biochemical indices, and the electronic nose response to the fish muscle, were measured over the range of effectiveanaestheticc concentrations. An increased concentration of vanillin shortened the time taken to achieve deep anaesthesia but increased the recovery time. The levels of white blood cells, red blood cells, haemoglobinn, platelets, alanine aminotransferase, alkaline phosphatase, lactate dehydrogenase, phosphorus, potassium, magnesium, total protein, and serum albumin were lower than the control in the vanillin treatment group. Triglycerides and total cholesterol were not significantly affected. Histology showed no effect of vanillin on the liver, except at 1.00 g/L vanillin. Vanillin resulted in a nondose-responsive effect on the gill tissue, increasing the width and spacing of the gill lamellae. E-Nose analysis of the carp-muscle flavour volatiles was able to distinguish between different vanillin treatment concentrations. GC-IMS identified 40 flavour compounds, including 8 aldehydes, 11 alcohols, 10 ketones, 2 esters, and 1 furan. Vanillin had aanaestheticic effect on crucian carp and these findings provide a theoretical basis for improving the transport and experimental manipulation of crucian carp.

Keywords: vanillin; crucian carp; anesthetic effect; physiological and biochemical indices; pathology; volatile components

Citation: Jiang, L.; Tang, J.; Huang, B.; Zhang, C.; Jiang, P.; Chen, D. Effect of Vanillin on the Anaesthesia of Crucian Carp: Effects on Physiological and Biochemical Indices, Pathology, and Volatile Aroma Components. *Foods* **2023**, *12*, 1614. https://doi.org/10.3390/foods12081614

Academic Editor: Jean-Christophe Jacquier

Received: 10 March 2023
Revised: 7 April 2023
Accepted: 7 April 2023
Published: 11 April 2023

1. Introduction

Freshwater fish are popular with consumers for their high-protein and low-fat contents, and there is increasing consumer demand for high-quality fish. During the production and transportation of fish, trauma and stress often occur, affecting their quality and survival rate. Fish anaesthetics are widely used, as they have a sedative effect during transport, and effectively reduce fish mortality. Commonly used fish anaesthetics in China are ethyl m-amino benzoate (MS-222), eugenol, and 2-phenoxyethanol, however, all three anaesthetics have undesirable side effects. MS-222 solution is weakly acidic, resulting in the plasma-cortisol content increasing after the fish recover from deep anaesthesia; exposure to direct sunlight may produce toxicity to fish when using MS-222 [1]. Eugenol is volatile and its effect diminishes during anaesthesia [2,3]; 2-Phenoxyethanol is hazardous to fishery workers, causing neurological syndromes after prolonged exposure, has a severe, long-lasting residual effect, and remains active for three days after the fish recover [4]. Therefore, there is a need for research into safer anaesthetics for fishery use, such as those of a plant origin, which have antimicrobial and antioxidant activities, and carry a low risk to human health [5,6]. Plant extracts, such as alfalfa essential oil [7], thyme essential oil [8], and menthol [9] all have anaesthetic effects on fish.

Vanillin is the common name for 3-methoxy-4-hydroxybenzaldehyde, which is extracted from the seeds of orchids of the genus *Vanilla* [10]. Annual consumption of vanillin in China is 2000–2500 t, with consumption increasing every year [11]. Vanillin has biological activities, such as antibacterial, anti-inflammatory, and sedative. Vanillin improved the resistance of papaya to invasion by pathogenic bacteria, thereby improving fruit quality and improved resistance to papaya postharvest rot; vanillin can be used as a safe, nontoxic inducer and preservative to inhibit the growth of rot pathogens [12]. Vanillin encapsulated in nanofiber membranes extended the shelf life of turbot fillets at 4 °C [13]. Inhalation of vanillin by mice had no effect on activity and cognitive function, though inhalation of essential oils containing vanillin produced effective sedation [14]. Vanillin alleviated rotational behavioural symptoms, reduced dopamine neuron damage, inhibited microglial activation in a rat model, and had a therapeutic effect in an animal model of lipopolysaccharide-induced Parkinson's disease [15]. The antimicrobial activity of vanillin reduced the heat resistance of *E. coli* [16]. Vanillin has human health-promoting and medicinal properties and is also an important raw material for the synthesis of many drugs. Vanillin has been reported to reduce infarct volume after hypoxic-ischaemic brain injury, by reducing blood–brain barrier damage and oxidative damage [17]. Vanillin has neuroprotective effects by regulating the expression of inflammatory cytokines in mice with transient middle cerebral artery occlusion, demonstrating that vanillin can inhibit the TLR4/NF-κB signalling pathway by an anti-inflammatory mechanism [18]. Vanillin improved survival and reduced oxidative stress in rats suffering from sepsis [19]. In this study, the anaesthetic effect of vanillin was tested on crucian carp and the resulting changes in physiological and biochemical parameters, pathology and volatile aroma compounds were investigated.

2. Materials and Methods

2.1. Materials and Reagents

Live crucian carp (*Carassius auratus*) from the Jinan seafood market (Jinan, China); fish were selected for good health and active movement. The fish were maintained in a temperature-controlled circulating water filtration system (dissolved oxygen $\geq$ 7 mg/L, pH 7–8), at the National Agricultural Products Modern Logistics Engineering Technology Research Center. Vanillin (purity $\geq$ 99%), was from Sinopharm Chemical Reagents (Shanghai, China). Potassium (K^+), sodium (Na^+), glucose (GLU), total protein (TP), albumin (ALB), total cholesterol (CHO), and triglyceride (TG) assay kits were all from Shandong Boke Biological Industries (Jinan, China). Xylene and neutral resin were from National Group Chemical Reagents (Shanghai, China). Hematoxylin and eosin (H&E) staining solution, differentiation solution, and blue return solution were from Wuhan Xavier Biotechnology (Wuhan, China). Ethylene glycol ethyl ether acetate was from Shanghai Macklin Biochemical Technology (Shanghai, China).

2.2. Experimental Methods

2.2.1. Determination of Effective Concentration of Vanillin for Anaesthesia of Crucian Carp

The effective concentration for this determination was defined as the onset of anaesthesia within 3 min and awakening within 5 min [20]. Six treatment concentrations (0.25, 0.50, 0.75, 1.00, 1.25, and 1.50 g/L) were used and 60 carp were randomly selected in total for the treatment and control groups. When the fish entered the anaesthesia phase (A_3), the corresponding time was recorded as anaesthesia onset, then the fish were immediately transferred to fresh water for recovery and the time was recorded as the onset of recovery phase 3 (R_3).

2.2.2. Determination of Behavioural Characteristics during the Recovery Phases of Anaesthesia, at the Optimal Vanillin Concentration

When the anaesthetic concentration is low, the fish cannot enter anaesthesia quickly; when the anaesthetic concentration is high, the fish enter anaesthesia too quickly and the behavioural characteristics of the anaesthesia are difficult to distinguish. To observe stable and distinct anaesthetic staging characteristics, an appropriate anaesthetic concentration is required. Based on the results from Section 2.2.1, the behavioural characteristics of the carp during anaesthesia were observed in a 1.00 g/L vanillin solution. The anaesthesia (A_0–A_4) and recovery (R_1–R_4) periods were classified according to the criteria of Mirghaed et al. [21] and correlated with the behavioural characteristics of the different periods.

2.2.3. Effect of Vanillin Concentration on Blood and Serum Parameters of Crucian Carp

Three crucian carp were randomly placed in treatment (different vanillin concentrations) and control tanks, then each fish was tested for anaesthesia (each fish was tested only once). The fish were kept in a state of complete anaesthesia for 10 min, and then blood samples (5 mL) were drawn with a syringe from the tail vein. Part of the sample was mixed with an anticoagulant and used for the determination of the physiological parameters, red blood cells (RBC), white blood cells (WBC), haemoglobin (HGB), and platelets (PLT); the other part was centrifuged at 4 °C for 20 min and 4000 rpm. Then, the supernatant was stored at −80 °C until needed for the determination of serum alanine aminotransferase (ALT), aspartate aminotransferase (AST), alkaline phosphatase (ALP), lactate dehydrogenase (LDH), inorganic phosphate (P_i), magnesium (Mg^{2+}), potassium (K^+), sodium (Na^+), glucose (GLU), total protein (TP), albumin (ALB), cholesterol (CHO), and triglycerides (TG).

2.2.4. H&E Staining and Histological Examination of Liver and Gill Tissues

The carp were put under complete anaesthesia with different concentrations of vanillin, maintained for 10 min and then gently removed and placed on a dissecting table to remove the liver and gill parts for H&E staining. All samples were cut into 5–6 μm thick slices. Then H&E was stained for microscopic observation. Dehydration and transparency were performed in ethanol and xylene, and then cover slipped. The stained sections were observed and photographed using a light microscope (Nikon, DS-Fi2) and spliced into a complete image using ImageJ software.

2.2.5. E-nose Analysis

The carp were put under complete anaesthesia with different concentrations of vanillin, maintained for 10 min, and then gently removed and placed on a dissecting table to remove the muscle for the electronic nose test. The E-Nose (FOX4000, Alpha MOS, Toulouse, France) sampling was performed as described previously [22], with some modifications. Muscle tissue (5.0 g) was homogenized, then sealed into 10 mL sample vials and analyzed each sample 4 times. The samples were analyzed as follows: carrier gas velocity, 150 mL/min; headspace generation temperature, 40 °C; injection volume, 2 mL; injection speed, 2 mL/s; headspace generation time, 600 s; data acquisition time, 120 s; delay time, 400 s. Radar charts and linear discriminant analysis (LDA) were used to analyse the data and remove outliers. The response characteristics of each E-Nose sensor are shown in Table 1.

2.2.6. Gas Chromatography-Ion Mobility Spectrometry (GC-IMS) Analysis of Flavour Volatile Compounds

The carp were put under complete anaesthesia with different concentrations of vanillin, maintained for 10 min and then gently removed and placed on a dissecting table to extract muscle parts for GC-IMS studies. The analysis was performed with a FlavourSpec flavour analyser (G.A.S., Dortmund, Germany). The control software has built-in NIST and IMS databases, which were used for analyte identification. A 4.0 g sample was placed in a 20.0 mL headspace vial and incubated for 15 min at 60 °C. Headspace sampling conditions were as follows: headspace temperature, 60 °C; incubation time, 15 min; heating mode, oscillation; headspace injection needle temperature, 85 °C; injection volume, 500 µL; non-split mode; carrier gas, N_2 (purity ≥ 99.999%); and wash time 0.50 min. Chromatography conditions were as follows: column temperature, 60 °C; run time, 15 min; carrier gas, N_2 (≥99.999%); flow rate: initially 5.0 mL/min, held for 10 min, and then linearly increased to 150 mL/min over 5 min. Detection conditions were as follows: drift tube length, 5 cm linear voltage in the tube, 400 V/cm; drift tube temperature, 40 °C; drift gas, N_2, (≥99.999%); flow rate, 150 mL/min; and IMS detector temperature, 45 °C.

Table 1. Response characteristics of E-Nose sensors.

Serial Number	Sensor Name	Sensor Response Characteristics
1	LY2/LG	chlorine, fluorine, nitrogen oxides, sulfides
2	LY2/G	ammonia, amine compounds, carbon oxides
3	LY2/AA	ethanol, acetone, ammonia
4	LY2/GH	ammonia, amine compounds
5	LY2/gCTL	hydrogen sulfide
6	LY2/gCT	propane, butane
7	T30/1	polar compounds, hydrogen chloride
8	P10/1	nonpolar; hydrocarbons, ammonia, chlorine
9	P10/2	nonpolar; methane, ethane
10	P40/1	fluorine, chlorine
11	T70/2	toluene, xylene, carbon monoxide
12	PA/2	ethanol, ammonia, amine compounds
13	P30/1	hydrocarbons, ammonia, ethanol
14	P40/2	chlorine, hydrogen sulfide, hydrogen fluoride
15	P30/2	hydrogen sulfide, ketones
16	T40/2	chlorine
17	T40/1	fluorine
18	TA/2	ethanol

2.2.7. Statistical Analysis

SPSS 26.0 (SPSS Inc., Chicago, IL, USA) was used for data analysis. Data are expressed as the mean ± standard deviation, according to an analysis of variance (ANOVA) and Duncan's multiple-range test ($p < 0.05$). The figures were plotted using OriginPro 2021 (OriginLab, Northampton, MA, USA). GC-IMS data were collected and analyzed from different perspectives using the Laboratory Analytical Viewer (LAV) software and three plug-ins (Reporter, Gallery Plot, Dynamic PCA) as well as the GC-IMS Library Search qualitative software.

3. Results

3.1. Determination of the Effective Vanillin Anaesthesia Concentration Range for Crucian Carp

As the concentration of vanillin increased, the time taken to achieve deep anaesthesia (stage A5; anaesthesia time) decreased and the time for complete recovery increased (Table 2, Figure 1); the anaesthesia time with 0.25 g/L vanillin was markedly longer than at other concentrations. The average anaesthesia time at 1.50 g/L was 50% of that at 0.50 g/L, whereas the average recovery time was 281% of that at 0.50 g/L. The recovery rate of carp was 100% at all six vanillin concentrations, indicating that, within the concentration range of 0.50–1.50 g/L, the vanillin bath was safe for the fish. As the vanillin concentration increased, the anaesthesia time decreased, in a dose-responsive, though nonlinear manner (Table 2). Taking market application as the starting point, the short average anaesthesia time is the focus of market demand, and it can be seen that the anaesthesia efficiency is higher when the test concentration is 0.50 g/L~1.50 g/L.

Table 2. Effect of different vanillin concentrations on the time taken to achieve anaesthesia and time taken for complete recovery for crucian carp.

Vanillin Concentration/(g/L)	Test the Number of Fish/Tail	Length/cm	Body Mass/g	Anaesthesia Time/min	Recovery Time/min	Recovery Rate/100%
0.25	10	16.1 ± 0.35	271.7 ± 37.86	33.00 ± 8.54	2.58 ± 0.13	100
0.50	10	16.3 ± 0.95	273.3 ± 79.11	2.48 ± 0.03	2.13 ± 0.12	100
0.75	10	16.6 ± 0.96	291.6 ± 68.07	1.64 ± 0.35	2.21 ± 0.22	100
1.00	10	15.8 ± 0.41	263.4 ± 70.71	1.42 ± 0.52	2.46 ± 0.07	100
1.25	10	16.6 ± 0.80	288.3 ± 25.65	1.36 ± 0.13	2.48 ± 0.02	100
1.50	10	17.8 ± 0.72	310.1 ± 13.22	1.24 ± 0.09	4.59 ± 0.27	100

Figure 1. Appearance of crucian carp at the five stages of anaesthesia.

Behavioural Characteristics of Crucian Carp during the Five Stages of Anaesthesia Onset and the Four Stages of Recovery

The timing and behavioural characteristics at each stage of anaesthesia (A_0–A_4) and recovery (R_1–R_4) of crucian carp were determined using 1.00 g/L vanillin (Table 3). When a fish stopped swimming, completely lost the ability to respond, and the operculum opened weakly and slowly, it was defined as being at the A_3 stage of the anaesthesia period, and when the fish body was balanced and responsive, and the operculum opened normally, it was defined as being at the R_3 stage of recovery.

Table 3. Behavioural characteristics of crucian carp during the successive stages of anaesthesia and recovery.

Stages	Behavioural Characteristics	Minute
A_0 stress period	A stress response occurs, swimming is accelerated, and operculum opening and closing are accelerated	0.40 ± 0.06
A_1 sedation period	The response to external stimuli is weakened, the ability to swim is weakened, the body is slightly out of balance, and the breathing rate is further increased	0.75 ± 0.12
A_2 Mild anaesthesia phase	The body rolls on its side, the ability to respond to external stimuli continues to weaken, swimming slowly, and the rate of operculum opening and closing decreases	1.09 ± 0.15
A_3 anaesthesia period	The body of the fish is out of balance, ventral face up, stationary, and the operculum opening is reduced but continuous	2.60 ± 0.13
A_4 deep anaesthesia period	The body of the fish is stationary, and the operculum opens and closes extremely slowly and irregularly	4.45 ± 0.42
R_1 recovery stage 1	The ventral side of the fish is stationary and breathing begins to slowly resume continuously	1.44 ± 0.12
R_2 recovery stage 2	The fish can swim slowly laterally, but the sense of direction is not clear, and the frequency of operculum opening is close to that before anaesthesia	1.91 ± 0.10
R_3 recovery stage 3	The fish body was completely restored to its preanaesthesia state, and the operculum and upper and lower jaw opening frequency returned to normal	3.61 ± 0.18
R_4 recovery stage 4	The operculum opens and closes normally, fully returns to normal swimming, and responds rapidly to stimuli	4.78 ± 0.27

3.2. *Effect of Vanillin Concentration on Blood and Serum Parameters of Crucian Carp*

3.2.1. Effect of Vanillin Concentration on the Blood Composition Index of Crucian Carp

The control WBC count was 809.76 $\pm$ 21.51 $\times$ 10^9/L; the vanillin treatment WBC count initially markedly decreased, then slightly increased, with the increased vanillin concentration (Figure 2A). The control RBC count was 0.17 $\pm$ 0.10 $\times$ 10^{12}/L and markedly decreased at all vanillin treatment concentrations, with a minimum of 0.75 g/L (Figure 2B). The control HGB content was 95.11 $\pm$ 4.59 g/L and moderately decreased, then slightly increased, with increased vanillin treatment concentration (Figure 2C). The control blood PLT content was 27.22 $\pm$ 4.63 $\times$ 10^9/L and markedly decreased at all vanillin treatment concentrations; the PLT content was not significantly different between the five vanillin concentrations (Figure 2D).

3.2.2. Effect of Vanillin Concentration on Blood Serum Enzymes in Crucian Carp

The control serum ALT concentration was 97.73 $\pm$ 58.18 U/L; the vanillin treatment ALT concentration initially markedly decreased, then slightly, though not significantly, increased, with increased vanillin concentration, except for a marked increase at 1.50 g/L vanillin, to 111.77 $\pm$ 21.97 U/L (Figure 3A). The control serum AST concentration was 440.4 $\pm$ 177.39 U/L; the vanillin treatment AST concentration markedly increased, though remained relatively constant, at all vanillin treatment concentrations (Figure 3B). The control serum ALP concentration was 26.67 $\pm$ 12.66 U/L; the vanillin treatment ALT concentration initially markedly decreased, then gradually increased, with the increased vanillin concentration (Figure 3C). The control serum LDH concentration was 2560.02 $\pm$ 7.23 U/L and moderately decreased at all vanillin treatment concentrations, with a minimum of 1.00 g/L (Figure 3D).

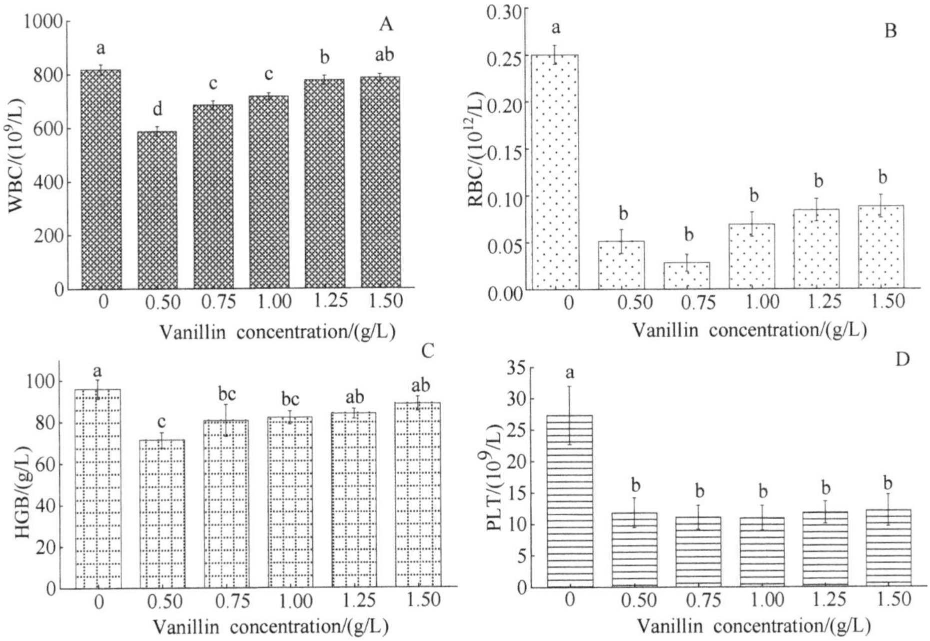

Figure 2. Effect of vanillin concentration on the blood composition of crucian carp: (**A**) white blood cells (WBC); (**B**) red blood cells (RBC); (**C**) haemoglobin (HGB); and (**D**) platelets (PLT). Different letters above the columns indicate a significant difference ($p < 0.05$); the same letter indicates no significant difference ($p > 0.05$).

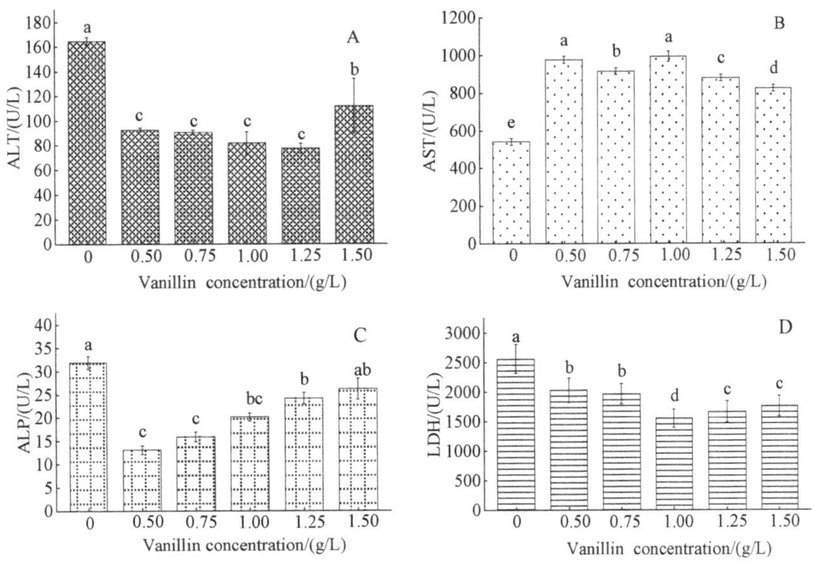

Figure 3. Effect of vanillin concentration on serum enzyme concentrations in crucian carp: (**A**) alanine aminotransferase (ALT); (**B**) aspartate aminotransferase (AST); (**C**) alkaline phosphatase (ALP); and (**D**) lactate dehydrogenase (LDH). Different letters above the columns indicate a significant difference ($p < 0.05$); the same letter indicates no significant difference ($p > 0.05$).

3.2.3. Effect of Vanillin Concentration on the Blood Serum Ion Concentrations of Crucian Carp

The PO_4^{3-} concentration gradually decreased from the control value with the increased vanillin concentration, with a minimum of 1.25 g/L, and a significant increase at 1.50 g/L (Figure 4A). The K^+ concentration markedly decreased from the control value with increased vanillin concentration, with a minimum of 1.00 g/L, and a significant increase at 1.50 g/L (Figure 4B). The Mg^{2+} concentration gradually decreased from the control value with increased vanillin concentration, with a minimum of 1.25 g/L, and a slight increase of 1.50 g/L (Figure 4C). The Na^+ concentration significantly increased from the control value at all vanillin treatment concentrations, with a gradual increase with increased vanillin concentration to a maximum of 1.25 g/L, and a small decrease at 1.50 g/L (Figure 4D).

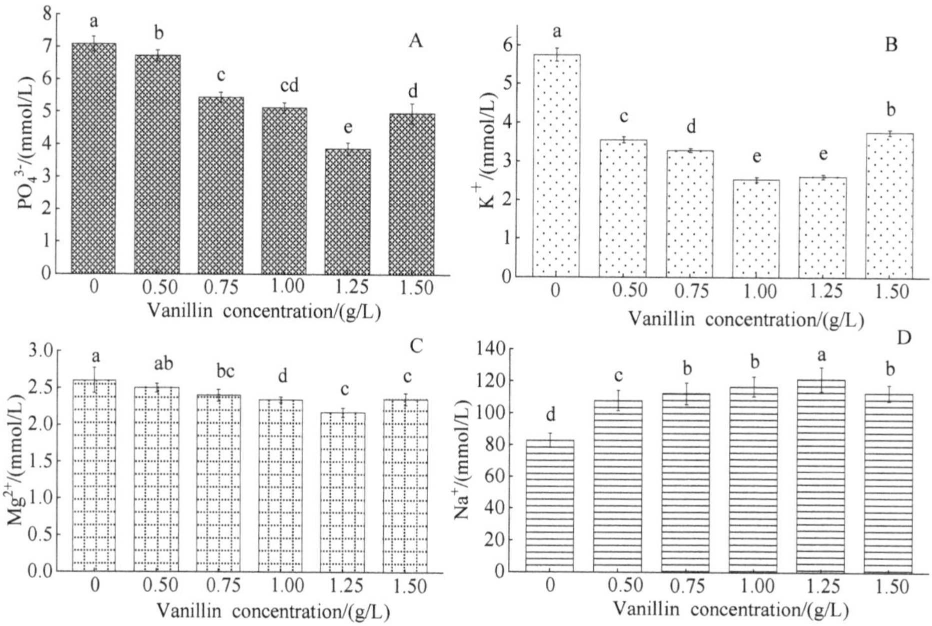

Figure 4. Effect of vanillin concentration on the blood serum ion content of crucian carp: (**A**) phosphate (PO_4^{3-}); (**B**) potassium (K^-); (**C**) magnesium (Mg^{2+}); and (**D**) sodium (Na^+). Different letters above the columns indicate a significant difference ($p < 0.05$); the same letter indicates no significant difference ($p > 0.05$).

3.2.4. Effect of Vanillin Concentration on the Blood Serum Concentrations of Organic Components in Crucian Carp

The control TP concentration of 76.33 ± 6.79 g/L decreased gradually with increasing vanillin treatment concentration to a minimum of 1.00 g/L vanillin, then gradually increased; the variation in ALB content was very similar to that of the TP content (Figure 5A). The CHO concentration of the vanillin treatments was slightly higher than the control (Figure 5B). The GLU concentration varied widely with increased vanillin concentration, increasing markedly at 0.50 g/L with a further increase at 0.75 g/L, followed by a marked decrease to a minimum at 1.25 g/L, and then a small increase at 1.50 g/L (Figure 5C). The control TG concentration slightly increased at 0.50 g/L vanillin, then slightly decreased with increasing vanillin concentration (Figure 5D).

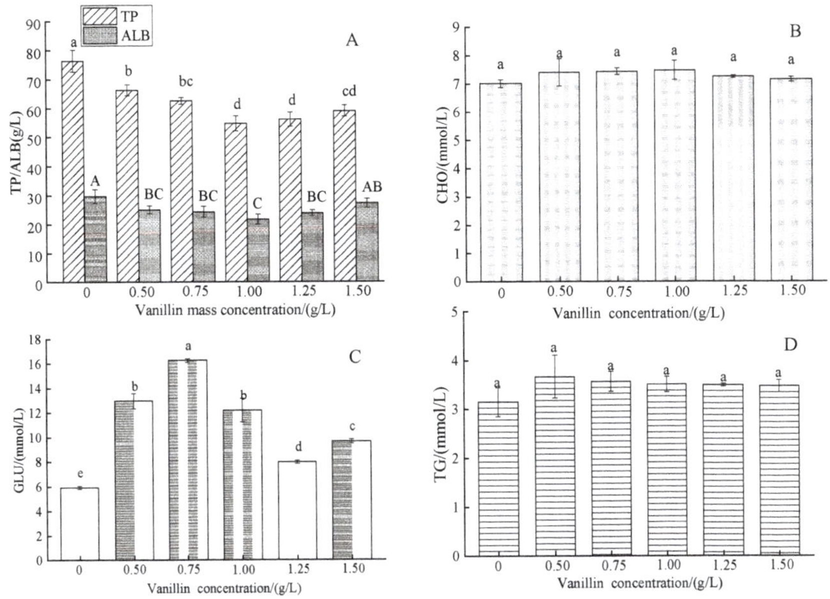

Figure 5. Effect of vanillin concentration on the blood serum concentrations of organic components in crucian carp: (**A**) total protein (TP) and albumin (ALB); (**B**) cholesterol (CHO); (**C**) glucose (GLU); and (**D**) triglycerides (TG). Different letters above the columns indicate a significant difference ($p < 0.05$); the same letter indicates no significant difference ($p > 0.05$).

3.3. Histopathological Examination

Liver tissue sections (Figure 6) showed that the hepatocytes in the control and treatment groups were evenly arranged, intact and clear, except at 1 g/L vanillin, and vacuolation in the treatment groups was more marked than in the control. At 1.50 g/L vanillin, the accumulation of hepatocyte nuclei was greater than in the control, though their integrity was better.

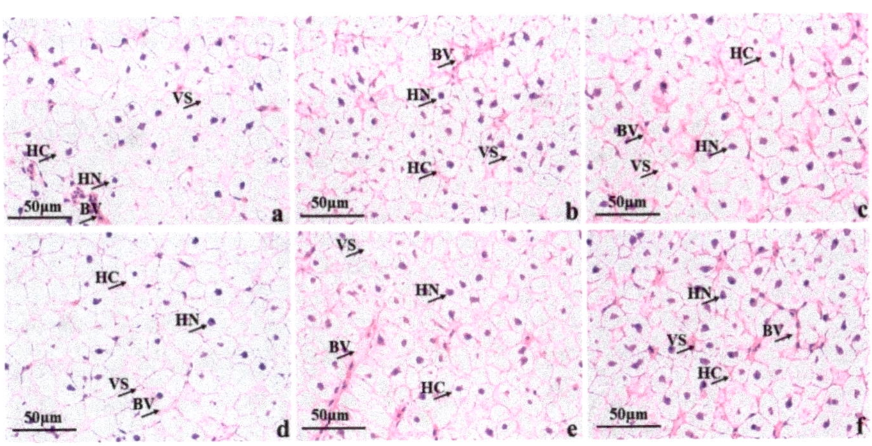

Figure 6. H&E-stained crucian carp liver tissue sections of (**a**) control group; (**b–f**) vanillin treatments at 0.50, 0.75, 1.00, 1.25, and 1.50 mg/L, respectively. Note: VS: vacuoles; HC: Hepatocytes; HN: Hepatocyte nucleus; BV: blood vessels.

Gill tissue sections (Figure 7) showed that the control gill lamellae (GL) were flattened, vesicular, elongated, and arranged in a comb-like pattern, they consisted of a single layer of epithelial cells, capillaries, and columnar support cells. The epithelial cells were flattened with spindle-shaped nuclei, and the columnar support cells had large, round nuclei and were attached to the basement membrane. The filaments and lamellae contained small numbers of chlorine-secreting cells. Compared to the control, the filaments and lamellae were swollen at 1.50 g/L vanillin; there was a proliferation of flattened and columnar cells in the vascular walls of the filaments, an increase in chlorine-secreting cells, curling and fusion of the lamellae, and detachment of the supragillar cells. Microstructural measurements showed a slight swelling of the chlorine-secreting cells at all treatment concentrations, which was not significantly different from the control, except at 1.00 g/L, at which the width and spacing of the gill lamellae increased significantly.

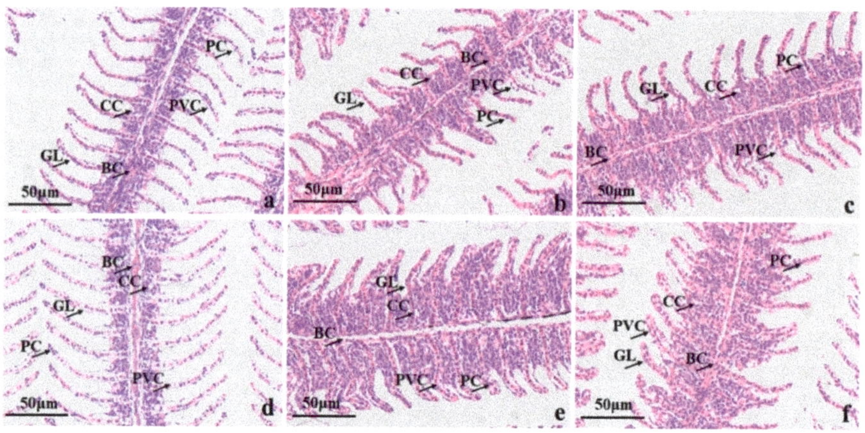

Figure 7. H&E-stained crucian carp gill tissue sections of (**a**) control group; (**b–f**) vanillin treatments at 0.50, 0.75, 1.00, 1.25, and 1.50 mg/L, respectively. Note: BC: blood corpuscle; CC: chlorine cells; GL: gill lamellae; PC: column cells; PVC: flat epithelial cells.

3.4. *Effect of Vanillin Concentration on the Volatile Flavour Compound Profile of Carp Muscle, Determined by E-Nose Analysis*

3.4.1. Radar Fingerprinting of the E-Nose Sensor Responses to Carp-Muscle Flavour Volatiles

The muscle aroma of the anaesthetized crucian carp was analyzed at different vanillin treatment concentrations. The 18 sensors of the FOX4000 E-Nose are sensitive to different chemical classes of volatile compounds and the ratio of relative conductivity G/G_0 is proportional to the concentration of the corresponding compound class. When $G/G_0 > 1$, the volatile concentration is detectable and when $G/G_0 \leq 1$ the volatile concentration is below the detection limit [23]. There was a wide variation in sensor response (Figure 8), with six sensors giving no response and those of T30/1, P10/1, PA/2, and P30/1 being relatively strong. However, the differences between the control and the six vanillin treatment concentrations were very small. It appears that vanillin treatment has a negligible effect on fish flavour, which is highly desirable from a commercial viewpoint.

3.4.2. Linear Discriminant Analysis (LDA) of E-Nose Data

LDA is a statistical method that uses samples of known categories to establish a discriminant model and discriminate samples of unknown categories [24]. The E-Nose data were subjected to an LDA dimension reduction analysis (Figure 9). The contributions of LDA1 and LDA2 were 77.29% and 19.69%, respectively, and the combined contribution was 96.98%. LDA can distinguish the treatment and control samples, though there was no dose response to the vanillin treatment.

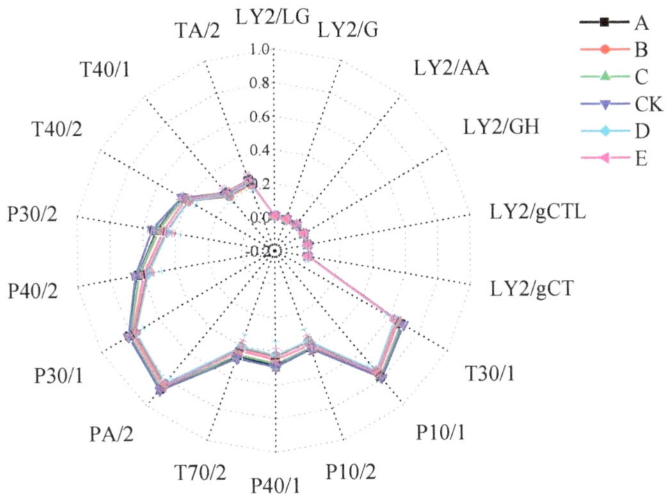

Figure 8. Effect of vanillin concentration on the radar fingerprint of E−Nose responses to carp-muscle flavour volatiles. Note: A, B, C, CK, D, and E represent vanillin concentrations of 0.50, 0.75, 1.00, 0 (control), 1.25, and 1.50 g/L, respectively.

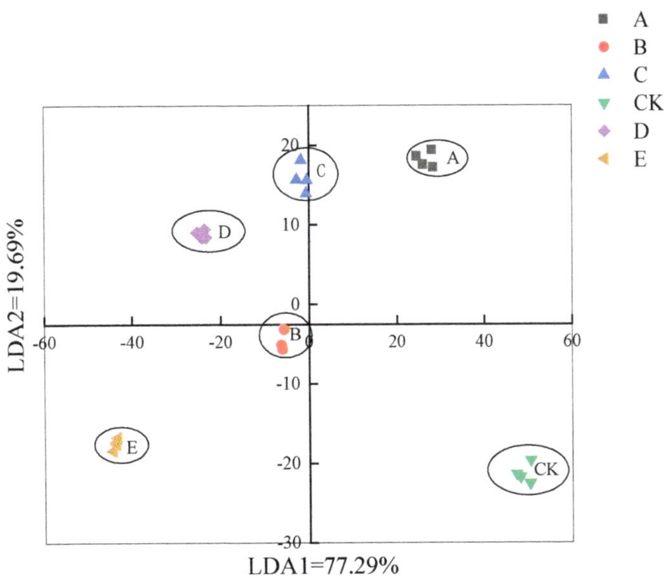

Figure 9. LDA of E−Nose response to crucian carp muscle at different vanillin treatment concentrations. Note: A, B, C, CK, D, and E represent vanillin concentrations of 0.50, 0.75, 1.00, 0 (control), 1.25, and 1.50 g/L, respectively.

3.5. GC-IMS Analysis of Aroma Volatiles from Crucian Carp Muscle at Different Vanillin Treatment Concentrations

The GC-IMS 3D spectra of carp-muscle flavour volatiles resulting from different vanillin treatment concentrations (Figure 10) are not easily distinguished by the eye.

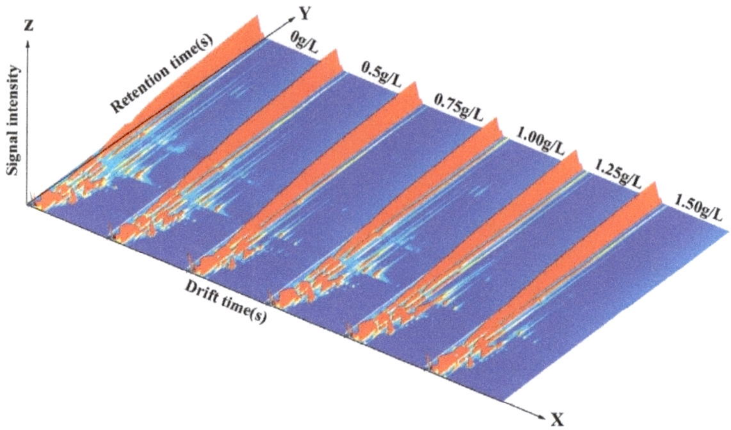

Figure 10. GC-IMS 3D spectra of crucian carp muscle after anaesthesia. From left to right: 0, 0.50, 0.75, 1.00, 1.25, and 1.50 g/L vanillin.

The 2D top-view GC-IMS plots (Figure 11A,B) allow visual comparison of the volatile flavour profiles of carp muscle resulting from different vanillin treatment concentrations and significant differences between samples are visible.

The GC retention time and IMS migration time of the volatile flavour compounds from carp muscle after treatment, with different concentrations of vanillin, were compared with the GC-IMS database on crucian carp muscle volatiles. The library search identified 40 flavour compounds (Figure 12, Table 4), including eight aldehydes, 11 alcohols, 10 ketones, two esters, and one furan.

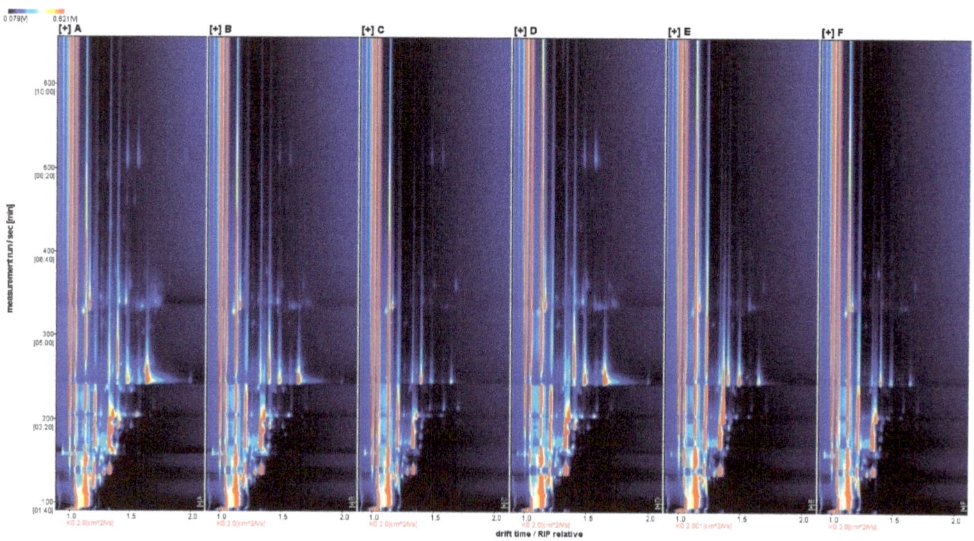

(A)

Figure 11. *Cont.*

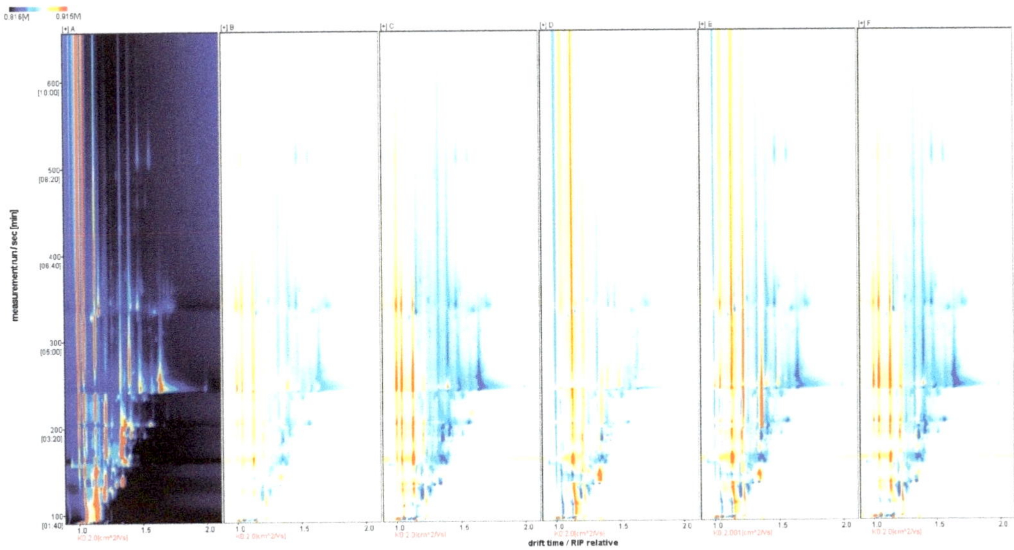

(B)

Figure 11. GC-IMS two-dimensional spectra of flavour volatiles resulting from different vanillin treatment concentrations on crucian carp muscle. Note: (**A**): top view; (**B**): difference plot. From left to right: 0, 0.50, 0.75, 1.00, 1.25, and 1.50 g/L vanillin.

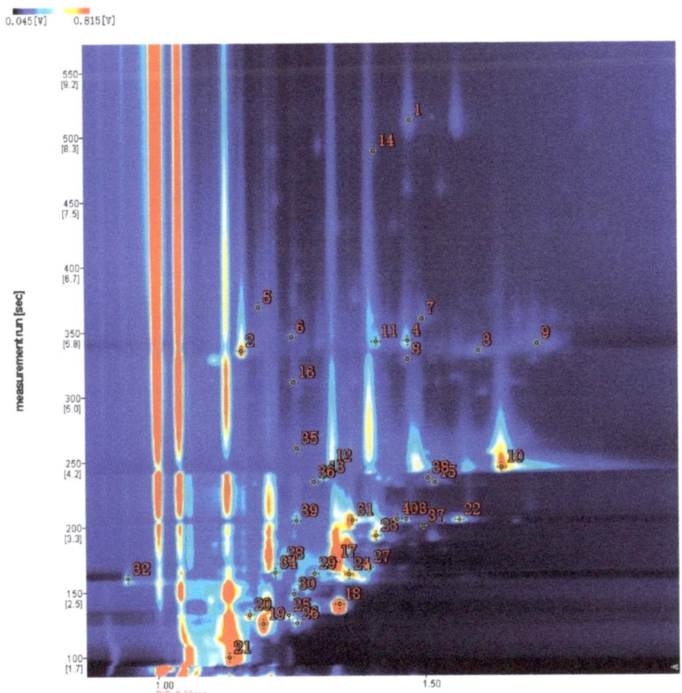

Figure 12. The 2D GC-IMS chromatogram of volatile organic compounds in crucian carp muscle, after treatment with vanillin. The numbers correspond to the identities in Table 4.

Table 4. Volatile organic compounds were identified from crucian carp muscle, after treatment with vanillin.

Category	Characteristic Peak Number	Name of the Compound	CAS#	Retention Index	Retention Time/min	Migration Time/min	Description of the Incense
Aldehydes	1	Nonanal	C124196	1107.9	512.244	1.47608	grease, cucumber and sweet orange flavours
	5	(E,E)-2,4-heptadienal	C4313035	1012.9	368.968	1.19232	aroma of grass and chicken
	8	5-methylfurfural	C620020	974.9	328.377	1.47055	cocoa, almonds
	11	Octanal	C124130	989.8	342.276	1.41258	fat, soap
	15	(E)-2-hexenal	C6728263	844.7	233.629	1.5202	fruity, green and vegetable
	16	(E)-hept-2-enal	C18829555	955.1	310.888	1.25698	aroma of grass and oil
	22	Hexanal-D	C66251	788.4	204.748	1.56613	grassy flavour
	39	Hexanal-M	C66251	787.5	204.354	1.26096	grassy flavour
Alcohols	2	1-octen-3-ol- M	C3391864	982.7	335.632	1.15976	mushroom, lavender, rose and hay aromas
	3	1-octen-3-ol-D	C3391864	982.2	335.107	1.60295	mushroom, lavender, rose and hay aromas
	10	*n*-Hexanol-D	C111273	865.2	245.133	1.64596	herbal flavour
	12	*n*-Hexanol-M	C111273	867	246.147	1.32588	herbal flavour
	13	3-Methyl-1-pentanol	C589355	852.3	237.822	1.31307	fermented taste
	21	ethanol	C64175	421.1	99.501	1.13351	alcohol
	28	2-methylbutan-1-ol	C137326	714.7	171.67	1.23403	aromatic with wine and ether
	31	2,3-Butanediol	C513859	788.4	204.748	1.36608	fermented taste
	32	pent-1-en-3-ol	C616251	684.2	159.959	0.94764	fruity aroma
	36	1-Pentanol,2-methyl	C105306	845.8	234.228	1.29357	fermented taste
	38	(E)-2-hexen-1-ol	C928950	851.7	237.511	1.50757	grassy, fruity
Ketones	9	3-Octanone	C106683	987.9	340.428	1.71197	fruity aroma
	14	2-nonanone	C821556	1094.4	488.937	1.40882	fruity, sweet and green notes
	17	3-hydroxybutan-2-one	C513860	717.3	172.744	1.33455	aromatic smell
	20	2,3-butanedione	C431038	580.2	132.579	1.17293	Fermented aroma, sweet aroma
	24	3-Pentanone	C96220	693.3	163.079	1.36017	sweet scent
	25	2-Butanone	C78933	576.6	131.72	1.24487	aromatic smell
	34	2,3-pentanedione	C600146	697	164.522	1.22074	Caramel aroma, diluted with a creamy smell
	35	2-heptanone	C110430	890.4	260.017	1.26204	fruity aroma
	37	2-Hexanone	C591786	779	200.268	1.50009	spicy smell
	40	3-Penten-2-one, 4-methyl	C141797	790.2	205.63	1.45009	sweet scent
Esters	18	ethyl acetate	C141786	610.7	140.096	1.34145	pineapple flavour
	19	methyl acetate	C79209	548.8	125.276	1.19757	pineapple flavour
Furans	6	2-pentyl furan	C3777693	993.6	345.924	1.25402	fruity, grassy

Note: M indicates monomer and D indicates dimer.

To analyze differences in aroma profiles resulting from different vanillin treatment concentrations, a flavour profile fingerprint was generated from the triplicate analyses of the control and vanillin treatment samples (Figure 13). Some compounds, such as 3-hydroxy-2-butanone, 2,3-butanedione, 2,3-pentanedione, 2,3-butanediol, ethanol, and 2-methylbutanol were present in similar amounts in all samples. Some other compounds varied in content greatly between different samples, such as hexanal, E-2-hexenal, methyl acetate, 2-heptanone, 3-octanone, 4-methyl-3-penten-2-one, 2-pentylfuran, and 5-methylfuran, which were higher in A or D and lower in C, E, and F. The difference in muscle volatile flavour substances between group A and group D was not significant, indicating that vanillin with a mass concentration of 1 g/L had little effect on the muscle flavour of the crucian carp.

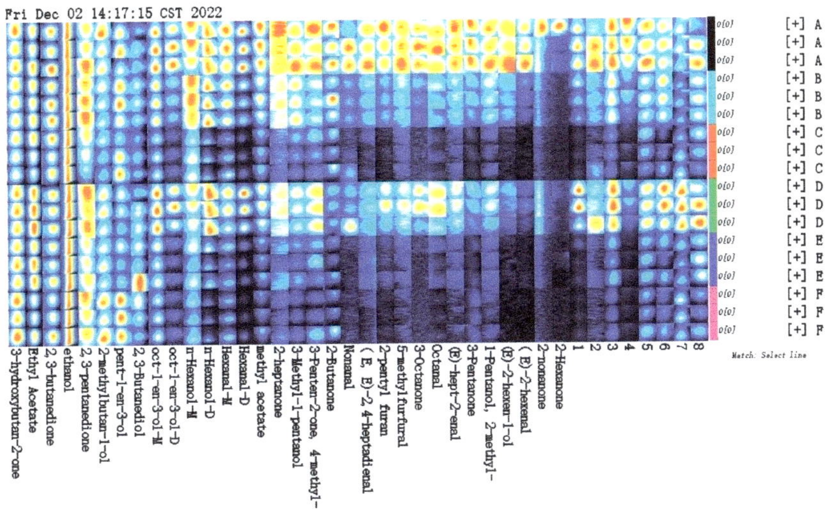

Figure 13. Fingerprint of volatile flavour compounds detected in crucian carp muscle after treatment with different concentrations of vanillin. Note: A-control; B-0.50 g/L; C-0.75 g/L; D-1.00 g/L; E-1.25 g/L; and F-1.50 g/L.

4. Discussion and Conclusions

Vanillin had an anaesthetic effect on crucian carp when added to the water in which the fish were present. With an increasing vanillin concentration, the time taken to achieve deep anaesthesia decreased in an exponential, dose-responsive manner and the recovery time of the fish, when removed to vanillin-free water, increased. All the test fish recovered completely and there was no evidence of adverse effects. This is consistent with reports on the anaesthetic effect of eugenol on juvenile yellow-spotted basketfish (*Siganus oramin*) [25], magnolia essential oil on spotted seabass (*Lateolabrax maculatus*) [26], and carbon dioxide on juvenile oval pompano trevally (*Trachinotus ovatus*) [27].

Blood physiological indices in fish are good indicators of their metabolic state, nutritional status, and health. White blood cells (WBC) are an essential part of the immune system [28]; red blood cells (RBC) deliver oxygen to body tissues [29]; platelets (PLT) are mainly involved in blood clotting, though they also have immunomodulatory effects; and haemoglobin HGB is the oxygen-binding protein in RBCs [30]. Liang et al. [31] analysed the anaesthetic effect of clove oil on the blood indexes of tilapia; the content of WBC, RBC, HGB, and PLT increased after anaesthesia. In this study, vanillin-induced anaesthesia moderately reduced the contents of WBC and HGB, though more than halved the contents of RBC and PLT.

Blood serum indices in fish are also good indicators of their metabolic state, nutritional status, health, and of cell membrane integrity [32]. Alanine and aspartate aminotransferases (ALT and AST) are important mitochondrial enzymes and are abundant in hepatocytes [33]. Under normal conditions, only small amounts of aminotransferases are released into the blood from hepatocytes and, therefore, serum aminotransferase activity is low. Elevated serum levels of ALT and AST are indicators of liver damage. Alkaline phosphatase (ALP) is involved in metabolism and the immune system [34]. Lactate dehydrogenase (LDH) is abundant in cardiac muscle cells and increased activity is an indicator of damage to the cardiomyocytes [35]. In this study, the ALT activity of carp anaesthetized with vanillin was reduced, whereas the AST activity markedly increased, indicating that there may have been some liver damage in the fish, which is consistent with the effect of anaesthesia with *Buddleja lindleyana* on serum AST in crucian carp [36]. The LDH activity was moderately reduced, suggesting that anaesthesia reduced myocardium damage, consistent with the response of MS-222 anaesthesia on bream (*Parabramis pekinensis*) [37]. The ALP activity of crucian carp anaesthetized by the various concentrations of vanillin solution was lower than that of the control group, indicating that the vanillin aqueous solution contained components that inhibited ALP activity, indicating that vanillin had less effect on ALP in crucian carp.

Inorganic ions such as PO_4^{3-}, Mg^{2+}, K^+, and Na^+ in serum are important for maintaining osmolality, acid-base balance, and overall homeostasis in fish, and their content changes after environmental stress [38]. Elevated serum PO_4^{3-} and Mg^{2+} levels in fish can indicate kidney damage [39]. In this study, the levels of PO_4^{3-} and Mg^{2+} in anaesthetized crucian carp were reduced, indicating that vanillin did not cause kidney damage. The increase in serum Na^+ and the decrease in serum K^+ suggest increased permeability of the gill epithelium, which allows leakage of Na^+ and ingress of K^+. In this study, with the increase in the vanillin mass concentration, the Na^+ content of each concentration of the anaesthesia group increased compared with the control group, and the K^+ content decreased compared with the control group, and it was speculated that the higher concentrations of vanillin caused increased gill tissue cell deformation. This is consistent with the anaesthesia of tilapia with MS-222, which decreased serum K^+ in a dose-responsive manner, whereas the Na^+ concentration remained relatively stable [40].

Glucose in the bloodstream supplies energy for various vital activities in fish [41] and its concentration is regulated by insulin and epinephrine and fluctuates in response to environmental factors [42]. In this study, the treatment group's blood glucose levels were higher than the control in different proportions, possibly due to less dissolved oxygen in the water, which lowered the metabolic rate and/or because of the reduced physical activity of the anaesthetized fish. The increases in GLU concentration after anaesthesia may result from a need for increased energy associated with the stress response, which is met by increased glycogenolysis. At higher vanillin concentrations, there was an irregular decrease in blood glucose, consistent with more rapid anaesthesia, which is consistent with a previous report on electroanesthetized pearl gentian grouper [43]. In contrast, blood glucose concentrations decreased after CO_2 anaesthesia of the grouper [44]. Serum albumin (ALB) is the most abundant serum protein and acts as a transporter for insoluble fatty acids [45]. The ALB concentration was slightly decreased by vanillin treatment. Cholesterol (CHO) is an important component of the cell membrane and is transported in the bloodstream by lipoproteins in fish [46]. Triglycerides, cholesterol, and total protein levels are affected by protein catabolism and hepatic glycogenolysis. In this study, except for triglycerides, the TP and ALB contents of the crucian carp after anaesthesia were lower than those in the control group, indicating that the protein breakdown and liver glycogen decomposition of crucian carp decreased after anaesthesia.

Fishery anaesthetics are a class of substances that inhibit the sensory centres of the fish brain to varying degrees, causing the fish to lose the ability for reflex action. The mechanism of action is to first inhibit the brain cortex (tactile loss phase), then act on the basal ganglia and cerebellum (excitation phase), and finally on the spinal cord (anaesthesia phase) [47]. An appropriate dose of anaesthetic reduces oxygen consumption and ammonia excretion, inhibits excessive stress in the fish, and effectively reduces injury, or mortality during handling. The liver is an important glandular and digestive-metabolic organ in fish and is involved in bile secretion, metabolism, detoxification, and defence [48]. The gills are the main respiratory organ of fish and excrete metabolic wastes, such as ammonia and nitrogen [49]. Anaesthesia, with 40 mg/L MS-222 for 24 h, may cause some damage to the liver tissue of the *Larimichthys crocea* [50]; eugenol anaesthesia of the carp (*Cyprinus carpio*) did not result in detectable liver or kidney damage [51]; deep anaesthesia of rainbow trout with *Coriandrum sativum* essential oil did not result in gill or liver damage [52]. In this study, vanillin caused no detectable pathological effects on the liver. All concentrations of vanillin, except for 1.00 g/L, had some effect on the gill tissue, increasing the width and spacing of the gill lamellae. Further research will be needed to determine whether this is harmful to the fish and the mechanism of any harm.

The electronic nose (E-Nose) is an array of gas sensors that mimics the human olfactory system and characterizes the aroma of a sample, which has low cost, ease of operation, and high accuracy [53]. E-Nose analysis of carp muscle after vanillin treatment indicated that the most abundant flavour volatiles are alcohols, amines, and hydrocarbons. Linear discriminant analysis separated the control and the different vanillin concentration treatments, though there was no clear dose response in the separation, indicating that, although the E-Nose could discriminate the control and treatment samples, there was little difference in the flavour profiles between the different vanillin treatment concentrations. The response of different concentrations of vanillin to the muscle of crucian carp. Gas-phase ion mobility spectrometry (GC-IMS) combines gas chromatography and ion mobility spectrometry [54] and has the advantages of the high resolution of ion mobility spectrometry and the high sensitivity of gas chromatography, resulting in richer chemical information than GC-mass spectroscopy [55]. In this study, the GC-IMS was used to analyze differences in carp-muscle flavour profile after anaesthesia with different vanillin concentrations. A total of 40 flavour compounds were identified, including eight aldehydes, 11 alcohols, 10 ketones, two esters, and one furan. Of these, 3-hydroxy-2-butanone, 2,3-butanedione, 2,3-pentanedione, 2,3-butanediol, ethanol, and 2-methylbutanol were present in all samples with similar signal intensities, suggesting that these compounds are the main contributors to the overall flavour of crucian carp muscle tissue after anaesthesia. Moreover, the difference between the control group and the vanillin concentration of 1 g/L on muscle volatile flavour substances was not significant, indicating that vanillin with a mass concentration of 1 g/L had little effect on the muscle flavour of crucian carp.

In conclusion, the effective concentration range of vanillin to anaesthetize crucian carp is 0.50–1.50 g/L, which resulted in relatively small changes to blood biochemistry, and no detectable liver damage, though minor, nonlethal damage to the gills. Future research on the anaesthetic effect of vanillin should include an examination of molecular toxicology and protein metabolism. GC-IMS proved very useful for the identification of carp-muscle flavour volatiles, however, the GC-IMS database is not yet complete for fish muscle flavour compounds, so more information on these compounds will be needed.

Author Contributions: Data Management, L.J., J.T. and D.C.; Formal Analysis, L.J.; Funding acquisition, B.H., C.Z. and P.J.; Methodology, B.H. and C.Z.; Resources, P.J. and D.C.; Writing—Original Draft, L.J.; Writing—Review and Editing, L.J. All authors have read and agreed to the published version of the manuscript.

Funding: This work was supported by the National Key R&D Program of China (2022YFD2100604, 2021YFD2100504).

Data Availability Statement: The data are available from the corresponding author.

Conflicts of Interest: The authors declare no conflict of interest.

References

1. Bodur, T.; Leon, B.S.; Navarro, A.; Tort, L.; Afonso, J.M.; Montero, D. Effects of new plant based anesthetics *Origanum* Sp. and Eucalyptus Sp. oils on stress and welfare parameters in *Dicentrarchus Labrax* and their comparison with clove oil. *Aquaculture* **2018**, *495*, 402–408. [CrossRef]
2. De Oliveira, C.P.B.; da Lemos, C.H.P.; Vidal, L.V.O. Anesthesia with eugenol in hybrid amazon catfish (*Pseudoplatystoma Reticulatum* × *Leiarius Marmoratus*) handling: Biochemical and haematological responses. *Aquaculture* **2019**, *501*, 255–259. [CrossRef]
3. De Oliviera, C.P.B.; Lemosda Paixão Lemos, C.H.; e Silva, A.F.; De Souza, S.A.; Albinati, A.C.L.; Lima, A.O.; Copatti, C.E. Use of eugenol for the anesthesia and transportation of freshwater angelfish (*Pterophyllum scalare*). *Aquaculture* **2019**, *513*, 734409. [CrossRef]
4. Fu, Y.Y.; Xu, S.L. The research status of the application of three fishery anesthetics was analyzed. *China Food Saf. Mag.* **2019**, *9*, 148–150.
5. Aydın, B.; Barbas, L.A.L. Sedative and anesthetic properties of essential oils and their active compounds in fish: A review. *Aquaculture* **2020**, *520*, 734999. [CrossRef]
6. Purbosari, N.; Warsiki, E.; Syamsu, K.; Santoso, J.; Effendi, I. Evaluation of the application of seaweed (*Eucheuma cottonii*) extract as fish anesthetic agent. *Aquac. Int.* **2021**, *29*, 1545–1560. [CrossRef]

7. Santos, A.C.D.; Junior, G.B.; Zago, D.C.; Zeppenfeld, C.C.; Silva, D.T.D.; Heinzmann, B.M.; Baldisserotto, B.; Cunha, M.A.D. Anesthesia and anesthetic action mechanism of essential oils of *Aloysia triphylla* and *Cymbopogon flexuosus* in silver catfish (*Rhamdia quelen*). *Vet. Anesth. Analg.* **2017**, *44*, 106–113. [CrossRef]
8. Boaventura, T.P.; dos Santos, F.A.C.; Souza, A.D.S.; Batista, F.S.; Julio, G.S.C.; Luz, R.K. Thymol and linalool chemotypes of the essential oil of thymus vulgaris (thyme) as anesthetic for *Colossoma macropomum*: Physiology and feed consumption. *Aquaculture* **2022**, *554*, 738161. [CrossRef]
9. Ananias, I.D.M.C.; de Melo, C.L.; Costa, D.C.; Ferreira, A.L.; Martins, E.D.F.F.; Takata, R.; Luz, R.K. Menthol as anesthetic for juvenile lophiosilurus alexandri: Induction and recovery time, ventilatory frequency, hematology and blood biochemistry. *Aquaculture* **2022**, *546*, 737373. [CrossRef]
10. Han, Z.C.; Long, L.; Ding, S. Expression and characterization of carotenoid cleavage oxygenases from herbaspirillum seropedicae and rhodobacteraceae bacterium capable of biotransforming isoeugenol and 4-vinylguaiacol to vanillin. *Front. Microbiol.* **2019**, *10*, 1869. [CrossRef] [PubMed]
11. Lv, X.J. Development status of vanillin industry. *Modern Food* **2019**, *7*, 14–16.
12. Wang, J.; Chen, P.S.; Meng, X.C. Effect of composite coating combined with UV treatment on the chilling quality of fresh-cut papaya. *J. Agric. Eng.* **2012**, *2*, 273–278.
13. Mei, G.L.; Liu, W.Q.; Li, T.T.; Li, J.R.; Mou, W.L.; Guo, X.H. Characteristics of chitosan/vanillin/polyvinyl alcohol composite electrospun nanofiber film and its application in the preservation of turbot. *Food Sci.* **2021**, *5*, 221–227.
14. Ueno, H.; Shimada, A.; Suemitsu, S.; Murakami, S.; Kitamura, N.; Wani, K.; Takahashi, Y.; Matsumoto, Y.; Okamoto, M.; Fujiwara, Y.; et al. Comprehensive behavioral study of the effects of vanillin inhalation in mice. *Biomed. Pharmacother.* **2019**, *115*, 108879. [CrossRef]
15. Yan, X.; Liu, D.F.; Zhang, X.Y.; Liu, D.; Xu, S.Y.; Chen, G.X.; Huang, B.X.; Ren, W.Z.; Wang, W.; Fu, S.P.; et al. Vanillin protects dopaminergic neurons against inflammation-mediated cell death by inhibiting ERK1/2, P38 and the NF-κB signaling pathway. *Int. J. Mol. Sci.* **2017**, *18*, 389. [CrossRef]
16. Orizano, P.E.; Char, C.; Sepulveda, F.; Ortiz, V.J. Heat sensitization of Escherichia coli by the natural antimicrobials vanillin and emulsified citral in blended carrot-orange juice. *Food Microbiol.* **2022**, *107*, 104058. [CrossRef]
17. Lan, X.B.; Wang, Q.; Yang, J.M.; Ma, L.; Zhang, W.J.; Zheng, P.; Sun, T.; Niu, J.G.; Liu, N.; Yu, J.Q. Neuroprotective effect of vanillin on hypoxic-ischemic brain damage in neonatal rats. *Biomed. Pharmacother.* **2019**, *118*, 109196. [CrossRef]
18. Wang, P.; Li, C.; Liao, G.; Huang, Y.; Lv, X.; Liu, X.; Chen, W.; Zhang, L. Vanillin attenuates proinflammatory factors in a TMCAO mouse model via inhibition of TLR_4/NF-KB signaling pathway. *Neuroscience* **2022**, *491*, 65–74. [CrossRef]
19. Tripathi, A.S.; Awasthi, S.; Maurya, R.K.; Yasir, M.; Mohapatra, L.; Srivastav, V. Protective effect of vanillin on the management of cecal ligation and puncture induced sepsis rat model. *Microb. Pathog.* **2022**, *165*, 105493. [CrossRef]
20. Marking, L.L.; Meyer, F.P. Are better anesthetics needed in fisheries? *Fisheries* **1985**, *10*, 2–5. [CrossRef]
21. Mirghaed, A.T.; Ghelichpour, M.; Hoseini, S.M. Myrcene and linalool as new anesthetic and sedative agents in common carp, Cyprinus carpio-comparison with eugenol. *Aquaculture* **2016**, *464*, 165–170. [CrossRef]
22. Chen, D.J.; Zhang, M.G.; Nie, X.B.; Jiang, P.H.; Zhang, Y.H.; Zhang, C.F.; Ren, F. Quality detection of turbot (*Scophtalmus maximus*) treated with electrostatic field using gas chromatography-ion mobility spectrometry. *Food Sci.* **2019**, *24*, 313–319.
23. Jia, Z.; Chen, X.T.; Pan, N.; Cai, S.L.; Zhang, Y.; Liu, Z.Y. Changes of volatile flavor compounds in *Takifugu bimaculatus* during refrigeration storage. *Food Sci.* **2021**, *20*, 188–196.
24. Li, H.D.; Xu, Q.S.; Liang, Y.Z. LibPLS: An integrated library for partial least squares regression and linear discriminant analysis. *Chemom. Intell. Lab. Syst.* **2018**, *176*, 34–43. [CrossRef]
25. Cao, X.C.; Huang, X.L.; Sun, X.Y.; Lin, H.Z.; Shu, H.; Yang, Y.K.; Huang, Z. Anesthesia effects of eugenol on juvenile *Siganus oramin*. *South China Fish. Sci.* **2019**, *3*, 50–56.
26. Zeng, X.; Dong, H.; Wu, J.; Wang, W.; Duan, Y.; Chen, J.; Zhang, J. Essential oil of Magnolia denudata is an effective anesthetic for spotted seabass (*Lateolabrax maculatus*): A test for its effect on blood biochemistry, physiology and gill morphology. *Fish Physiol. Biochem.* **2022**, *48*, 1349–1363. [CrossRef]
27. Huang, X.L.; Dai, C.; Yu, W.; Yang, J.; Yang, Y.K.; Li, T.; Lin, H.Z.; Huang, Z.; Sun, X.Y.; Shu, H. Anesthetic effect of eugenol on juvenile *Trachinotus ovatus*. *J. Guangdong Ocean. Univ.* **2020**, *4*, 124–131.
28. Safuan, S.N.M.; Tomari, M.R.M.; Zakaria, W.N.W. White blood cell (WBC) counting analysis in blood smear images using various color segmentation methods. *Measurement* **2018**, *116*, 543–555. [CrossRef]
29. Kumar, M.; Matoba, O.; Quan, X.; Rajput, S.K.; Morita, M.; Awatsuji, Y. Quantitative dynamic evolution of physiological parameters of RBC by highly stable digital holographic microscopy. *Opt. Lasers Eng.* **2022**, *151*, 106887. [CrossRef]
30. Wang, K.; Bian, X.; Zheng, M.; Liu, P.; Lin, L.; Tan, X. Rapid determination of hemoglobin concentration by a novel ensemble extreme learning machine method combined with near-infrared spectroscopy. *Spectrochim. Acta Part A Mol. Biomol. Spectrosc.* **2021**, *263*, 120138. [CrossRef]
31. Liang, Z.Y.; An, L.N.; Dong, Z.J.; Miao, L.H.; Xu, P.; Xie, Z. The anesthetic effects of clove oil on tilapia (*Oreochromis* spp.) and its influence on haematological indices and hormone level. *J. Shanghai Ocean. Univ.* **2009**, *5*, 629–635.
32. Amenyogbe, E.; Yang, E.; Xie, R.; Huang, J.; Chen, G. Influences of indigenous isolates Pantoea agglomerans RCS_2 on growth, proximate analysis, haematological parameters, digestive enzyme activities, serum biochemical parameters, antioxidants activities,

intestinal morphology, disease resistance, and molecular immune response in juvenile's cobia fish (*Rachycentron canadum*). *Aquaculture* **2022**, *551*, 737942.
33. Feng, C.J.; Wang, Y.L.; Zhang, Y.; Yang, H.L.; Wang, N.M.; Lyu, W.H.; Cao, D.C. Effects of different culture density on growth indices, serum biochemical indices and lipase activity of juvenile *Huso dauricus*. *Heilongjiang Anim. Sci. Vet. Med.* **2022**, *6*, 125–129.
34. Zhong, Y.M. *Clone and Expression Analysis of Alkaline Phosphatase Gene from S. Constricta under the Stress of Cu and Zn*; Shanghai Ocean University: Shanghai, China, 2012.
35. Yong, W.; Huang, Y.; Li, X.; Wu, X. Changes of AST, ALP, CK and LDH level in biological injury models caused by 18.4 mm rubber bullet. *Int. J. Lab. Med.* **2012**, *33*, 3.
36. Li, Y.; Wang, Z.X.; Li, S.Y.; Gu, T.Y.; Lin, S.; Nie, X.B.; Huang, B.S.; Zhang, C.F. Effects of Buddleja lindleyana Fortune on anesthesia and serum biochemical indices in crucian carp (*Carassius auratus*). *J. Dalian Ocean. Univ.* **2020**, *4*, 491–495.
37. Ding, Y.T.; Wang, Z.H.; Wang, L.L.; Shi, W.Z. Effect of MS-222 on survival of bream fish during anesthesia transportation. *Fish. Sci.* **2019**, *3*, 296–304.
38. Zhang, Y.C.; Wen, H.S.; Li, L.M.; Feng, Q.C. Effect of acute temperature stress on serum cortisol and hematological physiology of gestated *Sebastes schlegelii*. *J. Fish. China* **2015**, *12*, 1872–1882.
39. Bijvelds, M.J.; Van, D.V.; Kolar, Z.I. Magnesium transport in freshwater teleosts. *J. Exp. Biol.* **1998**, *201*, 1981–1990. [CrossRef]
40. Tang, B.G.; Yang, S.K.; Zhen, S.B.; Ni, F.R.; Zhou, X.X.; Xu, Y.R.; Xie, W.Y. Effects of MS-222 and benzocaine on electrolyte levels of *Oreochromis niloticus* and its correlation with content of erythrocyte. *J. Guangdong Ocean. Univ.* **2015**, *6*, 25–29.
41. Wu, B.; Xie, J. Optimization of water temperature and salinity for live transportation of grouper. *Food Sci.* **2019**, *16*, 235–241.
42. Toni, C.; Becker, A.G.; Simoes, L.N.; Pinheiro, C.G.; de Lima Silva, L.; Heinzmann, B.M.; Caron, B.O.; Baldisserotto, B. Fish anesthesia: Effects of the essential oils of *Hesperozygis ringens* and *Lippia alba* on the biochemistry and physiology of silver catfish (*Rhamdia quelen*). *Fish Physiol. Biochem.* **2014**, *40*, 701–714. [CrossRef]
43. Bai, Z.; Shen, J.; Xu, W.Q.; Zhao, X.Y.; Guo, S.F.; Ma, T.T. Effects of electronarcosis on the behavior and blood biochemical indices of ♀*Epinephelus fuscoguttatus* × ♂*E. lanceolatus*. *Fish. Mod.* **2022**, *1*, 89–96.
44. Du, H.; Qin, X.M.; Fan, X.P.; Zhang, J.S.; Li, S.J. Synergistic effect of CO_2 and low water temperature on the anesthesia and survival of *Trachinotus ovatus*. *J. Guangdong Ocean. Univ.* **2022**, *01*, 35–43.
45. Cicek, S.; Ozogul, F. Effects of selenium nanoparticles on growth performance, hematological, serum biochemical parameters, and antioxidant status in fish. *Anim. Feed. Sci. Technol.* **2021**, *281*, 115099. [CrossRef]
46. Zhu, T.; Ai, Q.; Mai, K. Feed intake, growth performance and cholesterol metabolism in juvenile turbot (*Scophthalmus maximus* L.) fed defatted fish meal diets with graded levels of cholesterol. *Aquaculture* **2014**, *428*, 290–296. [CrossRef]
47. Cooke, S.J.; Suski, C.D.; Ostrand, K.G.; Tufts, B.L.; Wahl, D.H. Behavioral and physiological assessment of low concentrations of clove oil anesthetic for handling and transporting largemouth bass (*Micropterus salmoides*). *Aquaculture* **2004**, *239*, 509–529. [CrossRef]
48. Zheng, J.L.; Luo, Z.; Liu, C.X.; Chen, Q.L.; Tan, X.Y.; Zhu, Q.; Gong, Y. Differential effects of acute and chronic zinc (Zn) exposure on hepatic lipid deposition and metabolism in yellow catfish *Pelteobagrus fulvidraco*. *Aquat. Toxicol.* **2013**, *132*, 173–181. [CrossRef] [PubMed]
49. Ou, Y.j.; Liu, Q.Q.; Wen, J.F.; Li, J.E.; Li, H. The effects of acute low temperature stress on liver, muscle and gill tissues of juvenile *Eleutheronema tetradactylum*. *Ecol. Sci.* **2018**, *5*, 53–59.
50. Zhang, X.; Xu, Z.J.; Li, W.Y.; Yin, X.; Wang, Y.F.; Chen, S.A.; Ma, X.B. Effects of MS-222 on anesthesia, tissue structure and antioxidant enzyme activity of *Larimichthys crocea*. *J. Dalian Ocean. Univ.* **2022**, 1–13. [CrossRef]
51. Velisek, J.; Svobodova, Z.; Piackova, V.; Groch, L.; Nepejchalova, L. Effects of clove oil anesthesia on common carp (*Cyprinus Carpio* L.). *Vet. Med.* **2005**, *50*, 269–275. [CrossRef]
52. Yigit, N.O.; Kocaayan, H. Efficiency of thyme (*Origanum onites*) and coriander (*Coriandrum sativum*) essential oils on anesthesia and histopathology of rainbow trout (*Oncorhynchus mykiss*). *Aquaculture* **2023**, *562*, 738813. [CrossRef]
53. Wilson, A.D. Application of electronic-nose technologies and VOC-biomarkers for the noninvasive early diagnosis of gastrointestinal diseases. *Sensors* **2018**, *18*, 2613. [CrossRef]
54. Wen, R.; Kong, B.; Yin, X.; Zhang, H.; Chen, Q. Characterisation of flavor profile of beef jerky inoculated with different autochthonous lactic acid bacteria using electronic nose and gas chromatography-ion mobility spectrometry. *Meat Sci.* **2022**, *183*, 108658. [CrossRef] [PubMed]
55. Qian, M.; Zheng, M.; Zhao, W.; Liu, Q.; Zeng, X.; Bai, W. Effect of marinating and frying on the flavor of braised pigeon. *J. Food Process. Preserv.* **2021**, *45*, 15219. [CrossRef]

foods

MDPI

Article

Hydrogen-Rich Water Treatment of Fresh-Cut Kiwifruit with Slightly Acidic Electrolytic Water: Influence on Antioxidant Metabolism and Cell Wall Stability

Yanan Sun [1,2], Weiyu Qiu [1], Xiaoqi Fang [1], Xiaomei Zhao [1], Xingfeng Xu [1] and Wenxiang Li [1,2,*]

1 School of Food Science and Engineering, Qingdao Agricultural University, Qingdao 266109, China
2 Qingdao Special Food Research Institute, Qingdao 266109, China
* Correspondence: xiang7332@126.com

Abstract: The synergistic impact of hydrogen-rich water (HRW, 394 ppb) and slightly acidic electrolyzed water (SAEW, pH of 6.25 ± 0.19) on the antioxidant metabolism of fresh-cut kiwifruit during storage was investigated (temperature: (3 ± 1) °C, humidity: 80%–85%). Compared with control group, H+S treatment increased the contents of active oxygen-scavenging enzymes (SOD, CAT, POD, and APX) and inhibited the increase of $O_2^{\bullet -}$ and H_2O_2 contents during the storage of fresh-cut kiwifruit. Meanwhile, H+S treatment could reduce the activities of the cell wall-degrading enzymes PG, PME, PL, Cx, and β-Gal, inhibit the formation of soluble pectin, delay the degradation rate of propectin, cellulose, and pseudocellulose, and maintain higher fruit hardness and chewability. The results showed that H+S treatment could enhance free radical scavenging ability and reduce the cell wall metabolism of fresh-cut kiwifruit, maintaining the good texture found in fresh-cut fruit.

Keywords: HRW; SAEW; fresh-cut; kiwifruit; antioxidant metabolism

Citation: Sun, Y.; Qiu, W.; Fang, X.; Zhao, X.; Xu, X.; Li, W. Hydrogen-Rich Water Treatment of Fresh-Cut Kiwifruit with Slightly Acidic Electrolytic Water: Influence on Antioxidant Metabolism and Cell Wall Stability. *Foods* **2023**, *12*, 426. https://doi.org/10.3390/foods12020426

Academic Editor: Sergio Torres-Giner

Received: 22 November 2022
Revised: 5 January 2023
Accepted: 12 January 2023
Published: 16 January 2023

1. Introduction

Kiwifruit (*Actinidia chinensis*) is not only rich in vitamin C but also contains nutrients needed by the human body, such as dietary fiber, potassium, vitamin E, and folic acid, as well as antioxidants, phytonutrients, and a variety of biologically active ingredients that offer great benefits to health. The vitamin C content in kiwifruit is higher than that in other fruits; one study found that the vitamin C content in Sunshine Golden kiwifruit was almost three times that found in oranges and strawberries [1], which can improve the bioavailability of micronutrient iron. Kiwifruit has a higher antioxidant capacity than apples, grapefruit, and pears, and regular consumption of kiwifruit can help reduce the risk of cardiovascular disease. Compared with whole fruits, fresh-cut fruits and vegetables are more favored by consumers because of their convenience, freshness, and lack of waste. However, it is easy to cause cutting damage to kiwifruit in the process of fresh cutting, such as tissue softening, browning, and loss of nutrients, seriously shortening its shelf life [2,3]. In addition, due to the high water content, juice leakage from fresh-cut kiwifruit would stimulate microbial growth, which may lead to foodborne hazards, seriously hindering the successful development of the industry.

Hydrogen-rich water (HRW) has received increasing attention in the field of fresh food processing and preservation. In animals, H_2 plays a beneficial role in many physiological and pathological processes, including antioxidant activity and anti-apoptotic protection. In plants, H_2 can be used as a bioactive gas molecule that promotes seed germination. Chen et al. [4] treated paddy and mung bean with HRW and found that low HRW concentrations could improve the seed germination rate, while high HRW concentrations inhibited germination [4]. The HRW concentrations required for affecting the germination rate of different kinds of seeds are different. H_2 is also considered to be an anti-stress molecule that is capable of alleviating various abiotic stresses by regulating the antioxidant defense system.

For example, Guan et al. [5] found that HRW treatment could improve the germination rate of black barley at lower temperatures (10–15 °C), improve the bioactive ingredients content in black barley germination (including free vanillic acid, conjugate coumaric acid, sinapic acid, and vanillic acid), and enhance the in vitro antioxidant activity [5]. In recent years, the application of H_2 in fruit and vegetable preservation has attracted widespread attention due to its strong antioxidant activity and permeability. Chen et al. [6] explored the theory that 25% of HRW pretreatment could effectively inhibit the rot of *Eusculus eusculus*, maintain high hardness, improve the activity of antioxidant enzymes, and maintain freshness [6]. High permeability and residue-free properties make H_2 popular; more importantly, H_2 is a non-toxic gas compared to NO and H_2S and does not react with most compounds at room temperature.

Electrolytic water (EW) can be divided into strongly acidic electrolytic water (AEW, pH 2.4–2.9, ACC 31–201 mg/L) and slightly acidic electrolyzed water (SAEW, pH 5.1–6.6, ACC 11–31 mg/L) according to the differences in pH and effective chlorine concentration (ACC) [7]. In recent years, compared with AEW, SAEW has attracted much attention due to its strong antibacterial activity and lack of adverse effects on human health and the environment [8,9]. Today, SAEW is used as a disinfectant for food safety and in preservation [10]. The effective form of the chlorine compounds in SAEW is hypochlorous acid (HClO), which has strong bacteriostatic activity. Issa-Zacharia et al. [11] confirmed that the bacteriostatic activity of SAEW is mainly caused by the potential oxidative damage of HClO to biomolecules [12]. Compared with acid preservatives, SAEW could reduce the environmental damage and corrosive effects of the food industry [13].

Due to the extremely unstable chemical properties of SAEW, exposure to air and light may affect the ability of SAEW to kill microorganisms, seriously affecting its timeliness [14,15]. However, when SAEW is combined with other technologies, this may have additional influences on the bactericidal effect. At present, the stability and feasibility of SAEW when combined with ascorbic acid [16] and plasma water [17] have been reported. There are many studies on the application of SAEW in aquatic products, most of which focus on the reduction of pathogenic microorganisms, but there are few studies on the effect of SAEW treatment on the storage quality of fresh-cut fruits. Therefore, in this paper, the effects of HRW and SAEW on reactive oxygen species (ROS) metabolism and the cell-wall stability of fresh-cut kiwifruit were studied, and the mechanism of HRW and SAEW in fresh-cut kiwifruit preservation was discussed, providing a theoretical basis for maintaining fresh-cut kiwifruit preservation quality.

2. Materials and Methods

2.1. Preparation of SAEW and HRW

SAEW (pH 6.25, ACC 30 mg/L) was obtained by electrolysis of a 10% hydrochloric acid solution prepared using aqueous electrolysis equipment, with the parameters controlled using pH measurement equipment (PHS-3C, SH Leici, Shanghai, China) and an effective chlorine detector (RC-3F, KRK Kasahara, Kuki, Japan). HRW with an H_2 concentration of 394 ppb was obtained via electrolysis in a hydrogen-rich water cup.

2.2. Materials and Processing

Kiwifruits were purchased from Qingdao Haofeng Food Group Co., Ltd. (Jiaozhou, China); they were selected for their uniform size, greenness, and freedom from diseases and pests. The whole fruits were washed, dried, and peeled, then cut into 6–8 mm slices (about 10–15 g) with a stainless-steel knife. All the slices were randomly divided into 4 groups, each group being 1000 g, and were treated as follows: (1) soaking in sterile distilled water for 5 min (CK); (2) soaking in SAEW for 5 min (SAEW); (3) soaking in HRW for 5 min (HRW); (4) soaking for 5 min in SAEW and HRW, mixed at 1:1 (H+S). After the above four groups of treatment, the processed fresh-cut kiwifruit slices were taken out, placed on the shelf to remove moisture, then stored in a 17.5 × 13.5 × 7.5 cm crisper, under the conditions of 4 ± 1 °C storage. After 0 d, 2 d, 4 d, 6 d, 8 d, and 10 d of storage, slices from each group were randomly sampled to determine the indexes.

2.3. Quantification of ROS Antioxidant Metabolism

2.3.1. Determination of $O_2^{\bullet -}$ and H_2O_2 Contents

The determination of H_2O_2 content was modified according to Cheng et al. [18]. The pre-cooled acetone was added to the 2 g sample, then ground and centrifuged to obtain the supernatant. First, 1 mL of supernatant, 0.1 mL of $TiCl_4$ solution (10%), and 0.2 mL of concentrated ammonia water were shaken several times, centrifuged, and then left to precipitate. Then, 3 mL of 2 mol/L concentrated H_2SO_4 was added to dissolve the precipitation, and the supernatant was obtained via centrifugation. By measuring the absorbance value of the supernatant at a wavelength of 412 nm, the content of H_2O_2 in the sample was calculated; the results are expressed in μmol/g.

2.3.2. Determination of SOD, CAT, POD, APX Activities

A superoxide dismutase (SOD) assay kit, catalase (CAT) assay kit, peroxidase (POD) assay kit, and ascorbate peroxidase (APX) assay kit were obtained from the Nanjing Jiancheng Institute of Biological Engineering, China, for measuring enzyme activity.

2.4. Cell Wall Antioxidant Metabolism

2.4.1. Texture Analysis

A cylindrical (P/75) probe with a diameter of 75 mm was used for the texture analysis (TPA) (TAXT2i, Stable Micro Systems Ltd., London, UK) in accordance with the procedure followed by Cheng et al. [18]. The test conditions were as follows: the pre-speed was 5.0 mm/s, the test speed was 2.0 mm/s, the post-speed was 1.0 mm/s, the penetration distance was 3 mm, the interval between the two cycles was 5 s, and the trigger force was 1.0 N.

2.4.2. Determination of Pectin, Cellulose, and Hemicellulose Contents

The carbazole colorimetric method was adopted, in accordance with the study by Lohani et al. [19], slightly modified for the determination of pectin; the results were expressed in mg/g. The content of hemicellulose in the fruit cell wall was determined by the anthrone colorimetric method, and the result was in mg/g. The content of cellulose in the fruit cell wall was determined by a weighing method, and the result was expressed as a percentage.

2.4.3. Determination of Activity of Cell Wall-Degrading Enzymes (PG, PME, PL, Cx, and β-Gal)

For the enzyme solution preparation, 12 mL of buffer solution was added to 5 g of sample, ground, and homogenized at 4 °C, then centrifuged at 12,000× *g* at 4 °C for 30 min, and, finally, the supernatant was retained.

PG activity and Cx activity were determined by the DNS colorimetric method; the result was expressed as mg/(g·h). PME activity was determined via NaOH titration, and the results were expressed as μg/(g·min).

PL activity determination: After 2 mL of 0.5% pectin solution was heated at 40 °C for 5 min, 0.5 mL of enzyme extract was added and held at 40 °C for 10 min. The mixture of 0.5 mL and 4.5 mL 0.01 mol/L of HCl was mixed and shaken. The absorbance value of the mixture was measured at 235 nm, then the result was expressed as OD235/(g·min).

β-Gal activity determination: 1.5 mL of 1% salicin solution and 0.5 mL of enzyme extract were shaken several times and then thoroughly mixed. After the mixture was fully reacted at 37 °C, 1.5 mL of DNS was quickly incorporated into the sample mixture and then it was boiled in a water bath for 5 min. After that, the cuvette was rinsed with cold water to achieve rapid cooling; the mixture absorbance was measured at 540 nm and the outcome was expressed as μg/(g·min).

2.5. Statistics

A one-way analysis of variance (ANOVA) was performed on the experimental data. Median and normal deviations were reported. The differences between means were tested using Duncan's new multiple range test, with a confidence level of 95%. SPSS 20 (CoStat Version 6.451, CoHort Software, Pacific Grove, CA, USA) was utilized in the calculation of all statistics. All experiments were performed in triplicate unless otherwise specified.

3. Results

3.1. Changes in the $O_2^{\bullet-}$ and H_2O_2 Contents of Fresh-Cut Kiwifruit by Different Treatments

The influence of different treatments on the $O_2^{\bullet-}$ content of fresh-cut kiwifruit during storage is shown in Figure 1A. The $O_2^{\bullet-}$ content of fresh-cut kiwifruit in each treatment group tended to increase as the storage period was extended. The $O_2^{\bullet-}$ content of samples processed with HRW, SAEW, and H+S were lower than CK, and the $O_2^{\bullet-}$ content of fresh-cut kiwifruit treated with H+S was lower than that of HRW and SAEW alone. On the 8th day, the $O_2^{\bullet-}$ content of fresh-cut kiwifruit treated with HRW, SAEW, and H+S was 92.7%, 95.4%, and 86.0% of CK, respectively. This showed that all three treatments could inhibit the increase in the $O_2^{\bullet-}$ content of fresh-cut kiwi fruit, and the H+S treatment had the most significant effect ($p < 0.05$).

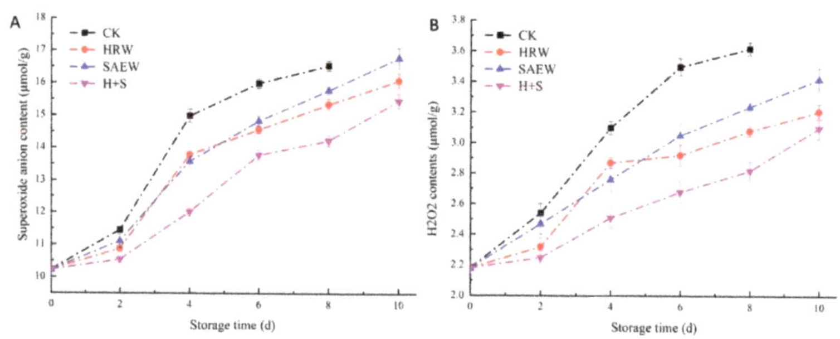

Figure 1. Changes in superoxide anion content (**A**) and H_2O_2 content (**B**) in fresh-cut kiwifruit after different treatments, during storage.

The consequences of the different disposal methods on the H_2O_2 content of fresh-cut kiwifruit during storage is shown in Figure 1B. It can be seen that the H_2O_2 content of fresh-cut kiwifruit continued rising with storage time in each treatment group. On the 8th day of storage, the H_2O_2 content of fresh-cut kiwifruit treated with HRW, SAEW, and H+S were 85.1%, 89.5%, and 77.9% of CK, respectively. This result showed that the H+S treatment could significantly inhibit the increase in H_2O_2 content ($p < 0.05$) and reduce its damage to fresh-cut fruit.

3.2. Changes in SOD, CAT, POD, and APX Activities after Different Treatments

The impact of various treatments on the SOD activity of fresh-cut kiwifruit during storage is shown in Figure 2A. Throughout the storage period, the SOD activity of the samples in each treatment group reached a peak value on day 2 and then gradually decreased. The SOD activity of fresh-cut kiwifruit treated with HRW, SAEW, and H+S was significantly higher ($p < 0.05$) than that of CK during storage. The SOD activity of samples treated with H+S was higher than that of samples treated with HRW and SAEW, except for the 6th day. The SOD activity of fruits treated with HRW, SAEW, and H+S was 1.24, 1.18, and 1.30 times higher than that of CK, respectively, after storage for the 8th day. The above analysis showed that the H+S treatment can effectively improve the SOD activity and delay the decay of fresh-cut kiwifruit.

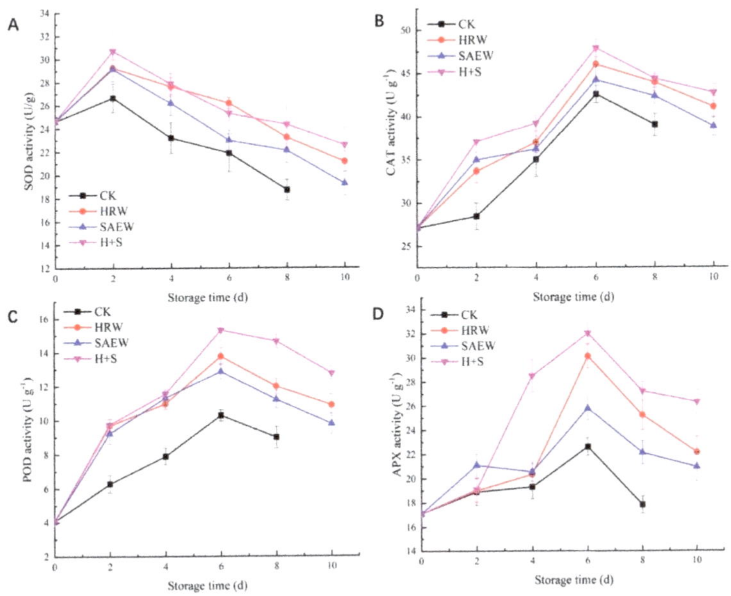

Figure 2. SOD activity (**A**), CAT activity (**B**), POD activity (**C**), and the APX activity (**D**) of fresh-cut kiwifruit after different treatments during storage.

The change in the CAT activity of fresh-cut kiwifruit during storage is shown in Figure 2B. During storage, the CAT activity of the sample in each group gradually increased in the early stages and began to decline gradually after reaching the maximum value on the 6th day. During the storage, the CAT activity of the CK group of fresh-cut kiwifruit was always lower than that of other treatment groups, and the HRW, SAEW, and H+S treatments significantly increased the CAT activity of fresh-cut kiwifruit ($p < 0.05$). On the 6th day, the CAT activity of the samples treated with HRW, SAEW, and H+S was 45.98 U/g, 44.21 U/g, and 47.87 U/g, respectively, which was significantly higher than that of the CK group ($p < 0.05$). It indicates that the CAT activity of fresh-cut kiwifruit was effectively improved after HRW, SEW, and H+S treatment, and the H+S treatment was the best.

As shown in Figure 2C, throughout the storage process, the POD activity of fresh-cut kiwifruit in each group increased rapidly in the early storage period and reached a peak on the 6th day. Throughout the storage process, the POD activity of the samples in the CK group was significantly less than those in the other treated groups ($p < 0.05$). The POD activity of fresh-cut kiwifruit treated with HRW, SAEW, and H+S showed no significant difference during the first six days of storage ($p > 0.05$). From the 6th day, the effect of H+S treatment on the POD activity of the sample was significantly higher than that of the HRW and SAEW treatment groups ($p < 0.05$). On the 8th day of storage, the POD activity of fresh-cut kiwifruit treated with HRW, SAEW, and H+S were 1.33 times, 1.25 times, and 1.63 times that of the CK group samples, respectively.

As seen in Figure 2D, the APX activity of fresh-cut kiwifruit first increased internally and then decreased with storage time, reaching a peak value at day 6. During the storage, except for the second day, the APX activity of samples treated with HRW, SAEW, and H+S was significantly higher than that of the CK group sample ($p < 0.05$). Since then, the APX activity of samples treated with H+S has been at a high level. Until the 6th day of storage, the APX activity of samples treated with HRW, SAEW, and H+S was 1.33 times, 1.17 times, and 1.42 times that of the CK group samples, respectively. The APX activity of fresh-cut kiwifruit treated with H+S was significantly increased ($p < 0.05$), which is beneficial for maintaining fruit quality.

3.3. Changes in the Hardness, Elasticity, and Chewability

The firmness, elasticity, and chewiness of fresh-cut kiwifruits showed a declining trend during storage (Table 1), which was due to fruit softening caused by tissue aging. It can be seen from the table that the firmness and chewiness of the fruit decreased significantly after 8 days of storage, while the elasticity of all samples remained stable. The sample hardness of the H+S treatment group was always at a higher level compared with other treatment groups, and the fruit hardness of the H+S treatment group was 5.1 N, which was 2.2 times that of the CK treatment group when it was stored for 8 days. The elasticity value remained stable during storage and fluctuated slightly between the different treatment groups, but there was no significant difference. The chewiness value reflected the anti-masticatory property of the fruit. The results showed that the masticatory property of the fruit in the CK group decreased significantly during storage. The SAEW, HRW, and H+S treatments could significantly delay ($p < 0.05$) the decline in the masticatory properties to varying degrees, among which the H+S treatment had the best effect.

Table 1. Changes in the hardness, elasticity, and chewiness of fresh-cut kiwifruit during storage after different treatments.

	Storage Time (d)	H+S	HRW	SAEW	Control
Hardness (N)	0	9.21 ± 0.28 [a]	9.21 ± 0.28 [a]	9.21 ± 0.28 [a]	9.21 ± 0.28 [a]
	2	8.21 ± 0.32 [d]	7.94 ± 0.25 [c]	7.02 ± 0.31 [b]	6.31 ± 0.15 [a]
	4	7.72 ± 0.24 [d]	7.14 ± 0.15 [c]	6.20 ± 0.13 [b]	5.67 ± 0.15 [a]
	6	6.91 ± 0.18 [c]	6.10 ± 0.47 [b]	4.54 ± 0.28 [a]	3.98 ± 0.10 [a]
	8	5.12 ± 0.19 [c]	4.12 ± 0.32 [b]	2.65 ± 0.11 [a]	2.32 ± 0.12 [a]
	10	3.89 ± 0.22 [c]	3.32 ± 0.16 [b]	2.33 ± 0.17 [a]	
Elasticity	0	0.7 ± 0.03 [a]	0.7 ± 0.03 [a]	0.7 ± 0.03 [a]	0.7 ± 0.03 [a]
	2	0.69 ± 0.07 [a]	0.65 ± 0.05 [a]	0.66 ± 0.03 [a]	0.66 ± 0.06 [a]
	4	0.67 ± 0.01 [a]	0.66 ± 0.09 [a]	0.67 ± 0.02 [a]	0.66 ± 0.03 [a]
	6	0.65 ± 0.07 [a]	0.64 ± 0.01 [a]	0.64 ± 0.04 [a]	0.62 ± 0.02 [a]
	8	0.64 ± 0.03 [a]	0.62 ± 0.03 [a]	0.63 ± 0.02 [a]	0.61 ± 0.01 [a]
	10				
Chewiness (N)	0	9.78 ± 0.55 [a]	9.78 ± 0.55 [a]	9.78 ± 0.55 [a]	9.78 ± 0.55 [a]
	2	9.31 ± 0.56 [c]	8.99 ± 0.56 [b]	8.43 ± 0.38 [a]	8.12 ± 0.47 [a]
	4	7.89 ± 0.47 [c]	7.46 ± 0.61 [b]	7.20 ± 0.31 [b]	6.21 ± 0.44 [a]
	6	7.46 ± 0.39 [d]	6.99 ± 0.22 [c]	6.41 ± 0.34 [b]	5.89 ± 0.36 [a]
	8	7.21 ± 0.33 [c]	6.45 ± 0.38 [b]	5.58 ± 0.51 [a]	5.21 ± 0.37 [a]
	10	6.89 ± 0.24 [b]	5.69 ± 0.25 [a]	5.45 ± 0.22 [a]	

[a–d] Different letters, marked on values of the same sampling time, indicate significant differences between treatments ($p < 0.05$).

3.4. Changes in the Contents of Pectin, Soluble Pectin, Cellulose and Hemicellulose

The influence of the different processes on the content of pectin in fresh-cut kiwifruit is shown in Figure 3A. During storage, the content of the pectin in the sample was gradually reduced, due to the continuous decomposition. HRW, SAEW, and H+S treatments could delay the decline in the pectin content of fresh-cut kiwifruit, to varying degrees. The pectin content of the H+S treatment sample was always at a high level, and the pectin content of fresh-cut kiwifruit after the H+S treatment was 1.33 mg/g after storage until the 8th day, which figure was 2.46 times higher than that in the CK group.

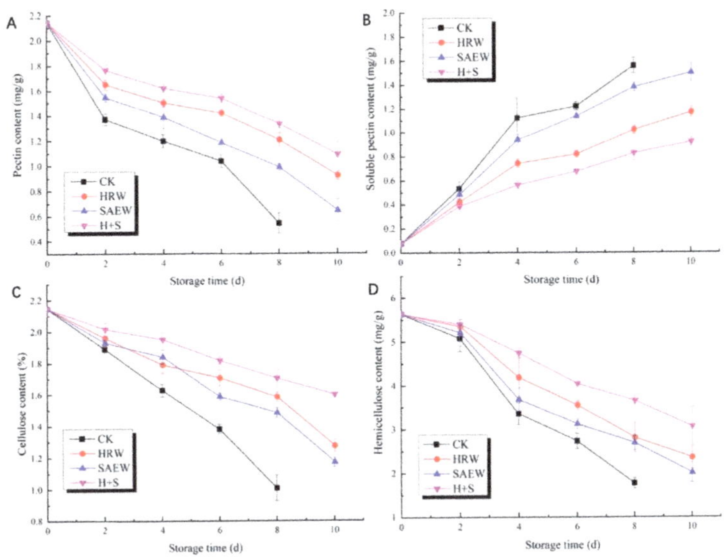

Figure 3. Pectin content (**A**), soluble pectin content (**B**), cellulose content (**C**), and hemicellulose content (**D**) of fresh-cut kiwifruit, with different treatments during storage.

As indicated in Figure 3B, the changing trend of the soluble pectin content of fresh-cut kiwifruit was opposite to that of the protopectin content, and the soluble pectin content increased continuously during storage. Compared with the CK samples, HRW, SAEW, and H+S treatments could significantly ($p < 0.05$) delay the increase in the soluble pectin content in fresh-cut kiwifruit. Until the 8th day, the soluble pectin content in fresh-cut kiwifruit treated with HRW, SAEW, and H+S were 65.4%, 88.5%, and 53.2% of the CK samples, respectively. In all these ways, H+S treatment demonstrated the best effect, and the pectin content in the fruit was only half that of the CK samples.

It can be seen from Figure 3C that the cellulose content of fresh-cut kiwifruit was constantly decreasing during the storage period. At the early stage of storage, there was no significant difference in the cellulose content of fruit among the treatments of each group ($p > 0.05$). With the extension of storage time, the cellulose content of the sample under the H+S treatment was at a higher level. By the 8th day of storage, the cellulose content of fresh-cut kiwifruit under HRW, SAEW, and H+S treatments was 1.58 times, 1.48 times, and 1.70 times higher than that under CK treatment, respectively.

The hemicellulose content of fresh-cut kiwifruit decreased continuously during the whole storage period (Figure 3D). HRW, SAEW, and H+S treatments can delay the decrease in fruit hemicellulose to varying degrees, among which H+S treatment has the best effect. On the 8th day after storage, the cellulose content of fresh-cut kiwifruit treated with HRW, SAEW, and H+S was 1.59 times, 1.52 times, and 2.07 times higher than that of CK, respectively.

3.5. Effects of Different Treatments on PG, PME, PL, and Cx and β-Gal Activity

Fresh-cut kiwifruit PG activity under different treatments is shown in Figure 4A. The PG activity of the fruit first increased throughout the storage period and then decreased with the increasing storage time. However, the PG activity of the sample in HRW, SAEW, and H+S treatments was below that of the CK group throughout the storage period. The PG activity of fresh-cut kiwifruit in the CK group reached its peak on the 6th day, being 1.9 times, 1.64 times, and 2.22 times higher than that in the HRW, SAEW, and H+S treatments, respectively. The H+S treatment samples reached a peak on the 8th day, which was 2 days later than in the CK group, which showed that H+S treatments could significantly reduce the PG activity of fresh-cut kiwifruit ($p < 0.05$).

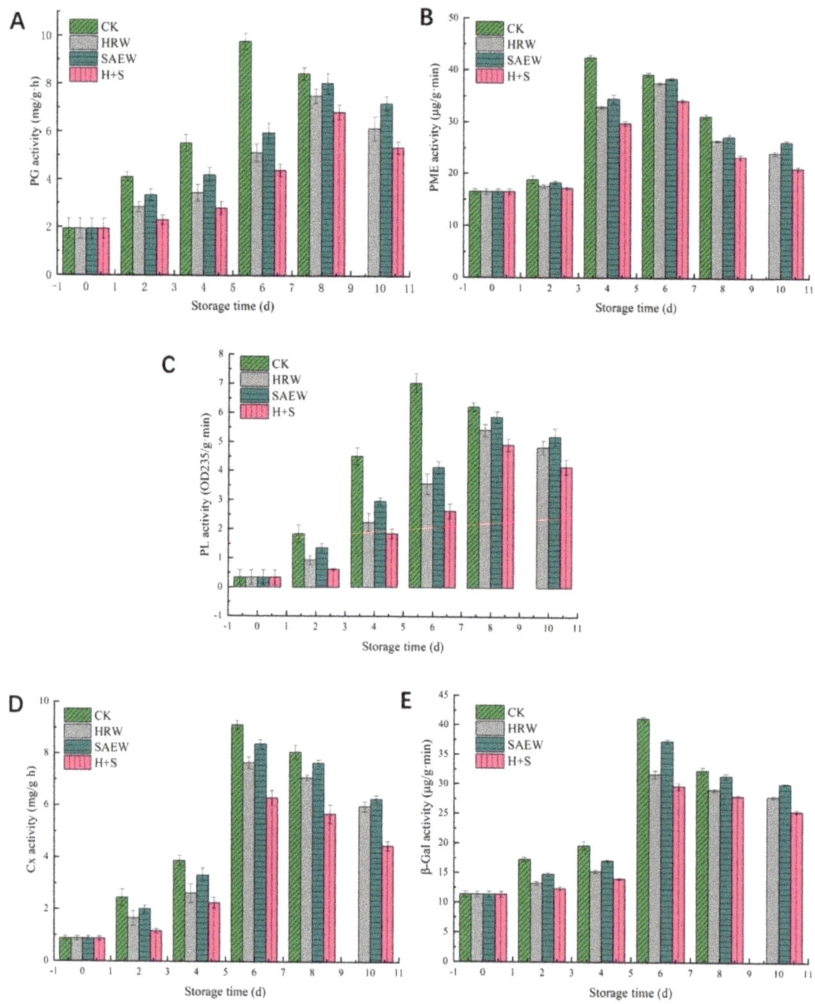

Figure 4. PG activity (**A**), PEM activity (**B**), PL activity (**C**), Cx activity (**D**), and β-Gal activity (**E**) of fresh-cut kiwifruit given different treatments during storage.

It is evident from Figure 4B that the PME activity of fresh-cut kiwifruit first increased and then decreased with increasing storage time. After 2 days of storage, the PME activity of fruit treated with HRW, SAEW, and H+S were 17.53 μg/g·min, 18.23 μg/g·min, and 17.17 μg/g min, respectively. No significant differences were found between them ($p > 0.05$). The PME activity of fruit in the CK treatment reached its maximum value on day 4, compared with other treatment groups, The H+S treatment significantly suppressed the increase in PME activity ($p < 0.05$), and the fruit PME activity of the H+S treatment group on day 4 was only 70.3% of that of the CK group. These results suggest that H+S treatment could delay the degradation of pectin substances by reducing PME activity in fresh-cut kiwifruit.

Figure 4C shows the effect of different treatments on the PL activity of fresh-cut kiwifruit. The PL activity of the samples increased and then decreased as the storage time increased, and the PL activity of the samples treated with HRW, SAEW, and H+S was lower than that of the samples treated with CK throughout the storage period. The PL activity of the fruits in the CK group peaked on day 6, while that of the other three treated samples peaked on day 8. The peak PL activity of fresh-cut kiwifruit treated with CK, HRW, SAEW, and H+S was 7.03 OD235/g·min, 5.43 OD235/g·min, 5.88 OD235/g·min and 4.92 OD235/g·min, respectively.

The influence of different treatments on the Cx activity of fresh-cut kiwifruit is shown in Figure 4D. During storage, the Cx activity of the fruit first increased and then decreased, reaching its peak value on the 6th day. The Cx activity of the sample in the CK group was significantly higher than that in other treatment groups ($p < 0.05$), which was 1.19 times, 1.09 times, and 1.45 times higher than that in the HRW, SAEW, and H+S treatment groups, respectively, at the 6th day of storage.

Fresh-cut kiwifruit β-Gal activity first increased and then decreased with the prolongation of storage time, as shown in Figure 4E. The sample β-Gal activity when treated with HRW, SAEW, and H+S was lower than in the CK group, and in fruit treated with H+S, the β-Gal activity was higher than that of the other two treatments during storage. Fresh-cut kiwifruits in each group were stored until the 6th day, when β-Gal activity reached its peak, and the fruit was treated with HRW and SAEW. H+S β-Gal activity was 77.2%, 90.66%, and 72.1% that of the CK group, respectively.

4. Discussion

The ripening and senescence of fruits and vegetables are closely related to ROS; the production rate and clearance rate of ROS in organisms can reach a relative balance under undamaged conditions. However, in the senescence of fruits and vegetables, the clearance rate of free radicals could be lower than the production rate of free radicals in the body. At this time, the balance would be broken, which leads to the accumulation of ROS in fruit [9]. The main reason for the ripening and senescence of fruits and vegetables is that an increase in ROS, especially the accumulation of $O_2^{\bullet-}$ and H_2O_2, causes oxidative damage to the biomolecules, such as lipid peroxidation, and will ultimately lead to cell death [20]. This study found that the contents of $O_2^{\bullet-}$ and H_2O_2 in fresh-cut kiwifruit increased rapidly in the early stages of storage (Figure 1), which is related to the decreasing function of the ROS scavenging system in the fruit. Compared with the CK group, HRW and SAEW have a positive effect on ROS removal and should keep them at a low level; the H+S treatment could reduce the content of $O_2^{\bullet-}$ and H_2O_2 in fresh-cut kiwifruit to a large extent, alleviating the oxidative stress of fruit.

Enhancing the activity of the ROS-scavenging enzymes (SOD, CAT, POD, and APX) is a common mechanism for eliminating ROS in fruit cells. SOD plays an important role in the ROS clearance of fruit and vegetable tissue, which is the first line of defense for eliminating ROS. It catalyzes the disproportionate nature of $O_2^{\bullet-}$ into H_2O_2 and molecular oxygen, and it is the only enzyme for $O_2^{\bullet-}$ scavenging [21]. The CAT, POD and APX enzymes work together to decompose H2O2 into H_2O and O_2 in fruits and vegetables [22]. In this study, H+S treatment significantly improved the SOD, CAT, POD, and APX activities of fresh-cut kiwifruit (Figure 2), indicating that the antioxidant activity was enhanced. One of the functions of HRW is probably that H_2 can easily penetrate the cell membrane, thus increasing the antioxidant gene expression of the encoding SOD, CAT, POD, and APX [23,24]. The effect of SAEW may be that the chloride in SAEW stimulates the oxidation of fruit [25].

The firmness of the fruit decreases continuously during storage, which is closely related to structural changes in the cell wall, mainly due to the hydrolysis of the pectin polymer in the pulp cell wall. This softening of the fruit is caused by an increase in the cell-wall metabolizing enzymes (PME, PG, PL, Cx, and β-Gal), resulting in the gradual degradation of cell-wall material components [26,27]. The fresh-cut fruit's firmness decreased continuously during storage; the H+S treatment group sample showed the highest firmness and chewiness value (Table 1), while the elasticity value remained stable during storage. At the early stages of storage, fresh-cut kiwifruit had a high hardness, and its cell wall was mainly composed of protopectin, hemicellulose, and cellulose. With the increase in storage time, the fruit appeared post-ripening and softening, and the protopectin in the fruit was continuously degraded (Figure 3), while the soluble pectin was constantly increasing in this process. In addition, the content of cellulose and hemicellulose also continued to decline, which is consistent with the research results of Langer et al. [28].

The main reason for the texture-softening of fruit is that the cell wall-degrading enzyme acts on the cell wall, reducing the connectivity degree between cells, and resulting in cell separation [29]. PG and PME are closely related to the decomposition of pectin in the fruit cell wall. The acting substrate of PG is the polygalacturonic acid in the cell wall, which is hydrolyzed into galacturonic acid or its oligomers. PME acts on the carboxyl group of pectin molecules to convert it into polygalacturonic acid, which is used as the hydrolysis substrate of PG. Peak PME activity appeared on day 4 for control samples and on day 6 for the HRW, SAEW, and H+S treatment groups, while the peak PG activity appeared on day 6 for the control group and on day 8 for the other treatment groups, indicating that the peak PME activity appeared earlier than PG, which is consistent with the results of previous studies [30,31].

5. Conclusions

(1) Compared with the CK group, the H+S treatment significantly ($p < 0.05$) reduced the content of free radicals ($O_2^{\bullet -}$ and H_2O_2) in fresh-cut kiwifruit during storage, increased the activities of the ROS scavenging enzymes, SOD, CAT, POD, and APX, and improved the antioxidant capacity of fruit.

(2) During storage, the firmness and chewability of fresh-cut kiwifruit decreased continuously, and the protopectin and cellulose were degraded. Compared with the CK group, the H+S treatment significantly increased ($p < 0.05$) the content of protopectin, cellulose, and hemicellulose in fresh-cut kiwifruit during storage, inhibited the increased rate of soluble pectin, delayed the decomposition of cell-wall substances, and maintained the integrity of the fruit cell wall.

(3) The decline in fruit firmness is related to the activity of cell wall-degrading enzymes. The H+S treatment reduced the PG, PME, PL, Cx, and β-Gal enzyme activity, and inhibited the degradation of pectin, cellulose, and hemicellulose, which are the main components of the cell wall so that the softening of fresh-cut kiwifruit tissues was delayed.

Author Contributions: Conceptualization, Y.S. and W.L.; methodology, X.Z. and W.Q.; software, X.Z.; validation, X.F. and X.Z.; formal analysis, X.Z.; investigation, Y.S.; resources, X.Z. and W.Q.; data curation, W.Q.; writing—original draft preparation, Y.S. and W.Q.; writing—review and editing, Y.S. and W.Q.; visualization, X.F.; supervision, Y.S.; project administration, W.L and X.X. All authors have read and agreed to the published version of the manuscript.

Funding: This research was funded by the National Key R&D Program of China (2018YFD0700303), the Shandong Province Technology Innovation Guidance Program (66812211118), the Shandong Province Modern Agricultural Industry Technology System (SDAIT-07-08), the Breeding Plan of Shandong Provincial Qingchuang Research Team-Innovation Team of Functional Plant Protein-Based Food (2021), Natural Science Foundation of Shandong Province (ZR2022QC104).

Institutional Review Board Statement: Not applicable.

Informed Consent Statement: Not applicable.

Data Availability Statement: Data is contained within the article.

Acknowledgments: Thanks to the National Key R&D Program of China (2018YFD0700303), the Shandong Province Technology Innovation Guidance Program (66812211118), the Shandong Province Modern Agricultural Industry Technology System (SDAIT-07-08), the Breeding Plan of Shandong Provincial Qingchuang Research Team-Innovation Team of Functional Plant Protein-Based Food (2021), Natural Science Foundation of Shandong Province (ZR2022QC104).

Conflicts of Interest: The authors declare no conflict of interest.

References

1. Richardson, D.P.; Ansell, J.; Drummond, L.N. The nutritional and health attributes of kiwifruit: A review. *Eur. J. Nutr.* **2018**, *57*, 2659–2676. [CrossRef] [PubMed]
2. Curutchet, A.; Dellacassa, E.; Ringuelet, J.A.; Chaves, A.R.; Viña, S. Nutritional and sensory quality during refrigerated storage of fresh-cut mints (*Mentha* × *piperita* and *M. spicata*). *Food Chem.* **2014**, *143*, 231–238. [CrossRef] [PubMed]
3. Carbone, K.; Macchioni, V.; Petrella, G.; Cicero, D.O.; Micheli, L. *Humulus lupulus* Cone Extract Efficacy in Alginate-Based Edible Coatings on the Quality and Nutraceutical Traits of Fresh-Cut Kiwifruit. *Antioxidants* **2021**, *10*, 1395. [CrossRef] [PubMed]
4. Chen, M.; Cui, W.; Zhu, K.; Xie, Y.; Zhang, C.; Shen, W. Hydrogen-rich water alleviates aluminum-induced inhibition of root elongation in alfalfa via decreasing nitric oxide production. *J. Hazard. Mater.* **2014**, *267*, 40–47. [CrossRef] [PubMed]
5. Guan, Q.; Ding, X.-W.; Jiang, R.; Ouyang, P.-L.; Gui, J.; Feng, L.; Yang, L.; Song, L.-H. Effects of hydrogen-rich water on the nutrient composition and antioxidative characteristics of sprouted black barley. *Food Chem.* **2019**, *299*, 125095. [CrossRef]
6. Chen, H.; Zhang, J.; Hao, H.; Feng, Z.; Chen, M.; Wang, H.; Ye, M. Hydrogen-rich water increases postharvest quality by enhancing antioxidant capacity in Hypsizygus marmoreus. *AMB Express* **2017**, *7*, 221. [CrossRef]
7. Ni, L.; Zheng, W.; Zhang, Q.; Cao, W.; Li, B. Application of slightly acidic electrolyzed water for decontamination of stainless steel surfaces in animal transport vehicles. *Prev. Veter. Med.* **2016**, *133*, 42–51. [CrossRef]
8. Cao, W.; Zhu, Z.W.; Shi, Z.X.; Wang, C.Y.; Li, B.M. Efficiency of slightly acidic electrolyzed water for inactivation of Salmonella enteritidis and its contaminated shell eggs. *Int. J. Food Microbiol.* **2009**, *130*, 88–93. [CrossRef]
9. Li, X.; Yue, H.; Xu, S.; Tian, J.; Zhao, Y.; Xu, J. The effect of electrolyzed water on fresh-cut eggplant in storage period. *LWT* **2020**, *123*, 109080. [CrossRef]
10. Xuan, X.-T.; Ding, T.; Li, J.; Ahn, J.-H.; Zhao, Y.; Chen, S.-G.; Ye, X.-Q.; Liu, D.-H. Estimation of growth parameters of Listeria monocytogenes after sublethal heat and slightly acidic electrolyzed water (SAEW) treatment. *Food Control* **2017**, *71*, 17–25. [CrossRef]
11. Issa-Zacharia, A.; Kamitani, Y.; Tiisekwa, A.; Morita, K.; Iwasaki, K. In vitro inactivation of Escherichia coli, Staphylococcus aureus and Salmonella spp. using slightly acidic electrolyzed water. *J. Biosci. Bioeng.* **2010**, *110*, 308–313. [CrossRef]
12. Ding, T.; Ge, Z.; Shi, J.; Xu, Y.-T.; Jones, C.L.; Liu, D.-H. Impact of slightly acidic electrolyzed water (SAEW) and ultrasound on microbial loads and quality of fresh fruits. *LWT* **2015**, *60*, 1195–1199. [CrossRef]
13. Zhang, C.; Lu, Z.; Li, Y.; Shang, Y.; Zhang, G.; Cao, W. Reduction of Escherichia coli O157:H7 and Salmonella enteritidis on mung bean seeds and sprouts by slightly acidic electrolyzed water. *Food Control* **2011**, *22*, 792–796. [CrossRef]
14. Zhang, C.; Zhang, Y.; Zhao, Z.; Liu, W.; Chen, Y.; Yang, G.; Xia, X.; Cao, Y. The application of slightly acidic electrolyzed water in pea sprout production to ensure food safety, biological and nutritional quality of the sprout. *Food Control* **2019**, *104*, 83–90. [CrossRef]
15. Tomás-Callejas, A.; Martínez-Hernández, G.B.; Artés, F.; Artés-Hernández, F. Neutral and acidic electrolyzed water as emergent sanitizers for fresh-cut mizuna baby leaves. *Postharvest Biol. Technol.* **2011**, *59*, 298–306. [CrossRef]
16. Yan, W.; Zhang, Y.; Yang, R.; Zhao, W. Combined effect of slightly acidic electrolyzed water and ascorbic acid to improve quality of whole chilled freshwater prawn (*Macrobrachium rosenbergii*). *Food Control* **2020**, *108*, 106820. [CrossRef]
17. Liao, X.; Xiang, Q.; Cullen, P.J.; Su, Y.; Chen, S.; Ye, X.; Liu, D.; Ding, T. Plasma-activated water (PAW) and slightly acidic electrolyzed water (SAEW) as beef thawing media for enhancing microbiological safety. *LWT* **2019**, *117*, 108649. [CrossRef]
18. Cheng, R.; Li, W.; Wang, Y.; Cheng, F.; Wu, H.; Sun, Y. Low voltage electrostatic field treatment of fresh-cut pineapples with slightly acidic electrolytic water: Influence on physicochemical changes and membrane stability. *Sci. Hortic.* **2023**, *308*, 111602. [CrossRef]
19. Lohani, S.; Trivedi, P.K.; Nath, P. Changes in activities of cell wall hydrolases during ethylene-induced ripening in banana: Effect of 1-MCP, ABA and IAA. *Postharvest Biol. Technol.* **2004**, *31*, 119–126. [CrossRef]
20. Bangar, S.P.; Trif, M.; Ozogul, F.; Kumar, M.; Chaudhary, V.; Vukic, M.; Tomar, M.; Changan, S. Recent developments in cold plasma-based enzyme activity (browning, cell wall degradation, and antioxidant) in fruits and vegetables. *Compr. Rev. Food Sci. Food Saf.* **2022**, *21*, 1958–1978. [CrossRef]
21. Dissanayake, I.H.; Zak, V.; Kaur, K.; Jaye, K.; Ayati, Z.; Chang, D.; Li, C.G.; Bhuyan, D.J. Australian native fruits and vegetables: Chemical composition, nutritional profile, bioactivity and potential valorization by industries. *Crit. Rev. Food Sci. Nutr.* **2022**, 1–34. [CrossRef] [PubMed]
22. Xu, F.; Liu, S. Control of Postharvest Quality in Blueberry Fruit by Combined 1-Methylcyclopropene (1-MCP) and UV-C Irradiation. *Food Bioprocess Technol.* **2017**, *10*, 1695–1703. [CrossRef]
23. Xu, F.; Liu, S.; Liu, Y.; Xu, J.; Liu, T.; Dong, S. Effectiveness of lysozyme coatings and 1-MCP treatments on storage and preservation of kiwifruit. *Food Chem.* **2019**, *288*, 201–207. [CrossRef] [PubMed]
24. Cui, W.; Gao, C.; Fang, P.; Lin, G.; Shen, W. Alleviation of cadmium toxicity in Medicago sativa by hydrogen-rich water. *J. Hazard. Mater.* **2013**, *260*, 715–724. [CrossRef] [PubMed]
25. Ohsawa, I.; Ishikawa, M.; Takahashi, K.; Watanabe, M.; Nishimaki, K.; Yamagata, K.; Katsura, K.-I.; Katayama, Y.; Asoh, S.; Ohta, S. Hydrogen acts as a therapeutic antioxidant by selectively reducing cytotoxic oxygen radicals. *Nat. Med.* **2007**, *13*, 688–694. [CrossRef]

26. Hu, X.; Chen, Y.; Wu, X.; Liu, W.; Jing, X.; Liu, Y.; Yan, J.; Liu, S.; Qin, W. Combination of calcium lactate impregnation with UV-C irradiation maintains quality and improves antioxidant capacity of fresh-cut kiwifruit slices. *Food Chem. X* **2022**, *14*, 100329. [CrossRef]
27. Su, N.; Wu, Q.; Liu, Y.; Cai, J.; Shen, W.; Xia, K.; Cui, J. Hydrogen-Rich Water Reestablishes ROS Homeostasis but Exerts Differential Effects on Anthocyanin Synthesis in Two Varieties of Radish Sprouts under UV-A Irradiation. *J. Agric. Food Chem.* **2014**, *62*, 6454–6462. [CrossRef]
28. Langer, S.E.; Marina, M.; Francese, P.; Civello, P.M.; Martínez, G.A.; Villarreal, N.M. New insights into the cell wall preservation by 1-methylcyclopropene treatment in harvest-ripe strawberry fruit. *Sci. Hortic.* **2022**, *299*, 111032. [CrossRef]
29. Huang, J.; Wu, W.; Fang, X.; Chen, H.; Han, Y.; Niu, B.; Gao, H. *Zizania latifolia* Cell Wall Polysaccharide Metabolism and Changes of Related Enzyme Activities during Postharvest Storage. *Foods* **2022**, *11*, 392. [CrossRef]
30. Zhao, X.; Meng, X.; Li, W.; Cheng, R.; Wu, H.; Liu, P.; Ma, M. Effect of hydrogen-rich water and slightly acidic electrolyzed water treatments on storage and preservation of fresh-cut kiwifruit. *J. Food Meas. Charact.* **2021**, *15*, 5203–5210. [CrossRef]
31. Ma, Q.; Cong, Y.; Wang, J.; Liu, C.; Feng, L.; Chen, K. Pre-harvest treatment of kiwifruit trees with mixed culture fermentation broth of Trichoderma pseudokoningii and Rhizopus nigricans prolonged the shelf life and improved the quality of fruit. *Postharvest Biol. Technol.* **2020**, *162*, 111099. [CrossRef]

foods MDPI

Review

Perspectives on Novel Technologies of Processing and Monitoring the Safety and Quality of Prepared Food Products

Jinjin Huang [1,2], Min Zhang [1,3,*] and Zhongxiang Fang [4]

1 State Key Laboratory of Food Science and Resources, Jiangnan University, Wuxi 214122, China; leisurehuang1108@163.com
2 Jiangsu Province International Joint Laboratory on Fresh Food Smart Processing and Quality Monitoring, Jiangnan University, Wuxi 214122, China
3 China General Chamber of Commerce Key Laboratory on Fresh Food Processing & Preservation, Jiangnan University, Wuxi 214122, China
4 School of Agriculture and Food, The University of Melbourne, Parkville, VIC 3010, Australia; zhongxiang.fang@unimelb.edu.au
* Correspondence: min@jiangnan.edu.cn; Tel./Fax: +86-510-85877225

Abstract: With the changes of lifestyles and rapid growth of prepared food industry, prepared fried rice that meets the consumption patterns of contemporary young people has become popular in China. Although prepared fried rice is convenient and nutritious, it has the following concerns in the supply chain: (1) susceptible to contamination by microorganisms; (2) rich in starch and prone to stall; and (3) vegetables in the ingredients have the issues of water loss and discoloration, and meat substances are vulnerable to oxidation and deterioration. As different ingredients are used in prepared fried rice, their food processing and quality monitoring techniques are also different. This paper reviews the key factors that cause changes in the quality of prepared fried rice, and the advantages and limitations of technologies in the processing and monitoring processes. The processing technologies for prepared fried rice include irradiation, high-voltage electric field, microwave, radio frequency, and ohmic heating, while the quality monitoring technologies include Raman spectral imaging, near-infrared spectral imaging, and low-field nuclear magnetic resonance technology. These technologies will serve as the foundation for enhancing the quality and safety of prepared fried rice and are essential to the further development of prepared fried rice in the emerging market.

Keywords: prepared fried rice; food processing; quality monitoring; food safety

Citation: Huang, J.; Zhang, M.; Fang, Z. Perspectives on Novel Technologies of Processing and Monitoring the Safety and Quality of Prepared Food Products. *Foods* **2023**, *12*, 3052. https://doi.org/10.3390/foods12163052

Academic Editor: Ken'ichi Ohtsubo

Received: 15 July 2023
Revised: 8 August 2023
Accepted: 12 August 2023
Published: 15 August 2023

1. Introduction

In recent years, especially after the COVID-19 pandemic, people's lifestyles have been changed tremendously, including the food consumption pattern. With the popularity of simple household appliances such as refrigerators and microwave ovens, the markets for prepared foods have grown significantly in North America, Europe, and China. In the convenience food sector alone, revenues reached $562.2 billion in 2022, where China contributed a significant proportion ($126.5 billion), and the convenience food industry is predicted to expand at a rate of 6.21% annually [1]. Prepared fried rice is mainly made from rice, which is stir-fried in cooking oil, seasoned with salt and other ingredients (with pre-processed dishes or ingredients such as eggs, seafood, green beans, carrots, sausages, etc.), and then sterilized and packaged for distribution. Consumers only need a simply reheating process before eating. With a more balanced nutritional profile, prepared fried rice caters to young customers who are busy and have no time to cook, and fits a more efficient and healthier lifestyle. The emergence of smart kitchens and future food manufacturing models have produced a new pattern of Chinese cuisine. In 2016, the State Council of China published the opinions to broaden the supply channel of staple food and create a high standardized and high-quality staple food brand [2]. It promoted industrial production of traditional Chinese staple foods and dishes. As a result, many representatively prepared fried rice brands have emerged in China markets (Table 1). Moreover, the cold chain logistics market of China has maintained an annual growth rate of more than 11% from 2016. The improvement of cold-chain logistics level and the development of food process technologies (including quick-freezing and preservation technologies) provided technical supports for the long-distance supply of prepared fried rice [3]. Due to different food ingredients in making prepared fried rice, appropriate processing and preservation and quality monitoring techniques should be applied to ensure food safety and quality [4]. Most companies currently choose −18 °C for low-temperature preservation during transportation. Convenience and health, low carbon emissions are the most critical development direction of prepared foods worldwide. Some companies are already experimenting with 4 °C refrigerated preservation with a view to reducing cold-chain transportation costs.

Prepared fried rice is a type of emerging food with Chinese characteristics (taste, flavor etc.), and there are few reported pieces of research on it. Similar to Chinese fried rice, Thailand has rice cracker snacks that require deep-frying [5], and Korea has the traditional oil-puffed snack Yukwa [6,7]. Although there are representative brands of fried rice in Japan, there is no classification of the types of fried rice. Most European and American consumers are used to eating risotto. Research has mainly focused on the pathogenicity [8] and growth kinetics [9,10] of harmful microorganisms in fried rice, with little discussion on the potential applications of novel physical processing technologies and nondestructive testing techniques on prepared fried rice. The processing of prepared fried rice is divided into the treatment of raw materials before cooking; the way of flipping during cooking; and the sterilization, packaging, retarding aging, and reheating after cooking. Uneven reheating can affect the flavor and texture of prepared fried rice. Therefore, the post-cooking processing technology is critical. Traditional heat treatment sterilization techniques (pasteurization: temperature below 100 °C; high temperature sterilization: temperature higher than 100 °C; ultra-high temperature instant sterilization: temperature at 135~150 °C) could destroy the flavor substances and color of the prepared fried rice. Physical field technology and anti-bacterial coating technology consume less energy and can retain the inherent nutrients and freshness of raw materials, thus minimizing the deterioration of quality caused by conventional sterilization processes. Moreover, physical field technologies have wide applications in enzymatic reaction [11], fermentation [12], and retarding aging [13]. Their advantages of no additives and continuous inhibition of harmful microbial growth and reproduction are more readily accepted by consumers. Therefore, this paper focuses on the potential sterilization technologies (irradiation, high-voltage electric field, microwave, radio frequency, ohmic heating), and quality monitoring (Raman spectral imaging, near infrared spectral imaging, and low-field nuclear magnetic resonance) technologies for prepared fried rice (Figure 1). These technologies are crucial to ensure the quality and safety of prepared fried rice and will provide guidance for further research and development for this group of emerging food products.

Table 1. Selective prepared fried rice brands and products on the Chinese market.

Brands	Introduction	Representative Products	Features	Main Ingredients	Official Website	Other Information (Reprieved from https://nlc.chinanutri.cn/fq/, accessed on: 8 July 2023)
好米道 炒饭	• Haitong Foods • Founded in 1985, Haitong Foods Group Co., Ltd., Cixi, China. • Green food research and development • Food production freezing and refrigeration of fruits and vegetables and aquatic products	Mixed fried rice	• Precise temperature control • Keep fresh at −35 °C	• Preserved pork • Ox tripe • Ham • Bacon	https://www.haitonggroup.com/, accessed on: 8 July 2023	• Storage conditions and shelf life: −18 °C or 4 °C, 3–12 months • Additives: modified soy phospholipids, condiment, pectin, acacia senegal, sodium bicarbonate • Capacity: 500–1000 kJ/100 g • Protein: 3–5 g/100 g • Mainly from rice, eggs, meat, and beans • Fat content: 5–15 g/100 g • Carbohydrate content: 20–40 g/100 g, mainly from rice and vegetables • Sodium content: 400–500 mg/100 g • Taste: similar to fresh fried rice; • salty, spicy, or sweet and spicy • Flavor: emits a strong aroma after heating
		Korean kimchi grilled sausage fried rice		• Sausage • Pickle		
		Japanese eel fried rice		• Eel		
三全	• Sanquan Foods • One of the companies producing frozen foods in China, which was founded in 1992. Prepared fried rice is a new type of prepared food produced by Sanquan.	Yangzhou fried rice	• Don't need to thaw • Don't need to open the bag • Microwave for 3 min in bag	• Egg • Green bean • Corn • Carrot • Shrimp meat	http://www.sanquan.com/sanquan, accessed on: 8 July 2023	
		Cantonese fried rice		• Sausage • Green vegetable		
		Thai fried rice		• Seafood • Pineapple		
北戴河 BDH	• HUANGDAO OCEAN FOOD • Founded in 1960, and is a comprehensive food processing enterprise integrating R&D, production, and sales. • Canned food • Compressed dry food • Self-heating food • Ready-to-eat food • Soup products	Mixed fried rice	• Unplugged • No fire • Heated for 15 min to eat • Easy to carry • Heated with water • Convenient and fast • Separated for easy storage	• Pork • Green bean • Carrot • Mushroom	http://www.hysp1960.com/, accessed on: 8 July 2023	
肯德基自在厨房。	• Kentucky Fried Chicken • Produced prepared foods since 2020 • Products stored in refrigerated cabinets • Steak • Fried rice/noodles • Coffee • Chicken soup • Chicken breast	New Orleans-style chicken fried rice	• Continuous stir-frying above 200 °C • Quick-freezing process • No firing • No washing dishes • Microwave heating	• Chicken • Carrot • Green bean	http://www.yumchina.com/Brand, accessed on: 8 July 2023	
		Sichuan tender beef Japanese fried rice		• Beef • Green bean • Carrot		
伍の料理 Wuの LIAOLI	• Wu Cuisine is a brand of Wu Group in Quanzhou, China. • Wu Group is a catering enterprise integrating the development and production of frozen seasoning kits, brand chain restaurants, and meal hosting with a "central kitchen factory". • Chinese and western frozen seasoning packages • Quick frozen food	Laver salmon fried rice	• Dishes go from 98 °C to −35 °C in 25 min	• Salmon • Nori • Garlic sprouts	http://www.wusjt.com, accessed on: 8 July 2023	
		European-style country beef fried rice		• Beef • Green bean		
		Bacon fried rice		• Corn • Green bean • Bacon		
		Tuna fried rice		• Tuna • Corn		

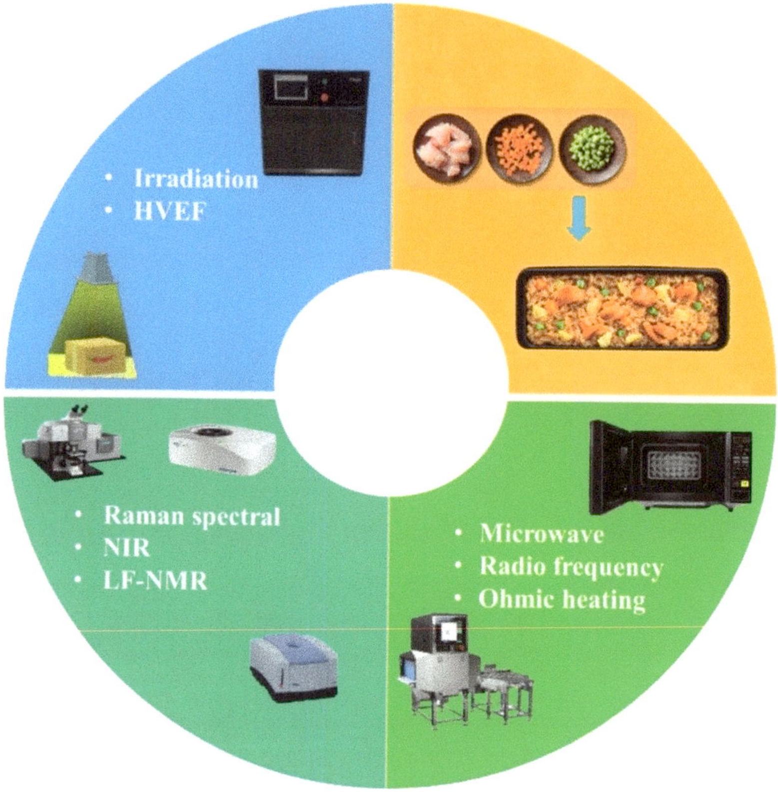

Figure 1. Schematic diagram of processing and quality testing technologies for prepared fried rice.

2. Potential Processing Technologies for Prepared Fried Rice

2.1. Sterilization Technologies

Unfavorable conditions during prepared fried-rice processing or storage can lead to the growth of pathogenic bacteria and hazardous substances produced by microbial metabolism [9]. As a raw material, rice can become contaminated during cultivation and storage by dust, water, plants, insects, soil, fertilizers, and animal dung, carrying harmful substances such as aflatoxin and *Bacillus cereus* [14]. There are potential biohazards associated with the use of contaminated rice for fried rice preparation. Moreover, the processing methods, factory sanitary conditions, packaging methods, and storage environment also affect the quality and safety of prepared fried rice [9]. *Vibrio parahaemolyticus* had a high growth rate in egg fried rice at room temperature due to inadequate sterilization and cold chain systems [15]. In prepared fried rice, microbial contamination mainly appeared on the surface of fried rice. Heat-sensitive foods such as egg whites or prepared fried rice were prone to contain *Bacillus cereus*, which is highly resistant to high temperatures and acidity during cooking and may cause acute liver failure and encephalopathy if ingested [16]. Typically, rapid cooling is applied after the fried rice is cooked, but it is also the most vulnerable period to microbial contamination. The correct refrigeration temperature was used during storage to control the harmful microorganisms, but those with spores remained alive [17]. *Bacillus cereus* may grow exponentially in fried rice when cooked in ham fried rice, scrambled egg fried rice, and pea fried rice, which were slowly cooled and stored at 15 °C [18]. Diarrhea and vomiting in accidental eaters of fried rice in a Korean restaurant were reported, due to the contamination of *Bacillus cereus* with a total colony count of 1.48–3.47 log CFU/g, presumably due to spoilage from improper storage [19]. In addition, eggs and eggshells were susceptible to *Salmonella* contamination, and *Salmonella* serovar can cause severe gastrointestinal disease [20]. Some researchers have studied the growth and hemolysin production mechanism of two strains of *Vibrio alginolyticus* and reported their toxicity in fried rice to raise public awareness of the safety of fried rice [21]. The examples shown above demonstrated a high possibility of microbiological contamination in prepared fried rice. Prepared fried rice should be primarily inhibited from harmful microorganisms such as *Bacillus cereus*, *Vibrio parahaemolyticus*, and aflatoxins produced by *Aspergillus flavus* and *Aspergillus parasiticus* during processing. Ensuring fresh raw materials and good sanitary environment during processing are critical to the safety and quality of prepared fried rice. The following are some important methods that could be applied to further improve the product safety and quality.

2.1.1. Irradiation

Irradiation is a non-thermal method that is frequently used to preserve food, where food is subjected to a specific level of ionizing radiation generated by mechanical or natural sources; and the DNA of microorganisms is destroyed, cell division is prevented, and the cells stop activities, thus providing a sterilizing effect. Generally, electron beam, X-ray, and γ-ray are used for food irradiation. Mechanical sources produce X-rays and electron beam, which can be turned on and off. γ-rays are generated by radionuclides ^{60}Co and ^{137}Cs. They have different effects on microorganisms, with penetration depths of about 8 cm, 20 cm, and 40 cm, respectively [1]. However, low consumer acceptance to irradiated food is an issue for the rapid application of this technology, although the Joint Expert Committee on Food Irradiation of the World Health Organization has demonstrated the safety of food irradiation at doses below 10 kGy and that these sources are not radioactive to cause safety hazards [22]. Moreover, appropriate irradiation doses are beneficial to preserve the color and nutrition of the food. In an investigation into the shelf-life of ready-to-eat fried diced chicken with chili, researchers found that 10 kGy γ-rays effectively killed pathogenic bacteria, such as *Salmonella*, *Escherichia coli*, *Bacillus* spp., and *Staphylococci*, with no loss of sensory quality and protein content [23]. Prepared fried rice usually contains rice, meat, and green vegetables, such as green beans and cabbage. It is crucial to take into account how the sterilization methods will affect the protein, fat, and starch in the fried rice and reduce its impact on the flavor and quality of the rice and vegetables during the sterilization process. Researchers found that irradiating brown rice with an intensity of 1 kGy γ-rays dramatically reduced its microbial population (molds and yeasts) after two-month storage without affecting the overall sensory tolerability. The irradiated samples became softer and the members of the sensory evaluation panel considered that the cooked brown rice was of better quality under 1 kGy irradiation [24]. Furthermore, a dose of 1 kGy γ-rays was discovered to be adequate to significantly lower the level of *Escherichia coli* (ATCC# 25922) in all mixed vegetables and increase the shelf life by up to 4 days [24].

Although irradiation benefits from a quick processing period and environmental friendliness, it also has limitations. For electron beam, the energy and dose limit are 10 MeV and 10 kGy. The maximum energy and dosage for X-rays are 5 MeV (US 7.5 MeV) and 10 kGy. The maximum energy and dosage for γ-rays is 5 MeV and 10 kGy [1]. High doses (4.5 kGy) of electron beam irradiation intensity can oxidize lipids and vitamins in foods [25]. Moreover, it has been shown that irradiation levels of up to 44 kGy are necessary to inactivate microbial spores, but the current commonly used 10 kGy is far below this level. Therefore, it is necessary to combine irradiation with other techniques to achieve effective bactericidal effects [26]. For example, combining different doses of γ-irradiation with 4 °C cold storage significantly reduced the number of toxic *Bacillus cereus* [27]. γ-irradiation combined with active coating at 0.4 kGy had a synergistic inhibition effect on *Escherichia coli* and *Aspergillus niger* [28]. Therefore, combining radiation with other techniques could be an effective method in the preservation of prepared fried rice.

2.1.2. High-Voltage Electric Field

High-voltage electric field sterilization technology is an effective, low energy-consuming, and non-polluting method that can preserve food quality. The high-voltage electric field is divided into high-voltage pulse electric field and high-voltage electrostatic field. High-voltage pulsed electric fields apply a strong electric field to food within a relatively short time. Compared to monopolar pulses, bipolar pulses have a greater impact on the porosity of cell membranes [29], and increasing the number of pulses and pulse frequency can also enhance the sterilization effect [30]. High-voltage pulsed electric field sterilization can be carried out at room temperature. It causes damage to the bacterium's enzymatic activity, DNA, and protein structure, impairing its typical physiological function and resulting in cellular breakage or death [31]. Prepared fried rice with ingredients including rice, green beans, and carrots may have been infected with *Staphylococcus aureus* and *Bacillus licheniformis* on the farm or during production. High-voltage pulsed electric fields have significant lethal effects on common foodborne pathogens such as *Staphylococcus aureus*, *Salmonella typhimurium*, *Escherichia coli*, and *Listeria monocytogenes* [32]. Fernandez-Diaz et al. (2000) concluded that pasteurization at 55.6 °C for 6.2 min or 56.7 °C for 3.5 min may not inactivate *Salmonella* spp. and *Listeria monocytogenes* [33]. While applying 26 kV/cm and 37 °C with 100 exponential decay pulses of 4 μs to continuously processed the liquid eggs, the proteins do not coagulate and the flavor substances of the eggs are retained; however, the harmful microorganisms decrease with the number and width of pulses increasing. They can be combined with other techniques to obtain better bactericidal effects. When high-voltage pulse electric field technology and cryoconcentration are used together, sub-lethal microbes can be suppressed without causing heat damage to the food [34]. Similarly, with the addition of nisin, a synergistic effect occurred and the bactericidal effects of the pulsed electric field was more effective [33].

High-voltage electrostatic field equipment is relatively simpler and lower cost compared with that of a high-voltage pulsed electric field (Figure 2). The electric field strength is also lower. High-voltage electrostatic fields kill mold and bacteria on food surfaces by causing small amounts of charged particles present in the air to constantly collide with neutral molecules or atoms. High-voltage electrostatic fields also cause oxidative stress and DNA sub-damage in bacteria, which have a sterilization effect on both solid and liquid foods, reducing the number of chemical preservatives and preserving the nutrients of the food [35]. Qi et al. (2021) studied the bactericidal effect of high-voltage electrostatic fields on salmon, pork, and sausage, in which the lethality of *Staphylococcus aureus* reached 92.1–99.8% [36]. These ingredients are often used in different types of prepared fried rice. Hsieh et al. (2011) measured the sensory quality index of tilapia at an electric field intensity of 1 kV/cm, and the treated group was fresher compared with the untreated ones, where the electric field intensity inhibited the protein denaturation of fish and significantly reduced the number of microorganisms [37]. Ko et al. (2016) obtained the same conclusions. After successive treatments with high voltage electrostatic fields at 3, 6, and 9 kV/cm for 8 days, it maintained the protein content, solubility, and Ca_2-ATPase activity [38]. *Acinetobacter johnsonii*, which is commonly existed in the water and soil around fish culture, is also present in fresh or spoiled processed foods and possibly in prepared fried rice with complex ingredients. After 15 min of treatment with a 30 kV high-voltage electrostatic field, the number of *Acinetobacter johnsonii* decreased, intracellular nucleic acids and proteins leaked, conductivity and reactive oxygen species (ROS) content increased 16.88-fold [39]. The sterilizing effect can be improved by combining high-pressure electrostatic field technology with other technologies. For example, the combination of high-pressure electrostatic field and controlled freezing point technique (temperature range from 0 °C to the freezing temperature of the organism) can successfully suppress microbial development and reproduction and postpone the loss of food quality [38].

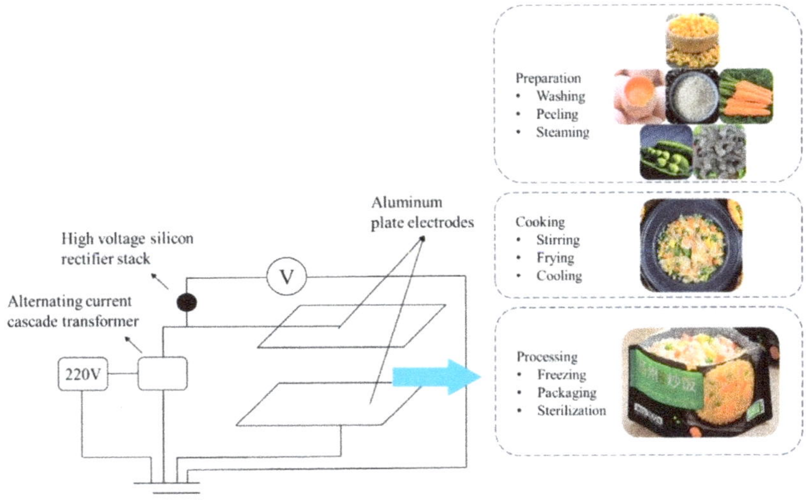

Figure 2. A processing diagram of prepared fried rice under high-voltage electrostatic field, adapted from [36], with permission from ELSEVIER, 2023.

The prepared fried rice needs to be reheated before consumption, i.e., the second heating of the cooked food. Although reheating prepared fried rice is a common consumer behavior, many restaurants are now introducing prepared fried rice to improve the efficiency of meal delivery, and convenience stores are selling prepared fried rice and providing reheating services. Therefore, new processing technology of reheating should be taken into consideration for prepared fried rice. Reheating can cause changes in sensory quality such as moisture, texture, color, starch pasting, digestibility, and other physical and chemical properties of food. The current research on reheating technologies focuses on the overcooked taste and the food quality after reheating. For prepared fried rice, moving closer to the quality of restaurant fried rice after reheating has attracted more interest. Traditional reheating, such as steaming, can maintain certain qualities of the food, but has the disadvantage of taking a long time and adding moisture to the food. Recently, microwave reheating and radio frequency reheating have become popular in reheating processing.

2.1.3. Microwave

Microwave heating has both a sterilization and reheating effect on the food. Microwave-assisted pasteurization is the process of pasteurization by using microwaves to act directly on the material, using the movement of the food's own polar molecules and conductive ions, friction, to generate heat. Compared to traditional heating methods, microwave heating is three to five times faster [40]. In addition, it reduces the loss of nutrients, flavor substances, and texture deterioration during food heating [41]; reduces thermal degradation of heat-sensitive ingredients; improves food quality; and extends shelf life [42]. For instance, *Bacillus cereus* was inoculated in fried rice and then heated in a frying pan at high temperature (internal temperature of fried rice: 103.8–121.4 °C), medium temperature (internal temperature of fried rice: 69.2°C), or microwave (internal temperature of fried rice: 86.3–90.6 °C) for 3 min, respectively; and the results showed that heating in microwave oven was the most effective to control the *Bacillus cereus* cells and spores. With microwave heating, the fried rice can be kept safe at 25 °C for 6 h, 35 °C for 3 h, and 45 °C for 2 h [43]. After microwave-assisted pasteurization, Montero et al. (2020) examined the physical, chemical, sensory, and microbiological properties of fried egg rice. The treated product had a shelf life of up to 6 weeks at 7 °C storage, where no sensory related microbial-induced spoilage was detected; and the eggs had an acceptable firm texture, showing the great potential of microwave-assisted pasteurization for this group of food [44]. It is noteworthy that 10% of American consumers store their food at a temperature of 7 °C most of the time [42], indicating the practical application of this technology.

Microwave not only has a good sterilization effect, but also a reheating effect. When microwave applies, the whole food will be heated up instantly and homogeneously. Researchers have investigated the quality loss and flavor characteristics of food products during microwave reheating. It was reported that meat products have little difference in cooking loss during microwave reheating compared to other reheating methods such as steaming, because of microwave's high heating efficiency [45]. Microwave can also well-retain active ingredients such as ascorbic acid in vegetables; and the loss of total phenols, total anthocyanins, and chlorogenic acid is much lower than that of stir-frying and deep-frying [46,47]. However, it is important to note that hot and cold spots caused by uneven microwave heating and moisture loss can lead to bland sensory quality of foods after microwave reheating. Geedipalli, Rakesh, and Datta (2007) placed 3.6 × 4.7 × 2.1 cm potato cubes in the center of a turntable inside a 2459 Hz microwave oven and heated for 35 s by completing a rotational cycle every 10 s [48]. Figure 3 compared the effect of the same oven with and without the turntable on the uniformity of potato temperature distribution. The static-heated potatoes had concentrated high-temperature areas, while the potatoes heated by the combined microwave rotary heating had more uniform temperature contours on their surfaces and were heated more efficiently with low-quality loss. Moreover, the electric field applied to the microwave cavity and temperature distribution can be influenced by the food dielectric characteristics and microwave frequency [49–52]. Therefore, choosing the appropriate microwave time and power is important to achieve the best results for reheating prepared foods. Researchers compared the sensory score and temperature distribution of lasagna with different combinations of frying time (5.5, 7, and 8.5 min) and microwave reheating power (low, medium low, medium, medium high, and high power), and determined that a frying time of 7.0 min and microwave power of medium-low power (4.20 W/g) and 80 s reheating resulted in the best quality. Wang et al. (2018) studied the uneven temperature distribution of convenience rice during microwave reheating and proposed an intermittent reheating power and packaging container design method to improve the microwave heating uniformity [53]. In addition, combining microwave and infrared technology was considered more effective than the single microwave heating [54]. This is because polar molecules and conductive ions inside the dish produced heat as a result of microwave; and infrared tubes radiate heated the food from the external environment to reduce the temperature difference between the interior and exterior, therefore inhibiting the migration of polar molecules such as moisture and enhancing temperature uniformity [55]. Lasagna, a pre-made pasta, is similar to prepared fried rice. It is made in a similar process and the ingredients are rich in starch, with a supplementary of oil, salt, and others; in addition, both need to be kept refrigerated [56]. The way lasagna is reheated can also be considered as a potential prepared fried rice reheating method. Comparing traditional reheating methods (steam, baking) and new reheating methods (air fryer, microwave, and combined infrared microwave), lasagna reheated by infrared combined with microwave had better temperature uniformity than microwave alone and increased internal moisture to 33.85%. This method also inhibited the outward migration of moisture, reduced the hardness, and improved the sensory property [57].

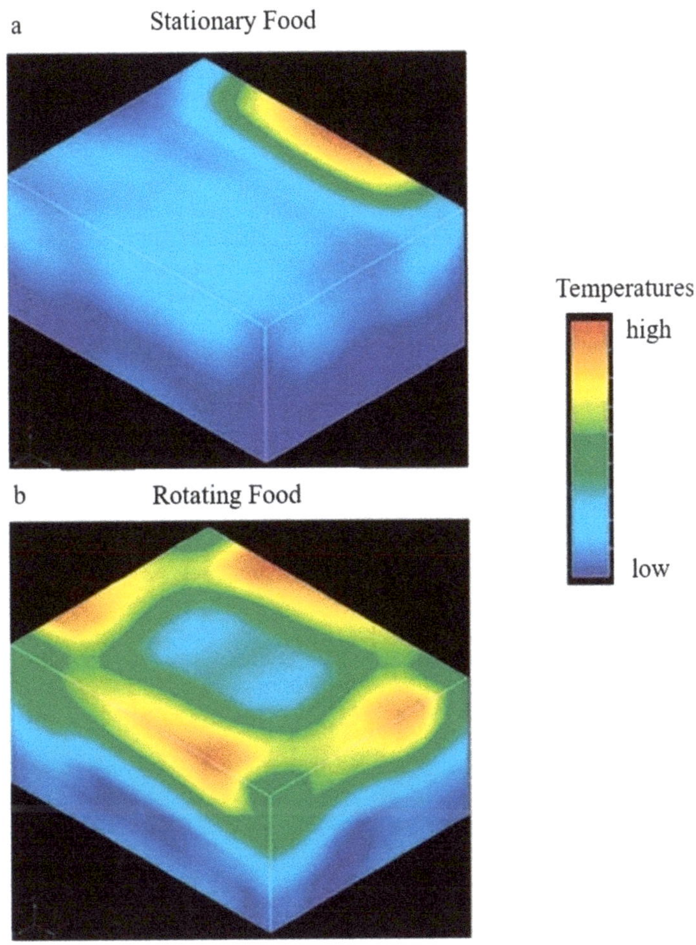

Figure 3. Microwave heating temperature distribution in a non-rotation (**a**) and rotation mode (**b**), adapted from [48], with permission from ELSEVIER, 2023.

2.1.4. Radio Frequency

Radio frequency (RF) is a high-frequency alternating current electromagnetic wave with frequencies between 300 kHz~300 MHz; and only 13.56 MHz, 27.12 MHz, and 40.68 MHz are used in industry [4]. RF treatment not only has a sterilizing effect on the individual components of prepared fried rice, but also has the effect of reheating and regulating the structure, digestibility, and physicochemical properties of rice starch [58].

In terms of the sterilization effect, RF penetrates the material interior and causes the rhythmic migration of charged ions, converting electrical energy into heat, and rapidly increasing the bulk temperature of the prepared fried rice. With a deep penetration rate and high energy efficiency, it overcomes the defects of long traditional heating time and excessive heating at the edges of packaged food [59]. The RF sterilization process is milder, but the effect is similar with the conventional retorting method; in addition, its effect on the quality deterioration of the product, such as nutrient content, texture, taste, and odor, is significantly reduced. RF has a high potential to be applied for prepared fried rice processing. It was observed that freshly cooked eggs were better preserved with a significantly longer shelf life after RF sterilization, so the U.S. military used RF technology to improve the shelf life of eggs to meet soldiers' needs [60]. After the sterilization of *Nostoc sphaeroides* using 6 kW, 27 MHz RF for 20 min, microbial counts were significantly reduced and the effect on its color and flavor was significantly lower than that of conventional autoclaving [61]. In addition, researchers have conducted extensive research on the sterilization effect of RF in combination with other techniques on food products. It was found that the effects of ZnO nanoparticles combined with RF heating on the sterilization of carrots and product quality characteristics was superior to either ZnO nanoparticles or RF heating treatment alone and increased the shelf life of prepared carrots up to 60 days [62].

For the research on the reheating effect of RF, most previous experiments only sterilized or reheated one ingredient; and it is difficult to evaluate the effect on prepared fried rice, as it has multiple ingredients. There are few studies using RF to examine the reheating behavior of non-homogeneous food materials. Lan et al. (2020) performed RF reheating of pizza, which is also a heterogeneous food. With selected RF parameters of 6 kW, 27.12 MHz, a maximum reheating temperature of 85 °C, and an electrode gap of 110 mm, a good consistency of pizza bases, cheese, salami, and onions temperatures after 7.33 min of reheating was obtained. The results suggested that heating non-homogeneous foods with similar loss factors and dielectric constants may require higher RF heating rates and that the uniformity of RF heating can be achieved by the moving or rotating of the raw materials [63]. Wang et al. (2012) also reheated meat lasagna noodles, a kind of highly heterogeneous food, with RF. When the beef, mozzarella cheese, noodles, and sauce were properly distributed, they showed little difference in temperature and maintained good product quality after the reheating process [64].

2.1.5. Ohmic Heating

Ohmic heating is also known as energized heating, where the food is acted as a resistance unit and electrical energy is applied through two electrodes located on both sides of the heating chamber in direct contact with the food material (Figure 4). When an electric current is passed through a conductive material, the internal temperature of the material increases due to the conversion of electrical energy into thermal energy by the Joule effect. Although convective heat-transfer modes are still the most popular heating technologies in the food industry, ohmic heating with its time and energy intensive advantages could serve as a potential alternative to traditional methods in prepared fried rice [65]. Ohmic heating is used in food industries for sterilization, thawing, extraction, and others [66], which has benefits of a high rate of electric heat conversion, simple and easy-to-control equipment, and no scaling, etc. [67]. It can save heating time by 48% and is more energy efficient than steam heating. Additionally, unlike microwave heating, there is no need for radiation isolation material during processing to protect the operator [68]. The effectiveness of ohmic heating is directly impacted by the electrical conductivity of the food substance. Electric field strength is proportional to the square of electrical conductivity; however, as particle size rises, electrical conductivity decreases [69]. As shown by Benabderrahmane and Pain (2000) and Icier and Ilicali (2005), conductivity decreases with increasing stacking density [70,71]. After salting, the electrical conductivity of the food material increases [72]. The conductivity is also influenced by particle position and orientation, with the solid phase heating up more quickly than the liquid phase under series settings and the opposite being seen under parallel conditions [69]. The material state of the food, the contact area between the electrodes and the food material, and the distance between these electrodes also affect the heating rate.

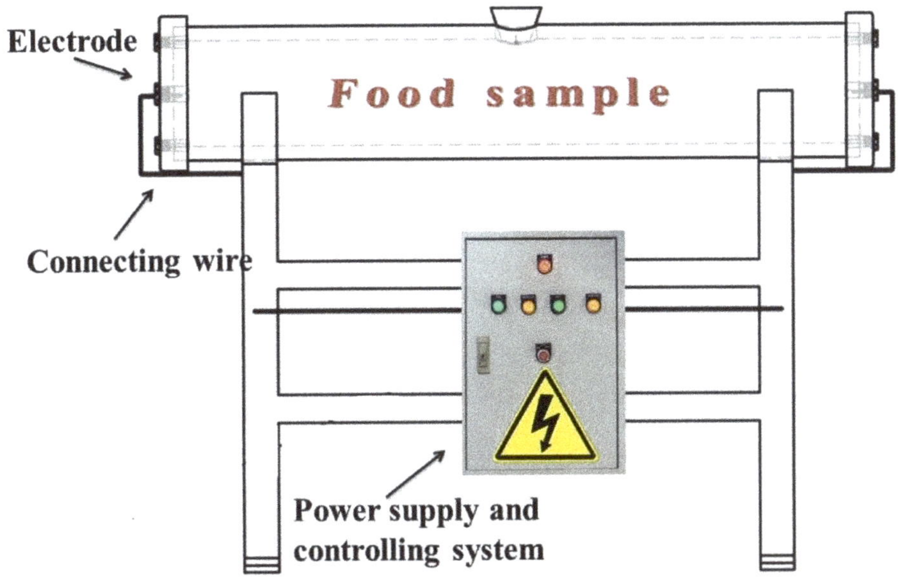

Figure 4. An ohmic food processing system, adapted from [65], with permission from ELSEVIER, 2023 (Gavahian et al., 2019).

Ohmic heating has a non-thermal effect (50–60 Hz) due to the presence of an electric field. The cell membrane has a specific dielectric strength and the presence of an electric field exceeding the electric field strength leads to changes in the cellular tissue structure. The cell wall accumulates charge and forms pores, cell membrane permeability is enhanced, and intracellular components diffuse, thus inactivating microorganisms [73]. Ohmic heating has a bactericidal effect on *Bacillus licheniformis*, *Escherichia coli*, *Bacillus subtilis*, and *Streptococcus thermophilus* that may be present during the processing of prepared fried rice [74]. It can even be used to inactivate heat-resistant bacterial spores, such as the most heat-resistant *Bacillus thermophilus* spores [75]. However, Somavat et al. (2012) argued that additional non-thermal effects can only be observed under specific treatment conditions or in specific food samples [75]. If a stronger non-thermal effect in prepared fried rice is required, its electrical conductivity should be increased, for example, by changing the stacking density of the rice grains. The thermal effects frequently conceal the non-thermal effects of ohmic heating that are used to inactivate foodborne bacteria. However, when other methods are utilized in addition to ohmic heating, the non-thermal effect is more pronounced and the pathogen inactivation rate rises dramatically. Combined treatments of non-homogenized salsa (tomatoes, onions, small peppers, salt, black pepper powder, lemon juice) with ohmic heating and parsley phenol had a synergistic bactericidal effect on foodborne pathogens such as *Escherichia coli O157:H7*, *Salmonella Typhimurium*, and *Listeria monocytogenes*. Compared to single ohmic heat sterilization, the combined treatment is less destructive to the composition of the food and does not affect the color and quality of the salsa [76]. Similarly, the combination of ohmic heating with citral or muscimol showed a synergistic bactericidal effect against the above-mentioned food-borne microorganisms [77]. Because foodborne pathogens can survive for long periods in cold spots, De Alwis and Fryer (1990) applied a mathematical modeling approach to simulate the ohmic heating process to improve food safety [78]. Choi et al. (2020) also applied mathematical modeling methods to accurately simulate the temperature distribution and inactivation of *Escherichia coli O157:H7*, resulting in a 5-log reduction [79].

Normally, thermal processing affects the texture of food, causing water loss and hardening. Ohmic heating reduces the damage to food quality. Compared to traditional steaming methods, ohmic heating consumes less energy and accelerated softening of white rice, brown rice, and germinated brown rice. It also offers greater advantages by reducing the loss of hardness by 50% after the pretreatment of foods containing starch [66]. These characteristics of ohmic heating are extremely important in the reheating of prepared fried rice. Combining ohmic heating with microwave and Near infrared radiation can improve the uniformity of heating [80]. The combination of cold plasma, RF and ohmic heating significantly reduces the processing time and the target temperature for ohmic heating. By using combined heating, food quality degradation can also be minimized and food safety improved. The elderly have special needs for food nutrient content, and processing techniques should maximize the retention of the nutrients and flavor. Joe et al. (2021) developed a combined ohmic and vacuum heating system to process elderly foods. The effects of vacuum pressure intensity and action time on the physical and electrical properties of foods with solid–liquid mixtures were discussed. It was found that the combined technique shortened the heating time of foods and solved the problem of uneven temperature distribution [81]. This provides an idea for the processing of prepared fried rice. Prepared fried rice is a kind of heterogeneous food, in which the fat wraps around the rice grains and auxiliary ingredients to slow down the aging of starch and reduce the viscosity change of fried rice.

At present, ohmic heating technology is still in its infancy and has some challenges. If electrode materials are not selected properly, ohmic heating may suffer from the electrode corrosion of toxic chemicals and overheating of the foods. Ohmic heating requires corrosion-resistant, non-polluting electrodes, and controlled heating speed to avoid local overheating [82]. In addition, for non-homogeneous foods such as prepared fried rice, the particle density affects the heating effect. The energization should be combined with other techniques to avoid uneven currents due to different resistances of the components inside the prepared fried rice. Since the electrical conductivity of the mixture of rice and water is low, it has been suggested that salt can be added to the rice mixture to increase the electrical conductivity [72]. It has also been suggested that soaking rice before making fried rice can effectively increase the electrical conductivity of the mixture, thus making it possible to steam rice with ohmic heating. It is noted that ohmic heating has been rarely reported in foods containing fat and oil, and further research is needed.

3. Potential Quality and Safety Monitoring Technologies for Prepared Fried Rice

Most of the traditional techniques used to analyze harmful chemical components and contaminants in food such as high-performance liquid chromatography (HPLC), gas chromatography (GC), and HPLC or GC combined with mass spectrometry are expensive, time-consuming, and invasive to the samples, requiring specialized technician operating [83]. Non-destructive, portable, and environmentally friendly Raman spectroscopy [84], near-infrared spectroscopic imaging [85], and low-field nuclear magnetic resonance [86] have been developed rapidly in recent years and have the potential for rapid and efficient analysis in the field of food quality monitoring. In particular, the combination of chemometric methods for predicting food quality has received wide attention [87]. However, there are few literature reviews on the quality monitoring techniques for prepared fried rice.

Prepared fried rice is rich in proteins and fats, which are highly susceptible to microbial attack and lead to food spoilage. Microorganisms convert nitrates to nitrites in prepared fried rice with Chinese sausage. Fried rice grains are rich in carbohydrates; and after cooking at high temperatures, free amino acids (asparagine) in the food react with reducing sugars or other carboxyl compounds to produce acrylamide [88]. Both nitrites and acrylamide are harmful to human health [89]. Food additives may also undergo transformation and produce toxins when interacting with other ingredients in the food or during heat treatment and freezing. Prepared fried rice packaging materials in an oil and salt environment may also migrate some plastic components into the food, creating potential food safety problems [90]. Therefore, the use of fast and effective detection techniques can reduce harmful substances in these food products.

3.1. Raman Spectroscopy Imaging Technology

Raman spectroscopy is an analytical technique used in the study of molecular structure to determine the vibration and rotation of molecules by examining the scattering spectra of various frequencies of incident light. Raman spectroscopy imaging is a combination of Raman spectroscopy and imaging methods that can simultaneously acquire structural morphological data and distribution images of the prepared fried rice. Different hazardous substances may have different spectral characteristics. After obtaining the Raman spectral data of prepared fried rice, the data can be analyzed and interpreted by using professional software. By comparing the spectral profiles of known hazardous substances, the presence of hazardous substances in fried rice can be determined [82]. Prepared fried rice needs to be stir-fried in oil and may contain about 500 $\mu g\ kg^{-1}$ of acrylamide [91]. Insertion of the aggregating agent into the metal nanoparticle aggregates to generate more hotspots, which theoretically produces high SERS surface resonance to further enhance the Raman signal [92]. Cheng et al. (2019), combining QuECHERS extraction and SERS, and Ye et al. (2023), using AgNP substrate-assisted SERS technique with 0.5 M NaCl as the agglomerating agent, both demonstrated that Raman spectroscopy can rapidly detect acrylamide in fried foods; and the results are consistent with LC-MS/MS, which can be used as a technical tool for on-site screening [92,93]. It is worth noting that the SERS technique has limitations. The experimental conditions and control parameters need to be optimized because of various samples with different surface structure and preparation. It is currently only used in typical fried foods such as potato chips, needing to further expand the scope of application. Raman spectroscopy techniques can also provide data on the spatial structure of proteins based on the backbone structure and side-chain microenvironment of the protein polypeptide, as well as reliable identification and imaging of starch and gluten [94]. According to the different characteristic peaks corresponding to different structural information, starch molecular structure, protein and fat composition, and additives content were predicted by comparing the variations in the Raman spectra of the prepared fried rice. This information is necessary to evaluate the prepared fried rice quality and safety as well as to detect microorganisms on food surfaces [95]. This technique does not require the complex pretreatment of the prepared fried rice; it is simple to operate and combines with fingerprinting to simultaneously identify multiple molecules for effectively regulating the safety of fried rice ingredients [96], such as rapid online monitoring or on-site detection of pork ractopamine and clenbuterol hydrochloride residues using SERS with chemometric methods [97], non-destructive detection of fungal spore counts and textural features of corn [98], distinguishing of aflatoxin contaminated corn kernels [99], and classification of the fake and real eggs (Figure 5) [100]. These are all raw materials of prepared fried rice. In addition, Raman spectroscopy was used to detect mycotoxins, metal ions, and drug residues in food products (Table 2) [101]. Although the direct application of Raman spectroscopy detection technology on fried rice is not widely reported, its accuracy, rapidity, and wide detection range have great potential for monitoring the quality of this group of foods.

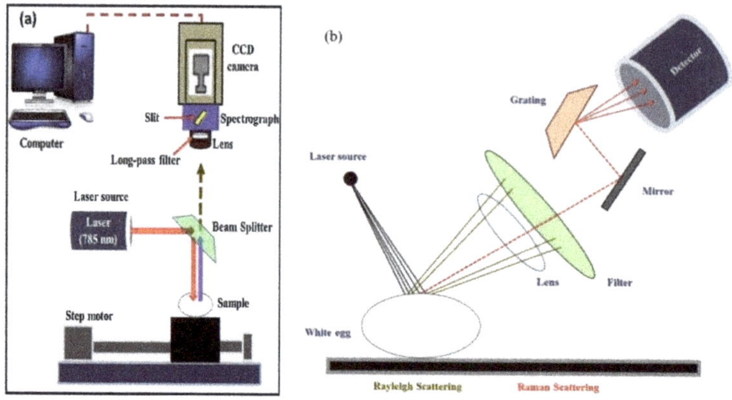

Figure 5. Schematic diagram (**a**) and details (**b**) of Raman hyperspectral imaging system used for measuring real and fake egg samples, adapted from [100], with permission from ELSEVIER, 2023 (Joshi et al., 2020).

Table 2. Application of Raman spectral imaging technology to the detection of harmful substances in food, adapted from [101], with permission from ELSEVIER, 2023 (Sun et al., 2022).

Analytes		Excitation Laser and Spectral Range	Results	References
Microorganisms	Viruses: influenza AH1N1 virus and human adenovirus	785 nm 1100–1600 cm^{-1}	LOD_{H1N1} = 50 pfu/mL LOD_{HAdV} = 50 pfu/mL	[102]
	Bacteria: *Escherichia coli* (*E. coli* ATCC 25922 and K-12 strains) on culture plate	633 nm 600–1800 cm^{-1}	LOD = 6×10^4 CFU/mL	[103]
	Fungi: *Aspergillus flavus* (AF36, AF13) infect corn kernels	785 nm 103–2831 cm^{-1}	All prediction accuracy ≥ 75.55%	[103]
Mycotoxins	AFB1 in peanut extracts	785 nm 400–1800 cm^{-1}	LOD = 0.5μg/L	[103]
Biotoxins	Saxitoxin (STX)	785 nm 400–2000 cm^{-1}	LOD = 1×10^{-7} M	[104]
Pesticides	Carbamate: thiram in apple juice	785 nm 500–1600 cm^{-1}	LOD = 86.1μg/L	[105]
Food authenticity	Authenticating Australian grain-fed and grass-fed beef products	785 nm 600–1900 cm^{-1}	All discrimination accuracy ≥ 83%	[105]
Metal ions	Hg^{2+} in natural ground and lake water	785 nm 200–1800 cm^{-1}	LOD = 0.1 nm	[106]
Drug residues	Tetracycline in water	532 nm 200–1800 cm^{-1}	LOD = 10^{-9} m	[106]
	Malachite green in grass carp, bream fish, and crucian fillets	633 nm 400–1800 cm^{-1}	LOD = 0.5μg/L	[107]
	Clenbuterol in pork, chicken, and sausage	785 nm 400–1600 cm^{-1}	LOD = 0.05 ng/mL	[108]

Several companies and institutions such as Thermo Fisher Scientific and HORIBA Scientific have established Raman spectroscopy databases to compare unknown molecules in food products. Dias, Jussiani, and Appoloni (2019) established a database using Raman spectra of 78 major commercial pesticides. In the analysis of pesticides, they used a portable delta nu Raman spectrometer with a laser wavelength of 785 nm and a resolution of 8 cm^{-1} to rapidly detect the pesticide species [109]. The establishment of the database provides a reference for the detection of foods like prepared fried rice that are susceptible to pesticide or microbial contamination.

3.2. Near-Infrared Spectroscopy Imaging Technology

The electromagnetic spectrum in the near-infrared (NIR) range (780–2526 nm) is used for spectral analysis. Most commercially available NIR spectrometers are focused on the short-wavelength region of 750–1100 nm [110], while NIR spectral wavelengths in the range of 1100 nm and 2400 nm are associated with branched starch and can be analyzed for starch aging [111]. NIR causes the fried rice molecules to absorb some of the NIR light and transition, reducing the intensity of the light due to different functional groups containing different chemical bonds (O-H, C-H, N-H, S-H bonds, etc.) and hydrogen-containing groups corresponding to different NIR absorption peaks. It is worth noting that this technique is mainly used for macronutrient analysis, of which the content is generally 0.1% higher than that of the sample weight. For prepared fried rice, NIR spectra are not a simple superposition of the individual spectra of each component [101]. Using chemometrics to extract the correct information from the NIR spectra allows qualitative and quantitative detection of water, proteins, and lipids in the pre-made fried rice because of their different structure and composition [112].

The quality of rice decreases with storage time, but it is difficult to quickly distinguish aged rice from the fresh one. Using the advantages of nondestructive and rapid detection by NIR spectroscopy, Shi et al. (2023) combined partial least-squares-discriminant analysis (PLS-DA), support vector machines (SVM), and classification regression trees (CART) algorithms to distinguish the freshness of rice; and obtained that PLS-DA and SVM have excellent classification ability and sensitivity [113]. This is similar to a design of Lapchareonsuk and Sirisomboon (2015) and Onmankhong and Sirisomboon (2021), who proposed to build mathematical models by NIR spectroscopy for quality control in the rice industry [114,115]. Before shipping, prepared fried rice needs to be frozen to reduce the moisture. To detect changes in the quality of pre-made fried rice after freezing, it is usually necessary to measure ice crystal size, cell microstructure, and nutrient content, which is complicated and time-consuming using conventional methods [116]. Since it is technically non-destructive, inexpensive, and real-time, NIR spectroscopy detection has the potential to be used by intelligent applications to quickly identify freezing parameters in food goods and track the freezing process [117]. It is feasible to combine NIR imaging with machine learning to predict the quality of frozen foods because the physical properties of the material components are strongly correlated with the chemical composition and can be reflected by NIR spectral images. This provides ideas for the intelligent monitoring of quality changes in prepared fried rice and predicting shelf life. Jiang et al. (2023) demonstrated the potential of combining NIR spectroscopy with BP-ANN modeling (Figure 6) for the rapid non-destructive detection of frozen foods by modeling the relationship between NIR spectra and quality indicators of frozen samples after thawing using PCR, PLSR, SVR, and a back-propagation artificial neural network (BP-ANN) [118]. In another study, researchers used NIR spectroscopy to determine the fat, moisture, and protein content of frozen beef online during transportation; and this technique allows better control of the production process, resulting in a uniform and consistent quality [119]. There is potential to use this technology for the real-time monitoring of prepared fried rice during the frozen storage. NIR spectroscopy enables the reliable assessment of protein and fat content, and the detection of pH values of beef [120], TVBN, and pH in pork [121]; and moisture and pH in chicken breast [122]. Peyvasteh et al. (2020) used NIR coupled with PCA to determine the relative absorbance changes of oxidized myosin and myoglobin, fat, water, and collagen in the visible spectrum of pork for the rapid evaluation of pork freshness [123]. These ingredients are common sources of protein in prepared fried rice and directly affect its quality.

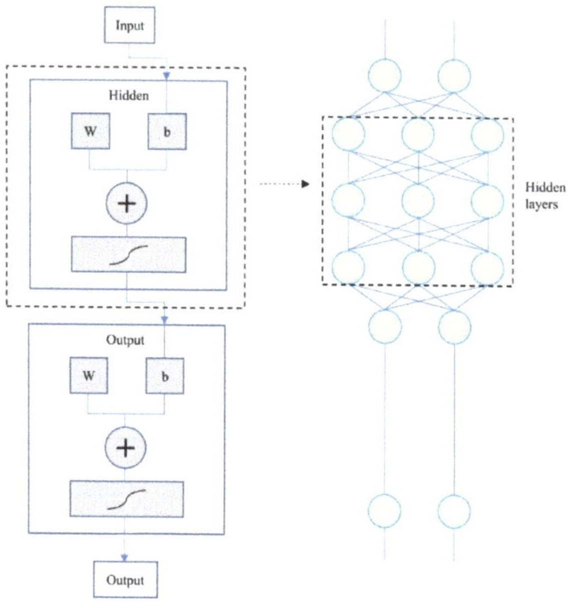

Figure 6. Schematic diagrams of a back-propagation artificial neural network (BP-ANN), adapted from [118], with permission from ELSEVIER, 2023 (Jiang et al., 2023).

Currently, the applications of NIR spectroscopy are mostly limited due to the insensitivity of it to impurities. In addition, in dynamic online detection mode, the stability of the instrument and the repeatability of the data are poor and the measurement accuracy is susceptible to environmental factors such as temperature [124]. Therefore, the ability of the models or software to process NIR data need to be further improved to increase the prediction accuracy.

3.3. *Low-Field Nuclear Magnetic Resonance Technology*

Low-field nuclear magnetic resonance (LF-NMR) technology uses atomic nuclei with fixed magnetic moments that exchange energy with alternating magnetic fields in the presence of static and alternating magnetic fields [125]. It is an emerging technology in the field of food quality testing, which is fast, efficient, and non-destructive. Materials of prepared fried rice have different internal compositions and physical properties. The quality classification can be achieved by transverse relaxation time T_2 and signal intensity. The peaks of LF-NMR spectra are T_{21}, T_{22}, and T_{23} from left to right, which represent strongly bound water, weakly bound water, and free water in order; in addition, the changes of the position and area of the characteristic peaks reflect the changes of moisture state of the prepared fried rice after processing [126]. This technique has been used in the determination of water content in fruits and vegetables, oil content in plant seeds [127], and water and oil content in fried starches [128].

The quality of prepared fried rice is affected by the temperature fluctuations during storage. High temperatures can damage the resistant starch as well as increase the water absorption of the rice grains, resulting in increased volume, reduced cohesion and adhesion, and floppier rice grains [129]. Prepared fried rice loses moisture content after freezing, thawing, and reheating. After repeated freezing and thawing, the water-holding capacity of food components in the fried rice were affected [130], which can be monitored by the LF-NMR technique. For example, Li, et al. (2012) obtained LF-NMR parameters under different conditions; and determined the color, shear, water retention, pH, and related steaming indicators, concluding that LF-NMR and color measurements are good methods to distinguish the water-holding capacity [131](Li et al., 2012). Ali et al. (2015) reported the analysis of the water state using LF-NMR and showed that after repeated freeze–thaw cycles, lipid and protein oxidation increased with the increasing number of cycles and affected the water-holding capacity [130]. Zhang, Zhang, and Mujumdar (2021) concluded that ice crystals formed after freezing cause damage to the structure of muscle cells, leading to the migration of water molecules. Water loss in chicken breast, beef and seafood, lipid types on fermented sausage and myofibrillar protein gels can be sensitively detected by LF-NMR technique [112]. Moreover, the integrity of proteins is revealed by the decrease in the lateral relaxation time of their protons due to protein aggregation, which further leads to the decrease in the lateral relaxation time of water protons. In addition to monitoring the state and distribution of water, LF-NMR is also effective in the study of detecting the quality of vegetables after water loss. Sun, Zhang, and Yang (2019) studied the trends of NMR signals (A_{21}, A_{22}, A_{23}, and total) of vegetables at different drying powers (200, 300, and 400 W) to obtain the characteristic variables, and used the BP-ANN model for the nondestructive testing of the nutritional quality [132]. Similarly, Zhao et al. (2022) established a method for the non-destructive detection of free-fatty-acid content in frying oil samples based on LF-NMR and BP-ANN [133]. The water and oil contents of prepared fried rice samples have now been examined in the laboratory by LF-NMR. The signals of oil and water were clearly differentiated. The LF-NMR approach is more precise and quicker than the Soxhlet method for the identification of oil-rich fried rice [128].

Interestingly, LF-NMR also played a role in the identification of adulteration in vegetable oils. LF-NMR detection revealed that the imaging difference between vegetable oils and those with additions was the distinction in the transverse relaxation distribution (T_2 distribution) of the third peak (A) and the shift of the peak T_2 value, which is likely to be the characteristic peak of polymorphs produced during the frying process of the oils. The A peak area grew linearly as a percentage of the total peak area as the addition of other materials increased. Based on the change in peak area, a linear equation can be established to discriminate adulterations [134]. In another investigation, LF-NMR detected contamination with various hydrocolloids in prepared fried rice ingredients, and found prominent hydrocolloid accumulation locations as well as typical T_2 fitting curves [135]. Prepared fried rice, as a heterogeneous type of food, is prone to quality deterioration due to uneven moisture after reheating and irregularities in the oil used by producers. The quick, non-intrusive, and affordable LF-NMR approach offers a potent tool for the real-time quality monitoring of various foods, including prepared fried rice.

4. Conclusions and Future Trends

Prepared fried rice is produced through a central kitchen, appropriately packaged to maintain a sterile environment and stored and transported at −18 °C using rapid freezing technologies. However, the use of metamorphic ingredients, low standard production processes, high oil and salt content, and unsafe packaging materials lead to poor quality of prepared fried rice. Compared with traditional processing technologies, irradiation technology and high-voltage electric field technology have the potential to control the microorganisms in prepared fried rice. Microwave, radio frequency and ohmic heating have the potential to sterilize and reheat prepared fried rice. These processing technologies consume less energy and are friendly to the environment. However, these technologies still have limitations. Irradiation and high-voltage electric field sterilization technology require high-capital investment. Microwave, RF, and ohmic heating normally need to be combined with other techniques to improve the uniformity of heating. The processing technologies will develop towards a greener, low-carbon, and environment-friendly direction.

Raman spectral imaging technology, NIR, and LF-NMR, as the new analytical procedures for monitoring, have some advantages. They can non-destructively and efficiently monitor the moisture, protein, and lipid content in food products; and detect quality changes such as color and pH value. These technologies also improve the detection efficiency of enterprises and law-enforcement officers. However, they also have some limitations. The sensitivity of portable Raman detectors needs to be improved. The accuracy of NIR is readily affected by environmental factors such as temperature, which requires correction or compensation to ensure the accuracy and reliability of the data. LF-NMR requires a high-liquid fraction in the food sample to be able to generate a detection signal. Factors such as the viscosity and temperature of the sample may also affect the accuracy. For thicker or multilayer samples, these detection techniques may need to be combined with other techniques to obtain accurate results. With the development of machine learning and artificial intelligence, data processing and model building for Raman spectral imaging technology, NIR, and LF-NMR will be available in the industry. This will significantly improve the accuracy and efficiency in monitoring the quality of prepared fried rice and ensure the safety and consistency of ingredients. The emergence of portable equipment will make it easier to perform rapid testing and analysis of prepared fried rice in the field or at sales terminals. In the future, the development of detecting technology will require the improvement of appropriate software. This allows food researchers to operate equipment and obtain experimental data more easily. Depending on the characteristics of the detection technology, a comprehensive database for a wide range of food products will also need to be established.

In addition, with the rapid development of prepared foods in the market, the relevant laws and regulations should be in place. The raw material standards, processing flows, and product standards (physical and chemical indicators, sensory quality, microbiological indicators, etc.) should be established. Although there are few examples of direct application of processing and testing techniques to prepared fried rice, there have been many studies on the application to individual components of fried rice. In summary, the market of prepared fried rice is expanding, and emerging processing and detection technologies have great potential to be applied to prepared fried rice to ensure a high standard of food quality and safety.

Author Contributions: Literature research and drafting were performed by J.H. Oversight of the project and field experience were provided by M.Z. Critical review of the manuscript was performed by M.Z. and Z.F. All authors have read and agreed to the published version of the manuscript.

Funding: This research was funded by National Key R & D Program of China (Contract No. 2022YFD2100601), the 111 Project (BP0719028), and the Jiangsu Province Key Laboratory Project of Advanced Food Manufacturing Equipment and Technology (No. FMZ202003).

Data Availability Statement: Not applicable.

Acknowledgments: The authors acknowledge the financial supports from the National Key R & D Program of China (Contract No. 2022YFD2100601), the 111 Project (BP0719028), and the Jiangsu Province Key Laboratory Project of Advanced Food Manufacturing Equipment and Technology (No. FMZ202003).

Conflicts of Interest: The authors declare no conflict of interest.

References

1. Yu, Q.; Zhang, M.; Ju, R.; Mujumdar, A.S.; Wang, H. Advances in prepared dish processing using efficient physical fields: A review. *Crit. Rev. Food Sci. Nutr.* **2022**, 1–15.
2. The State Council of China. Opinions of the General Office of the State Council on Further Promoting the Development of the Agricultural Product Processing Industry. Available online: http://www.gov.cn/zhengce/content/2016-12/28/content_5153844.htm (accessed on 28 December 2016).
3. Aimedia Consulting. *China Prepared Dishes Industry Analysis Report 2021*; Ai Media Restaurant Research Institute: Guangzhou, China, 2021.
4. Huang, M.; Zhang, M.; Bhandari, B. Recent development in the application of alternative sterilization technologies to prepared dishes: A review. *Crit. Rev. Food Sci. Nutr.* **2019**, *59*, 1188–1196. [CrossRef]
5. Maneerote, J.; Noomhorm, A.; Takhar, P.S. Optimization of processing conditions to reduce oil uptake and enhance physico-chemical properties of deep fried rice crackers. *LWT Food Sci. Technol.* **2009**, *42*, 805–812. [CrossRef]
6. Lim, S.-T.; Han, J.-A. Improvement in antioxidant functionality and shelf life of yukwa (fried rice snack) by turmeric (*Curcuma longa* L.) powder addition. *Food Chem.* **2016**, *199*, 590–596. [CrossRef] [PubMed]
7. Cho, S.-J.; Jo, M.N. Changes in physical properties of Yukwa (fried rice snack) by soybean extract concentration and incubation time. *Food Sci. Biotechnol.* **2018**, *27*, 433–440. [CrossRef]

8. Schreiber, N.; Hackl, G.; Reisinger, A.C.; Zollner-Schwetz, I.; Eller, K.; Schlagenhaufen, C.; Pietzka, A.; Czerwenka, C.; Stark, T.D.; Kranzler, M.; et al. Acute Liver Failure after Ingestion of Fried Rice Balls: A Case Series of Bacillus cereus Food Poisonings. *Toxins.* **2022**, *14*, 12. [CrossRef] [PubMed]
9. Huang, L.; Hwang, C.A. One-step dynamic analysis of growth kinetics of Bacillus cereus from spores in simulated fried rice—Model development, validation, and Marko Chain Monte Carlo simulation. *Food Microbiol.* **2022**, *103*, 103935. [CrossRef] [PubMed]
10. Jiang, H.; Zhang, M.; Chen, J. Effect of Processing Methods on Flavor Quality of Vegetable Balls after Boiling. *J. Food Sci. Biotechnol.* **2021**, *40*, 58–64.
11. Ding, Y.; Xiao, Y.; Ouyang, Q.; Luo, F.; Lin, Q. Modulating the in vitro digestibility of chemically modified starch ingredient by a non-thermal processing technology of ultrasonic treatment. *Ultrason. Sonochemistry* **2021**, *70*, 105350. [CrossRef]
12. Asaithambi, N.; Singh, S.K.; Singha, P. Current status of non-thermal processing of probiotic foods: A review. *J. Food Eng.* **2021**, *303*, 110567. [CrossRef]
13. Ribeiro, N.G.; Xavier-Santos, D.; Campelo, P.H.; Guimarães, J.T.; Pimentel, T.C.; Duarte, M.C.K.H.; Freitas, M.Q.; Esmerino, E.A.; Silva, M.C.; Cruz, A.G. Dairy foods and novel thermal and non-thermal processing: A bibliometric analysis. *Innov. Food Sci. Emerg. Technol.* **2022**, *76*, 102934. [CrossRef]
14. Ankolekar, C.; Rahmati, T.; Labbe, R.G. Detection of toxigenic Bacillus cereus and Bacillus thuringiensis spores in US rice. *Int. J. Food Microbiol.* **2009**, *128*, 460–466. [CrossRef] [PubMed]
15. Wang, R.; Sun, L.; Wang, Y.; Deng, Y.; Fang, Z.; Liu, Y.; Liu, Y.; Sun, D.; Deng, Q.; Gooneratne, R. Growth and Hemolysin Production Behavior of Vibrio parahaemolyticus in Different Food Matrices. *J. Food Prot.* **2018**, *81*, 246–253. [CrossRef] [PubMed]
16. Soni, A.; Oey, I.; Silcock, P.; Bremer, P.J. Impact of temperature, nutrients, pH and cold storage on the germination, growth and resistance of Bacillus cereus spores in egg white. *Food Res. Int.* **2018**, *106*, 394–403. [CrossRef]
17. Guinebretiere, M.H.; Girardin, H.; Dargaignaratz, C.; Carlin, F.; Nguyen-The, C. Contamination flows of Bacillus cereus and spore-forming aerobic bacteria in a cooked, pasteurized and chilled zucchini puree processing line. *Int. J. Food Microbiol.* **2003**, *82*, 223–232. [CrossRef] [PubMed]
18. Tirloni, E.; Bernardi, C.; Ghelardi, E.; Celandroni, F.; Cattaneo, P.; Stella, S. Bacillus cereus in fried rice meals: Natural occurrence, strain dependent growth and haemolysin (HBL) production. *LWT Food Sci. Technol.* **2019**, *114*, 108393. [CrossRef]
19. Chang, H.-J.; Lee, J.-H.; Han, B.-R.; Kwak, T.-K.; Kim, J. Prevalence of the levels of Bacillus cereus in fried rice dishes and its exposure assessment from Chinese-style restaurants. *Food Sci. Biotechnol.* **2011**, *20*, 1351–1359. [CrossRef]
20. Zhang, Y.; Liu, K.; Zhang, Z.; Tian, S.; Liu, M.; Li, X.; Han, Y.; Zhu, K.; Liu, H.; Yang, C.; et al. A Severe Gastroenteritis Outbreak of Salmonella enterica Serovar Enteritidis Linked to Contaminated Egg Fried Rice, China, 2021. *Front. Microbiol.* **2021**, *12*, 779749. [CrossRef]
21. Wang, R.; Hu, X.; Deng, Y.; Gooneratne, R. Effect of Food Matrix Type on Growth Characteristics of and Hemolysin Production by Vibrio alginolyticus. *J. Food Prot.* **2021**, *84*, 1411–1420. [CrossRef]
22. Tahergorabi, R.; Matak, K.E.; Jaczynski, J. Application of electron beam to inactivate Salmonella in food: Recent developments. *Food Res. Int.* **2012**, *45*, 685–694. [CrossRef]
23. Chen, Q.; Cao, M.; Chen, H.; Gao, P.; Fu, Y.; Liu, M.; Wang, Y.; Huang, M. Effects of gamma irradiation on microbial safety and quality of stir fry chicken dices with hot chili during storage. *Radiat. Phys. Chem.* **2016**, *127*, 122–126. [CrossRef]
24. Feliciano, C.P.; de Guzman, Z.M.; Tolentino, L.M.M.; Asaad, C.O.; Cobar, M.L.C.; Abrera, G.B.; Baldos, D.T.; Diano, G.T. Microbiological quality of brown rice, ready-to-eat pre-cut fresh fruits, and mixed vegetables irradiated for immuno-compromised patients. *Radiat. Phys. Chem.* **2017**, *130*, 397–399. [CrossRef]
25. Feng, X.; Ahn, D.U. Volatile profile, lipid oxidation and protein oxidation of irradiated ready-to-eat cured turkey meat products. *Radiat. Phys. Chem.* **2016**, *127*, 27–33. [CrossRef]
26. Li, X.; Farid, M. A review on recent development in non-conventional food sterilization technologies. *J. Food Eng.* **2016**, *182*, 33–45. [CrossRef]
27. Munshi, M.K.; Sukhi, F.A.; Huque, R.; Hossain, A.; Mahbub, S.; Hoque, S.A.; Hossain, T.; Khan, M.S.I.; Hossain, M.F. Combination impacts of gamma radiation and low temperature on the toxin-producing Bacillus cereus isolated from fried and steam rice. *J. Food Process. Preserv.* **2021**, *45*, 192–200. [CrossRef]
28. Ben-Fadhel, Y.; Saltaji, S.; Khlifi, M.A.; Salmieri, S.; Dang Vu, K.; Lacroix, M. Active edible coating and gamma-irradiation as cold combined treatments to assure the safety of broccoli florets (*Brassica oleracea* L.). *Int. J. Food Microbiol.* **2017**, *241*, 30–38. [CrossRef]
29. Wang, Q.; Li, Y.; Sun, D.-W.; Zhu, Z. Enhancing Food Processing by Pulsed and High Voltage Electric Fields: Principles and Applications. *Crit. Rev. Food Sci. Nutr.* **2018**, *58*, 2285–2298. [CrossRef]
30. Mosqueda-Melgar, J.; Raybaudi-Massilia, R.M.; Martin-Belloso, O. Influence of treatment time and pulse frequency on Salmonella Enteritidis, Escherichia coli and Listeria monocytogenes populations inoculated in. melon and watermelon juices treated by pulsed electric fields. *Int. J. Food Microbiol.* **2007**, *117*, 192–200. [CrossRef] [PubMed]
31. Zhu, N.; Yu, N.; Zhu, Y.; Wei, Y.; Zhang, H.; Sun, A.D. Inactivation of Pichia rhodanensis in relation to membrane and intracellular compounds due to microchip pulsed electric field (MPEF) treatment. *PLoS ONE* **2018**, *13*, e0198467. [CrossRef] [PubMed]
32. Zhao, W.; Yang, R.; Zhang, H.Q. Recent advances in the action of pulsed electric fields on enzymes and food component proteins. *Trends Food Sci. Technol.* **2012**, *27*, 83–96. [CrossRef]

33. Fernandez-Diaz, M.D.; Barsotti, L.; Dumay, E.; Cheftel, J.C. Effects of Pulsed Electric Fields on Ovalbumin Solutions and Dialyzed Egg White. *J. Agric. Food Chem.* **2000**, *48*, 2332–2339. [CrossRef]
34. Martin, O.; Qin, B.L.; Chang, F.J.; BarbosaCanovas, G.V.; Swanson, B.G. Inactivation of Escherichia coli in skim milk by high intensity pulsed electric fields. *J. Food Process Eng.* **1997**, *20*, 317–336. [CrossRef]
35. Huang, H.; Sun, W.; Xiong, G.; Shi, L.; Jiao, C.; Wu, W.; Li, X.; Qiao, Y.; Liao, L.; Ding, A.; et al. Effects of HVEF treatment on microbial communities and physicochemical properties of catfish fillets during chilled storage. *LWT* **2020**, *131*, 109667. [CrossRef]
36. Qi, M.; Zhao, R.; Liu, Q.; Yan, H.; Zhang, Y.; Wang, S.; Yuan, Y. Antibacterial activity and mechanism of high voltage electrostatic field (HVEF) against Staphylococcus aureus in medium plates and food systems. *Food Control* **2021**, *120*, 107566. [CrossRef]
37. Hsieh, C.-W.; Lai, C.-H.; Lee, C.-H.; Ko, W.-C. Effects of High-Voltage Electrostatic Fields on the Quality of Tilapia Meat during Refrigeration. *J. Food Sci.* **2011**, *76*, M312–M317. [CrossRef] [PubMed]
38. Ko, W.-C.; Yang, S.-Y.; Chang, C.-K.; Hsieh, C.-W. Effects of adjustable parallel high voltage electrostatic field on the freshness of tilapia (*Orechromis niloticus*) during refrigeration. *LWT Food Sci. Technol.* **2016**, *66*, 151–157. [CrossRef]
39. Huang, H.; Gao, T.; Qian, X.; Wu, W.; Fan, X.; Shi, L.; Xiong, G.; Ding, A.; Li, X.; Qiao, Y.; et al. In Vitro Antibacterial Mechanism of High-Voltage Electrostatic Field against *Acinetobacter johnsonii*. *Foods* **2022**, *11*, 955. [CrossRef]
40. Auksornsri, T.; Songsermpong, S. Lethality and quality evaluation of in-packaged ready-to-eat cooked Jasmine rice subjected to industrial continuous microwave pasteurization. *Int. J. Food Prop.* **2016**, *20*, 1856–1865. [CrossRef]
41. Viji, P.; Rao, B.M.; Debbarma, J.; Ravishankar, C.N. Research developments in the applications of microwave energy in fish processing: A review. *Trends Food Sci. Technol.* **2022**, *123*, 222–232. [CrossRef]
42. Peng, J.; Tang, J.; Barrett, D.M.; Sablani, S.S.; Anderson, N.; Powers, J.R. Thermal pasteurization of ready-to-eat foods and vegetables: Critical factors for process design and effects on quality. *Crit. Rev. Food Sci. Nutr.* **2017**, *57*, 2970–2995. [CrossRef]
43. Kim, H.S.; Choi, S.J.; Yoon, K.S. Efficacy Evaluation of Control Measures on the Reduction of Staphylococcus aureus in Salad and Bacillus cereus in Fried Rice Served at Restaurants. *Foodborne Pathog. Dis.* **2018**, *15*, 198–209. [CrossRef]
44. Montero, M.L.; Sablani, S.; Tang, J.; Ross, C.F. Characterization of the sensory, chemical, and microbial quality of microwave-assisted, thermally pasteurized fried rice during storage. *J. Food Sci.* **2020**, *85*, 2711–2719. [CrossRef]
45. Guo, Q.; Sun, D.-W.; Cheng, J.-H.; Han, Z. Microwave processing techniques and their recent applications in the food industry. *Trends Food Sci. Technol.* **2017**, *67*, 236–247. [CrossRef]
46. Tian, J.; Chen, J.; Lv, F.; Chen, S.; Chen, J.; Liu, D.; Ye, X. Domestic cooking methods affect the phytochemical composition and antioxidant activity of purple-fleshed potatoes. *Food Chem.* **2016**, *197*, 1264–1270. [CrossRef] [PubMed]
47. Fan, L.P.; Zhang, M.; Mujumdar, A.S. Vacuum frying of carrot chips. *Dry. Technol.* **2005**, *23*, 645–656. [CrossRef]
48. Geedipalli, S.S.R.; Rakesh, V.; Datta, A.K. Modeling the heating uniformity contributed by a rotating turntable in microwave ovens. *J. Food Eng.* **2007**, *82*, 359–368. [CrossRef]
49. Regier, M.; Knoerzer, K.; Schubert, H. The Microwave Processing of Foods 2. In *Microwave Heating and the Dielectric Properties of Foods*, 2nd ed.; Woodhead Publishing: Sawston, UK, 2017; pp. 23–43.
50. Wang, R.; Zhang, M.; Mujumdar, A.S.; Jiang, H. Effect of salt and sucrose content on dielectric properties and microwave freeze drying behavior of re-structured potato slices. *J. Food Eng.* **2011**, *106*, 290–297. [CrossRef]
51. Wang, R.; Zhang, M.; Mujumdar, A.S. Effect of Osmotic Dehydration on Microwave Freeze-Drying Characteristics and Quality of Potato Chips. *Dry. Technol.* **2010**, *28*, 798–806. [CrossRef]
52. Zhang, J.; Zhang, M.; Shan, L.; Fang, Z.X. Microwave-vacuum heating parameters for processing savory crisp bighead carp (*Hypophthalmichthys nobilis*) slices. *J. Food Eng.* **2007**, *79*, 885–891. [CrossRef]
53. Wang, X.; Xu, H.; Zhu, Y.; Liu, C.; Liu, H.; Liu, C.; Fu, H.; Zheng, X. Improvement of Temperature Uniformity of Instant Rice Inside Plastic Rectangular Container Under Microwave Reheating. *J. Northeast. Agric. Univ.* **2018**, *25*, 79–89.
54. Wang, Y.C.; Zhang, M.; Mujumdar, A.S.; Mothibe, K.J.; Azam, S.M.R. Study of Drying Uniformity in Pulsed Spouted Microwave-Vacuum Drying of Stem Lettuce Slices with Regard to Product Quality. *Dry. Technol.* **2013**, *31*, 91–101. [CrossRef]
55. Piza, L.V.; Von Atzingen, G.V.; Costa, E.J.X. Electronic Instrumentation and Computational Simulation to Evaluate the Combined Use of Microwave and Infrared Technologies for Reheating Biphasic Foods. *Int. J. Food Eng.* **2019**, *15*, 20180192. [CrossRef]
56. Chen, C.; Zhang, M.; Liu, W.C.; Lin, Z.H. Baking characteristic improvement and starch retrogradation inhibition of Chinese pancakes by hydrocolloids. *J. Food Process. Preserv.* **2022**, *46*, e16529. [CrossRef]
57. Chen, C. Study on Edible Quality Improvement of Fresh/Cooling and Reheating of Pre-Made Pancake. Master's Thesis, Jiangnan University, Wuxi, China, 2022. Available online: https://kns.cnki.net/KCMS/detail/detail.aspx?dbname=CMFD202301&filename=1022642003.nh (accessed on 8 July 2023).
58. Zhang, Z.N.; Zhang, M.Q.; Zhang, B.; Wang, Y.Y.; Zhao, W. Radio frequency energy regulates the multi-scale structure, digestive and physicochemical properties of rice starch. *Food Biosci.* **2022**, *47*, 101616. [CrossRef]
59. Jiao, Y.; Tang, J.; Wang, Y.; Koral, T.L. Radio-Frequency Applications for Food Processing and Safety. *Annu. Rev. Food Sci. Technol.* **2018**, *9*, 105–127. [CrossRef]
60. Luechapattanaporn, K.; Wang, Y.; Wang, J.; Tang, J.; Hallberg, L.M.; Dunne, C.P. Sterilization of Scrambled Eggs in Military Polymeric Trays by Radio Frequency Energy. *J. Food Sci.* **2006**, *70*, E288–E294. [CrossRef]
61. Xu, J.; Zhang, M.; An, Y.; Roknul, A.S.; Adhikari, B. Effects of radio frequency and high pressure steam sterilisation on the colour and flavour of prepared Nostoc sphaeroides. *J. Sci. Food Agric.* **2018**, *98*, 1719–1724. [CrossRef]

62. Xu, J.C.; Zhang, M.; Bhandari, B.; Kachele, R. ZnO nanoparticles combined radio frequency heating: A novel method to control microorganism and improve product quality of prepared carrots. *Innov. Food Sci. Emerg. Technol.* **2017**, *44*, 46–53. [CrossRef]
63. Lan, R.; Qu, Y.; Ramaswamy, H.S.; Wang, S. Radio frequency reheating behavior in a heterogeneous food: A case study of pizza. *Innov. Food Sci. Emerg. Technol.* **2020**, *65*, 102478. [CrossRef]
64. Wang, J.; Luechapattanaporn, K.; Wang, Y.; Tang, J. Radio-frequency heating of heterogeneous food—Meat lasagna. *J. Food Eng.* **2012**, *108*, 183–193. [CrossRef]
65. Gavahian, M.; Tiwari, B.K.; Chu, Y.H.; Ting, Y.W.; Farahnaky, A. Food texture as affected by ohmic heating: Mechanisms involved, recent findings, benefits, and limitations. *Trends Food Sci. Technol.* **2019**, *86*, 328–339. [CrossRef]
66. Jittanit, W.; Khuenpet, K.; Kaewsri, P.; Dumrongpongpaiboon, N.; Hayamin, P.; Jantarangsri, K. Ohmic heating for cooking rice: Electrical conductivity measurements, textural quality determination and energy analysis. *Innov. Food Sci. Emerg. Technol.* **2017**, *42*, 16–24. [CrossRef]
67. Mannozzi, C.; Fauster, T.; Haas, K.; Tylewicz, U.; Romani, S.; Dalla Rosa, M.; Jaeger, H. Role of thermal and electric field effects during the pre-treatment of fruit and vegetable mash by pulsed electric fields (PEF) and ohmic heating (OH). *Innov. Food Sci. Emerg. Technol.* **2018**, *48*, 131–137. [CrossRef]
68. Gavahian, M.; Chu, Y.H.; Farahnaky, A. Effects of ohmic and microwave cooking on textural softening and physical properties of rice. *J. Food Eng.* **2019**, *243*, 114–124. [CrossRef]
69. Zareifard, M.R.; Ramaswamy, H.S.; Trigui, M.; Marcotte, M. Ohmic heating behaviour and electrical conductivity of two-phase food systems. *Innov. Food Sci. Emerg. Technol.* **2003**, *4*, 45–55. [CrossRef]
70. Benabderrahmane, Y.; Pain, J.P. Thermal behaviour of a solid/liquid mixture in an ohmic heating sterilizer—Slip phase model. *Chem. Eng. Sci.* **2000**, *55*, 1371–1384. [CrossRef]
71. Icier, F.; Ilicali, C. Temperature dependent electrical conductivities of fruit purees during ohmic heating. *Food Res. Int.* **2005**, *38*, 1135–1142. [CrossRef]
72. Shirsat, N.; Lyng, J.G.; Brunton, N.P.; McKenna, B. Ohmic processing: Electrical conductivities of pork cuts. *Meat Sci.* **2004**, *67*, 507–514. [CrossRef] [PubMed]
73. Gavahian, M.; Chu, Y.-H.; Sastry, S. Extraction from Food and Natural Products by Moderate Electric Field: Mechanisms, Benefits, and Potential Industrial Applications. *Compr. Rev. Food Sci. Food Saf.* **2018**, *17*, 1040–1052. [CrossRef]
74. Kaur, N.; Singh, A.K. Ohmic Heating: Concept and Applications-A Review. *Crit. Rev. Food Sci. Nutr.* **2016**, *56*, 2338–2351. [CrossRef]
75. Somavat, R.; Mohamed, H.M.H.; Chung, Y.K.; Yousef, A.E.; Sastry, S.K. Accelerated inactivation of Geobacillus stearothermophilus spores by ohmic heating. *J. Food Eng.* **2012**, *108*, 69–76. [CrossRef]
76. De Alwis, A.A.P.; Fryer, P.J. A finite-element analysis of heat generation and transfer during ohmic heating of food. *Chem. Eng. Sci.* **1990**, *45*, 1547–1559. [CrossRef]
77. Kim, S.S.; Kang, D.H. Synergistic effect of carvacrol and ohmic heating for inactivation of E-coli O157:H7, S. Typhimurium, L-monocytogenes, and MS-2 bacteriophage in salsa. *Food Control* **2017**, *73*, 300–305. [CrossRef]
78. Kim, S.S.; Kang, D.H. Combination treatment of ohmic heating with various essential oil components for inactivation of food-borne pathogens in buffered peptone water and salsa. *Food Control* **2017**, *80*, 29–36. [CrossRef]
79. Choi, W.; Kim, S.S.; Park, S.H.; Ahn, J.B.; Kang, D.H. Numerical analysis of rectangular type batch ohmic heater to identify the cold point. *Food Sci. Nutr.* **2020**, *8*, 648–658. [CrossRef] [PubMed]
80. Choi, W.; Lee, S.H.; Kim, C.T.; Jun, S. A finite element method based flow and heat transfer model of continuous flow microwave and ohmic combination heating for particulate foods. *J. Food Eng.* **2015**, *149*, 159–170. [CrossRef]
81. Joe, S.Y.; So, J.H.; Hwang, S.H.; Cho, B.K.; Lee, W.H.; Kang, T.Y.; Lee, S.H. Application of Ohmic-Vacuum Combination Heating for the Processing of Senior-Friendly Food (Multiphase Food): Experimental Studies and Numerical Simulation. *Foods* **2021**, *10*, 138. [CrossRef]
82. Samaranayake, C.P.; Sastry, S.K. Electrode and pH effects on electrochemical reactions during ohmic heating. *J. Electroanal. Chem.* **2005**, *577*, 125–135. [CrossRef]
83. Byanju, B.; Rahman, M.M.; Hojilla-Evangelista, M.P.; Lamsal, B.P. Effect of high-power sonication pretreatment on extraction and some physicochemical properties of proteins from chickpea, kidney bean, and soybean. *Int. J. Biol. Macromol.* **2020**, *145*, 712–721. [CrossRef] [PubMed]
84. Wang, K.; Li, Z.; Li, J.; Lin, H. Raman spectroscopic techniques for nondestructive analysis of agri-foods: A state-of-the-art review. *Trends Food Sci. Technol.* **2021**, *118*, 490–504. [CrossRef]
85. Lopez-Maestresalas, A.; Arazuri, S.; Garcia, I.; Mangado, J.; Jaren, C. A Review of the Application of Near-Infrared Spectroscopy for the Analysis of Potatoes. *J. Agric. Food Chem.* **2013**, *61*, 5413–5424. [CrossRef]
86. Li, X.; Xing, C.; Wang, Z.; Sun, W.; Wu, C.; Xu, G.; Wang, X. LF-NMR intelligent evaluation for lipid oxidation indices of polar compound distribution, fatty acid unsaturation, and dynamic viscosity: Preference and mechanism. *Food Res. Int.* **2022**, *161*, 111807. [CrossRef]
87. Olivares Díaz, E.; Kawamura, S.; Matsuo, M.; Kato, M.; Koseki, S. Combined analysis of near-infrared spectra, colour, and physicochemical information of brown rice to develop accurate calibration models for determining amylose content. *Food Chem.* **2019**, *286*, 297–306. [CrossRef]

88. Li, X.Q.; Kahlon, T.; Wang, S.; Friedman, M. Low Acrylamide Flatbreads Prepared from Colored Rice Flours and Relationship to Asparagine and Proximate Content of Flours and Flatbreads. *Foods* **2021**, *10*, 2909. [CrossRef]
89. Flores, M.; Mora, L.; Reig, M.; Toldra, F. Risk assessment of chemical substances of safety concern generated in processed meats. *Food Sci. Hum. Wellness* **2019**, *8*, 244–251. [CrossRef]
90. Gao, P.Y.; Cha, R.T.; Luo, H.Z.; Xu, Y.R.; Zhang, P.; Han, L.; Wang, X.H.; Zhang, Z.L.; Jiang, X.Y. Development of antimicrobial oxidized cellulose film for active food packaging. *Carbohydr. Polym.* **2022**, 278. [CrossRef] [PubMed]
91. Hai, Y.D.; Tran-Lam, T.T.; Nguyen, T.Q.; Vu, N.D.; Ma, K.H.; Le, G.T. Acrylamide in daily food in the metropolitan area of Hanoi, Vietnam. *Food Addit. Contam. Part B Surveill.* **2019**, *12*, 159–166. [CrossRef] [PubMed]
92. Ye, Z.H.; Chen, X.T.; Zhu, H.Y.; Liu, X.Q.; Deng, W.H.; Song, W.; Li, D.X.; Hou, R.Y.; Cai, H.M.; Peng, C.Y. Aggregating-agent-assisted surface-enhanced Raman spectroscopy-based detection of acrylamide in fried foods: A case study with potato chips. *Food Chem.* **2023**, *403*, 134377. [CrossRef]
93. Cheng, J.; Zhang, S.; Wan, S.; Wang, P.L.; Su, X.O.; Xie, J.C. Rapid and sensitive detection of acrylamide in fried food using dispersive solid-phase extraction combined with surface-enhanced Raman spectroscopy. *Food Chem.* **2019**, *276*, 157–163. [CrossRef] [PubMed]
94. Zhang, W.Y.; Ma, J.; Sun, D.W. Raman spectroscopic techniques for detecting structure and quality of frozen foods: Principles and applications. *Crit. Rev. Food Sci. Nutr.* **2021**, *61*, 2623–2639. [CrossRef]
95. Yaseen, T.; Sun, D.-W.; Cheng, J.-H. Raman imaging for food quality and safety evaluation: Fundamentals and applications. *Trends Food Sci. Technol.* **2017**, *62*, 177–189. [CrossRef]
96. Zhang, W.; Tang, S.; Jin, Y.; Yang, C.; He, L.; Wang, J.; Chen, Y. Multiplex SERS-based lateral flow immunosensor for the detection of major mycotoxins in maize utilizing dual Raman labels and triple test lines. *J. Hazard. Mater.* **2020**, *393*, 122348. [CrossRef]
97. Zhao, J.H.; Yuan, H.C.; Peng, Y.J.; Hong, Q.; Liu, M.H. Detection of Ractopamine and Clenbuterol Hydrochloride Residues in Pork Using Surface Enhanced Raman Spectroscopy. *J. Appl. Spectrosc.* **2017**, *84*, 76–81. [CrossRef]
98. Long, Y.; Huang, W.; Wang, Q.; Fan, S.; Tian, X. Integration of textural and spectral features of Raman hyperspectral imaging for quantitative determination of a single maize kernel mildew coupled with chemometrics. *Food Chem.* **2022**, *372*, 131246. [CrossRef]
99. Tao, F.; Yao, H.; Hruska, Z.; Rajasekaran, K.; Qin, J.; Kim, M. Use of line-scan Raman hyperspectral imaging to identify corn kernels infected with Aspergillus flavus. *J. Cereal Sci.* **2021**, *102*, 103364. [CrossRef]
100. Joshi, R.; Lohumi, S.; Joshi, R.; Kim, M.S.; Qin, J.; Baek, I.; Cho, B.-K. Raman spectral analysis for non-invasive detection of external and internal parameters of fake eggs. *Sens. Actuators B Chem.* **2020**, *303*, 127243. [CrossRef]
101. Sun, Q. A Study on Intelligent Drying Quality Detection of Typical Fruits and Vegetables Based on the Combination of LF-NMR and NIR. Ph.D. Dissertation, Jiangnan University, Wuxi, China, 2022. Available online: https://kns.cnki.net/KCMS/detail/detail.aspx?dbname=CDFDLAST2023&filename=1022642213.nh (accessed on 8 July 2023).
102. Wang, C.; Wang, C.; Wang, X.; Wang, K.; Zhu, Y.; Rong, Z.; Wang, W.; Xiao, R.; Wang, S. Magnetic SERS Strip for Sensitive and Simultaneous Detection of Respiratory Viruses. *ACS Appl. Mater. Interfaces* **2019**, *11*, 19495–19505. [CrossRef]
103. Tai, Y.H.; Lo, S.C.; Montagne, K.; Tsai, P.C.; Liao, C.C.; Wang, S.H.; Chin, I.S.; Xing, D.; Ho, Y.L.; Huang, N.T.; et al. Enhancing Raman signals from bacteria using dielectrophoretic force between conductive lensed fiber and black silicon. *Biosens. Bioelectron.* **2021**, *191*, 113463. [CrossRef]
104. Cao, C.; Li, P.; Liao, H.; Wang, J.; Tang, X.; Yang, L. Cys-functionalized AuNP substrates for improved sensing of the marine toxin STX by dynamic surface-enhanced Raman spectroscopy. *Anal. Bioanal. Chem.* **2020**, *412*, 4609–4617. [CrossRef]
105. Sun, Y.; Zhai, X.; Xu, Y.; Liu, C.; Zou, X.; Li, Z.; Shi, J.; Huang, X. Facile fabrication of three-dimensional gold nanodendrites decorated by silver nanoparticles as hybrid SERS-active substrate for the detection of food contaminants. *Food Control* **2021**, *122*, 107772. [CrossRef]
106. Zhao, Q.; Zhang, H.; Fu, H.; Wei, Y.; Cai, W. Raman reporter-assisted Au nanorod arrays SERS nanoprobe for ultrasensitive detection of mercuric ion (Hg (2+)) with superior anti-interference performances. *J. Hazard. Mater.* **2020**, *398*, 122890. [CrossRef]
107. Zhang, Y.; Huang, Y.; Kang, Y.; Miao, J.; Lai, K. Selective recognition and determination of malachite green in fish muscles via surface-enhanced Raman scattering coupled with molecularly imprinted polymers. *Food Control* **2021**, *130*, 108367. [CrossRef]
108. Su, L.; Hu, H.; Tian, Y.; Jia, C.; Wang, L.; Zhang, H.; Wang, J.; Zhang, D. Highly Sensitive Colorimetric/Surface-Enhanced Raman Spectroscopy Immunoassay Relying on a Metallic Core-Shell Au/Au Nanostar with Clenbuterol as a Target Analyte. *Anal. Chem.* **2021**, *93*, 8362–8369. [CrossRef]
109. Dias, L.A.F.; Jussiani, E.I.; Appoloni, C.R. Reference Raman Spectral Database of Commercial Pesticides. *J. Appl. Spectrosc.* **2019**, *86*, 166–175. [CrossRef]
110. Walsh, K.B.; McGlone, V.A.; Han, D.H. The uses of near infra-red spectroscopy in postharvest decision support: A review. *Postharvest Biol. Technol.* **2020**, *163*, 111139. [CrossRef]
111. Nallan Chakravartula, S.S.; Cevoli, C.; Balestra, F.; Fabbri, A.; Rosa, M.D. Evaluation of the effect of edible coating on mini-buns during storage by using NIR spectroscopy. *J. Food Eng.* **2019**, *263*, 46–52. [CrossRef]
112. Zhang, L.; Zhang, M.; Mujumdar, A.S. Technological innovations or advancement in detecting frozen and thawed meat quality: A review. *Crit. Rev. Food Sci. Nutr.* **2021**, *63*, 1483–1499. [CrossRef] [PubMed]
113. Shi, S.; Feng, J.; Yang, L.; Xing, J.; Pan, G.; Tang, J.; Wang, J.; Liu, J.; Cao, C.; Jiang, Y. Combination of NIR spectroscopy and algorithms for rapid differentiation between one-year and two-year stored rice. *Spectrochim. Acta Part A Mol. Biomol. Spectrosc.* **2023**, *291*, 122343. [CrossRef]

114. Lapchareonsuk, R.; Sirisomboon, P. Sensory Quality Evaluation of Rice Using Visible and Shortwave Near-Infrared Spectroscopy. *Int. J. Food Prop.* **2015**, *18*, 1128–1138. [CrossRef]
115. Onmankhong, J.; Sirisomboon, P. Texture evaluation of cooked parboiled rice using nondestructive milled whole grain near infrared spectroscopy. *J. Cereal Sci.* **2021**, *97*, 103151. [CrossRef]
116. Li, L.; Zhang, M.; Bhandari, B.; Zhou, L. LF-NMR online detection of water dynamics in apple cubes during microwave vacuum drying. *Dry. Technol.* **2018**, *36*, 2006–2015. [CrossRef]
117. Deidda, R.; Sacre, P.-Y.; Clavaud, M.; Coïc, L.; Avohou, H.; Hubert, P.; Ziemons, E. Vibrational spectroscopy in analysis of pharmaceuticals: Critical review of innovative portable and handheld NIR and Raman spectrophotometers. *TrAC Trends Anal. Chem.* **2019**, *114*, 251–259. [CrossRef]
118. Jiang, Q.; Zhang, M.; Mujumdar, A.S.; Wang, D. Non-destructive quality determination of frozen food using NIR spectroscopy-based machine learning and predictive modelling. *J. Food Eng.* **2023**, *343*, 111374. [CrossRef]
119. Togersen, G.; Arnesen, J.F.; Nilsen, B.N.; Hildrum, K.I. On-line prediction of chemical composition of semi-frozen ground beef by non-invasive NIR spectroscopy. *Meat Sci.* **2003**, *63*, 515–523. [CrossRef]
120. Reis, M.M.; Rosenvold, K. Early on-line classification of beef carcasses based on ultimate pH by near infrared spectroscopy. *Meat Sci.* **2014**, *96*, 862–869. [CrossRef] [PubMed]
121. Qu, F.; Ren, D.; He, Y.; Nie, P.; Lin, L.; Cai, C.; Dong, T. Predicting pork freshness using multi-index statistical information fusion method based on near infrared spectroscopy. *Meat Sci.* **2018**, *146*, 59–67. [CrossRef] [PubMed]
122. Yang, Y.; Zhuang, H.; Yoon, S.-C.; Wang, W.; Jiang, H.; Jia, B.; Li, C. Quality Assessment of Intact Chicken Breast Fillets Using Factor Analysis with Vis/NIR Spectroscopy. *Food Anal. Methods* **2017**, *11*, 1356–1366. [CrossRef]
123. Peyvasteh, M.; Popov, A.; Bykov, A.; Meglinski, I. Meat freshness revealed by visible to near-infrared spectroscopy and principal component analysis. *J. Phys. Commun.* **2020**, *4*, 095011. [CrossRef]
124. Porep, J.U.; Kammerer, D.R.; Carle, R. On-line application of near infrared (NIR) spectroscopy in food production. *Trends Food Sci. Technol.* **2015**, *46*, 211–230. [CrossRef]
125. Zhu, F. NMR spectroscopy of starch systems. *Food Hydrocoll.* **2017**, *63*, 611–624. [CrossRef]
126. Yu, X.; Wang, Z.; Zhang, Y.; Wadood, S.A.; Wei, Y. Study on the water state and distribution of Chinese dried noodles during the drying process. *J. Food Eng.* **2018**, *233*, 81–87. [CrossRef]
127. Niu, L.; Li, J.; Chen, M.-S.; Xu, Z.-F. Determination of oil contents in Sacha inchi (*Plukenetia volubilis*) seeds at different developmental stages by two methods: Soxhlet extraction and time-domain nuclear magnetic resonance. *Ind. Crops Prod.* **2014**, *56*, 187–190. [CrossRef]
128. Chen, L.; Tian, Y.; Sun, B.; Wang, J.; Tong, Q.; Jin, Z. Rapid, accurate, and simultaneous measurement of water and oil contents in the fried starchy system using low-field NMR. *Food Chem.* **2017**, *233*, 525–529. [CrossRef] [PubMed]
129. Paesani, C.; Gómez, M. Effects of the pre-frying process on the cooking quality of rice. *Lwt* **2021**, *140*, 110743. [CrossRef]
130. Ali, S.; Zhang, W.; Rajput, N.; Khan, M.A.; Li, C.B.; Zhou, G.H. Effect of multiple freeze-thaw cycles on the quality of chicken breast meat. *Food Chem.* **2015**, *173*, 808–814. [CrossRef] [PubMed]
131. Li, C.; Liu, D.; Zhou, G.; Xu, X.; Qi, J.; Shi, P.; Xia, T. Meat quality and cooking attributes of thawed pork with different low field NMR T21. *Meat Sci.* **2011**, *92*, 79–83. [CrossRef]
132. Sun, Q.; Zhang, M.; Yang, P. Combination of LF-NMR and BP-ANN to monitor water states of typical fruits and vegetables during microwave vacuum drying. *LWT* **2019**, *116*, 108548. [CrossRef]
133. Zhao, L.; Zhang, M.; Wang, H.; Mujumdar, A.S. Monitoring of free fatty acid content in mixed frying oils by means of LF-NMR and NIR combined with BP-ANN. *Food Control* **2022**, *133*, 108599. [CrossRef]
134. Zhang, Q.; Saleh, A.S.M.; Shen, Q. Discrimination of Edible Vegetable Oil Adulteration with Used Frying Oil by Low Field Nuclear Magnetic Resonance. *Food Bioprocess Technol.* **2012**, *6*, 2562–2570. [CrossRef]
135. Li, M.; Li, B.; Zhang, W. Rapid and non-invasive detection and imaging of the hydrocolloid-injected prawns with low-field NMR and MRI. *Food Chem.* **2018**, *242*, 16–21. [CrossRef] [PubMed]

foods

MDPI

Review

Research Progress of Nitrite Metabolism in Fermented Meat Products

Qiyuan Shen [1,2,3], Xiaoqun Zeng [1,2,3,*], Lingyu Kong [1,2,3], Xiaoqian Sun [1,2,3], Jingjing Shi [1,2,3], Zhen Wu [1,2,3], Yuxing Guo [4] and Daodong Pan [1,2,3]

1 State Key Laboratory for Managing Biotic and Chemical Threats to the Quality and Safety of Agro-Products, Ningbo 315211, China
2 Key Laboratory of Animal Protein Food Processing Technology of Zhejiang Province, College of Food and Pharmaceutical Sciences, Ningbo University, Ningbo 315800, China
3 Zhejiang-Malaysia Joint Research Laboratory for Agricultural Product Processing and Nutrition, Ningbo 315800, China
4 School of Food Science and Pharmaceutical Engineering, Nanjing Normal University, Nanjing 210097, China
* Correspondence: zengxiaoqun@nbu.edu.cn

Abstract: Nitrite is a common color and flavor enhancer in fermented meat products, but its secondary amines may transfer to the carcinogen *N*-nitrosamines. This review focuses on the sources, degradation, limitations, and alteration techniques of nitrite. The transition among NO_3^- and NO_2^-, NH_4^+, and N_2 constitutes the balance of nitrogen. Exogenous addition is the most common source of nitrite in fermented meat products, but it can also be produced by contamination and endogenous microbial synthesis. While nitrite is degraded by acids, enzymes, and other metabolites produced by lactic acid bacteria (LAB), four nitrite reductase enzymes play a leading role. At a deeper level, nitrite metabolism is primarily regulated by the genes found in these bacteria. By incorporating antioxidants, chromogenic agents, bacteriostats, LAB, or non-thermal plasma sterilization, the amount of nitrite supplied can be decreased, or even eliminated. Finally, the aim of producing low-nitrite fermented meat products is expected to be achieved.

Keywords: nitrogen synthesis; nitrogen degradation; lactic acid bacteria; nitrite reductase; antioxidant; nitrite degradation

Citation: Shen, Q.; Zeng, X.; Kong, L.; Sun, X.; Shi, J.; Wu, Z.; Guo, Y.; Pan, D. Research Progress of Nitrite Metabolism in Fermented Meat Products. *Foods* **2023**, *12*, 1485. https://doi.org/10.3390/foods12071485

Academic Editor: Valérie Monteils

Received: 21 February 2023
Revised: 14 March 2023
Accepted: 21 March 2023
Published: 1 April 2023

1. Introduction

Fermented meat products were developed in response to a demand for meat storage. During the fermentation process, a series of biochemical and physical changes caused by microbial fermentation or enzymes provide the meat products with a unique flavor, color, texture, and antioxidant properties, so as to improve the edible quality of the meat [1]. Subsequently, nitrite, comprising salty white or light yellow particles [2], which primarily contain sodium or potassium nitrite, is commonly used in meat curing [3] as a color protectant, antioxidant, and preservative, inhibiting the growth of spoilage and pathogenic bacteria, such as *Clostridium botulinum* and *Listeria monocytogenes* [4]. In long-term studies, nitrite was investigated as a mammalian vasodilator, which released a protective substance that can save a mammal's life during hypoxia [5]. Nitrite has antibacterial, antioxidant, color development, and flavor production properties, making it almost irreplaceable [3]. Therefore, nitrite is commonly utilized in the processing of meat products.

Regulation No.1333/2008, which applies to sausages, pig hoofs, and other products in the European Union (EU), previously regulated the use of nitrite to a maximum dosage of 150 mg/kg [6]. Excessive nitrite might pose a threat to food safety. Firstly, once nitrite enters the body, it causes hypoxic poisoning by binding to hemoglobin in the blood. Secondly, after ingestion of animal products, nitrite reacts with secondary amines to form nitrosamine [7]. Increasing awareness of health care advances meant that people gradually realized the harm of excessive nitrite to the body, causing concerns regarding fermented meat product ingredients such as nitrite, which has had a massive effect on cured meat product sales. Therefore, preferences for the consumption of safe, low-salt, and low-fat meat products without chemical additives have increased considerably [8]. Several investigations have been undertaken to determine whether antioxidants, chromogenic compounds, and bacteriostatic agents can substitute for nitrite in the curing process [9]. However, cured meat products are complex products, in which each ingredient plays a special role, making reformulation difficult. Challenges arise as a result of the reduction of salt and chemicals in cured meat products, which alter their sensory qualities and might have an impact on the microbial ecosystem, resulting in uncharacteristic and unsafe products [8].

Recent research reviewed the role of nitrite in fermented meat products, and numerous nitrite metabolic pathways have been documented and collated. The LAB in fermented meat products are the key factors affecting nitrite metabolism, and the acids, enzymes, and other compounds produced by LAB contribute to nitrite synthesis and degradation. Consequently, this review summarizes research from the previous years, the majority of which were published within the last five years, and discusses nitrite's source, degradation pathway, and conversion regulation from the four perspectives of nitrogen synthesis and degradation, nitrite metabolism, nitrite safety, and methods and application of fermented meat products. This review mainly describes the role of LAB in the degradation of nitrite in fermented meat products to provide a theoretical foundation for future research into nitrite and its applications.

2. Microorganisms Related to Nitrite Metabolism

Nitrite is one of the nitrogen cycle intermediates on Earth [10]. Nitrogen in nature is mainly present as NH_4^+, as the inert N_2, the lowest oxidized state, and as NO_3^-, the highest oxidized state. The transition between NO_3^- and NO_2^-, NH_4^+, and N_2 in the biogeochemical cycle constitutes the balance of nitrogen synthesis and degradation through nitrate and denitrification bacteria (Figure 1).

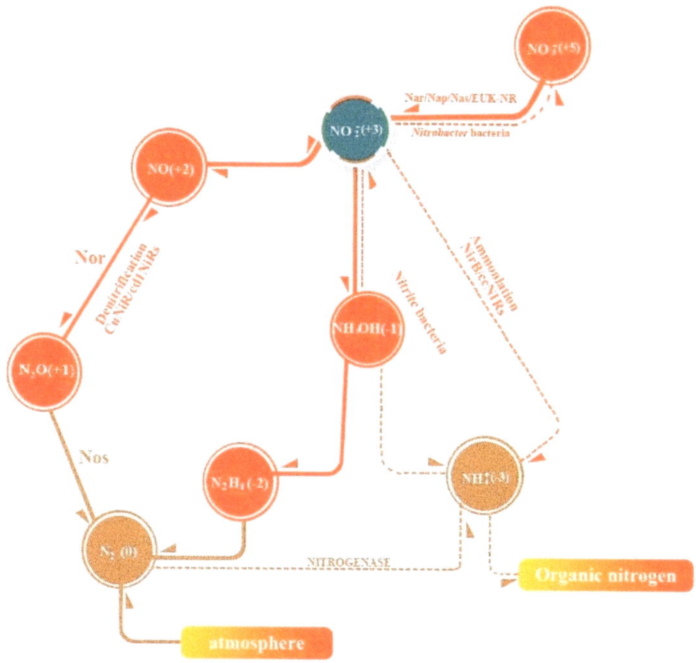

Figure 1. Biochemical cycling of nitrogen.

Nitrifying bacteria are a class of aerobic bacteria with two physiological subgroups, *Nitrobacter* and *Nitrosomonas*. *Nitrobacter* (also known as nitrite-oxidizing bacteria), containing nitrite oxidase, oxidize nitrite to nitrate (reaction (1)) and reproduce for a generation in 18 h. *Nitrosomonas* (nitrite bacteria, also known as ammonia-oxidizing bacteria), oxidize NH_4^+ to nitrite (reaction (2)) and reproduce for a generation in 18 min [11]. The growth and propagation rates of *Nitrobacter* are significantly lower than those of *Nitrosomonas*. The conversion rate of ammonium nitrogen to nitroso nitrogen is significantly slower than that of nitrate-nitrogen to nitrite-nitrogen, leading to the accumulation of nitrite-nitrogen.

$$2NO_2^- + O_2 \rightarrow 2NO_3^- + \text{Energy} \tag{1}$$

$$2NH_4^+ + 3O_2 \rightarrow 2NO_2^- + 4H^+ + 2H_2O + \text{Energy} \tag{2}$$

Denitrifying bacteria are a group of bacteria that reduce nitrate-nitrogen to N_2 and NH_4^+. Mostly heterotrophic, facultative anaerobic bacteria, such as *Bacillus stephensi* and *Trichomonas aeruginosa*, contain a variety of nitrite reductase (NiR) enzymes to degrade nitrite. Copper-type nitrite reductases (CuNiRs) and cytochrome cd1 nitrite reductases (cd1NiRs) convert NO_3^- to N_2 (reactions (3) and (4)) by denitrification, and polyheme c nitrite reductases (ccNiRs) convert NO_3^- to NH_4^+ by amination. Denitrifying bacteria maintain the nitrogenous nitrogen content at a steady low-concentration level and the nitrogen cycle remains in dynamic equilibrium [12].

$$C_6H_{12}O_6 + 12NO_3^- \rightarrow 6H_2O + 6CO_2 + 12NO_2^- + \text{Energy} \tag{3}$$

$$5CH_3COOH + 8NO_3^- \rightarrow 6H_2O + 10CO_2 + 4N_2 + 8OH^- + \text{Energy} \tag{4}$$

Bacteria with degradation effects mainly include *Lactiplantibacillus plantarum*, *Levilactobacillus brevis*, *Leumesenteroides*, *Pediococcus cerevisiae*, *Streptococcus faecalis*, and others. Paik et al. found that *Lactiplantibacillus plantarum* KGR5105, *Levilactobacillus brevis* KGR3111, *Latilactobacillus curvatus* KGR2103, and *Lactobacillus serans* KGR4108 also produce NiR. LAB showed significant degradation ability in fermented meat products under optimal conditions [13].

Fermented meat products, which typically contain nitrite, a nitrogen-containing dietary additive, follow the same natural rhythm of nitrogen synthesis and breakdown. In general, the pH of meat products ranges between 5.5 and 6.2, while the pH of dry fermented sausage ranges between 4.5 and 5.5 [14], both of which are optimum conditions (pH 4.5–5.5) that allow LAB to grow. However, *N*-nitrosamine formation may be more easily achieved in dry fermented sausages as the pH of the product approaches the optimum pH (pH 3.5) of the nitrosation reaction [14,15]. Nitrite is synthesized and degraded by bacteria in the endogenous system of fermented meat products, completing the nitrite metabolism cycle in these items [10].

3. The Role of Nitrite in Fermented Meat Products

3.1. Color Formation Effect

NO reacts with myoglobin (Fe^{2+}) and methemoglobin (Fe^{3+}) in fermented meat products to form cured pink, which serves as a coloring and antioxidant. Both free and heme-bound iron is the principal pro-oxidant in meat products [16]. NO binds to myoglobin (Fe^{2+}) to form unstable NO-myoglobin, which is converted to a stable pink pigment, nitrosoheme, upon heating and prevents iron-induced oxidation [17]. Myoglobin can also react with nitrite and be oxidized to methemoglobin (Fe^{3+}), which in turn reacts with NO to form NO-methemoglobin, and reducing agents restore NO-methemoglobin to form NO-myoglobin and finally form cured pink under heating [18]. Postmortem, active cytochrome enzymes possess the ability to utilize oxygen, which is responsible for the red color on the surface of meat caused by the presence of oxymyoglobin [19]. Similarly, nitric oxide promotes the lipid peroxidation cycle due to its lipophilic nature by reacting with alkyl-, alkoxyl-, and peroxyl radicals [20]. Consequently, nitrite can inhibit both primary and secondary oxidation.

3.2. Flavor Improvement

The exact mechanism of the effect of nitrite on flavor is still unclear, but the antioxidant activity of nitrite is an important factor affecting its flavor, although other antioxidants added to meat products cannot show the unique flavor of fermented meat products. Nitrite will not directly produce a specific flavor substance in fermented meat products. However, nitrite can inhibit lipid oxidation, which inhibits the production of aldehydes, such as hexanal and pentanal, which masks the sulfur-containing compounds that make meat products produce a pickled flavor [21]. On the other hand, nitrite can induce the formation of Strecker aldehyde, which is formed via the degradation of amino acids by dicarbonyls that are produced from the Maillard reaction, due to the pro-oxidant effect of nitrite, and is related to food flavor [22].

3.3. Antioxidant Properties

Nitrite can also be used as an antioxidant by receiving oxygen from sensitive molecules or producing active nitrogen. Villaverde et al. reported that the formation of carbon-based compounds in fermented sausage increased with the increase in nitrite content [21]. Vossen and De Smet studied the effect of sodium nitrite on protein oxidation in isolated myofibrillar protein and porcine patties [23]. It was found that the TBARS value of sodium nitrite was significantly lower than that of the control group (without sodium nitrite) but had no effect on the content of carbon-based compounds as an indicator of protein oxidation. In addition, when sodium nitrite and sodium ascorbate are used together, the yield of carbon-based compounds will increase, but it will not increase when used alone, which is the result of nitrite as an oxidant of ascorbate. Dehydroascorbate is produced by the oxidation of ascorbate by nitrite, which is similar to the carbon group of reducing sugar. It produces non-protein carbon-based compounds via non-enzymatic glycosylation [24].

3.4. Antimicrobial Effect

Nitrite is the substance of interest in microbial inhibition, but it is not the nitrite itself that produces the inhibition, which is closely related to its ability to trigger the formation of NO. In vitro, NO can directly react with microbial proteins that contain an iron–sulfur enzyme cluster (Fe-S-NO) and form protein-bound dinitrosyl dithiolato iron complexes [25]. Ferredoxin in *Clostridium* spp. is involved in ATP synthesis by the microorganisms in this Fe-S broad group. Therefore, this Fe-S-NO mechanism is considered to be the main factor inhibiting the growth of *Clostridium* spp. vegetative cells, and it can also be observed in aerobic and facultative pathogens associated with cured meat products [26]. The dinitric iron complex formed by NO binding to proteins is present in prokaryotic and eukaryotic cells, and iron–sulfur proteins are the major source of protein-bound dinitrosyl iron complexes formed in *Escherichia coli* cells under nitric oxide stress [27]. At the same time, peroxynitrite, a highly oxidized and unstable compound, may be involved in the inhibition mechanism of NO, and the formation of peroxynitrite is related to the changes in the oxidation state of intermediate compounds during the reduction of myoglobin and nitrate. Due to its oxidation and nitrification ability, it will damage proteins, DNA, and lipids, and ultimately inhibit the growth of microorganisms [28].

4. Metabolic Pathways of Nitrite in Fermented Meat Products

Exogenous manufacture, contamination during processing, and endogenous microbial synthesis are the main sources of nitrite in fermented meat products. In addition, a small amount of nitrite is produced by contamination during rearing and endogenous microbial production. Acids, enzymes, and non-acid and non-enzyme compounds in bacteria can degrade nitrite in fermented meat products (Figure 2).

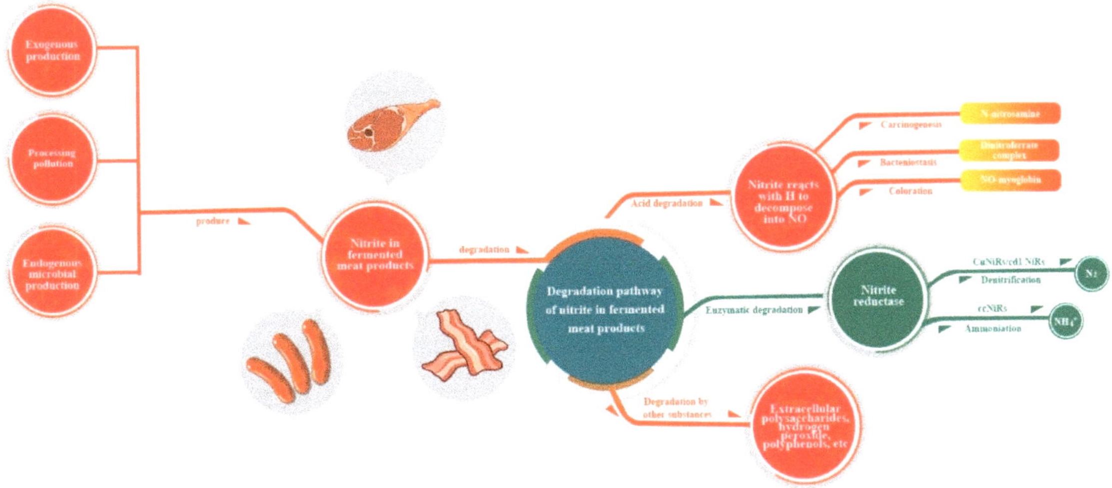

Figure 2. Metabolic pathway of nitrite in fermented meat products.

4.1. *Source of Nitrite in Fermented Meat Products*

4.1.1. Exogenous Manufacture during Processing

The addition of nitrite to meat products during processing improves their appearance, color, flavor, and safety. It is also the main source of nitrite in fermented meat products. In meat products, nitrite serves four purposes. The first is the reaction of nitric oxide (NO) with hemoglobin and myoglobin to form nitrohemoglobin and nitrosymyoglobin, which maintain the bright color of meat products [29]. Jarulertwattana found that 4×10^{-6} g/g nitrite was suitable for curing chicken when used with ginger sauce, but more than 4×10^{-6} g/g nitrite will cause pink defects in chicken legs [30]. Second, nitrite can inhibit the growth of *Clostridium botulinum* and *Staphylococcus aureus*. The minimum concentration of nitrite that inhibits the outgrowth of Clostridium botulinum is from 4×10^{-5} to 8×10^{-5} g/g [1]. Third, nitrite has an antioxidant effect and can suppress lipid peroxidation, lowering the production of foul odors. Ji found that sodium nitrite can effectively inhibit lipid oxidation at 1×10^{-4} g/g during the curing process of mutton [31]. Finally, nitrite can aid some fermentation bacteria in the production of fermented flavor components in cured meat products through metabolism or the hydrolysis of protease and lipase. [4,32].

Some vegetable extracts are added to fermented meat products as nitrite substitutes because they contain a large amount of nitrite: celery, watercress, lettuce, spinach, and rape (<2500 mg/100 g); Chinese cabbage, leek, and parsley (1000 to <2500 mg/100 g); radish and cabbage (500 to <1000 mg/100 g); carrots, cucumbers, pumpkins, and broccoli (200 to <500 mg/100 g); and potatoes, tomatoes, onions, eggplants, mushrooms, and asparagus (<200 mg/100 g) [33,34]. Meat products cured with vegetable extracts showed similar quality and sensory properties to nitrite-cured products [35]. Natural nitrite sources have a higher consumer preference than synthetic nitrite because they contain a variety of antioxidants and antibacterial compounds. Phenolic compounds in vegetable extracts may hinder the conversion of nitrite into *N*-nitrosamines, making the probability of nitrosamines production lower than that of synthetic nitrite. However, there is no relevant research to prove its conversion relationship. Nevertheless, these plant extracts have many application limitations. Firstly, only vegetables that contain enough nitrate can be used. Furthermore, the taste and color of vegetables will also affect the sensory quality of fermented meat products. Furthermore, the transformation of nitrate to nitrite in plant extracts is uncontrollable, and the content of nitrite finally transformed may not reach the desired level [36].

4.1.2. Contamination during Rearing

During animal rearing, domestic animals will ingest nitrite in the environment. Nitrogen compounds are widely used as fertilizers in agriculture. Crops produce plant chlorophyll in response to nitrogen fertilizer, which increases crop yield [37]. The use of nitrogen fertilizer has risen substantially in recent years, particularly in Southeast Asia [38]. Despite having a positive and significant impact on global food production, nitrogen fertilizer has a significant negative impact on the larger ecosystem. Under natural conditions, nitrifying microorganisms in the environment influence the nitrogen cycle of water, ensuring that nitrogen metabolism remains balanced. However, a large amount of high-nitrogen metabolites are added to aquaculture water, which outperforms the metabolism of natural bacteria in the reservoir, alters the dynamic balance of the nitrogen cycle, causes nitrite to deposit in the water, and eventually accumulates nitrite in animals [39].

Unintentionally added nitrate and nitrite in livestock products have been reported. Iammarino et al. reported a maximum endogenous nitrate content of 30 mg/kg in horse meat and beef, and up to 40 mg/kg in fresh pork meat [40]. Iacumin et al. also determined that the thresholds for nitrite and nitrate concentrations in raw meat were less than 4 mg/kg and 22 mg/kg, respectively, which are below the threshold; the authors believed that nitrite and nitrate in raw meat were not intentionally added [41].

4.1.3. Endogenous Microbial Production

In addition to the nitrite accumulated by exogenous addition and contamination of raw materials, a tiny quantity of nitrite is produced during meat product fermentation by endogenous microbial action. In ammonia-oxidizing bacteria, monoamine oxidase converts NH_4^+ to hydroxylamines, which are then converted to nitrite by hydroxylamine oxidase (reaction (2)). Denitrifying bacteria can also convert nitrate-nitrogen to N_2 and NH_4^+, with nitrite as an intermediary product. Iacumin et al. investigated dry pickled ham from San Daniele without adding nitrite for 14–19 months and found that nitrite in the ham was present at less than 4 mg/kg [42]. When products have a long curing process, adding nitrate instead of nitrite to meat products, the nitrate is reduced to nitrite by microbial nitrate reductases [43].

4.2. The Degradation Pathway of Nitrite in Fermented Meat Products

LAB and denitrifying bacteria degrade nitrite in fermented meat products, mostly through acid and enzyme degradation, while other non-acid and non-enzyme compounds can also degrade nitrite. LAB can also limit the metabolism of nitrate reduction bacteria, reducing the generation of nitrite, and inhibiting the growth of *Escherichia coli*, *Klebsiella pneumoniae*, *Pseudomonas fluorescens*, *Pseudomonas alkalogenes*, and other bacteria [44]. Some bacteria, such as *Latilactobacillus sakei*, *Latilactobacillus curvatus*, and *Levilactobacillus brevis*, have a strong ability to reduce nitrite by inhibiting the formation of *N*-nitro-sodimethylamine precursors [45]. Wu et al. used a co-culture of *Lactiplantibacillus plantarum* Shanghai brewing 1.08 and *Zygosaccharomyces rouxii* CGMCC 3791 to minimize nitrite and biogenic amine concentrations and increase the flavor components in Chinese sauerkraut [46].

The degradation by LAB is separated into two stages according to Zeng et al. [47]. In the early stage of fermentation, when the pH > 4.5, nitrite is mostly degraded enzymatically. Second, because LAB create acid, the breakdown of nitrite in late fermentation is primarily acid degradation after pH < 4.5.

In the curing process, LAB mostly focused in the processing step degrade nitrite rather than store it. According to Huang, *Limosilactobacillus fermentum* RC4 and *Lactiplantibacillus plantarum* B6 were added to cured meat. The content of nitrite was significantly reduced (0.75 mg/kg) during processing, which was significantly lower than that of regular cured meat (4.67 mg/kg); during 0–20 days of storage, the moisture content of bacon decreased significantly, the pH increased continuously, but the content of nitrite did not change significantly [48].

According to reports, nitrite is degraded in the following percentages in meat products: reactions with heme proteins (5–15%), non-heme proteins (20–30%), nitrous acid with free-amino acids via the Van Slyke reaction to generate nitrogen gas (1–5%), sulfhydryl groups (5–15%), lipids (1–5%), oxidation back to nitrate (1–10%), and free nitrite (5–20%) [49].

4.2.1. Acid Degradation

The acid degradation of nitrite demonstrates that nitrite serves as a colorant, antioxidant, unique flavoring agent, oxidant, and bacteriostatic agent in fermented meat products. Nitrite also can produce *N*-nitrosamines with a carcinogenic teratogenic effect.

Fermented meat products contain large amounts of H^+ after microbial fermentation, and nitrite and H^+ are decomposed into NO and H_2O (reactions (5)~(7)) [4].

$$NO_2^- + H^+ \leftrightarrow HNO_2 \tag{5}$$

$$2HNO_2 \leftrightarrow N_2O_3 + H_2O \tag{6}$$

$$N_2O_3 \leftrightarrow NO + NO_2 \tag{7}$$

Nitrite (NO_2^-) forms nitrite acid (HNO_2) when it combines with hydrogen ions (H^+). Nitric acid then gradually decomposes into nitrous trioxide (N_2O_3) and water molecules (H_2O) (reactions (6) and (7)). N_2O_3 further dissociates into NO and NO_2 (reaction (8)). NO can further form N-nitrosamines with carcinogenic and teratogenic effects [3]. In the endogenous environment, the intermediate between nitrite and NO, NO_2, establishes a dynamic balance.

4.2.2. Enzyme Degradation

Nitrite is degraded by NiR and reduced to NO or NH_3. NiR, which is a critical enzyme in the nitrogen cycle, catalyzes nitrite reduction and mainly comes from LAB and denitrifying bacteria. The majority of NiRs are intracellular enzymes that are mostly found in the periplasmic region or cell membrane, and some are free in the cytoplasm. Liu et al. reported that NiR produced by *Lacticaseibacillus rhamnosus* LCR6013 might, via the nitrate respiration pathway ($NO_2 \rightarrow NO \rightarrow N_2O \rightarrow N_2$), produce nitrous oxide (N_2O) and degrade nitrite [50].

Depending on the reactants and cofactors, NiRs can be classified into CuNiRs, cd1NiRs, ccNiRs, and ferredoxin-dependent nitrite reductases (FdNiRs) [51]. The genotypes of NiR mainly include *nir*K, *nir*S, *nrf*A, and *nir*B, etc., of which *nir*K and *nir*S are key genes in the denitrification process, encoding a CuNiR and cd1NiR, respectively [52]. *nrf*A and *nir*B are the key genes of nitrate ammonification, which reduce nitrite by transferring six electrons. Genes *nrf*A and *nrf*H encode the double-subunit complex ccNiR protein composed of two subunits, NrfA and NrfH, respectively.

CuNiRs are involved in the denitrification of nitrogen during metabolism. The reduction of NO_2^- by CuNiRs can be divided into five steps: the combination of NO_2^- with the enzyme, the reduction reaction, dehydration of bound intermediate products, NO release, and enzyme regeneration. NO_2^- combines with the T2Cu center in the oxidized form to replace a soluble molecule and forms a hydrogen bond between the Asp98 residue and an oxygen atom of NO_2^-. When the electron is transferred from T1Cu to T2Cu, the proton of the hydrogen bond is transferred from the Asp98 residue to the oxygen atom of the substrate to form an intermediate product O=N-O-H. The N-O bond of the oxygen atom is then broken and the product NO is released into the active center [53].

Similar to ccNiRs, the NiRB large and NiRD small subunits are encoded by the genes *nir*B and *nir*D, respectively, and are used to degrade nitrite by forming the NiRBD complex [54]. NiRB encoded by the *nir*B gene, catalyzes NO_2^- to NH_4^+, and is a soluble NADH-dependent NiR catalytic subunit-containing sirol heme. NiRB, under anaerobic conditions, uses NADH as an electron donor to reduce nitrite in the cytoplasm, while oxygen suppresses its activity [55]. Wang et al. used anaerobic chemostat culture technology, in which NiRB-lacZ was used to report fusion steady-state gene expression, revealing the differential expression pattern of gene *nir* in *E. coli* [56]. It was found that NiRB is an inducible enzyme, and its expression is induced by high nitrate and nitrite conditions. The synthesis of *nir*B-encoded NiRB is high only when the content of NiRB exceeds the consumption capacity of the cells.

Microorganisms can spontaneously regulate the content of nitrite in fermented meat products, which is mainly related to the genes in the microorganisms. The exogenous environment, the normal expression of genes, the deletion of gene fragments, and the interaction between multiple genes will affect the regulation of nitrite metabolism in microorganisms [57]. Zeng et al. used proteomic and bioinformatic analyses to identify 31, 87, and 190 differentially expressed genes in the process of nitrite degradation in *Limosilactobacillus fermentum* RC4, including *adh*E and *lpd*A which are involved in carbohydrate metabolism, *cys*K related to amino acid metabolism, *nir*B corresponding to nitrogen metabolism, *fab*I and *acc*D associated with lipid metabolism, and *gsk* involved in nucleotide metabolism [47]. These genes are involved in the metabolism of *Limosilactobacillus fermentum* RC4 during nitrite reduction. Chu et al. studied the effects of carbon sources (acetate and glucose) on the endogenous denitrification and ammoniation of *Candida* [58]. After adjusting the oxygen–phosphorus ratio, it was found that acetate (54.2%) had a higher efficiency of converting nitric acid into nitrite (90.2%), whereas glucose (51.3%) made the accumulation of nitrite more stable (85.3%). The total nitrogen removal efficiencies of acetate (88.8%) and glucose (91.3%) were similar (87.8% and 89.8%). Iino et al. studied the genes *ro06366* and *ro00862* of *Rhodococcus* rHA1 [59]. The single mutant with deletion of the above gene showed growth retardation in the environment when using nitrate or nitrite as the only nitrogen source. Iino allowed the double mutant to grow in the environment with a nitrate and nitrite nitrogen source. It was found that nitrate and nitrite were not the only nitrogen sources used and both *ro06366* and *ro00862* are involved in the regulation of nitrite and nitrate. Khlebodarova et al. studied the process of nitrite utilization by NiR in *E. coli* cells [60]. NO_2^- is reduced to NH_4^+ outside the cell via NrfA reductase, flows into the cell through the NirC transporter, participates in the degradation of proteins and their complexes, and is converted to NH_3 through NirB reductase, and then flows out of the cell through the NirC transporter.

4.2.3. Other Substances for Degradation

In addition to the aforementioned degradation pathways, several non-acid and non-enzyme substances, such as flavonoids, polyphenols, and ascorbate, also have effects in nitrite scavenging. Guo et al. measured the nitrite scavenging ability of flavonoids using the diazotization-coupling reaction in vitro, and found that the nitrite scavenging activity was closely related to phenolics ($r^2 = 0.990$, $p < 0.01$) and flavonoids ($r^2 = 0.923$, $p < 0.05$) [61]. Ben et al. found that polyphenols in sea buckthorn had a higher nitrite-degrading capacity than other compounds, with a nitrite-degrading rate of 75.9% [62]. Skibsted found that antioxidants such as ascorbate and polyphenols can induce the reduction of N_2O_3, promote the production of NO, and destroy the balance between nitrite, NO, and NO_2, ultimately reducing the content of nitrite [20]. Feng et al. reported the effect of nitrite on protein oxidation and nitrification of cooked sausage, in which the antioxidant effect of nitrite on protein oxidation was manifested as a significantly reduced base content, higher free amines, and lower surface hydrophobicity [63].

4.3. Factors Affecting Nitrite Degradation

4.3.1. Effects of the Food Matrix on the Degradation of Nitrite

Polysaccharides and inorganic salts in the food matrix mainly affect the degradation of nitrite. The food matrix influences the activity of catalase in vitro of LAB and produces hydrogen peroxide, altering the ability of autolysis in vitro of LAB and influencing the activity of nitrate reductase in vitro [64]. Seo et al. studied the extracellular polysaccharides produced by *Lactiplantibacillus plantarum* YML009 [65]. The nitrite clearance rate was 43.93%, which proved the extracellular polysaccharides have a strong ability to degrade nitrite, and the higher the sugar content, the higher the nitrite-free radical clearance rate. Inorganic salt can reduce the water content of fermented meat products and inhibit the growth of microorganisms. When excessive sodium chloride (NaCl) is used to pickle fermented meat products, bacteria will dehydrate and die, and the nitrite degradation effect will be significantly reduced. Delgado-Pando et al. evaluated bacon with different salt contents and found that the color of bacon with a high salt content was redder than that with a low salt content [66].

4.3.2. Effects of Fermentation Conditions on the Degradation of Nitrite

The lower the pH, the more appropriate the fermentation temperature, and the longer the fermentation duration, which leads to less nitrite production. Kilic et al. investigated the influence of pH on residual nitrite in cured meat products and discovered that lowering the pH greatly reduced residual nitrite, whereas increasing the pH increased residual nitrite [67]. Under an acidic environment, the microbial system in fermented meat products is regulated spontaneously, and *Lactobacillus acidophilic* has competitive advantages to inhibit other bacteria and possible pathogens [68]. Furthermore, the longer the fermentation time is, the more dominant the LAB will be.

5. Safety of Nitrite in Food

5.1. Toxicity of Nitrite

NO is a crucial compound generated from nitrite, which combines with myoglobin and methemoglobin, proteins in cells, and with amines to lower NO levels in the dynamic balance, resulting in nitrite degradation. In meat fermentation, NO has coloring, antioxidant, cured flavor, and antibacterial benefits; however, it can also be carcinogenic. At low concentrations (nm), NO acts as a signaling molecule for intercellular communication in neurons and the cardiovascular system [69]. At high concentrations (μM), NO can kill pathogens and cancer cells [70].

Through a nucleophilic substitution reaction, NO can combine with non-protonated secondary amines to form *N*-nitrosamines, which are carcinogenic and teratogenic [3]. The most common volatile *N*-nitrosamines in meat products are *N*-nitrosodimethylamine, *N*-nitrosodiethylamine, *N*-nitrosopiperidine, *N*-nitrosopyrrolidine, and *N*-nitrobenzylmorpholine. Among them, in terms of oncogenicity and genotoxicity, *N*-nitrosodimethylamine and *N*-nitrosodiethylamine are regarded as the most volatile *N*-nitrosamines [71]. The amount of *N*-nitrosamines production mainly depends on the intake of nitrite and the processing conditions in meat products, which also increases in concentration under conditions of pH 2.5~3.5, longer storage time, and high-temperature and high-acid conditions [72]. The presence of amines is one of the primary prerequisites for the creation of *N*-nitrosamines. There are not many *N*-nitrosamines in raw meat since amino acids are only decarboxylated to produce amine. Nevertheless, procedures including maturation, fermentation, and curing may increase their synthesis of *N*-nitrosamines. It was discovered during processing that the quantity of precursors was directly correlated with the amount of *N*-nitrosamines in meat products, but even if precursors were present, low water activity and an unfavorable pH would prevent the production of nitrosamines in meat products [73]. Additionally, several cooking techniques that reach temperatures greater than 130 °C, such as frying or barbecuing, may make it more likely that *N*-nitrosamines will occur [3]. To prevent nitrosation, chemicals such as ascorbate and tocopherol are added during processing. The presence of sulfhydryl compounds, some phenols, and tannins in meat products might also inhibit the formation of *N*-nitrosamines [15].

5.2. Addition Limit of Nitrite in Fermented Meat Products in Different Countries

According to current knowledge, the use of nitrite mainly considers two risks—the final formation of *N*-nitrosamines and the final growth of serious pathogens. Nitrite has potential carcinogenicity in the human body, and its intake should be limited [7]. In addition to the processing steps of curing meat products, consumers' cooking methods with temperatures higher than 130 °C, such as frying or grilling meat products, will also increase the possibility of *N*-nitrosamines formation [3]. By contrast, other studies have shown that nitrite, as a vasodilator in mammals, is a life-saving medication that releases a protective substance during hypoxic events [5]. The European Food Safety Authority (EFSA) assessed the acceptable daily intake (ADI) in 2017, excluding infants under 3 months of age, and determined that the ADI of nitrate is 3.7 mg/kg body weight (bw)/d and that of nitrite is 0.07 mg/kg bw/d. Most countries' management regulations allow for the use of nitrite (Table 1).

Table 1. Application requirements of nitrite in different countries (only fermented meat products).

Country	Institution	Application Requirements of Nitrite	Reference
America	Code of Federal Regulations	The maximum permitted amount of nitrite use ≤ 200 mg/kg	[74]
Japan	Ministry of Health, Labor and Welfare of Japan	The maximum residue ≤ 70 mg/kg	[75]
Korea	Ministry of Food and Drug Safety	The maximum residue ≤ 70 mg/kg	[76]
Europe	European Commission	The maximum residue ≤ 50 mg/kg	[6]
China	State Health and Family Planning Commission	The maximum permitted amount ≤ 150 mg/kg (The maximum residue of sodium nitrite ≤ 30 mg/kg)	[77]

5.3. Clean Label Movement

In some countries, there are plant extracts such as pre-converted nitrites added to fermented meat products, and these products are labeled "natural", "organic", and "uncured" [78], which may lead to confusion or even mislead consumers. In addition, the vegetable extract's precise chemical makeup need not be disclosed, which also encourages some businesses to use it in place of some chemical additions [8]. On the other hand, the use of plant extracts rich in nitrite does not avoid the production of *N*-nitrosamines because the residual nitrite in the product probably reacts under high-temperature conditions and may produce *N*-nitrosamines [79]. According to the Standing Committee on Plants, Animals, Food, and Feed (ScoPAFF), the use of vegetable extracts with a high nitrite content is regarded as an addition in the food processing process, and regulation N°1333/2008 also ensures the use of vegetable extracts in food processing [80]. Therefore, adding vegetable extracts to fermented meat products as nitrite to label items green is a false tactic used by food producers and operators [81]. Rivera et al. thought that the elimination of the "uncured" labeling policy for meats processed with pre-converted nitrites from vegetable sources would improve transparency for consumers [1].

6. Methods and Applications of Substituting Nitrite in Fermented Meat Products

Nitrite can inhibit the growth of bacteria due to its oxidation and nitrification ability. The survival or multiplication of microorganisms under reduced or in the absence of nitrite concentrations is increased substantially and represents a challenge for the meat industry to guarantee the safety of fermented meat products. Commercially speaking, it may provide value to draw customers if nitrite is not listed on the label of fermented meat products.

Researchers continue to find safe substitutes for nitrite, including plant substitutes, microbial substitutes, and organic acid substitutes. However, at present, no substitute can currently match the entire action of nitrite and completely replace it. Therefore, the most successful option is to use low-dose nitrite in the curing process of meat products in combination with other substances or technologies to provide anti-corrosion, bacteriostasis, color, and taste, while also preserving the quality of meat products as much as possible.

Recently, many studies have been conducted on the inhibition nitrite synthesis (Table 2). It is required to prevent the synthesis of nitrite in order to lower the level of nitrite in fermented products, such as by adding antioxidants and LAB, and replacing the role of nitrite with chromogenic agents and antibacterial agents.

Table 2. Studies on reducing the nitrite content in fermented meat products.

Methods		Products	Effects	Reference
Antioxidants substances	Sodium ascorbate (500 ppm) and α-Tocopherol acetate (10 ppm)	Fermented meat products	Reduced the content of nitrite, prevented the conversion of nitrite to nitrosamine, produced unique flavors, and improved the sensory quality.	[82]
	Ascorbic acid (0.0075%) and other curing additives	Beef sausage	Increased the antilisterial activity of *Enterococcus mundtii* to 2 log cfu/g.	[83]
	Phenolic	Fermented meat products	Phenolic compounds in bovine essential oils had antioxidant and antibacterial properties.	[84]
Lactic acid bacteria and its enzymes	*Limosilactobacillus fermentum* RC4 (1.06%) and *Lactiplantibacillus plantarum* B6 (0.53%)	No-nitrite-added cured meat	*Limosilactobacillus fermentum* RC4 had an effective nitrite degradation ability and *Lactiplantibacillus plantarum* B6 inhibited bacteria (0.53%).	[48]
	Enzymes	Pickle	Reduced nitrosamines.	[44]
Chromogenic agent	Beetroot and berries extracts	Fermented dry sausages	Imparted the red color of nitrite-free meat products to develop and stabilize color.	[85]
	Plasma-activated water	Beef jerky	Promoted the formation and fixation of color.	[86]
	Chili red and lycopene	Meat batters	Partially replaced nitrite from 150 mg/kg to 100 mg/kg, and improved the texture characteristics of the product.	[87]
Bacteriostat	Radish powder (0.5%) and chitosan (0.25%)	Fermented cooked sausages	Had good effects on physical, chemical and microbial stability of fermented cooked sausage.	[88]
	Garlic essential oil (125 mg/kg), allyl isothiocyanate (62.5 μL/kg), and nisin (20 mg/kg)	Fresh sausages	Effectively reduced *E. coli* O157H7 and LAB, and maintained the physical and chemical properties.	[89]

6.1. Antioxidant Substitutes

As an antioxidant, nitrite can largely influence flavor to prevent lipid oxidation and the formation of undesirable lipid oxidation byproducts such as hexanal and 2,4-decadienal [21]. Its antioxidant substitutes can be divided into synthetic antioxidants and natural antioxidants. Synthetic antioxidants include propyl gallate, tert butyl hydroquinone, and butyl hydroxyanisole. Natural antioxidants include phytic acid, vitamin E mixed concentration, guaiac resin, flavonoids, amino acids, and others [90]. Some natural ingredients not only have an antioxidant effect, but also have a potential inhibitory effect on controlling microbiological hazards, such as cloves, essential oils of aromatic and medicinal plants, plant extracts with high polyphenols concentration, acidified whey, honey, and other bee products [34].

6.2. LAB and Its Enzymes

LAB are the ideal bacteria in food processing. The vast majority of LAB have a status generally recognized as safe and they have good salt resistance. In addition, the European Food Safety Authority has granted the status of qualified presumption of safety to many LAB, included in the genera *Carnobacterium*, *Lactococcus*, *Leuconostoc*, *Oenococcus*, *Pediococcus*, *Streptococcus*, and the former *Lactobacillus* genus, recently reclassified into twenty-five new genera [91,92]. In the environment of pickled food, LAB can ferment food and increase flavor, and also produce acetic acid, bacterial peptides, and hydrogen peroxide to inhibit harmful bacteria [93,94]. LAB contain special enzymes that can reduce nitrosamine, and the lactate can also reduce nitrite content [44]. When LAB and other substitutes are used at the same time, it is very important to select the most appropriate combination, so as to produce favorable results and even synergy.

6.3. Chromogenic Agent Substitutes

Nitrite has a chromogenic effect in the pickling process. The natural pigments that can replace nitrite include *monascus* red, nitrosohemoglobin pigment, chili red, beet red, sorghum red, carboxyhemoglobin, and tomato red. The replacement synthetic pigments include carmine, erythritol, sunset yellow, lemon yellow, and amaranth. Some plant extracts, such as beetroot and berries extracts, can be used to impart the red color of nitrite-free meat products to develop and stabilize color [85]. However, the color produced by chromogenic agent substitutes is frequently different from that produced by traditional meat products. Patarata et al. reduced the nitrite content of cured loins made from wine- and water-based marinade, and studied the consumer's evaluation of product color [95]. Consumers prefer nitrite-free products even though the color of fermented meat products created with color additive alternatives is subpar if the label says "additive-free".

6.4. Bacteriostat Substitutes

Nitrite can effectively prevent food spoilage and inhibit the growth of *Clostridium botulinum* and other biological hazards, including *Listeria monocytogenes*, *bacillus cereus*, and *Staphylococcus aureus* [96,97]. The antipathogenic properties of nitrite are mainly attributed to three aspects. The first aspect is perturbing oxygen uptake and oxidative phosphorylation; the second is inhibiting critical enzymes, including aldolase, glyceraldehyde-3-phosphate, and nitrogenases; and the third is forming bactericidal nitrite derivatives [98]. Substitute bacteriostatic agents for nitrite include organic acids, tea polyphenols, spice extract, and bacteriocin [99]. On the other hand, some physical methods of food preservation can be mobilized to control microbial hazards, namely, the use of high isostatic pressures, activated plasma, pulse-field UV light, and active packaging [100–102].

However, these bacteriostat substitutes show several limitations. Firstly, their ability to inhibit microbial growth may be lower than nitrite. Secondly, some of these natural antimicrobials and plasma-treated water have a selective effect on Gram-positive or Gram-negative pathogens, vegetative cells, or spores. Thirdly, the components of some vegetable extracts will greatly change the aroma and color of products, which affects their application in the production and processing process [1].

6.5. Non-Thermal Plasma Sterilization

Some emerging food sterilization methods, such as non-thermal plasma sterilization, can be used for the preparation of fermented meat products without nitrite [30]. The interaction of plasma with water can result in the generation of nitrate and nitrite, as well as reactive oxygen species. Reactive oxygen and nitrogen compounds react with cellular macromolecules, such as proteins, lipids, enzymes, and DNA, to change the functional features of biofilms, affecting normal physiological functions, and ultimately leading to microbial cell death [96]. However, during non-thermal plasma sterilization, oxygen atoms react with vibration-excited nitrogen molecules to produce nitric oxide (NO) (reaction (8)).

$$O + N_2 \rightarrow NO + N \tag{8}$$

When NO combines with plasma, it forms nitrogen oxide (NO_2) and additional nitrogen oxides (NO_3, N_2O, N_2O_3, and N_2O_5) [103]. Nitrogen oxides diffuse and dissolve in liquids, where they react with water molecules to create nitrite, nitric acid, and nitrate acid, followed by nitrite acid degradation [36].

After non-thermal plasma sterilization, Yong et al. marinated tenderloin in brine without sodium nitrite and observed that the color was similar to that of sodium nitrite-treated tenderloin, with an increased red a* and no significant difference in yellow b* or brightness L* [104]. Lee et al. used plasma for 30 min to produce pork batter with 42.42 mg/kg of nitrite, with no significant differences in physicochemical or sensory performance when compared with pork batter prepared using sodium nitrite or celery powder [105].

7. Summary

Nitrite is a multifunctional food additive that is widely used in the fermentation of meat products. However, high levels of nitrite residues have become a significant concern that impacts the safety of fermented meat products and restricts the growth of the traditional fermented meat product industry. Exogenous addition, contamination during processing, generation by endogenous microorganisms, and non-thermal plasma sterilization technology accumulate nitrite in fermented meat products, which is degraded by acids, enzymes, and other non-acid and non-enzyme substances. Nitrite levels are primarily affected by the matrix, fermentation conditions, and antioxidants in fermented meat products. The content of nitrite in fermented meat products can be reduced by inhibiting nitrite formation and substituting nitrite's role in the fermentation process during meat processing. Future researchers should consider the following points: (i) The association between nitrite in plant extracts and nitrosamines based on their phenolic compounds has to be investigated using fermented meat products as the medium. (ii) More research is needed to evaluate the sensory acceptability of nitrite substitute products to the reformulated cured meat, taking into account consumer needs and concerns. (iii) Nitrite can be degraded by a variety of non-acid and non-enzyme compounds. Its composition and content, as well as the degradation effects of each component, and the synergistic degradation effects of each component and acid, should be analyzed further. (iv) Only CuNiRs have ever been explored in terms of their catalytic mechanism. Regarding the other three types of nitrite reductase (cd1NiRs, ccNiRs, and FdNiRs), their catalytic mechanisms have not been fully investigated. Excessive nitrite levels in cured meat products have become the principal issue influencing fermented meat product production. Further understanding of the nitrite metabolism in fermented meat products would support the appropriate application of nitrite in meat products and the production of more high-quality, low-nitrite fermented meat to meet consumer demand.

Author Contributions: All authors participated in the conception and design of this review article. Q.S. and L.K. conducted the literature search and screened the retrieved articles. Q.S., J.S. and X.S. performed the data extraction and analysis. Q.S. wrote the initial draft of the manuscript, which was then edited and revised by all authors. X.Z., Y.G., Z.W. and D.P. provided critical feedback and revised the manuscript. All authors have read and agreed to the published version of the manuscript.

Funding: We gratefully acknowledge the financial support of the National Key Research and Development Program of China (2022YFD2100603), the National Natural Science Foundation of China (32072195, 41406165, 41641052, 31972093, and 31972048), and the Zhejiang Provincial Natural Science Foundation of China (LR23C200001).

Data Availability Statement: The data presented in this study are available on request from the corresponding author.

Conflicts of Interest: The authors declare that they have no known competing financial interests or personal relationships that could have appeared to influence the work reported in this paper.

References

1. Rivera, N.; Bunning, M.; Martin, J. Uncured-labeled meat products produced using plant-derived nitrates and nitrites: Chemistry, safety, and regulatory considerations. *J. Agric. Food Chem.* **2019**, *67*, 8074–8084. [CrossRef] [PubMed]
2. Parthasarathy, D.K.; Bryan, N.S. Sodium nitrite: The "cure" for nitric oxide insufficiency. *Meat Sci.* **2012**, *92*, 274–279. [CrossRef]
3. Honikel, K.O. The use and control of nitrate and nitrite for the processing of meat products. *Meat Sci.* **2008**, *78*, 68–76. [CrossRef]
4. Sebranek, J.G. *Basic Curing Ingredients. Ingredients in Meat Products*; Springer: Berlin/Heidelberg, Germany, 2009; pp. 1–23.
5. Neto-Neves, E.M.; Sousa-Santos, O.; Ferraz, K.C.; Portella, R.L.; Sertório, J.T.; Tanus-Santos, J.E. Nitrite and tempol combination promotes synergic effects and alleviates right ventricular wall stress during acute pulmonary thromboembolism. *Nitric Oxide* **2021**, *115*, 23–29. [CrossRef] [PubMed]
6. European Union Council. Regulation (EC) No 1333/2008 of the European Parliament and of the Council. 2008. Available online: https://olympus.uniurb.it/images/stories/pdf/180%20reg.%201338.2008.pdf (accessed on 22 March 2023).

7. Hao, X.H.; Liang, Y.Q.; Zhen, H.; Sun, X.C.; Liu, X.L.; Li, M.W.; Shen, A.; Yang, Y.X. Fast and sensitive fluorescent detection of nitrite based on an amino-functionalized MOFs of UiO-66-NH2. *J. Solid State Chem.* **2020**, *287*, 121323. [CrossRef]
8. Fraqueza, M.J.; Laranjo, M.; Elias, M.; Patarata, L. Microbiological hazards associated with salt and nitrite reduction in cured meat products: Control strategies based on antimicrobial effect of natural ingredients and protective microbiota. *Curr. Opin. Food Sci.* **2021**, *38*, 32–39. [CrossRef]
9. Huang, Y.Y.; Liu, D.M.; Jia, X.Z.; Liang, M.H.; Lu, Y.Z.; Liu, J.S. Whole genome sequencing of *Lactobacillus plantarum* DMDL 9010 and its effect on growth phenotype under nitrite stress. *LWT-Food Sci. Technol.* **2021**, *149*, 111778. [CrossRef]
10. Nie, S.Q.; Zhang, Z.F.; Mo, S.M.; Li, J.H.; He, S.; Kashif, M.; Liang, Z.W.; Shen, P.H.; Yan, B.; Jiang, C.J. Desulfobacterales stimulates nitrate reduction in the mangrove ecosystem of a subtropical gulf. *Sci. Total Environ.* **2021**, *769*, 144562. [CrossRef]
11. Xia, L.; Li, X.M.; Fan, W.H.; Wang, J.L. Heterotrophic nitrification and aerobic denitrification by a novel *Acinetobacter* sp. ND7 isolated from municipal activated sludge. *Bioresour. Technol.* **2020**, *301*, 122749. [CrossRef]
12. Philips, S.; Laanbroek, H.J.; Verstraete, W. Origin, causes and effects of increased nitrite concentrations in aquatic environments. *Rev. Environ. Sci. Biotechnol.* **2002**, *1*, 115–141. [CrossRef]
13. Paik, H.D.; Lee, J.Y. Investigation of reduction and tolerance capability of LAB isolated from kimchi against nitrate and nitrite in fermented sausage condition. *Meat Sci.* **2014**, *97*, 609–614. [CrossRef] [PubMed]
14. De Mey, E.; De Maere, H.; Goemaere, O.; Steen, L.; Peeters, M.C.; Derdelinckx, G.; Paelinck, H.; Fraeye, I. Evaluation of *N*-nitrosopiperidine formation from biogenic amines during the production of dry fermented sausages. *Food Bioprocess Technol.* **2014**, *7*, 1269–1280. [CrossRef]
15. Pegg, R.B.; Shahidi, F. *Nitrite Curing of Meat: The N-Nitrosamine Problem and Nitrite Alternatives*; Food Nutrition Press, Inc.: Trumbull, CT, USA, 2004.
16. Sindelar, J.J.; Milkowski, A.L. Sodium Nitrite in Processed Meat and Poultry Meats: A Review of Curin and Examining the Risk and Benefit of Its Use. 2011. Available online: https://themeatsite.com/articles/contents/nitrite_report.pdf (accessed on 22 March 2023).
17. Zając, M.; Zając, K.; Dybaś, J. The effect of nitric oxide synthase and arginine on the color of cooked meat. *Food Chem.* **2022**, *373*, 131503. [CrossRef]
18. Alahakoon, A.U.; Jayasena, D.D.; Ramachandra, S.; Jo, C. Alternatives to nitrite in processed meat: Up to date. *Trends Food Sci. Technol.* **2015**, *45*, 37–49. [CrossRef]
19. Damodaran, S.; Parkin, K.L. *Fennema's Food Chemistry*, 5th ed.; CRC Press: Boca Raton, FL, USA, 2017.
20. Skibsted, L.H. Nitric oxide and quality and safety of muscle based foods. *Nitric Oxide* **2011**, *24*, 176–183. [CrossRef] [PubMed]
21. Thomas, C.; Mercier, F.; Tournayre, P.; Martin, J.L.; Berdagué, J.L. Effect of nitrite on the odourant volatile fraction of cooked ham. *Food Chem.* **2013**, *139*, 432–438. [CrossRef] [PubMed]
22. Villaverde, A.; Ventanas, J.; Estévez, M. Nitrite promotes protein carbonylation and Strecker aldehyde formation in experimental fermented sausages: Are both events connected? *Meat Sci.* **2014**, *98*, 665–672. [CrossRef] [PubMed]
23. Vossen, E.; De Smet, S. Protein oxidation and protein nitration influenced by sodium nitrite in two different meat model systems. *J. Agric. Food Chem.* **2015**, *63*, 2550–2556. [CrossRef]
24. Berardo, A.; De Maere, H.; Stavropoulou, D.A.; Rysman, T.; Leroy, F.; De Smet, S. Effect of sodium ascorbate and sodium nitrite on protein and lipid oxidation in dry fermented sausages. *Meat Sci.* **2016**, *121*, 359–364. [CrossRef]
25. Singh, N.; Bhatla, S.C. Heme oxygenase-nitric oxide crosstalk-mediated iron homeostasis in plants under oxidative stress. *Free Radic. Biol. Med.* **2022**, *182*, 192–205. [CrossRef]
26. Crack, J.C.; Brun, N.E.L. Biological iron-sulfur clusters: Mechanistic insights from mass spectrometry. *Coord. Chem. Rev.* **2021**, *448*, 214171. [CrossRef]
27. Landry, A.P.; Duan, X.W.; Huang, H.; Ding, H.G. Iron-sulfur proteins are the major source of protein-bound dinitrosyl iron complexes formed in *Escherichia coli* cells under nitric oxide stress. *Free Radic. Biol. Med.* **2012**, *50*, 1582–1590. [CrossRef] [PubMed]
28. Majou, D.; Christieans, S. Mechanisms of the bactericidal effects of nitrate and nitrite in cured meats. *Meat Sci.* **2018**, *145*, 273–284. [CrossRef] [PubMed]
29. Giguere, A.T.; Taylor, A.E.; Myrold, D.D.; Mellbye, B.L.; Sayavedra-Soto, L.A.; Bottomley, P.J. Nitrite-oxidizing activity responds to nitrite accumulation in soil. *FEMS Microbiol. Ecol.* **2018**, *94*, 1–9. [CrossRef] [PubMed]
30. Jarulertwattana, P.P.; Wanaloh, T.; Charurungsipong, P.; Pittarate, C.; Asavasanti, S. Influence of nitrate-nitrite contamination on pink color defect in ginger marinated steamed chicken drumsticks. *Appl. Sci. Eng. Prog.* **2021**, *14*, 417–424. [CrossRef]
31. Ji, X.; Ma, G.; Chen, Y.; Jin, J.; Yu, Q. Effect of sodium nitrite on fatty acid composition and lipid oxidation of mutton in marinating process. *J. Gansu Agric. Univ.* **2020**, *55*, 195–202. [CrossRef]
32. Wang, Z.L.; Wang, Z.X.; Ji, L.L.; Zhang, J.M.; Zhao, Z.P.; Zhang, R.; Chen, L. A review: Microbial diversity and function of fermented meat products in China. *Front. Microbiol.* **2021**, *12*, 1–8. [CrossRef]
33. Munekata, P.E.; Pateiro, M.; Domínguez, R.; Santos, E.M.; Lorenzo, J.M. Cruciferous vegetables as sources of nitrate in meat products. *Curr. Opin. Food Sci.* **2020**, *38*, 1–7. [CrossRef]
34. Ferysiuk, K.; Wójciak, K.M. Reduction of nitrite in meat products through the application of various plant-based ingredients. *Antioxidants* **2020**, *9*, 711. [CrossRef]
35. Sindelar, J.J.; Cordray, J.C.; Olson, D.G.; Sebranek, J.G.; Love, J.A. Investigating quality attributes and consumer acceptance of uncured, no-nitrate/nitrite-added commercial hams, bacons, and frankfurters. *J. Food Sci.* **2007**, *72*, S551–S559. [CrossRef]

36. Jo, K.; Lee, J.; Lee, S.; Lim, Y.; Choi, Y.S.; Jo, C.; Jung, S. Curing of ground ham by remote infusion of atmospheric non-thermal plasma. *Food Chem.* **2020**, *309*, 125643. [CrossRef] [PubMed]
37. Evans, J.R.; Clarke, V.C. The nitrogen cost of photosynthesis. *J. Exp. Bot.* **2019**, *70*, 7–15. [CrossRef] [PubMed]
38. Picetti, R.; Deeney, M.; Pastorino, S.; Miller, M.R.; Shah, A.; Leon, D.A.; Dangour, A.D.; Green, R. Nitrate and nitrite contamination in drinking water and cancer risk: A systematic review with meta-analysis. *Environ. Res.* **2022**, *210*, 112988. [CrossRef] [PubMed]
39. Park, S.; Bae, W. Modeling kinetics of ammonium oxidation and nitrite oxidation under simultaneous inhibition by free ammonia and free nitrous acid. *Process Biochem.* **2009**, *44*, 631–640. [CrossRef]
40. Iammarino, M.; Taranto, A.D.; Cristino, M. Endogenous levels of nitrites and nitrates in wide consumption foodstuffs: Results of five years of official controls and monitoring. *Food Chem.* **2013**, *140*, 763–771. [CrossRef]
41. Iacumin, L.; Cattaneo, P.; Zuccolo, C.; Galanetto, S.; Acquafredda, A.; Comi, G. Natural levels of nitrites and nitrates in San Daniele dry cured ham PDO, and in meat, salt and sugna used for its production. *Food Control* **2019**, *100*, 257–261. [CrossRef]
42. Fraqueza, M.J.; Borges, A.; Patarata, L. Strategies to Reduce the Formation of Carcinogenic Chemicals in Dry Cured Meat Products. In *Handbook of Food Bioengineering*; de Araujo Bandeira, N.R., Ed.; Food Control and Biosecurity; Academic Press: Cambridge, MA, USA, 2018; pp. 295–342. [CrossRef]
43. Jung, S.; Kim, H.J.; Park, S.; Yong, H.I.; Choe, J.H.; Jeon, H.J.; Choe, W.; Jo, C. Color developing capacity of plasma-treated water as a source of nitrite for meat curing. *Korean J. Food Sci. Anim. Resour.* **2015**, *35*, 703–706. [CrossRef]
44. Yu, L.; Guo, J.W.; Yi, G.B.; Yu, Q. Effects of LAB on nitrite degradation during pickle fermentation. *Adv. Mater. Res.* **2013**, *781*, 1656–1660. [CrossRef]
45. Kim, S.H.; Kang, K.H.; Kim, S.H.; Lee, S.; Lee, S.H.; Ha, E.; Sung, N.J.; Kim, J.G.; Chung, M.J. LAB directly degrade N-nitrosodimethylamine and increase the nitrite-scavenging ability in kimchi. *Food Control* **2017**, *71*, 101–109. [CrossRef]
46. Wu, C.D.; Zheng, J.; Huang, J.; Zhou, R.Q. Reduced nitrite and biogenic amine concentrations and improved flavor components of Chinese sauerkraut *via* co-culture of *Lactobacillus plantarum* and *Zygosaccharomyces rouxii*. *Ann. Microbiol.* **2014**, *64*, 847–857. [CrossRef]
47. Zeng, X.Q.; Pan, Q.; Guo, Y.X.; Wu, Z.; Sun, Y.Y.; Dang, L.Y.; Cao, J.X.; He, J.; Pan, D.D. Potential mechanism of nitrite degradation by Lactobacillus fermentum RC4 based on proteomic analysis. *J. Proteom.* **2019**, *194*, 70–78. [CrossRef] [PubMed]
48. Huang, L.; Zeng, X.Q.; Sun, Z.; Wu, A.J.; He, J.; Dang, Y.L.; Pan, D.D. Production of a safe cured meat with low residual nitrite using nitrite substitutes. *Meat Sci.* **2020**, *162*, 108027. [CrossRef] [PubMed]
49. Cassens, R.G.; Ito, T.; Lee, M.; Buege, D. The use of nitrite in meat. *BioScience* **1978**, *28*, 633–637. [CrossRef]
50. Liu, D.M.; Wang, P.; Zhang, X.Y.; Xu, X.L.; Wu, H.; Li, L. Characterization of nitrite degradation by *Lactobacillus casei* subsp. *rhamnosus LCR 6013*. *PLoS ONE* **2014**, *9*, e93308. [CrossRef]
51. Merkle, A.C.; Lehnert, N. Binding and activation of nitrite and nitric oxide by copper nitrite reductase and corresponding model complexes. *Dalton Trans.* **2011**, *41*, 3355–3368. [CrossRef] [PubMed]
52. Silveira, C.M.; Quintas, P.O.; Moura, I.; Moura, J.J.; Hildebrandt, P.; Almeida, M.G.; Todorovic, S. SERR spectroelectrochemical study of cytochrome cd1 nitrite reductase co-immobilized with physiological redox partner cytochrome c552 on biocompatible metal electrodes. *PLoS ONE* **2015**, *10*, e0129940. [CrossRef]
53. Horrell, S.; Kekilli, D.; Strange, R.W.; Hough, M.A. Recent structural insights into the function of copper nitrite reductases. *Metallomics* **2017**, *9*, 1470–1482. [CrossRef]
54. Izumi, A.; Schnell, R.; Schneider, G. Crystal structure of NirD, the small subunit of the nitrite reductase NirbD from *Mycobacterium tuberculosis* at 2.0 Å resolution. *Proteins* **2012**, *80*, 2799–2803. [CrossRef]
55. Rong, Z.M.; Wilson, M.T.; Cooper, C.E. A model for the nitric oxide producing nitrite reductase activity of hemoglobin as a function of oxygen saturation. *Nitric Oxide* **2013**, *33*, 74–80. [CrossRef]
56. Wang, H.N.; Gunsalus, R.P. The *nrf*A and *nir*B nitrite reductase operons in *Escherichia coli* are expressed differently in response to nitrate than to nitrite. *J. Bacteriol.* **2000**, *182*, 5813–5822. [CrossRef]
57. Shi, M.Z.; Liu, C.G.; Zhao, Y.; Wei, Z.M.; Zhao, M.Y.; Jia, L.M.; He, P.P. Insight into the effects of regulating denitrification on composting: Strategies to simultaneously reduce environmental pollution risk and promote aromatic humic substance formation. *Bioresour. Technol.* **2021**, *342*, 125901. [CrossRef] [PubMed]
58. Chu, G.Y.; Yu, D.S.; Wang, X.X.; Wang, Q.Y.; He, T.H.; Zhao, J. Comparison of nitrite accumulation performance and microbial community structure in endogenous partial denitrification process with acetate and glucose served as carbon source. *Bioresour. Technol.* **2021**, *320*, 124405. [CrossRef] [PubMed]
59. Iino, T.; Miyauchi, K.; Kasai, D.; Masai, E.; Fukuda, M. Characterization of nitrate and nitrite utilization system in *Rhodococcus jostii* RHA1. *J. Biosci. Bioeng.* **2013**, *115*, 600–606. [CrossRef] [PubMed]
60. Khlebodarova, T.M.; Ree, N.A.; Likhoshvai, V.A. On the control mechanisms of the nitrite level in *Escherichia coli* cells: The mathematical model. *BMC Microbiol.* **2016**, *16*, 15–30. [CrossRef] [PubMed]
61. Guo, H.F.; Saravanakumar, K.; Wang, M.H. Total phenolic, flavonoid contents and free radical scavenging capacity of extracts from tubers of *Stachys affinis*. *Biocatal. Agric. Biotechnol.* **2018**, *15*, 235–239. [CrossRef]
62. Ben, T.T.; Wu, P.; Zou, H.; Chen, L. Characterization of nitrite degradation by polyphenols in sea buckthorn (*Hippophaë rhamnoides* L.) by density function theory calculations. *LWT-Food Sci. Technol.* **2021**, *149*, 111884. [CrossRef]
63. Feng, X.C.; Li, C.Y.; Jia, X.; Guo, Y.; Lei, N.; Hackman, R.M.; Chen, L.; Zhou, G.H. Influence of sodium nitrite on protein oxidation and nitrosation of sausages subjected to processing and storage. *Meat Sci.* **2016**, *116*, 260–267. [CrossRef]

64. Aarti, C.; Khusro, A. Functional and technological properties of exopolysaccharide producing autochthonous *Lactobacillus plantarum* strain AAS3 from dry fish based fermented food. *LWT-Food Sci. Technol.* **2019**, *114*, 108387. [CrossRef]
65. Seo, B.J.; Bajpai, V.K.; Rather, I.A.; Park, Y.H. Partially purified exopolysaccharide from *Lactobacillus plantarum* YML009 with total phenolic content, antioxidant and free radical scavenging efficacy. *Indian J. Pharm. Educ. Res.* **2015**, *49*, 282–292. [CrossRef]
66. Delgado-Pando, G.; Fischer, E.; Allen, P.; Kerry, J.P.; O'Sullivan, M.G.; Hamill, R.M. Salt content and minimum acceptable levels in whole-muscle cured meat products. *Meat Sci.* **2018**, *139*, 179–186. [CrossRef]
67. Kilic, B.; Cassens, R.G.; Borchert, L.L. Influence of turkey meat on residual nitrite in cured meat products. *J. Food Prot.* **2001**, *64*, 235–239. [CrossRef] [PubMed]
68. Fraqueza, M.J.; Patarata, L.; Lauková, A. Protective cultures and bacteriocins in fermented meats. In *Fermented Meat Products: Health Aspects*; CRC Press: Boca Raton, FL, USA, 2017; pp. 228–269. [CrossRef]
69. Kuschman, H.P.; Palczewski, M.B.; Thomas, D.D. Nitric oxide and hydrogen sulfide: Sibling rivalry in the family of epigenetic regulators. *Free Radic. Biol. Med.* **2021**, *170*, 34–43. [CrossRef]
70. Baseler, W.A.; Davies, L.C.; Quigley, L.; Ridnour, L.A.; Weiss, J.M.; Hussain, S.P.; Wink, D.A.; McVicar, D.W. Autocrine IL-10 functions as a rheostat for M1 macrophage glycolytic commitment by tuning nitric oxide production. *Redox Biol.* **2016**, *10*, 12–23. [CrossRef] [PubMed]
71. Herrmann, S.S.; Granby, K.; Duedahl-Olesen, L. Formation and mitigation of *N*-nitrosamines in nitrite preserved cooked sausages. *Food Chem.* **2015**, *174*, 516–526. [CrossRef]
72. Mortensen, A.; Aguilar, F.; Crebelli, R.; Domenico, A.D.; Dusemund, B.; Frutos, M.J.; Galtier, P.; Gott, D.; Gundert-Remy, U.; Lambré, C.; et al. Re-evaluation of potassium nitrite (E249) and sodium nitrite (E250) as food additives. *EFSA J.* **2017**, *15*, 4786–4943. [CrossRef]
73. De Mey, E.; De Maere, H.; Paelinck, H.; Fraeye, I. Volatile *N*-nitrosamines in meat products: Potential precursors, influence of processing, and mitigation strategies. *Crit. Rev. Food Sci. Nutr.* **2017**, *57*, 2909–2923. [CrossRef]
74. United States Food and Drug Administration. Code of Federal Regulations Title 21 Food and Drugs. 2021. Available online: https://www.fda.gov/food (accessed on 22 March 2023).
75. Ministry of Health, Labor, and Welfare of Japan. Standard Specification of Food and Additives: No. 370 Notice of Ministry of Health and Welfare of Showa in 34. 2021. Available online: https://www.mhlw.go.jp/stf/seisakunitsuite/bunya/kenkou_iryou/shokuhin/zanryu/591228-1.html (accessed on 22 March 2023).
76. Ministry of Food and Drug Safety. Food Additives Code (#2021-19, 2021.03.09.). 2021. Available online: https://www.mfds.go.kr/eng/brd/m_15/view.do?seq=72436 (accessed on 22 March 2023).
77. State Health and Family Planning Commission of the People's Republic of China. National Standard for Food Safety Standard for Use of Food Additives. 2014. Available online: http://down.foodmate.net/standard/yulan.php?itemid=42543 (accessed on 22 March 2023).
78. Flores, M.; Toldrá, F. Chemistry, safety, and regulatory considerations in the use of nitrite and nitrate from natural origin in meat products—Invited review. *Meat Sci.* **2021**, *171*, 108272. [CrossRef]
79. Pegg, R.B.; Honikel, K.O. Principles of curing. In *Handbook of Fermented Meat and Poultry*, 2nd ed.; Toldrá, F., Hui, Y.H., Astiasarán, I., Sebranek, J.G., Talon, R., Eds.; Wiley-Blackwell: Chichester, West Sussex, UK, 2015; pp. 19–30.
80. Standing Committee on Plants, Animals, Food and Feed. Summary Report of the Standing Committee on Plants, Animals, Food and Feed Held in Brussels on 17 September 2018. 2018. Available online: https://ec.europa.eu/food/horizontal-topics/committees/paff-committees_en (accessed on 26 October 2021).
81. Bernardo, P.; Patarata, L.; Lorenzo, J.M.; Fraqueza, M.J. Nitrate is nitrate: The status quo of using nitrate through vegetable extracts in meat products. *Foods* **2021**, *10*, 3019. [CrossRef]
82. Gadekar, Y.P.; Sharma, B.D.; Shinde, A.K.; Verma, A.K.; Mendiratta, S.K. Effect of natural antioxidants on the quality of cured, restructured goat meat product during refrigerated storage (4 ± 1 °C). *Small Rumin. Res.* **2014**, *119*, 72–80. [CrossRef]
83. Orihuel, A.; Bonacina, J.; Vildoza, M.J.; Bru, E.; Vignolo, G.; Saavedra, L.; Fadda, S. Biocontrol of *Listeria monocytogenes* in a meat model using a combination of a bacteriocinogenic strain with curing additives. *Food Res. Int.* **2018**, *107*, 289–296. [CrossRef]
84. Dutra, T.V.; Castro, J.C.; Menezesa, J.L.; Ramos, T.R.; do Prado, I.N.; Junior, M.M.; Mikcha, J.M.G.; de Abreu Filho, B.A. Bioactivity of oregano (*Origanum vulgare*) essential oil against *Alicyclobacillus* spp. Industrial Crops and Products. *Ind. Crops Prod.* **2019**, *129*, 345–349. [CrossRef]
85. Ozaki, M.M.; Munekata, P.E.S.; Jacinto-Valderrama, R.A.; Efraim, P.; Pateiro, M.; Lorenzo, J.M.; Pollonio, M.A.R. Beetroot and radish powders as natural nitrite source for fermented dry sausages. *Meat Sci.* **2021**, *171*, 108275. [CrossRef] [PubMed]
86. Inguglia, E.S.; Oliveira, M.; Burgess, C.M.; Kerry, J.P.; Tiwari, B.K. Plasma activated water as an alternative nitrite source for the curing of beef jerky: Influence on quality and inactivation of *Listeria innocua*. *Innov. Food Sci. Emerg. Technol.* **2020**, *59*, 102276. [CrossRef]
87. Bázan-Lugo, E.; García-Martínez, I.; Alfaro-Rodríguez, R.H.; Totosaus, A. Color compensation in nitrite-reduced meat batters incorporating paprika or tomato paste. *J. Sci. Food Agric.* **2011**, *92*, 1627–1632. [CrossRef] [PubMed]
88. Ozaki, M.M.; Munekata, P.E.S.; Lopes, A.D.S.; do Nascimento, M.D.S.; Pollonio, M.A.R. Using chitosan and radish powder to improve stability of fermented cooked sausages. *Meat Sci.* **2020**, *167*, 108165. [CrossRef] [PubMed]
89. Araújo, M.K.; Gumiela, A.M.; Bordin, K.; Luciano, F.B.; Macedo, R.E.F. Combination of garlic essential oil, allyl isothiocyanate, and nisin Z as bio-preservatives in fresh sausage. *Meat Sci.* **2018**, *143*, 177–183. [CrossRef]

90. Oswell, N.J.; Thippareddi, H.; Pegg, R.B. Practical use of natural antioxidants in meat products in the U.S.: A review. *Meat Sci.* **2018**, *145*, 469–479. [CrossRef] [PubMed]
91. EFSA The 2019 updated list of QPS status recommended biological agents in support of EFSA risk assessments. *EFSA J.* **2020**, *17*, 1–5. [CrossRef]
92. Zheng, J.S.; Wittouck, S.; Salvetti, E.; Franz, C.M.A.P.; Harris, H.M.B.; Mattarelli, P.; O'Toole, P.W.; Pot, B.; Vandamme, P.; Walter, J.; et al. A taxonomic note on the genus *Lactobacillus*: Description of 23 novel genera, emended description of the genus *Lactobacillus* Beijerinck 1901, and union of *Lactobacillaceae* and *Leuconostocaceae*. *Int. J. Syst. Evol. Microbiol.* **2020**, *70*, 2782–2858. [CrossRef]
93. Wu, Y.Y.; Liu, F.J.; Li, L.H.; Yang, X.Q.; Deng, J.C.; Chen, S.J. Isolation and identification of nitrite-degrading LAB from salted fish. *Adv. Mater. Res.* **2012**, *393–395*, 828–834. [CrossRef]
94. Laranjo, M.; de Guía Córdoba, M.; Semedo-Lemsaddek, T.; Potes, M.E. Food Microbiology. *BioMed Res. Int.* **2019**, *2019*, 8039138. [CrossRef] [PubMed]
95. Patarata, L.; Carvalho, F.; Fraqueza, M.J. Nitrite-free implications on consumer acceptance and the behavior of pathogens in cured pork loins. *Foods* **2022**, *11*, 796. [CrossRef] [PubMed]
96. Alkawareek, M.Y.; Algwari, Q.T.; Gorman, S.P.; Graham, W.G.; O'Connell, D.; Gilmore, B.F. Application of atmospheric pressure nonthermal plasma for the in vitro eradication of bacterial biofilms. *FEMS Immunol. Med. Microbiol.* **2012**, *65*, 381–384. [CrossRef]
97. Hospital, X.F.; Hierro, E.; Arnau, J.; Carballo, J.; Aguirre, J.S.; Gratacós-Cubarsí, M.; Fernández, M. Effect of nitrate and nitrite on *Listeria* and selected spoilage bacteria inoculated in dry-cured ham. *Food Res. Int.* **2017**, *101*, 82–87. [CrossRef]
98. Weiss, J.; Gibis, M.; Schuh, V.; Salminen, H. Advances in ingredient and processing systems for meat and meat products. *Meat Sci.* **2010**, *86*, 196–213. [CrossRef]
99. Lavieri, N.A.; Sebranek, J.G.; Cordray, J.C.; Dickson, J.S.; Manu, D.K.; Mendonca, A.F.; Brehm-Stecher, B. Control of *Listeria monocytogenes* on alternatively cured ready-to-eat ham using natural antimicrobial ingredients in combination with post-lethality interventions. *Food Process. Technol.* **2015**, *6*, 1–7. [CrossRef]
100. Inguglia, E.S.; Zhang, Z.H.; Tiwari, B.K.; Kerry, J.P.; Burgess, C.M. Salt reduction strategies in processed meat products—A review. *Trends Food Sci. Technol.* **2017**, *59*, 70–78. [CrossRef]
101. Pinton, M.B.; dos Santos, B.A.; Lorenzo, J.M.; Cichoski, A.J.; Boeira, C.P.; Campagnol, P.C.B. Green technologies as a strategy to reduce NaCl and phosphate in meat products: An overview. *Curr. Opin. Food Sci.* **2021**, *40*, 1–5. [CrossRef]
102. Horita, C.N.; Baptista, R.C.; Caturla, M.Y.R.; Lorenzo, J.M.; Barba, F.J.; Sant'Ana, A.S. Combining reformulation, active packaging and non-thermal post-packaging decontamination technologies to increase the microbiological quality and safety of cooked ready-to-eat meat products. *Trends Food Sci. Technol.* **2018**, *72*, 45–61. [CrossRef]
103. Lee, J.; Lee, C.W.; Yong, H.I.; Lee, H.J.; Jo, C.; Jung, S. Use of atmospheric pressure cold plasma for meat industry. *Korean J. Food Sci. Anim. Resour.* **2017**, *37*, 477–485. [CrossRef]
104. Yong, H.I.; Park, J.; Kim, H.J.; Jung, S.; Park, S.; Lee, H.J.; Choe, W.; Jo, C. An innovative curing process with plasma-treated water for production of loin ham and for its quality and safety. *Plasma Process. Polym.* **2018**, *15*, e1700050. [CrossRef]
105. Lee, J.; Jo, K.; Lim, Y.; Jeon, H.J.; Choe, J.H.; Jo, C.; Jung, S. The use of atmospheric pressure plasma as a curing process for canned ground ham. *Food Chem.* **2018**, *240*, 430–436. [CrossRef] [PubMed]

MDPI
St. Alban-Anlage 66
4052 Basel
Switzerland
www.mdpi.com

Foods Editorial Office
E-mail: foods@mdpi.com
www.mdpi.com/journal/foods

www.ingramcontent.com/pod-product-compliance
Lightning Source LLC
Chambersburg PA
CBHW061617220926
51850CB00005B/533
* 9 7 8 3 0 3 6 5 9 9 8 7 8 *